DISINFECTION and DECONTAMINATION
Principles, Applications and Related Issues

DISINFECTION and DECONTAMINATION
Principles, Applications and Related Issues

Edited by
Gurusamy Manivannan

CRC Press
Taylor & Francis Group
Boca Raton   London   New York

CRC Press is an imprint of the
Taylor & Francis Group, an **informa** business

CRC Press
Taylor & Francis Group
6000 Broken Sound Parkway NW, Suite 300
Boca Raton, FL 33487-2742

© 2008 by Taylor & Francis Group, LLC
CRC Press is an imprint of Taylor & Francis Group, an Informa business

International Standard Book Number-13: 978-0-8493-9074-6 (Hardcover)

Library of Congress Cataloging-in-Publication Data

Disinfection and decontamination : principles, applications, and related issues / editor, Gurusamy Manivannan.
    p. ; cm.
"A CRC title."
Includes bibliographical references and index.
ISBN 978-0-8493-9074-6 (alk. paper)
    1. Disinfection and disinfectants. 2. Microbial contamination--Prevention. I. Manivannan, Gurusamy, 1960-
    [DNLM: 1. Disinfection--methods. 2. Anti-Infective Agents--therapeutic use. 3. Biofilms. 4. Decontamination--methods. 5. Infection Control--methods. WA 240 D611 2007]

RA761.D587 2007
614.4'8--dc22
                                 2007008723

Visit the Taylor & Francis Web site at
http://www.taylorandfrancis.com

and the CRC Press Web site at
http://www.crcpress.com

Table of Contents

vi

Preface

Where the mind is without fear and the head is held high
Where knowledge is free
Where the world has not been broken up into fragments
By narrow domestic walls
Where words come out from the depth of truth
Where tireless striving stretches its arms towards perfection
Where the clear stream of reason has not lost its way
Into the dreary desert sand of dead habit
Where the mind is led forward by thee
Into ever-widening thought and action
Into that heaven of freedom, my Father, let my country awake

—Rabindranath Tagore, *Gitanjali*

The ongoing battle of human against microbe has a legacy starting from the dawn of humankind and has been heightened by increased knowledge. Though we have many friendly interactions with microorganisms, there is little fondness for these invisible cohabitants.

We totally understood that the magnitude of issues connected to disinfection and decontamination could not be covered by a single book. Our objective in preparing this book was to further advance the knowledge of disinfection and decontamination in all walks of human life: through principles, applications in the form of devices, and governing issues such as regulatory and legal, and to pool various ideas under one platform. This book is unique in that it provides the knowledge of 34 experts in their fields rather than a single author. Moreover, it covers more sophisticated and practical topics.

Being a product development professional working with very capable and smart scientists, I always felt the need for a book that gave a broader perspective with in-depth knowledge. Such a book should also whet the appetite of readers to read more in this fascinating field. I sincerely believe that the contributing authors from across the globe, who are the leading experts in their fields, did more than justice to fulfilling that expectation.

I would like to personally thank all the contributors for their time and effort in sharing their hands-on experience and knowledge to advance the field of disinfection and decontamination through their well-thought out and well-written chapters. I had the pleasure of reading their chapters from draft to final version and we all hope that you will enjoy this book.

Acknowledgments

Were it not for the determination and unflagging commitment, patience, encouragement, and synergistic contributions of the remarkable team that worked with me in this exciting project, this book would have never seen the light of day.

So it is with deep gratitude that I express my appreciation to the following for their contributions:

- My sincere thanks to all contributing authors for offering me the great privilege of reading their outlines and multiple versions of their chapters.

- No book can be edited without the selfless support of loved ones. Thanks to my dear wife, Selvi, and our lovely children, Anand, Abinaya, and Anisha, who allowed me to spend a huge chunk of family time for the book, and offered constant encouragement with huge smiles, unconditional support, and nourishing love.

- I would like to thank my beloved and noble parents, Gurusamy and Themozhi, for the gifts of education, persistence, inspiration, and self-confidence, and for their continuous interest. Many thanks to my dad for leading a great model life with numerous achievements (author of more than 100 books and recipient of many awards) and for being a mentor—not only to me but also to his students through many years of teaching—and to my mom for her constant encouragement and support of him.

- My sincere thanks to my senior management, Dr. Jerry Newman, Dr. Wallace Puckett, and Dr. Peter Burke, and to my colleagues for their support, suggestions, and stimulating discussions.

- I would like to thank all my teachers, mentors, and international collaborators for their intellectual stimulus over these years.

- I express my sincere gratitude to those friends who honestly supported and pulled me through tough times. You know who you are.

- And one final large thanks to everybody I may have forgotten or was not mentioned here, but in some way contributed to this book.

Contributors

Terri A. Antonucci STERIS Corporation, Mentor, Ohio

Linda Corum Department of Pathology, West Virginia University, Morgantown, West Virginia

Joseph Kirby Farrington Eli Lilly, Indianapolis, Indiana

David P. Fidler Indiana University School of Law, Bloomington, Indiana

Nancy E. Kaiser STERIS Corporation, St. Louis, Missouri

Daniel A. Klein STERIS Corporation, St. Louis, Missouri

Steven P. Kutalek Division of Cardiac Electrophysiology, Drexel University College of Medicine, Philadelphia, Pennsylvania

Emilie Lermite Service de Chirurgie Digestive, Centre Hospitalier Universitaire d'Angers, Angers, France

Gurusamy Manivannan Milk Quality & Animal Health, DeLaval, Kansas, Missouri

Gerald McDonnell Research and European Affairs, STERIS Corporation, Basingstoke, Hampshire, United Kingdom

Lindsay Nakaishi Department of Pathology, West Virginia University, Morgantown, West Virginia

Kanavillil Nandakumar Department of Biology, Lakehead University, Thunder Bay, Ontario, Canada

Jerry L. Newman STERIS Corporation, St. Louis, Missouri

Ben M. J. Pereira Department of Biotechnology, Indian Institute of Technology Roorkee, Roorkee, Uttaranchal, India

Semali Perera   Department of Chemical Engineering, University of Bath, Bath, United Kingdom

Patrick Pessaux   Service de Chirurgie Digestive, Centre Hospitalier Universitaire d'Angers, Angers, France

Jim Polarine   Technical Services, STERIS Corporation, St. Louis, Missouri

Vikas Pruthi   Department of Biotechnology, Indian Institute of Technology Roorkee, Roorkee, Uttaranchal, India

José A. Ramirez   Johnson Diversey, Inc., Sturtevant, Wisconsin

Nancy A. Robinson   STERIS Corporation, Mentor, Ohio

Nora A. Sabbuba   Cardiff School of Biosciences, Cardiff University, Cardiff, Wales, United Kingdom

Michael G. Sarli   STERIS Corporation, St. Louis, Missouri

Syed A. Sattar   Centre for Research on Environmental Microbiology, Faculty of Medicine, University of Ottawa, Ontario, Canada

Scott W. Sinner   ID Associates, P.A., Hillsborough, New Jersey

Anthony W. Smith   The School of Pharmacy, University of London, United Kingdom

Susan Springthorpe   Centre for Research on Environmental Microbiology, Faculty of Medicine, University of Ottawa, Ontario, Canada

Kurissery R. Sreekumari   ALS Laboratory Group, Environmental Division, Thunder Bay, Ontario, Canada

David J. Stickler   Cardiff School of Biosciences, Cardiff University, Cardiff, Wales, United Kingdom

Vesna Šuljagić   Infection Control Department, Military Medical Academy, Institute of Preventive Medicine, Crnotravska, Belgrade, Serbia

Scott V. W. Sutton   Vectech Pharmaceutical Consultants, Inc., Farmington Hills, Michigan

John G. Thomas   Department of Pathology, West Virginia University, Morgantown, West Virginia

Allan R. Tunkel   Department of Medicine, Monmouth Medical Center, Long Branch, New Jersey

Nicole Williams   Repairs and Reprocessing, Richard Wolfe Medical Instruments Corporation, Vernon Hills, Illinois

George A. Yesenosky   Division of Cardiology, Drexel University College of Medicine, Philadelphia, Pennsylvania

1

Perspectives on Infection Challenges and Solutions

Gurusamy Manivannan

CONTENTS

1.1 Introduction

Emerging infectious diseases have long loomed like a black shadow over the human race and have successfully wiped out portions of it. People and pathogens have a long history of going hand in hand. It is not surprising that infections have been detected in the bones of human ancestors more than a million years old. Evidence from the mummy of the Egyptian pharaoh Ramses V suggested that smallpox infection might have caused his death. Similarly, widespread outbreaks of disease and their impact on the human race are also well documented from ancient literature to current Web sites. Between 1347 and 1351, roughly one-third of the population of medieval Europe was wiped out by the bubonic plague. In 1875, the son of a Fijian chief was afflicted with measles after a ceremonial trip to Australia. More than 20,000 Fijians died from this imported disease within four months following his return home [1].

The widespread availability of antibiotics led to a dramatic reduction of infectious diseases in the 1940s. Even nearly 40 years after penicillin was introduced, it has been used successfully to treat pneumonia. Now penicillin-resistant strains of pneumonia are dominant in many countries. These strains successfully mutate and acquire resistant genes from other microbes and develop resistance to antimicrobial drugs. It seems that microbes are finding their own way of responding to human intelligence. Table 1.1 shows the list of some of the major emerging diseases listed by the Centers of Disease Control and Prevention (CDC) [2]. Infectious diseases are the third leading cause of death in the United States, behind heart disease and cancer [3]. The magnitude of issues connected to disinfection and decontamination cannot be covered by a single book. While the intention of this chapter is to paint a broader picture on different issues of infection challenges, other chapters of this book focus on typical challenges and potential solutions covering some key aspects.

TABLE 1.1

Major Emerging Diseases [2]

Bovine spongiform encephalopathy (mad cow disease)
Variant Creutzfeldt–Jakob disease (VCJD)
Cholera and diphtheria
Dengue, Rift Valley, Ebola hemorrhagic, valley (coccidioidomycosis) and yellow fevers
Escherichia coli, Norovirus (formerly Norwalk virus), and streptococcal infection
Hantavirus pulmonary syndrome
Hepatitis C and HIV/AIDS
Influenza
Legionnaires' disease (legionellosis)
Malaria, measles, and meningitis
MRSA (methicillin-resistant *Staphylococcus aureus*)
Plague and polio (poliomyelitis)
Rabies and rotavirus infection
Salmonellosis and severe acute respiratory syndrome (SARS)
Smallpox and tuberculosis
VISA/VRSA—vancomycin-intermediate/resistant *Staphylococcus aureus*
West Nile virus infection

1.2 Fear of the Microorganisms and Response

How have humans responded to the threat posed by microorganisms? Penicillin, latex condoms, streptomycin, chloroquine, pyrimethamine, and vaccines are some of the effective prophylactics or treatments developed as an answer to some of the diseases. The combinations of widespread immunization campaigns and successful strategies and products developed by the scientific community have countered the threats posed to both developed and developing nations. While recent scientific achievements in infection control measures, especially antibiotics or antimicrobials, yield measurable successes, the misuse and underuse of them have diminished possible benefits in the developing countries. According to the World Health Organization (WHO), more than 10 million children die each year before their fifth birthday, most from preventable causes, and almost all such children from poor countries. Though the causes of death differ substantially from one country to another, diarrhea, pneumonia, and malaria remain the diseases most often associated with child deaths [4].

Antimicrobials, antiseptics, and new technologies are some of the answers of contemporary societies to these challenges. Do current technologies provide the answer for the ongoing health issues involving various aspects of human life and keep us healthier? Answers were provided by a wide choice of antimicrobial products, such as soaps, hand sanitizers, cleaners, wipes, appliances, medical devices, and even toys. Interest in these antimicrobial products has been clearly demonstrated by the steady increase in the consumption of such products. Such interest has been augmented by the striking headlines about killer viruses and deadly bacteria to hospital-acquired infection, infectious diseases outbreak, etc. The fact that these specialized products carry a premium of 10%–15% in price further bears witness to the consumers' awareness and concern. It is important to remember that clearly compelling literature needs to be created, demonstrating the need and benefit of routine use of such products in the home environment. However, the role of such products in the healthcare market is vitally important in areas such as surgical scrubs, surgical site preparations, and hand-hygiene products for healthcare providers and disinfection of critical areas. To understand the interest and effort on infection control measures, the literature for the last 15 years was evaluated using publicly available materials on infection control. The number of publications in 1990 was 1168 and in 2005, the number increased to 2328. (Figure 1.1)

The fear of microbes is well founded. History has seen repeated examples of incredible human suffering and death caused by microorganisms. In the late 1800s, recurrent outbreaks of deadly diseases such as cholera and diphtheria stimulated creative minds to understand the mechanisms of such outbreaks. Significant product development led to the design of toilets, sewer traps, window ventilators, and water filters. A century later, another new wave of infectious diseases rekindled fears, and shrewd marketers in

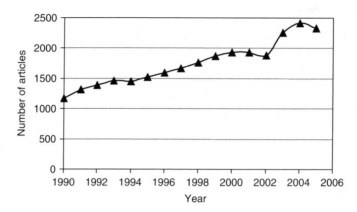

FIGURE 1.1
Growth of literature on infection control, 1990–2005.

combination with product developers responded with new germ-fighting strategies and brilliant products such as hand sanitizers and disinfectants. As shown in Table 1.1, new diseases have been reported (emerged during the past 25 years) such as the Ebola virus or Creutzfeldt–Jakob diseases, which can significantly affect health in a short period.

As per the WHO's report, there is a common misconception that frequent new drug discoveries are made to replace ineffective drugs [5]. In reality, while new versions of older drugs continue to be developed, there is a dearth of new classes of antibacterials [5]. With a period of 10–20 years taken for developing new anti-infective drugs and for the approval process, the emergence of new drugs or vaccines from the research and development pipeline is diminishing. Imagine the disaster if the eradication of smallpox in 1980 happened after few years delay; the unforeseen emergence of HIV (in the developing nations mostly) might have undermined safe smallpox vaccination in severely affected populations. The world learned new lessons when it faced the severe acute respiratory syndrome (SARS) epidemic in 2003. SARS affected a total of 8098 people worldwide, of which 774 died [6]. Avian influenza (bird flu), an infectious disease of birds caused by type A strains of the influenza virus, has received worldwide attention recently. The avian influenza viruses are readily transmitted from farm to farm by the movement of live birds, people, contaminated vehicles, equipment, feed, and cages [7]. According to the WHO's fact sheet about avian influenza, scientists are increasingly convinced that at least some migratory waterfowl are now carrying the H5N1 virus in its highly pathogenic form over long distances and introducing the virus to poultry flock in areas that lie along their migratory routes [7]. So far, the spread of H5N1 virus from person to person has not been observed. However, scientists are concerned that the H5N1 virus could infect humans and spread from one person to another. Since these viruses do not commonly infect humans, there is little or no immunity developed against them. The world is closely watching H5N1 virus because if it were to gain the capacity to spread easily from person to person, an influenza pandemic could begin [8].

Zelicoff and Bellomo have covered different outbreaks and their origins in their book, *Microbe: Are We Ready for the Next Plague*? [9] They have pointed out that while the United States is a world leader in science and medicine, its public health system is underequipped to respond to a major outbreak. If this is true, then, imagine the status of the rest of the world. Their book exposes conditions conducive to major biological crises such as gaps in information and communication chains, deployment and distribution strategies, and containment plans. Zelicoff and Bellomo also discuss the outbreaks of West Nile virus, SARS, Hantavirus, and other diseases in conjunction with the current health system [9]. In our society, the consequences of well-known risks have been accepted if there are beneficial tradeoffs. Humans deeply fear the unknown outbreaks because of the perception that it can go beyond the control of current measures. If our society is to arrive at a balanced solution to the problem of recognizing and managing new infectious diseases, we must each become better informed about the outbreaks and associated risks.

1.3 Antimicrobials

Antimicrobials are chemicals that kill or inhibit the growth of microorganisms, depending on the use concentration and conditions. They include chemical preservatives and antiseptics, as well as drugs used in the treatment of infectious diseases of plants and animals. Two major types of antimicrobial agents are antiseptics and disinfectants. Antiseptics are designed to be applied on the skin, but should not be taken internally. Common examples are iodine solution, alcohols, detergents, and silver nitrate. Disinfectants are agents that kill microorganisms, but not necessarily their spores on inanimate surfaces. They are not generally safe for application to living tissues, but they may be applied on objects such as tables, floors, utensils, etc. Some examples are chlorine and its compounds, hypochlorites, copper sulfate, phenols, and quaternary ammonium compounds. In general, antimicrobial or anti-infective materials encompass a wide variety of substances including heavy metals (silver, arsenic, bismuth), azo dye derivatives (sulfonamides), synthetic drugs (quinolones), and antibiotics (penicillins, aminoglycosides, etc.). Usage of antimicrobial agents such as alcohols or inorganic salts to treat infections dates back more than 2500 years. The most important property of these antimicrobial materials, from a host point of view, is their selective toxicity acting in a specific way that inhibits or kills bacterial pathogens, but has little or no toxic effect on the host. This indicates that the biochemical processes in bacteria are in some way different from those in the animal cells and that the advantage of this difference has been taken into account in actually eliminating them.

Most microbiologists classify antimicrobial agents used in infectious disease treatments based on their origin. Antibiotics are a major class of materials that are natural substances produced by certain groups of microorganisms, and an other group comprises chemotherapeutic agents, which are chemically synthesized. A hybrid substance is a semisynthetic antibiotic, wherein a molecular version produced by a microbe is subsequently altered to achieve desired properties. The modern era of antimicrobial chemotherapy began in 1929 with Fleming's discovery of a powerful bactericidal substance—penicillin—and Domagk's discovery of synthetic chemicals (sulfonamides) with broad antimicrobial activity in 1935. In the early 1940s, spurred partially by the need of World War II, the magical drug penicillin was isolated and administered to experimental animals. It was found not only to cure infections but also to possess an incomparable feature of very low toxicity for the animals. The rapid isolation of streptomycin, chloramphenicol, and tetracycline followed, and these and several other antibiotics were employed in clinical usage by the 1950s. In Chapter 6, McDonnell has covered various modes of action and mechanisms of resistance of various biocides in detail, and in Chapter 7 Šuljagić has discussed the pragmatic approach to the judicious selection and proper use of disinfectant and antiseptic agents in healthcare settings.

1.4 Host–Parasite Relationship

As dictated by nature, humans and microorganisms are always evolving together. The microorganisms become less virulent and humans gradually become more resistant, resulting in either mutualism (benefiting both species) or commensalism (one species benefits without any harm to the other). The microorganisms harming human species to obtain benefits result in a parasitic (positive–negative) relationship. The parasite is physiologically or metabolically dependent upon its host, and heavily infested hosts are killed by their parasites. Thinking logically, the best interests of an organism are not served if its host dies. The reverse of that cooperative paradigm could also be true: microorganisms and their hosts can exhaust their energies devising increasingly powerful weaponry and defenses. Again, paradoxically, as long as a microorganism has an effective mechanism of jumping from one host to another, it can afford to kill the host.

Thinking macroscopically, about the competition between microorganisms and human beings, humans should definitely have an upper hand if size matters: the average person is 10^{17} times the size of an average bacterium. On the contrary, humans must compete with more than 5,000 kinds of viruses and 300,000 species of bacteria. Furthermore, looking through the genetic hourglass, bacteria can reproduce a half-million times during the period of humans in achieving a single new generation. This disparity inherently offers distinct advantages to pathogens and helps to evolve adaptations that are ever more virulent and outstrip human responses. This elusive scenario is dramatically explained by Dawkins and Krebs as the "Red Queen Principle" (Alice has to run faster and faster just to stay in the same place), adapted from Lewis Carroll's *Through the Looking Glass* [10].

Egyptian demographer Abdel R. Omran offers another intriguing dimension. As observed through ages, the major killers in many modern industrial nations are no longer infectious diseases. Omran attributed the changes that happened in the middle of nineteenth century to the industrial revolution that took place in the United States and Europe. The industrial revolution changed the dynamics of the host–parasite relationship and tilted it in favor of humans. Better nutrition, improved public health measures, and awareness of medical advances resulted in greater control over microorganisms. An increase in life expectancy has been observed in the United States, from 49 to 77 years of age. In the developing countries, life expectancy has also increased to 63 from 40 years of age. These gains can be easily attributed to the increase in knowledge and affordability of healthcare, significant growth in infrastructure (such as availability of purified and treated water), and eradication or control of numerous infectious diseases [11,12].

1.5   Infectious Disease Challenges

The legacy of infectious diseases has been a history of microbes on the march, often in our wake taking advantage of the rich opportunities offered to them to thrive, prosper, and spread. Conditions of modern life such as faster communication, larger population bases, and better science ensure that the factors responsible for disease emergence are recognized faster than ever before. Emerging infections and selected outbreaks in the United States have been monitored and updated by various organizations in their Web sites such as those of CDC, WHO, Food and Drug Administration (FDA), National Institute of Health (NIH), and the Environmental Protection Agency (EPA). Emerging pathogens are those that have appeared in a human population for the first time or have occurred previously but are increasing in incidence or expanding into areas where they have not previously been reported usually over the last 20 years [13].

Despite predictions of healthcare workers and various organizations working toward public health, various microbial diseases have not been eliminated in total. The emerging pathogens have been broken down into the following categories: (1) emerging old infections due to antimicrobial resistance; (2) earlier unrecognized infections such as Lyme disease; (3) known infections such as West Nile encephalitis spreading to new geographic areas or populations; and (4) new infections due to change or evolution of existing organisms such as SARS [13]. Emergence of newly recognized diseases, reemergence of diseases thought to be controlled or eradicated, and the development of new drug-resistant organisms continue to be life- and welfare threatening in all parts of the world. However, proper surveillance networks, prompt epidemiologic investigation, and adequate diagnostic laboratory capacity in the affected areas can help to recognize and control these diseases.

The need for basic and applied research leading to practical solutions, training of healthcare professionals, caregivers, and informing the public is ever present. The recent outbreaks of bird flu in Asian countries reinforced the need for an effective control of infectious diseases. SARS triggered a wake up call and demonstrated the global havoc that can be wreaked by a newly emerging infectious disease. Also the report entitled "Emerging Infections: Microbial Threats to Health in the United States," issued by the Institute of Medicine in 1992, stressed the need for increased vigilance to ensure prompt recognition and rapid response to new and reemerging infectious diseases [14]. In the United States, the CDC, NIH, and FDA led a task force even in 1999 to create an action plan-providing blueprint for specific, coordinated federal actions to address the emerging threat of anti-microbial resistance. The plan contained 84 action items (including 13 prior-ity action steps) under 4 major components: surveillance, prevention and control, research, and product development. The 2005 report documents the domestic issues on the public health action plan to combat antimicrobial resistance [15]. These reports constitute prompt response to the infectious disease challenges.

1.5.1 Infectious Disease Emergence Factors

Among various major factors contributing to the emergence of infectious diseases, the important ones are human demographics and behavior, indus-try and technology, economic development and land use, globalization and international travel, microbial adaptation and change, breakdown of public health measures, and economic disparity of have and have-nots [16,17]. Major ecological changes due to economic development and land use such as change in water ecosystems; deforestation/reforestation; flood/drought; famine; and climate changes are also playing a role in addition to other factors. Some of the infectious diseases caused by such changes include schistosomiasis, Rift Valley fever, Argentine hemorrhagic fever, Hantaan, and the Hantavirus pulmonary syndrome. Human demographic and behavior changes led to the spread of HIV and other sexually transmitted diseases (Table 1.2).

Breakdown in public health measures such as curtailment or reduction in prevention programs, inadequate sanitation, and vector control measures contributed to the resurgence of tuberculosis in the United States, cholera in refugee camps in Africa and India, and resurgence of diphtheria in the former Soviet Union. These problems are more severe in the developing countries but are not confined to these areas. The United States outbreak of water-borne cryptosporidium infection in Milwaukee, Wisconsin, in the

TABLE 1.2

Examples of Various Infectious Diseases and Responsible Factors

Diseases	Responsible Factors
Airport malaria, rat-borne Hantaviruses	International travel and commerce
Hemolytic uremic syndrome, bovine spongiform encephalopathy, Creutzfeldt–Jakob disease	Globalization of food supplies, changes in food processing and packaging, organ or tissue transplantation, antibiotics usage
Antibiotic-resistant bacteria and antigenic drift in influenza virus	Microbial adaptation and changes as a response to the environment changes
Resurgence of tuberculosis, diphtheria, and cholera	Breakdown in public health measures such as curtailment or reduction in prevention programs, inadequate sanitation, and vector control measures
Rift Valley fever, Argentine hemorrhagic fever, Hantaan Korean hemorrhagic fever, Hantavirus pulmonary syndrome	Ecological changes such as agriculture, dams, deforestation, flood/drought, famine, or climate changes

spring of 1993, with over 400,000 estimated cases, was in part due to a nonfunctioning water filtration plant [18]. Similar deficiencies in water purification have also been found in other United States cities [19].

1.5.2 Food-Borne Diseases and Contributing Factors

Food-borne transmission of pathogenic and toxigenic microorganisms has been a well-documented and widely recognized hazard for decades. Even half a century ago, the dangers of botulism from underprocessed canned foods, staphylococcal poisoning from unrefrigerated food products, and salmonellosis from infected animal products were known. They pose major public health issues in developed as well as developing nations in different ways. It is a serious threat in developing nations where the availability of clean water and ideal food processing conditions are lacking. Microbial contamination, in the form of new pathogens, is the greatest concern in food safety in developed nations. Mass suppliers of food materials such as meat and dairy industries may inadvertently spread various microorganisms to unsuspecting consumers. Bacteria such as *Escherichia coli* and *Salmonella enteritidis* are more widespread as shown by the increase in the number of FDA-regulated products recalled for life-threatening microbial contamination [20]. In the United States alone, each year, food-borne illnesses affect up to 80 million people and cost an estimated $5 billion [21,22]. While many developed countries are using sophisticated data-collecting systems to report the incidence and causes of food-borne illness, the data may not represent all the infections. In many cases, infected individuals may not seek medical advice in case of minor infection. Even for those that seek treatment, their illness may not be recognized as food borne in origin or may not be properly reported.

Despite the forced protective measures of government regulations and demand due to economic reasons, changes in preservation techniques and lapses in following the recognized procedures have created unforeseen dangers. Salmonellosis is one of the most common and widely distributed food-borne diseases. It is generally contracted through the consumption of contaminated food of animal origin (mainly meat, poultry, eggs, and milk). The causative organisms pass through the food chain, from the primary production to consumption points such as households or food service establishments. Some of the new organisms responsible for food-borne disease are *E. coli* O157:H7, *Listeria monocytogenes*, *Campylobacter jejuni*, *Vibrio parahaemolyticus*, *Yersinia enterocolitica*, and others.

1.5.3   Human Behavioral Changes

Interestingly, changes in food consumption have brought some unrecognized hazards into the limelight. Consumption of fresh fruit and vegetables has increased dramatically due to the recent awareness toward carbohydrates and glycemic foods with the intention to control weight and follow a healthier life style. In general, the well-published value of high-fiber diet and an increase in health awareness have helped this trend. Fresh produce can be susceptible to microbial contamination during the various phases of transporting from the farm at one part of the globe to the dining table across the globe. Pathogens on the surface of produce such as mangoes or melons can contaminate the inner edible portion during cutting, and can easily multiply if the fruit slice is held at room temperature over a period.

Another notable change is in the general population's eating habits. The growing number of two-income families and the trend of eating away from home contribute fewer opportunities to discuss food safety information and follow best practices among family members. The percentage of income spent on food eaten away from home has significantly increased during recent decades. Fast food restaurants and salad bars are turning into primary sites for food consumption in today's fast-paced society [23]. The number of home-delivered meals—considered as the ultimate convenience food—has dramatically increased chiefly due to the rapid delivery mechanisms. Furthermore, aging has a profound effect on food safety considerations as the elderly are more susceptible to food-borne illnesses [24].

Developing nations are also experiencing food safety issues due to changes in food consumption practices. The booming economies of countries such as India and China, towing to the explosive growth in information technology (IT) and mass production, have led to changes in eating habits. The development of a variety of eating establishments in fast-growing cities to serve the increasingly affluent workforce poses concerns for proper food safety. Proper food-handling measures at such places are not implemented due to many reasons such as inadequate finance, less awareness, or inability of their governments and regulatory agencies to impose tight health regulations.

1.5.4 Industrialization and Technological Advancements

To provide greater demographic distribution of processed and unprocessed products, the food industry has changed toward the model of large, centralized processing centers, which facilitate the practice of modern food preservation techniques and offer greater economic advantages. This food industry model also carries huge risks of dispersing outbreaks to wider community, however unintentional. When mass-mediated food products are intermittently contaminated at a very low level during any segment of the complicated distribution channel, illnesses may appear as sporadic rather than part of an outbreak in an unexpected area. Consolidation of industries and mass distribution of foods in the industrialized nations could potentially lead to large outbreaks of food-borne diseases.

1.5.5 Travel and Commerce

International travel of humans and goods has increased dramatically during this century. Even with terrorist acts such as 9/11 affecting the extent of travel for a short duration, people resumed their travel habits shortly thereafter. It has been reported that international tourist arrivals in United States are expected to reach 937 million by 2010 [25]. Pathogens get a free ride with globetrotters and infect nontravelers as a result. Travelers may become infected with food-borne pathogens uncommon in their home country or their places of visit, thus complicating diagnosis and treatment when their symptoms begin after they return home or during their visit. Rapid increase in international travel and commerce brought diseases such as airport malaria and rat-borne Hantaviruses.

1.5.6 Prevention and Control

The primary objective of the public health system is to prevent disease before it occurs, since it is a common economic understanding that prevention is always better than cure. Surveillance and close monitoring are critical in meeting this objective. Outbreak response is also critical because even an ideal food safety system will not be able to prevent all possible food-borne illness. FoodNet Surveillance Network and PulseNet, National Antibiotic Resistance Monitoring System (NARMS), Foodborne Outbreak Response Coordinating Group (FORCG), Risk Assessment Consortium, and the Joint Institute for Food Safety Research are some of the organizations in promoting food safety [26].

Prevention of food-borne diseases depends on careful food production with proper handling of raw produce and preparation of finished foods. Health hazards can be introduced at any point from farm to table. Food preparations with a few practical food-handling precautions such as thorough heating and proper refrigeration can noticeably reduce the risk of food-borne diseases. Cross-contamination of foods can be avoided by separating cooked and raw foods. Food handlers should meticulously wash their hands, cutting boards, and contaminated surfaces as warranted to prevent cross-contamination. New safety standards and novel food-processing techniques should help to eliminate the deficiencies of archaic methods used.

1.6   Antimicrobial Drug Resistance

Drug-resistant pathogens are definitely a growing threat to all people, especially in healthcare settings. Each year about 90,000 people die because of their infection, among the nearly two million patients in the United States infected in a hospital. Unfortunately, more than 70% of the bacteria causing healthcare-acquired infections (HAI) are resistant to at least one of the drugs most commonly used to treat them [2,27]. As a consequence, persons infected with drug-resistant organisms are more likely to have longer hospital stays and require special treatment [27]. On the contrary, antimicrobials are one of the most commonly prescribed classes of drugs. The National Foundation for Infectious Diseases has estimated the annual cost of infection caused by drug-resistant organisms to be approximately $4 billion.

It is well understood that microorganisms have altered or are in the process of altering their genetics and physiology to become less susceptible to an antimicrobial agent because of Mother Nature's gift. They have been developing resistance to antimicrobial agents for several decades. In fact, as per a school of thought, there is no antimicrobial agent to which microbes cannot develop resistance [28]. Such developments in many pathogenic microbes pose one of the most serious problems in the arena of controlling infectious diseases [29].

Antimicrobial resistance is an inevitable consequence of the misuse, underuse, and abuse of antibiotics in animals and humans, exacerbating the natural selection pressures for decreased susceptibility to pharmacotherapies. Recent articles by Tenover [30] and Rice [31] covered the mechanisms of resistance to antimicrobial agents with case studies on *E. coli*'s resistance to third-generation cephalosporins, *Staphylococcus aureus'* resistance to high-level vancomycin, *Pseudomonas aeruginosa*'s resistance to multidrugs, and in depth account of methicillin-resistant *Staphylococcus aureus* (MRSA). As they pointed out, the prudent use of antibacterial drugs—using the appropriate drug at the appropriate dosage and for the appropriate duration—is one of the important means of reducing the selective pressure that helps resistant organisms emerge. Interested readers are referred to articles of Tenover [30], Rice [31], and Martinez [32] and the references cited therein. Antimicrobial drug misuse in both industrialized and developing countries is a problem in conjunction not only with human treatment but also with food production [33–35].

1.7 Nosocomial Infection and Related Issues

Nosocomial infection is deadly and prevalent. Organisms causing nosocomial infection are found virtually everywhere. Approximately 6% of all patients admitted to U.S. hospitals acquire nosocomial infections of which 4% prove fatal [36]. This issue is getting more complicated when the infection results due to drug-resistant organisms. According to the National Nosocomial Infection Surveillance (NNIS) statistics for infections acquired among ICU patients in the United States in 1999, 52% of infections resulting from *S. aureus* were identified as MRSA infections, and 25% of nosocomial infections were attributed to vancomycin-resistant enterococci (VRE). It reflects 37% and 43% increase from 1994 to 1998 [2], respectively. Nosocomial pneumonia, surgical wound infections, and vascular access–related bacteremia have caused the most illness and death in hospitalized patients with intensive care units (ICUs) identified as epicenters of antibiotic resistance [37]. Dettenkofer and Block provide a detailed description of the efficacy and safety issues of hospital disinfection [38,39]. A prevalence survey conducted under the auspices of WHO in 55 hospitals (14 countries) representing four WHO regions revealed that, on average, 8.7% of hospital patients suffer nosocomial infections.

Patients who are immunocompromised because of age, underlying diseases, or medical or surgical treatments are typically affected by nosocomial infections. Weinstein pointed out that aging of our population and increasingly aggressive medical and therapeutic interventions, including implanted foreign bodies, organ transplantations, and xenotransplantations, have created a cohort of particularly vulnerable persons [37]. Three major forces involved in nosocomial infections are (1) antimicrobial use in hospitals and long-term care facilities; (2) failure of hospital personnel to follow basic infection control, such as hand washing between patient contacts; and (3) patients in hospitals are immunocompromised increasingly. Infection control is making a paradigm shift. Given the choice of improving technology or improving human behavior, technology is the better choice. All infection control measures will need to continue to pass the test of the "four Ps": Biologically **P**lausible? Affordable and **P**ractical? Acceptable **P**olitically? Follow up by the **P**ersonnel? [40]

Currently, *Acinetobacter baumannii* is considered an important and emerging hospital-acquired pathogen [41]. It is responsible for 2%–10% of all gram-negative bacterial infections in ICUs in Europe and the United States [42]. One of the main concerns about *A. baumannii* is a characteristic that makes it a major threat to public health, namely, its remarkable ability to rapidly develop antibiotic resistance, mostly by acquisition of gene clusters carried by plasmids [43], transposons [44], or integrons [45,46], and arranged within the genome [47]. Despite vast research into antibiotic-resistant *A. baumannii*, it remains a puzzling bacterium that requires additional studies to solve all its mysteries, as commented by Richet and Fournier [48].

1.7.1   Initiatives in Hand Hygiene

Modern infection control is grounded in the work of Ignaz Semmelweis, who even in 1840s demonstrated the importance of hand hygiene for controlling transmission of infection in hospitals. He initiated a mandatory hand-washing policy for medical students and physicians [49]. In 1976, the Joint Commission on Accreditation of Healthcare Organizations published accreditation standards for infection control, creating the impetus and need for hospitals to provide administrative and financial support for infection control programs. In 1995, the CDC's study on the efficacy of nosocomial infection control reported that hospitals with four key infection control components—an effective epidemiologist, one infection control practitioner for every 250 beds, active surveillance mechanisms, and ongoing control efforts—reduced nosocomial infection rates by approximately one-third [50].

Hand hygiene is considered the most effective strategy in combating hospital-associated infection. According to the WHO guidelines, factors influencing poor adherence to recommended hand hygiene practices include risk factors such as irritating hand hygiene agents, lack of amenities (such as soap, paper, towels), lack of knowledge of guidelines and protocols, and additional perceived barriers such as lack of institutional priority. Compliance by healthcare workers for recommended hand hygiene procedures needs to be improved since compliance remains low [51–54]. Many factors have contributed to such poor compliance. They include a lack of knowledge on the importance of hand hygiene in reducing the spread of infection and on how hands become contaminated, lack of understanding of correct hand hygiene techniques, understaffing and overcrowding, poor access to hand hygiene facilities, irritant contact dermatitis associated with frequent exposure to poorly formulated soaps, and lack of institutional commitment to good hand hygiene [51–53,55–57].

To overcome these barriers, the CDC's Healthcare Infection Control Practices Advisory Committee (HICPAC) published a comprehensive "Guideline for hand hygiene in healthcare settings" in 2002. WHO launched its guidelines on hand hygiene in healthcare (advanced draft) in October 2005 and will issue a final version in 2007. These global consensus guidelines reinforce the need for multidimensional strategies involving key elements such as staff education and motivation, adoption of an alcohol-based hand rub as the primary method for hand hygiene, use of performance indicators, and strong commitment by all stakeholders. As pointed out by Kaiser and Newman, sophisticated formulation technologies and the proper selection of ingredients can provide products that encourage hand hygiene and can be a part of the solution for compliance improvement, an imperative in today's healthcare environments [58]. It is also interesting to note that even unsuspected areas such as patients' files in high-risk areas of ICUs can harbor multidrug-resistant pathogenic bacteria [59]. Conscientious healthcare workers (HCWs) must take steps to prevent the contamination of patients' files by strict hand hygiene compliance before entering the case notes [59].

1.7.2 Device-Related Infections

The insertion of prosthetic implants has become a clinically applied, often life-saving procedure in modern medicine. Whereas initially unlimited opportunities were anticipated, development of prosthetic devices for every conceivable application inherently carries several basic challenges. Bacterial contamination and consequent implant infection may have blurred the enthusiasm slightly. Therapy of such infection is difficult and most often irreversible, in many cases requiring complete removal of implanted material. The past few years have seen a dramatic increase in the types and numbers of medical devices. It is estimated that there are some 6,000 distinct types or generic groups of medical devices, and some 750,000 or more brands and models, ranging from simple to very complex devices. It is estimated that the costs of treating an infected joint prosthesis exceeded four- to sixfold the costs of the original prosthetic joint replacement [60]. Furthermore, the indirect costs can be significant as patients are precluded from social and occupational participation due to the prolonged hospitalization and rehabilitation.

Device-related infections add another well-understood dimension to the infectious disease challenges. Advances in medical devices (such as catheters, guidewires, stents, and pacemakers) and other invasive products have enormously improved or helped diagnostic and therapeutic practices, and thereby millions of human lives. However, the benefits of invasive devices are limited by the occurrence of device-related infections, even when the best aseptic techniques are practiced. Device-related infection occurs through many simple or complicated pathways such as the introduction of microorganisms during the device insertion or implantation procedure or from the attachment of blood-borne organisms to the newly inserted device and their subsequent propagation on its surface. Good clinical practices such as thoroughly cleaning and disinfecting the area before insertion; proper prepping by the clinical staff; and care in handling the device maintaining sterility prior to insertion, proper design, and cleaning/sterilization of surgical tools will significantly reduce infection occurrence.

The surfaces of indwelling medical devices, including prosthetic joints, are an excellent platform for the formation of life-threatening infections through biofilms [61]. Although biofilms that colonize indwelling medical devices have been studied extensively for the last 25 years [61–67], there still are some gaps in knowledge to be filled with respect to biofilm structure, community composition, and physiologic activity of biofilm organisms. Incorporation of antimicrobials in the bulk material that constitutes a device can be effective but costly, and various surface treatments are emerging as important and less expensive alternatives. Since more and more catheters are used for diverse applications such as chemotherapeutic, diagnostic, therapeutic, transfusional, and nutritional supportive care involving entirely different materials and surfaces, sophisticated approaches are needed to prevent infection.

1.7.2.1   Orthopedic Implant Infections

Orthopaedic implants help millions of people to stay active and maintain quality of life. However, these medical devices are prone to infection, forcing patients back to surgery for replacement or repair. Orthopaedic implant infection is a devastating disease because of the physical and emotional trauma the patient encounters associated with revision surgery, compounded by long-term postoperative treatment [64,68]. Current rates of infection for artificial joints may vary by hospital, surgeon, and scope of the study. But the best estimates suggest an infection rate of 1%–2%, which produces a figure of 4000–8000 infected arthroplastics requiring surgical revision annually [68]. It is estimated by the Center for Biofilm Engineering at Montana State University that there are 1.4 million prosthetic devices that become infected each year in the United States [68]. The cost, morbidity, and patient suffering caused by these persistent infections are enormous. The best strategy to combat orthopaedic implant biofilm infections is to prevent their occurrence through a combination of engineering, surgical, and micro-biological techniques.

1.7.2.2   Catheter-Associated Urinary Tract Infection

Catheter-associated urinary tract infection (CAUTI) is the most common nosocomial infection. It comprises >40% of all institutionally acquired infections [69]. Several catheter-care practices are universally recommended to prevent or at least delay the onset of CAUTI. These include avoiding unnecessary catheterizations, considering a condom or suprapubic catheter, having a trained professional insert the catheter aseptically, removing the catheter as soon as no longer needed, maintaining uncompromising closed drainage, minimizing manipulations of the system, separating catheterized patients, and considering adopting a novel anti-infective catheter [70–73]. However, few of these practices have been proven effective by randomized controlled trials. CAUTIs comprise perhaps the largest institutional reservoir of nosocomial antibiotic-resistant pathogens such as *E. coli, Enterococci, P. aeruginosa, Klebsiella, Enterobacter* spp., and *Candida* spp. Novel urinary catheters impregnated with nitrofurazone or minocycline and rifampin or coated with a silver alloy hydrogel exhibit anti-infective surface activity, which significantly reduces the risk of CAUTI for short-term catheterizations not exceeding two to three weeks [69]. Measures such as the use of anti-infective lubricants, sealed catheter-collection tubing junctions, microbe-impervious antireflux valves, continuous irrigation of the bladder with an anti-infective solution, conformable urethral catheters, instillation of anti-infectives into the collection bag, and anti-infective catheter material which use antimicrobial drugs such as nitrofurazone, minocycline–rifampin, and silver oxide are some of the current strategies employed in different devices [69–73].

1.8 Regulatory Constraints

Infection control professionals are facing an unprecedented wave of legislative and regulatory activity as individual states and federal agencies in the United States demand the public disclosure of infection rates. For example, legislation introduced in early part of 2006 in Illinois requires hospitals to screen all patients for MRSA in accordance with guidelines published by CDC. In addition to MRSA, emerging as a distinctly different community-acquired strain, VRE and *Clostridium difficile* may get the needed regulatory attention. As noted in the Web site for the Association for Professionals in Infection Control and Epidemiology (APIC), 14 states including Colorado, Connecticut, New Hampshire, South Carolina, Tennessee, and Vermont have adopted laws requiring mandatory reporting of hospital infection rates and many others are in some state of legislative study or discussion. At present, the impact of public reporting of infection data on the quality of patient care, consumers, or hospitals is not fully known. A unified, national infection rate reporting system by the National Quality Forum (NQF), APIC, and other stakeholders is expected to be completed in February 2007 and will provide nationwide infection rates on a single database. Regulatory agencies in Canada, United States, and Europe place huge constraints on antimicrobials, disinfectants, and medical device manufacturers on the proper usage of active materials and registering them at various levels. The manufacturers as well as consumers should have a clear understanding of the regulations.

1.9 Concluding Remarks

Overconfidence or ignorance has made humans vulnerable to emerging and reemerging diseases. These diseases are viewed as corrective actions of the evolutionary principles. Further, the history of infection control suggests that there is no quick fix in spite of all our medical and technological hubris. If humans are able to overcome the current issues, it will be through sensible changes in behavior, improved sanitization, and unwavering commitment to maintain the ecological balance. Although civilization can expose people to new pathogens, cultural progress also has a countervailing effect: it provides invaluable tools—medicines, sensible city planning, and education. Moreover, because biology favors microorganisms, people have no other choice except to depend on protective cultural practices such as vaccinations, adherence to the proper hand hygiene practices by healthcare providers and food handlers, and intelligent use of antimicrobials. All too often, simple protective measures, such as using only clean hypodermic needles or treating urban drinking water with chlorine, are neglected, whether out of ignorance or a misplaced emphasis on the short-term financial costs. Such lapses are showing the effects that the delicate balance between humans and invasive microorganisms can tip the other way. The increasing incidence of antimicrobial resistance has important implications for clinical management and infection control practices, disease surveillance and diagnosis, drug and vaccine development priorities, and professional and public education strategies.

Some of the identified basic concepts in disease emergence [74] listed below have not changed:

1. Emergence of infectious disease is complex.
2. Infectious diseases are dynamic and most new infections are not caused by genuinely new pathogens.
3. Agents involved in new and reemergent infections cross taxonomic lines.
4. Human activities are the most potent factors driving disease emergence.
5. Social, economic, political, climatic, technologic, and environmental factors shape the disease patterns and significantly influence emergence.
6. Understanding and responding to disease emergence require a global perspective independent of geographical and economic limitations.
7. Shrinking of the globe due to current globalization favors disease emergence.

It is not very difficult to predict what challenges the future holds. We are likely to confront a future influenza pandemic, continuing challenges posed by drug-resistant organisms, and additional chronic diseases. Recent outbreaks in different parts of the globe illustrate the global implications of a local problem [75–77]. The prevention and control of infectious diseases require prompt recognition by alert healthcare providers, clinicians, microbiologists, and public health personnel, and prompt application of sophisticated epidemiologic, molecular biologic, behavioral, and statistical approaches and technologies. Prevention is always cheaper than the treatment, and early identification of disease in a population gives us the luxury of using prevention for many. Earlier identification followed by isolation and treatment reduces the mortality rate significantly. The following statement of Nobel laureate Joshua Lederberg in 1997 still seems to be applicable: "Despite many potential defenses—vaccines, antibiotics, diagnostic tools—we are intrinsically more vulnerable than before, at least in terms of pandemic and communicable diseases" [78].

References

1. Armelagos, G.J., The viral superhighway, *The Sciences*, 24, 1998.
2. www.cdc.gov.
3. The agricultural use of antibiotics and its implications for human health, United States General Accounting Office, RCED-99-4 (April), 1999.
4. Black, R.E., Morris, S.S., and Bryce, J., Where and why are 10 million children dying every year? *Lancet*, 361 (June), 2226, 2003.
5. Drug resistance threatens to reverse medical progress, World Health Organization, 2000.
6. Schraag, J., Avian influenza: Why infection prevention and control maybe our only effective weapon, *Infection Control Today*, November, 32, 2006.
7. Avian influenza ("bird flu")—Fact sheet, World Health Organization, 2006.
8. Key facts about avian influenza (bird flu) and avian influenza A (H5N1) virus, Centers for Disease Control and Prevention, 2006.
9. Zelicoff, A.P. and Bellomo, M., Microbe: Are we ready for the next plague? *AMACOM/American Management Association*, 2006.
10. Dawkins, R. and Krebs, J.R., Arms races between and within species, *Proc. R. Soc. London*, B205, 489, 1979.
11. Older Americans 2000: Key indicators of well being, www.agingstats.gov/chartbook2000, Federal Interagency Forum on Aging Related Statistics, 2000.
12. The future of human life expectancy: Have we reached the ceiling or is the sky the limit? National Institute on Aging—Research Highlights in the Demography and Economics of Aging, March (8), 1, 2006.
13. Greene, L., Emerging infectious diseases, *Infection Control Today*, 10 (9), 46, 2006.
14. Lederberg, J., Shope, R.E., and Oaks, S.C.J., Emerging Infections: Microbial Threats to Health in the United States, Institute of Medicine, National Academy Press, Washington, DC, 1992.
15. A public health action plan to combat antimicrobial resistance Part I: Domestic issues, 2005 Annual Report, www.cdc.gov, 2006.
16. Hughes, J.M. and Tenover, F.C., Infectious disease challenges of the 1990s, *Infect. Med.*, 13 (9), 788, 1996.
17. Berkelman, R.L., Bryan, R.T., Osterholm, M.T., Le Duc, J.W., and Hughes, J.M., Infectious disease surveillance: A crumbling foundation, *Science*, 264 (5157), 368, 1994.
18. MacKenzie, W.R., Hoxie, N.J., Proctor, M.E., Gradus, M.S., Blair, K.A., Peterson, D.E., et al., A massive outbreak in Milwaukee of cryptosporidium infection transmitted through the public water supply, *N. Engl. J. Med.*, 331 (3), 161, 1994.
19. Assessment of inadequately filtered public drinking water—Washington, DC, December 1993, *Morb. Mortal. Wkly. Rep.*, 43 (36), 661, 1994.
20. Food safety from farm to table: A new strategy for the 21st century, U.S. Food and Drug Administration, Discussion draft, Washington DC, 1997.
21. Foodborne pathogens: Risks and consequences, Council for Agricultural Science and Technology, Task Force Report, 122, 1994.
22. Altekruse, S.K., Cohen, M.L., and Swerdlow, D.L, Emerging foodborne diseases, *Emerg. Infect. Dis.*, 3 (3), 285, 1997.
23. Manchester, A. and Clauson, A., 1994 Spending for food away from home outpaces food at home, *Food rev.*, 18, 12, 1995.

24. Zink, D.L., The impact of consumer demands and trends on food processing, *Emerg. Infect. Dis.*, 3 (4), 467, 1997.
25. Paci, E., Exploring new tourism marketing opportunities around the world, Proceedings of the 11th General Assembly of the World Tourism Organization, Cairo, Egypt, 1995.
26. Calvin, L., Response to US foodborne illness outbreaks associated with imported produce, USDA Bulletin number 789-5, 2004.
27. CDC, Antimicrobial resistance in healthcare settings, http://www.cdc.gov/ncidod/dhqp/ar.html#2 (accessed on December 11, 2006).
28. Levy, S.B., The Antibiotic Paradox: How Miracle Drugs Are Destroying the Miracle, Plenum Press, New York, 1992.
29. Chapman, J.S., Diehl, M.A., and Fearnside, K.B., Preservative tolerance and resistance, *Int. J. Cosmet. Sci.*, 20, 31, 1998.
30. Tenover, F.C., Mechanisms of antimicrobial resistance in bacteria, *Am. J. Infect. Control*, 34, S3, 2006.
31. Rice, L.B., Antimicrobial resistance in gram-positive bacteria, *Am. J. Infect. Control*, 34, S11, 2006.
32. Martinez, J.E., Resistance to disinfectants, Part 1, *Bioprocess Int.* (September), 32, 2005.
33. Fidler, D.P., Return of the fourth horseman: Emerging infectious diseases and international law, *Minn. Law Rev.*, 81 (3), 771, 1997.
34. Fidler, D.P., The globalization of public health: Emerging infectious diseases and international relations, *Ind. J. Global Leg. Stud.*, 5 (1), 11, 1997.
35. Fidler, D.P., Legal issues associated with antimicrobial drug resistance, *Emerg. Infect. Dis.*, 4 (2), 169, 1998.
36. DeLoach, C. and Brown, J., Cleaning for health: Understanding green cleaning, ASHES Conference, Phoenix, AZ, September, 25, 2005.
37. Weinstein, R.A., Nosocomial infection update, *Emerg. Infect. Dis.*, 4 (3), 416, 1998.
38. Dettenkofer, M., Wenzler, S., Amthor, S., Anter, G., Motschzall, E., and Daschner, F.D., Does disinfection of environmental surfaces influence nosocomial infection rates? A systematic review, *Am. J. Infect. Control*, 32 (2), 84, 2004.
39. Dettenkofer, M. and Block, C., Hospital disinfection: Efficacy and safety issues, *Curr. Opin. Infect. Dis.*, 18 (4), 320, 2005.
40. Weinstein, R.A., SHEA consensus panel report: A smooth takeoff, *Infect. Control Hosp. Epidemiol.*, 19, 91, 1998.
41. Villegas, M.V. and Hartstein, A.I., Acinetobacter outbreaks, 1977–2000, *Infect. Control Hosp. Epidemiol.*, 24, 284, 2003.
42. Hanberger, H., Garcia-Rodriguez, J.A., Gobernado, M., Goossens, H., Nilsson, L.E., and Struelens, M.J., Antibiotic susceptibility among aerobic gram-negative bacilli in intensive care units in 5 European countries. French and Portugese ICU study groups, *JAMA*, 281, 67, 1999.
43. Seifert, H., Boullion, B., Schulze, A., and Pulverer, G., Plasmid DNA profiles of *Acinetobacter baumannii*: Clinical application in a complex endemic setting, *Infect. Control Hosp. Epidemiol.*, 15, 520, 1994.
44. Devaud, M., Kayser, F.H., and Bachi, B., Transposon-mediated multiple antibiotic resistance in *Acinetobacter* strains, *Antimicrob. Agents Chemother.*, 22, 323, 1982.
45. Segal, H., Thomas, R., and Gay, E.B., Characterization of class 1 integron resistance gene cassettes and the identification of a novel IS-like element in *Acinetobacter baumannii*, *Plasmid*, 49, 169, 2003.

46. Poirel, L., Menuteau, O., Agoli, N., Cattoen, C., and Nordmann, P., Outbreak of extended spectrum beta-lactamase VEB-1 producing isolates of *Acinetobacter baumannii* in a french hospital, *J. Clin. Microbiol.*, 41, 3542, 2003.
47. Fournier, P.E., Vallenet, D., Barbe, V., Audic, S., Ogata, H., Poirel, L., et al., Comparative genomics of multidrug resistance in *Acinetobacter baumannii*, *PLoS Genet.*, 2:e7, 2006.
48. Richet, H. and Fournier, P.E., Nosocomial infections caused by *Acinetobacter baumannii*: A major threat worldwide, *Infect. Control Hosp. Epidemiol.*, 27 (7), 645, 2006.
49. Best, M. and Neuhauser, D., Ignaz Semmelweis and the birth of infection control, *Qual. Saf. Health Care*, 13, 233, 2004.
50. Haley, R.W., Culver, D.H., White, J.W., Morgan, W.M., Emori, T.G., Munn, V.P., and Hootan, T.M., The efficacy of infection surveillance and control programs in preventing nosocomial infections in US hospitals, *Am. J. Epidemiol.*, 121, 182, 1985.
51. Pittet, D., Mourouga, P., and Perneger, T.V., Compliance with handwashing in a teaching hospital. Infection control program, *Ann. Intern. Med.*, 130 (2), 126, 1999.
52. Pittet, D., Hugonnet, S., Harbarth, S., Mourouga, P., Sauvan, V., Touveneau, S., and Perneger, T.V., Effectiveness of a hospital-wide programme to improve compliance with hand hygiene. Infection control programme, *Lancet*, 356 (9238), 1307, 2000.
53. Pittet, D., Improving adherence to hand hygiene practice: A multidisciplinary approach, *Emerg. Infect. Dis.*, 7 (2), 234, 2001.
54. Boyce, J., It is time for action: Improving hand hygiene in hospitals, *Ann. Intern. Med.*, 130, 153, 1999.
55. Boyce, J.M. and Pittet, D., Guideline for hand hygiene in health-care settings. Recommendations of the Healthcare Infection Control Practices Advisory Committee and the HIPAC/SHEA/APIC/IDSA Hand Hygiene Task Force, *Am. J. Infect. Control*, 30 (8), S1, 2002.
56. Pyrek, K.M., Hand hygiene: New initiatives on the domestic and global fronts, *Infection Control Today*, 22 ff., 2006.
57. Pittet, D., Hugonnet, S., Harbarth, S., et al., Effectiveness of a hospital-wide programme to improve compliance with hand hygiene, *Lancet*, 356, 1307, 2000.
58. Kaiser, N.E. and Newman, J.L., Formulation technology as a key component in improving hand hygiene practices, *Am. J. Infect. Control*, 34 (10), S81, 2006.
59. Panhotra, B.R., Saxena, A.K., and Al-Mulhim, A.S., Contamination of patients' files in intensive care units: An indication of strict handwashing after entering case notes, *Am. J. Infect. Control*, 33 (7), 398, 2005.
60. Sugarman, B. and Young, E.J., Infections associated with prosthetic devices: Magnitude of the problem, *Infect. Dis. Clin. N. Am.*, 3, 187, 1989.
61. Donlan, R.M. and Costerton, J.W., Biofilms: Survival mechanisms of clinically relevant microorganisms, *Clin. Microbiol. Rev.*, 15 (2), 167, 2002.
62. Donlan, R.M., New approaches for the characterization of prosthetic joint biofilms, *Clin. Orthop. Relat. Res.* (437), 12, 2005.
63. Borriello, G., Werner, E., Roe, F., Kim, A.M., Ehrlich, G.D., and Stewart, P.S., Oxygen limitation contributes to antibiotic tolerance of *Pseudomonas aeruginosa* in biofilms, *Antimicrob. Agents Chemother.*, 48, 2659, 2004.
64. Stoodley, P., Kathju, S., Hu, F.Z., Erdos, G., Levenson, J.E., Mehta, N., Dice, B., Johnson, S., et al., Molecular and imaging techniques for bacterial biofilms in joint arthroplasty infections, *Clin. Orthoped. Relat. Res.* (437), 31, 2005.

65. Patel, R., Biofilms and antimicrobial resistance, *Clin. Orthoped. Relat. Res.* (437), 41, 2005.
66. Allon, M., Saving infected catheters: Why and how, *Blood Purif.*, 23 (1), 23, 2005.
67. Franco, F.F.S., Spratt, D., Leao, J.C., and Porter, S.R., Biofilm formation and control in dental unit waterlines, *Biofilms*, 2, 9, 2005.
68. Ehrlich, G.D., Stoodley, P., Kathju, S., Zhao, Y., McLeod, B., Balaban, N., Hu, F.Z., et al., Engineering approaches for the detection and control of orthopaedic biofilm infections, *Clin. Orthoped. Relat. Res.*, 437, 59, 2005.
69. Maki, D.G. and Tambyah, P.A., Engineering out the risk of infection with urinary catheters, *Emerg. Infect. Dis.*, 7, 1, 2001.
70. Stamm, W.E., Catheter associated urinary tract infections: Epidemiology, pathogenesis, and prevention, *Am. J. Med.*, 91 (suppl 3B), 65S, 1991.
71. Burke, J.P. and Riley, D.K., Nosocomial urinary tract infection, Mayhall, C.G. (ed.), *Hospital Epidemiology and Infection Control*, Baltimore, Williams and Wilkins, 139, 1996.
72. Warren, J.W., Catheter-associated urinary tract infections, *Infect. Dis. Clin. N. Am.*, 11, 609, 1997.
73. Kunin, C.M., Care of the urinary catheter, *Urinary Tract Infections: Detection, Prevention and Management*, Baltimore, Williams and Wilkins, Fifth edition, 227, 1997.
74. Wilson, M.E., Levins, R., and Spielman, A. (eds.), *Disease in Evolution: Global Changes and Emergence of Infectious Diseases*, New York Academy of Sciences, 740, 1, 1995.
75. Edler, A.A., Avian flu (H5N1): Its epidemiology, prevention, and implications for anesthesiology, *J. Clin. Anesth.*, 18 (1), 1, 2006.
76. Salzberger, B. and Ruf, B., Avian influenza and pandemic influenza threat, *Eur. J. Med. Res.*, 11 (1), 1, 2006.
77. Allen, P.J., Avian influenza pandemic: Not if, but when, *Pediatr. Nurs.*, 32 (1), 76, 2006.
78. Lederberg, J., Infectious disease as an evolutionary paradigm, *Emerg. Infect. Dis.*, 3 (4), 417, 1997.

2

Sisyphus in the Microbial World Revisited: Global Governance, Antimicrobial Strategies, and Humanity's Health

David P. Fidler

CONTENTS

2.1   Introduction

Chapter 1 analyzes some fascinating and frightening aspects of the challenges pathogenic microbes create. This chapter explores a different manifestation of the pathogenic challenge because it considers the radical changes in governance that have characterized global health activities against infectious diseases in the first years of the twenty-first century. In an earlier companion volume to this book published in 2000, the editors invited me, an international lawyer, to reflect on broader political, economic, cultural, and legal issues that affected antimicrobial strategies at the local, national, and global levels [1]. As described below, I posited that humankind was on the cusp of a fourth epidemiological transition, the nature of which would be determined not by science and technology but by how societies individually and collectively governed pathogenic threats. I used the image of Sisyphus to capture the labors the global community might have to endure to address the emerging and reemerging threats posed by infectious diseases.

The intervening years have seen governance on global pathogenic threats revolutionized. This revolution suggests that the fourth epidemiological transition has, in fact, begun. The changes seen in global health since 2000 provide me with an opportunity to revisit Sisyphus in the microbial world. In this chapter, I analyze the governance revolution that has taken place in global health with respect to infectious diseases. The governance transformations in global health are unprecedented and have produced, in the words of David Heymann of the World Health Organization (WHO), a "new way of working" on infectious diseases in the twenty-first century [2].

Sisyphus still labors in the microbial world, however, because all is not well in the wake of the governance revolution. This chapter looks particularly at the increased desire for "architecture" for global health governance. This desire reflects growing anxiety and frustration with the continued, and often worsening, burden infectious diseases impose on humanity, particularly those who are most vulnerable to infection. I reflect on the increased use of the architecture metaphor in discourse on global health and argue that a different metaphor or analogy is needed to describe and shape global governance on health in the first decades of the twenty-first century.

2.2   The Fourth Epidemiological Transition

In my chapter to the volume published in 2000, I used the theory developed by anthropologist George Armelagos and his colleagues that humankind has passed through three epidemiological transitions [3]. According to this theory, the first transition occurred when nomadic societies engaged in agriculture and domesticated animals as livestock. This transition provided pathogenic microbes with fertile territory into which to penetrate and spread locally and globally.

The second epidemiological transition happened when science and technology gave humankind the upper hand in the struggle with infectious diseases. The second transition witnessed the dawn of the "golden age" of scientific and technological advances against microbial threats. The third epidemiological transition reveals humanity confronted by "a disheartening set of changes that in many ways have reversed the effects of the second transition and coincide with the shift to globalism" ([3], p. 27). Powerful evidence of the third transition was found in the global crisis of emerging and reemerging infectious diseases detected in the 1990s [4]. Symbolic of the third transition was the relentless march of HIV/AIDS across the planet and the devastation this pandemic caused in developing nations, especially in sub-Saharan Africa.

In 2000, I wrote that "[t]he catalysts for any shift to a fourth epidemiological transition will be political, legal, economic, and cultural changes that allow humankind to organize its affairs to contain the threat from pathogenic microbes" ([1], p. 332). Rather than a second golden age of scientific and technological progress, I argued that multidimensional antimicrobial strategies, which integrated scientific progress with political, economic, and legal changes, would be the touchstones of the next epidemiological transition. Such strategies would constitute the building blocks of new approaches to governance nationally and globally concerning infectious disease threats.

Put another way, the fourth epidemiological transition was, I predicted, to be highly political and would be characterized by a new kind of *microbial-politik*. Paradoxically, the success achieved by scientific and technological progress during the second epidemiological transition contributed to many perceiving public health as merely a technical, humanitarian endeavor separated from politics, especially the politics of foreign policy and international relations. The lack of foreign policy and diplomatic interest in public health led to public health's political neglect and obscurity [5,6].

The crises of emerging and reemerging infectious diseases that unfolded in the third epidemiological transition thrust public health back into politics because political solutions to such crises had to be crafted and maintained within and among the nations of the world. The fourth epidemiological transition needed, more than anything, to be a governance transition of historic proportions.

2.3 The Governance Revolution in Global Health

The first 6 years of the twenty-first century have witnessed such dramatic changes in public health governance that I do not overstate matters by arguing that a governance revolution has taken place. Although indications of public health's political renaissance were appearing at the turn of the century, the extent of the transformation of public health's political and governance importance has surprised many experts. The past 6 years have seen public health emerge from obscurity and neglect to become important to many foreign policy and international political agendas [7–9]. Public health now features prominently in policy debates and activities concerning national and homeland security, international trade, development strategies, environmental protection, global governance, and human rights. The political attention health now receives is unprecedented. Ilona Kickbusch captured this change in public health's political status:

> The protection of health is no longer seen primarily as a humanitarian and technical issue relegated to a specialised UN agency, but more fully considered in relation to the economic, political, and security consequences for the complex post–Cold War system of interdependence. This has led to new policy and funding initiatives at many levels of governance and a new political space within which global health action is conducted. ([6], pp. 192–193)

More than any other factor, threats from the microbial world have caused public health's political revolution. Threats from pathogenic microbes have driven the rise of public health in the foreign policies of nations and in activities undertaken by intergovernmental organizations and non-state actors (e.g., nongovernmental organizations and multinational corporations). The political impact of the pathogen problem can be captured by considering two axes, the "axis of evil" and the "axis of illness." The axes help communicate the political dangers state and non-state actors believe microbes present.

2.3.1 The Axis of Evil: Biological Weapons, Biological Terrorism, and Public Health's New Security Importance

The axis of evil is, of course, President Bush's now famous description of a set of problems threatening U.S. national and homeland security after the tragic events of September 11, 2001 [10]. Although many think of Iran, Iraq, and North Korea as the axis of evil, the axis actually represents a post–9/11 strategic framework for thinking about U.S. security in the early twenty-first century. The axis of evil focuses U.S. security policy on three interdependent threats: repressive regimes, international terrorism, and weapons of mass destruction (WMD) (Figure 2.1).

The significance of the axis of evil for public health flows from the growing dangers policy makers and experts perceive exist with respect to the proliferation of biological weapons and the perpetration of biological terrorism. Biological weapons are not a new foreign policy problem [11], but the axis of evil framework communicates that the biological weapons threat has significantly changed. Critical to the changed nature of this threat is the need to prepare for the potential use of biological weapons against civilian

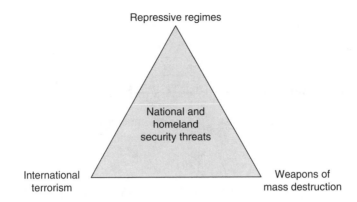

FIGURE 2.1
The axis of evil. (From Fidler, D.P., *McGeorge Law Rev.*, 35, 45, 2004.)

populations [12]. Such preparation makes national and international public health capabilities vital to homeland, national, and international security. The United States has been particularly focused on the threat of biological terrorism [13], and its hegemonic status in world politics has ensured that the relationship between public health and security has become important globally.

The linkage of public health with the pursuit of security against the malevolent use of microbes has taken public health into the "high politics" of foreign policy and international relations—the politics that address the state's security from violent attack. This linkage represents the collision of two policy areas—security and public health—previously separate from each other. The collision has produced controversies I cannot explore here; but the rise of public health on homeland, national, and international security agendas constitutes one of the most dramatic aspects of the governance revolution that has taken place in public health nationally and globally.

2.3.2　The Axis of Illness: Emerging and Reemerging Infectious Diseases

The second major pathogen-related vector transforming public health's political and governance status involves emerging and reemerging infectious diseases. In other writings, I developed the axis of illness to capture the key factors that converge to produce the dangers naturally occurring infectious diseases present [14,15] (Figure 2.2). The axis of illness attempts to synthesize lists of factors developed by public health experts to communicate the complexity of the threat posed by infectious diseases in an era of globalization [16,17]. The components of the axis represent key policy and governance challenges facing societies increasingly at risk from microbial emergence and spread.

Microbial resilience describes the importance of microbial, genetic, and biological factors that drive pathogenic evolution and its impact on human populations. The continuing global problem of antimicrobial resistance

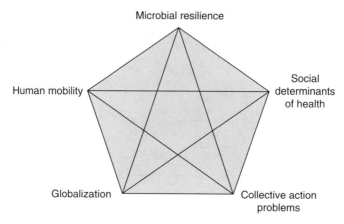

FIGURE 2.2
The axis of illness. (From Fidler, D.P., *SARS, Governance, and the Globalization of Disease*, Palgrave Macmillan, Basingstoke, United Kingdom, 2004.)

perhaps best illustrates the challenge microbial resilience presents [18]. Human mobility underscores how international trade, travel, and migration play critical roles in infectious disease emergence and spread. This factor includes the contributions that technology and industry make to the increased speed, scope, and impact of human mobility on pathogenic threats. The emergence and global spread of severe acute respiratory syndrome (SARS) in 2003 provides an example of the dangers human mobility presents in the context of a new transmissible pathogen.

Social determinants of health emphasizes the underlying problems that stimulate microbial emergence in, and penetration of, societies. These determinants include lack of access to clean water, sanitary housing, adequate food and nutrition, and education. More broadly, poverty continues to be a prolific generator of microbial-related danger and death.

Globalization focuses on many factors that speed up economic, technological, industrial, and cultural interconnectedness in ways that de-territorialize human behavior and the environment in which people and pathogens interact. The spread of HIV/AIDS, SARS, avian influenza, and resistant forms of tuberculosis have all been linked with the processes of globalization. Collective action problems describe the policy challenges created by infectious disease emergence and spread. These challenges include the breakdown or decay of public health capabilities at local, national, and global levels. Stronger and more effective collective political action and governance within and among states is required to address successfully the challenge of resurgent infectious diseases.

The threat the axis of illness tries to capture has generated activity in international relations that forms part of the political revolution transforming global health. Emerging and reemerging infectious diseases have risen in prominence across many policy agendas, but especially in the areas of security, trade, development policy, humanitarian action, and reform of global governance. Never before has control of naturally occurring infectious diseases had the foreign policy and international political importance it developed in the first half of the twenty-first century.

An efficient way to communicate this far-reaching and historic shift is to analyze how frequently control of infectious diseases featured in strategies to reform the United Nations (UN) for the challenges of the twenty-first century. UN reform is not a new governance issue in international relations, but never before has public health featured in UN reform proposals as significantly as it did in the December 2004 report of the UN Secretary-General's High-Level Panel [19] and in the UN Secretary-General's own March 2005 report [20]. These important documents on UN reform integrate the need for public health improvements across the range of problems the UN and its member states confront in the twenty-first century. The High-Level Panel argued that infectious disease control was a critical element of what it termed "comprehensive collective security" ([19], p. 14). The panel asserted that the weaknesses of public health nationally and globally threatened achievement of comprehensive collective security, and it called for global public health capabilities to be rebuilt.

Each of the UN Secretary-General's objectives for UN governance reform—freedom from fear, freedom from want, and freedom to live in dignity—depends on the international community making progress on the control of infectious diseases, especially HIV/AIDS. The connection between infectious disease control and development was also prominent in the UN Secretary-General's analysis. To combat poverty, the UN Secretary-General emphasized fulfillment of the Millennium Development Goals, three of which address specific health threats (child mortality; maternal health; and combating HIV/AIDS, malaria, and other diseases) and four of which seek improvement in key social determinants of health (extreme poverty and hunger, universal primary education, gender equality, and environmental sustainability). The UN Secretary-General also emphasized the need to strengthen global infectious disease surveillance to combat both naturally occurring infectious diseases and any acts of biological violence perpetrated by terrorists or states.

One message sent by public health's prominence in UN reform strategies is that public health, especially infectious disease control, represents a strategic "best buy" for national and global governance because public health is an integrated public good that contributes to the fight against security risks, disruptions of trade and commerce, poverty, disease, environmental degradation, and threats to human dignity. The importance public health has achieved suggests that, in *microbialpolitik*, public health itself has become an independent marker of good governance, alongside democracy, the rule of law, and human rights.

For example, the High-Level Panel argued that improved global surveillance was important not only for addressing intentional and naturally occurring pathogenic threats but also for "building effective, responsible

[s]tates" ([19], p. 29). Similarly, the Bush administration has identified effective public health systems as an indicator for larger governance attributes: "Pandemics require robust and fully transparent public health systems, which weak governments and those that fear freedom are unable or unwilling to provide" ([21], p. 48). These associations of effective public health with more transcendent notions of good governance and political ideology illuminate the dramatic sea change public health has undergone as a policy and governance concern in the early part of the twenty-first century.

2.3.3 Prominence in Perspective: The Axes of Evil and Illness in a Dangerous World

The analytical constructs of the axes of evil and illness help communicate the policy and governance revolution public health, and especially infectious disease control, has experienced in past few years. The transformation of public health from an obscure, neglected area of international relations into something that many believe sits at the core of the conduct of foreign policy and of reform of global governance in the early twenty-first century is nothing short of astonishing. Public health's new governance prominence should, however, be kept in perspective.

The world after 9/11 is a far more dangerous place than the first phase of the post–cold war era. Infectious diseases contribute to this danger, as the potential for avian influenza's spread to trigger pandemic human influenza suggests. But public health does not make the world go round, and the worsening of other problems, such as the crisis in Iraq, the emerging oil crisis, the nuclear standoff between Iran and the United States, and the ongoing war on global terrorism, can easily divert political capital and resources from public health to other policy and governance objectives. Other global problems are experiencing political and governance deterioration, creating the potential for public health to fall again into the "low politics" of foreign policy and international relations.

The dark side of the political revolution in global health is that the revolution's energy feeds off a perceived worsening public health situation around the world. The new political and governance prominence antimicrobial strategies have would not exist but for the continued rise of pathogenic dangers around the globe. Sisyphus may be reenergized in shouldering his burdens in the microbial world, but the boulder is getting heavier and the mountain steeper.

2.4 The New Way of Working: The New International Health Regulations

Section 2.3 provided an overview of the political revolution that has transpired in public health. Now I focus on a concrete example that illustrates the reality of the revolution and its impact on global health governance, particularly relating to infectious disease control. The example is the new International Health Regulations adopted by WHO in May 2005 (IHR 2005) [22]. The IHR 2005 radically depart from the traditional way in which the international system organized governance for the control of the international spread of infectious disease. The radical changes made in the IHR 2005 combine to create a new governance platform for antimicrobial strategies.

2.4.1 From the Classical Regime to the IHR 2005

The IHR 2005 replace the old IHR, first adopted in 1951 [23] and, before May 2005, last substantively revised in 1969 [24]. The old IHR's lineage extends further back in time because the old regulations continued an approach to governance of international infectious disease threats that first emerged in the latter half of the nineteenth century in the form of the international sanitary conventions. These conventions and the old IHR contained what I have called the "classical regime" of international governance on infectious diseases [25]. In brief, states designed and built the classical regime to reduce the burdens national quarantine systems placed on international trade and travel. The classical regime had two parts. First, states participated in an international surveillance system concerning a small number of infectious diseases the spread of which was traditionally associated with international trade and travel. After smallpox was eradicated in the late 1970s, the old IHR only applied to three infectious diseases, namely cholera, plague, and yellow fever. Second, the classical regime imposed obligations on states not to impose excessive or irrational measures on the trade and travel coming from other states affected by outbreaks of diseases subject to the regime. For these diseases, the classical regime prescribed the maximum measures that states could apply, and these measures were based on the best available scientific and public health understanding of the diseases in question.

As discussed in more detail in other works [25,26], the classical regime failed as a governance strategy. States generally failed to comply with their obligations on surveillance and on limiting their trade and travel restricting responses to what was epidemiologically appropriate. Under the old IHR, WHO's surveillance activities were limited to information provided by governments; and this state-centric approach to surveillance crippled the effectiveness of WHO's efforts. In addition, the crisis in emerging and reemerging infectious diseases in the 1990s made clear that limiting the classical regime to cholera, plague, and yellow fever made this approach anachronistic in terms of the infectious disease threats to the global system.

The IHR 2005 reflect WHO's attempt to cast aside the old IHR and create a governance regime sufficiently robust to handle the challenges of infectious disease emergence and spread in the era of globalization. Through this remarkable effort, we can see the governance revolution in global health as manifested in the axes of evil and illness. The IHR 2005 contain many important changes, but three changes most powerfully communicate why the IHR 2005 constitute a new way of working for public health governance in the early twenty-first century. These three transformative changes are (1) a dramatic expansion of the scope of the governance strategy, (2) granting WHO increased authority and power, and (3) integrating human rights principles into the implementation of the IHR 2005.

2.4.2   Expanding the Scope of the IHR's Governance Strategy

In many ways, the classical regime was a very limited governance approach. It only applied to a small number of diseases, contained obligations on states that were limited in nature, and only considered governments as legitimate participants of infectious disease governance. The IHR 2005 could not be more different because they expand (1) the scope of the IHR's disease application, (2) imposes serious and far-reaching surveillance and response obligations on states legally bound by the new rules (states parties), and (3) integrates non-state actors into the functioning and implementation to the governance strategy.

In terms of the scope of disease application, the IHR 2005 increase disease coverage in three ways. First, the IHR 2005 replace the static and closed disease-specific approach of the classical regime with an approach built on the broadly defined terms "disease," "event," "public health risk," and "public health emergency of international concern." This approach allows the IHR 2005 to catch not only already identified disease risks but also new, unexpected threats. The old IHR did not apply to the SARS outbreak because SARS was not on the list of diseases subject to the regulations. The IHR 2005 do, however, apply to unknown or unforeseen disease threats, which makes the IHR 2005 far more flexible, dynamic, and forward looking than any manifestation of the classical regime. The changes made in the IHR 2005 embody an attempt to design a governance regime to handle the myriad challenges presented by the axis of illness in a context of intensifying globalization.

Second, the broad definitions of key terms in the IHR 2005 mean that both infectious and noncommunicable disease threats are subject to the new governance approach. The classical regime never applied to anything but a handful of infectious diseases. The IHR 2005 acknowledge that international disease spread can involve agents that cause noncommunicable diseases. Thus, the governance strategy builds these threats into its operation. This radical departure from the past brings chemical, nuclear, and radiological events within the scope of application of the IHR 2005.

Third, the expanded scope of disease application of the IHR 2005 incorporates all possible sources of public health threats—naturally occurring, accidental, and intentionally caused. This change makes the IHR 2005 applicable to government or terrorist use of WMD, giving the direct relevance of the IHR 2005 to security strategies against the proliferation and use of WMD. The classical regime only applied to naturally occurring infectious diseases and never had any connection with concerns about nuclear, chemical, or biological weapons. The IHR 2005 reflect the linkage between public health and security produced by the security worries reflected in the axis of evil framework. This change formed part of why WHO stressed the importance of the IHR 2005 for achieving what it called "global health security" [27].

In terms of expanding the scope of obligations, the IHR 2005 are far more demanding of states parties in terms of surveillance and response than anything attempted with the classical regime. To begin, the expansion of the scope of the application of IHR 2005 discussed above changes significantly the nature of obligations that states parties confront under the new rules. More specifically, the obligations of the IHR 2005 on states parties to notify WHO of disease events and public health emergencies of international concern and respond to such public health risks require states to have robust surveillance and response capabilities. In fact, in another radical break from the classical regime, the IHR 2005 specifically impose on states parties duties to build and maintain minimum core surveillance and response capabilities. These changes connect to the linkages of public health capacity, development, and good governance analyzed earlier. The IHR 2005 also expand the nature of obligations on states parties through the authority and power they grant WHO and the integration of human rights principles, both of which are discussed below.

The IHR 2005 also increase the reach of the new governance strategy beyond the classical regime by incorporating non-state actors into the functioning of the new strategy. The IHR 2005 abandon the classical regime's state-centric approach by allowing WHO to collect and use surveillance information it receives from nongovernmental sources. Such sources of information have become valuable to global disease surveillance, and new information technologies, such as the Internet and e-mail, permit WHO to harness nongovernmental information more productively for global health. This approach also significantly reduces the prospects and incentives states parties have to try to hide or cover-up disease outbreaks in their territories, as China painfully discovered during SARS and as other governments realized during the spread of avian influenza. The IHR 2005 expand, thus, the scope of participation in the governance strategy for global health by making non-state actors critical components of governance efforts. This change connects to the increased activity and participation in global health of non-state actors that many experts believe characterizes global health governance in the early twenty-first century.

2.4.3 Granting WHO Increased Authority and Power

The second significant change achieved in the IHR 2005 concerns the granting to WHO of increased authority and power. Under the classical regime, WHO's authority and discretion to act were extremely limited. WHO essentially acted as a conduit of information that its member states could use or ignore as they pleased. By contrast, the IHR 2005 establish WHO as a governance actor with independent authority and power. The IHR 2005 accomplish this striking governance change in a number of ways. The incorporation of nongovernmental sources of surveillance information discussed above actually increases WHO's authority because it now can use and act on information that does not emanate from states parties. States parties have to answer to WHO's requests for verification of information the organization received from nongovernmental actors. In terms of surveillance, the IHR 2005 put WHO more in the driver's seat.

The IHR 2005 also grant WHO the power to decide whether a disease event constitutes a public health emergency of international concern. By moving from a disease-specific to broader surveillance coverage, the new governance strategy could have been vulnerable to states parties playing games with whether a disease event was serious enough to report to WHO. The IHR 2005 shut down this vulnerability. A state party may decide not to report an event in its territory because it does not believe that it may constitute a public health emergency of international concern. The final decision about whether the event actually constitutes such an emergency is, however, in the hands of the WHO Director-General and not the state party directly affected.

If the WHO Director-General declares that a public health emergency of international concern exists, the IHR 2005 authorize the Director-General to issue temporary recommendations on how states parties and non-state actors should respond to the threat. Although these recommendations are not legally binding, the ability to issue them accords the WHO Director-General tremendous power vis-à-vis states parties. The political and economic pain created for states by the travel advisories WHO issued during SARS demonstrates that the authority to issue temporary recommendations involves the exercise of real power. Examples of intergovernmental organizations wielding this kind of material power independently of states are virtually unheard of in international relations.

These changes in WHO's authority and power converge with WHO's access to nongovernmental sources of information to create a global health governance pincer that squeezes sovereign states to participate more openly and effectively in global health governance (Figure 2.3). This pincer, and its effects on sovereignty, reveals the IHR 2005 taking governance for public health in directions never seen before in this area of international law, and the unprecedented nature of this change demonstrates how the IHR 2005 stand as an excellent case study in the political revolution that has reshaped governance for global health.

2.4.4 Integrating Human Rights Principles

The third significant change that marks the status of the IHR 2005 as a radical governance departure from the past involves the integration of

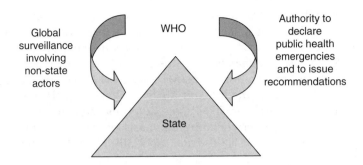

FIGURE 2.3
The global health governance pincer. (Fidler, D.P., *SARS, Governance, and the Globalization of Disease*, Palgrave Macmillan, Basingstoke, United Kingdom, 2004.)

human rights principles into the implementation of the regime. The classical regime originally developed before the post–World War II human rights revolution in international law and international relations. Revisions of the classical regime in the form of the IHR in 1951 and 1969 did not, however, link international control of infectious disease with the growing body of international human rights law. The IHR 2005 achieve this linkage. The IHR 2005 state that "[t]he implementation of these Regulations shall be with full respect for the dignity, human rights and fundamental freedoms of persons" (Article 3.1), and the regulations contain provisions that apply human rights principles to actions states parties might take.

Although the integration of human rights principles in the IHR 2005 is not complete [26,28], this change reveals the importance of human rights to global health governance. The IHR 2005 not only address security, economic, and development objectives but also address the enduring significance to public health of the protection and promotion of human dignity. The incorporation of human rights principles completes public health's nature as an integrated public good concerned with security, economic well-being, development, and human dignity. The IHR 2005 reveal a governance outlook on public health far removed from the classical regime's limited vision of public health as useful in reducing burdens on trade.

2.5 Is the New Way of Working Actually Working? The Quest for Governance Architecture for Global Health

2.5.1 What Do President Bush, Bill Gates, and Bono Have in Common?

The IHR 2005 represent only one example of the changes in governance the political revolution in global health has triggered. Global health experts have noted an explosion of new ideas, initiatives, projects, funding sources, and actors over the past decade that have transformed the landscape of global health permanently [29]. This explosion has involved many different types of participants with different agendas and motivations, all working at some level on improving global health. Traditional actors, such as states and WHO, have been joined by a growing cadre of non-state actors dedicating time and resources to global health causes. This cadre includes not only nongovernmental organizations (e.g., *Médecins Sans Frontières*) and philanthropic foundations (e.g., the Bill and Melinda Gates Foundation) but also individual issue entrepreneurs like Professor Jeffrey Sachs and the Irish rock star Bono. By far, the overwhelming emphasis of these new participants has been on infectious disease threats, especially HIV/AIDS, malaria, and tuberculosis.

The explosion of interest in global health, especially with respect to battling infectious diseases, is simultaneously cause for excitement and concern in terms of governance. The proliferation of actors in global health gives the field not only political and media prominence but also a fractured, disorganized feel perhaps best characterized as a situation of "unstructured plurality." Worries are frequently expressed that unstructured plurality does not, in the long run, serve the interests of global health governance well. This sentiment has led to a recent proliferation of calls for creating better governance architecture for global health. As the Dean of the Harvard School of Public Health argued, "There've been lots of creative ideas and lots of new people. But there's one missing piece. There's no architecture of global health" ([29], p. 162).

2.5.2   Beyond Unstructured Plurality: Whither Global Health Governance?

The increased calls for better governance architecture for global health suggests that the governance transition that will characterize the fourth epidemiological transition remains incomplete. Although groundbreaking and significant in so many ways, the IHR 2005 do not provide the overarching architecture experts perceive is necessary. In fact, many of the concerns arising from the development of unstructured plurality in global health involve infectious disease problems for which the IHR 2005 provide little, if any assistance. At the same time that the WHO Executive Board calls for voluntary accelerated implementation of the IHR 2005 with respect to potential pandemic influenza [30], most experts do not believe the IHR 2005 add much to the ongoing struggles with the three main infectious disease killers: HIV/AIDS, malaria, and tuberculosis. Each of these plagues is, depressingly, showing signs of becoming more deeply entrenched, and thus more difficult as a governance challenge.

The explosion in new interest, ideas, and funding for global infectious disease threats has provided momentum for people to begin to think about reshaping this movement into something more effective and sustainable. The architecture metaphor is attractive because it captures the desire to move beyond unstructured plurality toward organized unity. The architecture metaphor appeals because the architect engages in an effort to achieve functional vision—expressing normative values and interests in light of the practical constraints of the cost of materials and the forces of nature. In global health, the fourth epidemiological transition has perhaps reached a point at which the messy proliferation of governance initiatives requires reconstruction in order to address the still serious, and in some cases worsening, threats the microbial world creates. The IHR 2005 form part of this reconstruction of global health governance, but the new regulations mark only the end of the beginning of governance reform in global health.

Whether future governance reform will involve serious efforts to create organized unity in global health appears, however, questionable. The architecture metaphor communicates orderly and interdependent design and function that cannot perhaps be achieved in the early twenty-first century context of unstructured plurality. President Bush, Bill Gates, and Bono have all focused significant time, attention, and resources on global infectious disease problems. I have trouble, however, conceiving of a governance approach that will simultaneously organize and mesh together the foreign policy of the world's hegemonic power, the philanthropic strategies of one of the world's most influential foundations and businessmen, and the fierce independence of an Irish rock musician-turned global health advocate. Resistance to organized unity in global health governance will be strong not only from powerful states but also from the non-state actors that have increasingly become involved in the global fight against infectious diseases. I suspect that Bill Gates and Bono will no more march to the tune of WHO than President Bush will.

At best, the evolution of governance for global health, as the fourth epidemiological transition progresses, will involve a movement from unstructured plurality to purposeful plurality. In contrast to the former, the latter would foresee a continued plurality of actors, ideas, and initiatives in global health governance, but a plurality better focused on the epidemiological challenges the microbial world will continue to present. The result would not constitute architecture as much as a global health concert of increasingly coordinated and cooperating actors and activities. This vision for the fourth epidemiological transition may not be as dramatic as the one implicit in the desire for governance architecture for global health, but it may be more realistic.

2.6 Conclusion

The political revolution in global health that has transpired in the past few years reinforces the argument I made in 2000 that the labors of Sisyphus in the microbial world in the early twenty-first century would be most significant in the area of governance. Effectively responding to the axes of evil and illness requires the integration of science and technology with new forms of political motivation and action. The emergence of these axes reflects the increasing burden Sisyphus in the microbial world bears. This reality has raised the prominence and importance of governance to the basic tasks of epidemiology.

Evidence for a governance-centered fourth epidemiological transition has appeared abundantly in the past 6 years, and developments such as the IHR 2005 and the explosion of interest and initiative in global infectious disease activities suggest that Sisyphus has renewed vim and vigor in shouldering the burdens of global antimicrobial strategies. These developments act as a counterweight to the continuing threat pathogenic microbes present humankind. Whether Sisyphus is halfway up or halfway down the mountain is perhaps less important at this stage than the realization that the fourth epidemiological transition has begun in ways that lend some hope that the gap between the theory and practice of global health can be reduced without waiting, perhaps in vain, for deus ex machina salvation to be provided by science and technology.

References

1. Fidler, D.P., Sisyphus in the microbial world: Antimicrobial strategies and humanity's health, *Antimicrobial/Anti-Infective Materials: Principles, Applications and Devices*, Sawan, S.P. and Manivannan, G., (eds.), Technomic Publishing, Lancaster, PA, 2000, pp. 330–342.
2. Heymann, D.L., Testimony of David L. Heymann, Executive Director, Communicable Diseases, World Health Organization, Hearing on Severe Acute Respiratory Syndrome Threat, before the Committee on Health, Education, Labor, and Pensions of the U.S. Senate, 7 April 2003.
3. Armelagos, G., The viral superhighway, *The Sciences*, 24–29, Jan./Feb. 1998.
4. World Health Organization, World Health Report 1996: Fighting Disease, Fostering Development, WHO, Geneva, 1996.
5. Lee, K. and Zwi, A., A global political economy approach to AIDS: Ideology, interests and implications, *Health Impacts of Globalization: Towards Global Governance*, Lee, K., (ed.), Palgrave Macmillan, Basingstoke, United Kingdom, 2003, pp. 13–32.
6. Kickbusch, I., Global health governance: Some theoretical considerations on the new political space, *Health Impacts of Globalization: Towards Global Governance*, Lee, K., (ed.), Palgrave Macmillan, Basingstoke, United Kingdom, 2003, pp. 192–203.
7. Fidler, D.P., Germs, norms, and power: Global health's political revolution, *Journal of Law, Social Justice and Global Development*, 2004, Available at http://elj.warwick.ac.uk/global/04-1/fidler.html
8. *Health, Foreign Policy & Security: Towards a Conceptual Framework*, Ingram, A., (ed.), Nuffield Trust, London, 2004.
9. Kickbusch, I., Influence and opportunity: Reflections on the U.S. role in global public health, *Health Affairs*, 21, 13, 2001.
10. Bush, G., State of the Union address, 29 Jan. 2002, Available at http://www.whitehouse.gov/news/releases/2002/01/20020129-11.html

11. *Deadly Cultures: Biological Weapons since 1945,* Wheelis, M., Rózsa, L., and Dando, M., (eds.), Harvard University Press, Cambridge, MA, 2006.
12. Office of Homeland Security, National Strategy for Homeland Security, Washington, DC, 2002.
13. The White House, National Security Strategy of the United States, Washington, DC, 2002.
14. Fidler, D.P., Caught between paradise and power: Public health, pathogenic threats, and the axis of illness, *McGeorge Law Rev.,* 35, 45, 2004.
15. Fidler, D.P., Fighting the axis of illness: HIV/AIDS, human rights, and U.S. foreign policy, *Harvard Journal of Human Rights,* 17, 99, 2004.
16. Institute of Medicine, *Emerging Infections: Microbial Threats to Health in the United States,* National Academy Press, Washington, DC, 1992.
17. Institute of Medicine, *Microbial Threats to Health: Emergence, Detection and Response,* National Academy Press, Washington, DC, 2003.
18. World Health Assembly, Improving the Containment of Antimicrobial Resistance, WHO Doc. WHA58.27, 25 May 2005.
19. Report of the Secretary-General's High-level Panel on Threats, Challenges, and Change, A More Secure World: Our Shared Responsibility, United Nations, New York, 2004.
20. Annan, K., In Larger Freedom: Towards Development, Security and Human Rights for All—Report of the Secretary-General, UN Doc. A/59/2005, 21 Mar. 2005.
21. The White House, The National Security Strategy of the United States of America, Washington, DC, 2006.
22. World Health Assembly, Revision of the International Health Regulations, WHO Doc. WHA58.3, 23 May 2005.
23. International Sanitary Regulations, 25 May 1951, 175 United Nations Treaty Series 214.
24. World Health Organization, *International Health Regulations* (1969), 3rd annual edition, WHO, Geneva, 1983.
25. Fidler, D.P., *SARS, Governance, and the Globalization of Disease,* Palgrave Macmillan, Basingstoke, United Kingdom, 2004.
26. Fidler, D.P., From international sanitary conventions to global health security: The new international health regulations, *Chinese Journal of International Law,* 4, 325, 2005.
27. World Health Organization, Alert and Response Operations, Available at http://www.who.int/csr/alertresponse/en/
28. Fidler, D.P. and Gostin, L.O., The new international health regulations: An historic development for international law and public health, *Journal of Law, Medicine & Ethics,* 34, 85, 2006.
29. Cohen, J., The new world of global health, *Science,* 311, 162, 13 Jan. 2006.
30. WHO Executive Board, Application of the International Health Regulations (2005), WHO Doc. EB117.R7, 26 Jan. 2006.

3

The Need for Safer and Better Microbicides for Infection Control

Syed A. Sattar and Susan Springthorpe

CONTENTS

3.1 Introduction

Microbicides, by their very nature, are highly reactive with a wide range of chemical moieties, which is how they inactivate microorganisms. This very property also makes them relatively hazardous to handle and store, as well as accounting for their systemic and genotoxic effects. Effective use of microbicides thus requires an understanding of their relative properties, and demands that they be diligently and properly applied [1]. While judicious use of these potentially hazardous chemicals can be extremely effective in preventing and controlling the spread of infectious agents and nuisance organisms in critical settings, their widespread and often unnecessary application may be more deleterious than beneficial to humans as well as the environment. As microbicides inevitably, and often immediately, become diluted beyond the point of application, they no longer remain "killing agents," but instead become just sublethal toxins in the waste stream, where a complex mixture of pathogens and other microorganisms are exposed and potentially affected [2]. We are just beginning to understand the complexities and risks posed by such mixtures of chemicals and microbes [2–4]. More persistent microbicidal chemicals can cause wider contamination of the environment and enter higher life forms in low doses over extended periods through air, water, and food [5]. The longer-term consequences of such exposures, especially in the young, remain ill defined.

Should these and other risks from microbicides in current use be a matter of concern? If yes, are any safer alternatives and application strategies available? This chapter discusses the more salient issues with currently used classes of chemicals and then explores certain principles and approaches to produce safer and potentially more effective microbicidal products. While more fundamental research in this field still remains woefully inadequate, many specialty chemical producers have already taken on this challenge as a part of their R&D programs. The awareness of workplace and environmental safety of microbicide use in end users appears to progress in spurts and starts. A complex and dynamic topic such as this cannot be covered comprehensively in the limited space available. Therefore, what follows is more an emphasis on basic principles and a sampling of the technologies and approaches under consideration. Also, this discussion is focused primarily on liquid chemicals used on inanimate surfaces and objects in healthcare and domestic settings irrespective of whether they are applied as sprays, foams, and other means to inactivate microbial targets.

3.2   What Are Some of the Problems with Current Microbicides?

3.2.1   Safety and Handling Issues

First and foremost, microbicides in general must be handled with caution and stored out of the reach of young children, particularly in the case of domestic products [6]. Reports of poisoning from ingestion of microbicides, and the serious consequences that can result [7], are far outnumbered by the unpublished occurrences. While such dramatic events draw attention to the acute toxicity following microbicide ingestion, the majority of human exposures come from using microbicidal products on a regular basis. Occupational exposures can result in immune reactions including hypersensitivity [8–11] and contact dermatitis [12–16]. Data from California show that, between 1991 and 1995, four types of microbicidal chemicals (sodium hypochlorite, quats, chlorine gas, and glutaraldehyde) were responsible for the highest number of occupational illnesses [17]. This is not a trivial issue as many healthcare professionals can be affected [12,16,18]; while such sensitivities are rarely life threatening, they frequently affect performance and compromise ability or willingness to complete required procedures. This is an increased burden for the healthcare system, and occasionally the severity of effects lead to long-term disability. Working youth face higher risks from exposure to microbicides in occupational settings [6]. Safer microbicides are therefore highly desirable, but proper training is also needed for all who handle microbicides in an occupational capacity, including housekeeping personnel in both industry and healthcare. It is also very important that the label requirements for safe storage and disposal be simplified to be followed properly.

3.2.2 Environmental Toxicity

Environmental safety of spent microbicides is an issue that is not widely acknowledged because the use of such chemicals is deemed to benefit humans. However, now that microbicides are incorporated into many consumer products, both human and environmental exposures to these chemicals have already increased considerably. Most microbicides also contain detergents and inert ingredients that may potentiate their action and could also exacerbate their toxicity. There is little direct evidence of this because it has not been widely or systematically examined, but increasingly we are beginning to understand that complex chemical mixtures pose particularly significant environmental risks even when individual chemicals may be below their acknowledged toxicity threshold [4]. Unfortunately, toxicity of the active ingredients of microbicides and their by-products is poorly investigated under relevant conditions.

Liquid chemical microbicides are often used in relatively high concentrations and in large quantities in healthcare and industry. Sewer discharge is normal, even if prior neutralization has occurred, and after treatment any remaining chemical constituents are distributed between the sludge disposed off on land or in landfills and the effluent discharged directly into surface waters. The latter contain a variety of species that can be directly affected by sewage plant effluents, and it is important to remember that surface waters often serve as sources of drinking water for downstream treatment plants, which may not remove chemicals effectively. The half-lives of microbicidal chemicals and their by-products in the environment can be long, with breakdown taking weeks or months in some cases [19,20].

A significant issue not considered widely except for microbicides used in drinking water treatment is the formation of disinfection by-products (DBPs). For example, the levels of certain DBPs formed from chlorine are strictly regulated because of their known potential as mutagens or carcinogens [21–23]. Thus, a fine balance needs to be achieved between adequate disinfection and control of DBP formation in drinking water treatment. However, similar chemistries, often at much higher concentrations, are used in food production as well as in many consumer products, and one can therefore anticipate proportionately higher levels of DBP formation [24], although in smaller amounts.

Recognition of specific by-products from other drinking water disinfectants is an active area of research [25,26]. As more drinking water utilities switch to maintaining a disinfectant residual with monochloramine, there is a particular interest in understanding chloramine DBPs, including the potential for nitrosamine formation which may be higher than that with chlorine [26]. Even so, the potential formation of N-nitrosodimethylamine (NDMA) and other nitrosamines in water disinfection is likely much less than in disinfection of food or sewage. Moreover, ozone, an extremely powerful oxidant and microbicide when used in primary disinfection of drinking water, can result in decreased levels of many DBPs associated with chlorine, but potentially increased bromate and increased levels of metabolizable organic compounds, which can promote bacterial growth once ozone has declined below lethal levels. The nature, formation, toxicity, and half-lives of DBP formed from most other microbicides are poorly understood.

We have previously reviewed the potential for combined or sequential effects of chemical and pathogen exposure on the susceptible host [2], potentially through modulation of immune function. The likelihood that microbicides or their DBPs are involved in such processes would depend on their immuno- or genotoxicity and on the risks of exposure to them. However, it is clear that several demographic trends [27], including increased urbanization, raise the frequency of exposure to both chemicals and pathogens, and therefore the risks of such combined or sequential exposures.

3.2.3   Genotoxicity and Microbial Resistance

Many active ingredients in microbicides interact directly with nucleic acids and proteins (including enzymes). They have the potential, therefore, to affect DNA replication in a number of different ways, including directly causing mutations [28,29]. Several such chemicals in microbicides are also recognized as carcinogens [29,30]. Moreover, some components have been shown to have hormone-disrupting activities [31].

Sublethal exposure of microorganisms to microbicides always leads to selection of organisms with reduced susceptibility to the applied pressure. The terms "tolerance" and "resistance" as used below to describe changes in susceptibility to chemicals are not intended to denote specific mechanisms and are simply descriptive terms for observed phenomena. Although the mutagenic effect of microbicidal chemicals can equally result in strains with lower tolerance to the applied stressor, the latter are inevitably selected against if the selective pressure is repeatedly or continuously applied. Reduced susceptibility can arise from mutations that affect the incursion of the toxin into the cell at the membrane level. While such "keep-it-out" mechanisms may be expected to be relatively common, if membrane permeability to nutrient molecules and other beneficial chemicals are concurrently impaired, such cells may be at a competitive disadvantage and fail to thrive. However, bacterial cells also have inbuilt mechanisms to deal with environmental toxins, which include efflux pumps capable of rapidly exporting a range of chemicals from the cells. Observed mechanisms of microbicide resistance have frequently involved such efflux pumps [32–34]. Since many microbicides react with a wide range of intra-cellular chemicals, it is difficult or impossible for the microbial cell to alter multiple targets simultaneously; nevertheless, target alteration is a strategy that can be employed by gram-negative bacteria against more specific anti-microbial compounds including antibiotics [35,36]. Another mechanism for microbes to deal with toxins is production of excess of compounds that can react with and sequester the toxic materials, rendering them harmless. This can occur inside the cell and also by exporting such metabolites to the cell exterior. Such a function may be played in part by secondary metabolites exported from bacterial cells in stationary phase or the extracellular matrix in biofilms.

The range of strategies a microbe can employ against toxic chemicals is however limited, and thus a tolerance to one toxin may extend to others—the issue of most concern is that increased resistance to environ-mental chemicals, such as sublethal microbicide concentrations, may lead to reduced antibiotic susceptibility in exposed strains. Triclosan is the chemical for which there is the greatest evidence for cross-resistance [37–39], where a single enzyme system has been implicated as the main target [40], though this has recently been challenged as overly simplistic [41]. Cross-resistance can readily be demonstrated to occur in the laboratory for *Escherichia coli* and some other species [39], but has not yet been adequately demonstrated in the field. Even if resistant clones are selected, their significance is unclear in the larger ecological context [38,42]. Nevertheless, it is not only triclosan that should be of concern but also hypermutable regions of bacterial genomes that may be vulnerable to change from a variety of agents including microbicides and mutagenic DBPs, and there is every reason to consider that even if such events are comparatively rare, their selection under suitable environmental conditions could lead to their dominance. Exposure to quaternary ammonium compounds in the field has the real potential to lead to antibiotic resistance in environmental isolates because of the high co-selection of class 1 integrons with resistance to quaternary ammonium compounds [43].

Evidence for and against more widespread cross-resistance is limited. On the one hand some have failed to demonstrate the link between antibiotic resistance and tolerance to environmental chemicals [44], while other stud-ies have suggested that it occurs [43,45,46]. In some cases results can be seen with bacteria in pure culture that may not be replicated in natural populations [47].

Ultimately, the results obtained may depend on the experimental system employed and the methods used to screen the isolates. Since random mutational events to reduced susceptibility are of necessarily low frequency, strains that carry traits for tolerating exposure to a wide variety of chemicals are likely to arise rarely in a population. Therefore, studies that use only a small number of isolates may be predestined to provide data against cross-resistance, whereas those that examine large numbers of isol-ates may be more likely to demonstrate its occurrence. Moreover, use of microbicides at or near inhibitory levels is less likely to produce results for cross-resistance than those that act before any inhibition is observed, and therefore target the maximum populations, or that train the exposed bacteria over prolonged periods. In our own studies, we screened over 27,000 isolates of wild-type *E. coli* (K-12) and isogenic mutator strains for increases in antibiotic resistance with or without continuous exposure to drinking water disinfectants [48]. Not surprisingly, the mutator strain showed a higher frequency of resistance to multiple antibiotics upon exposure. The existence of bacterial mutator strains in the environment in demonstrable numbers [49,50] further raises the prospects of altered responses to toxic stressors and potential correlations with antibiotic resistance. Moreover, mutator strains may be more common among pathogenic isolates of *E. coli* and other bacteria [51].

3.2.4   Materials Compatibility

A desirable attribute of microbicides for use in healthcare is materials compatibility. For medical instruments and many pipes and surfaces that are disinfected regularly, the chemical concentrations needed to ensure effective microbicidal action can also cause corrosion or damage. Microbicides are therefore often formulated to enhance their materials compatibility. For example, acidic detergents used as sanitizers usually contain phosphoric acid, which is less corrosive than, say, hydrochloric acid, even though a sanitizer containing hydrochloric acid may be a more effective microbicide [52]. A second example is rubber that is continuously or intermittently exposed to reactive chemicals; the rubber components used in milking equipment are regularly washed with chlorine solutions, which tend to "crack" or craze the surfaces [53]. Such damage increases with exposure and if these items are not replaced regularly, the milk can contain small rubber/carbon particulates, causing it to be rejected.

Material incompatibility is indeed widespread to varying degrees. Even an apparently inert surface can react with the microbicide, thus exerting a demand and a consequent reduction in its microbicidal potential in proportion to the area of surface contacted. This is particularly significant when the product is applied in small quantities, perhaps by wiping with an applicator. Even if the surface is really inert to the microbicide, any accretion of surface soiling can contribute to this demand. Surface–microbicide interactions can be sufficient to cause a recommendation that a particular type of microbicide is not used on certain defined surface types. Tests of microbicide efficacy are performed on one or two types of inoculated but otherwise clean surfaces, so the onus may be on the user to query the suitability of a particular product for the application at hand.

Table 3.1 summarizes the key issues in the safe and effective use of microbicides in healthcare. Microbicide use is however much more widespread than only in healthcare, and so this list is not comprehensive.

TABLE 3.1

Key Considerations in the Safe and Effective Use of Microbicides in Healthcare

Factors	Examples of Problems That May Arise
Toxicological	
Healthcare worker	Microbicide vapors may cause respiratory sensitization [10] and exposure of bare skin may cause contact dermatitis [11,13,54]
Patient	Improperly rinsed endoscopes may release residues of microbicide into body cavity [55]
Microbiological	
Resistance	Pathogens may develop resistance to a given microbicide [56]
Cross-resistance	Pathogens with increased resistance to certain microbicides may also show higher resistance to antibiotics [43,57]
Product storage	Improper or prolonged storage of microbicides may lead to growth of bacteria in them [58]
Production dilution (if required)	Improper dilution or failure to clean container for diluted product may result in contamination of the solution [59]
Use of improper microbicide	Use of the wrong formulation can lead to cross-infections [60]
Chemical	
Improper storage	Improper storage of unstable microbicides may lead to explosions or fires. Reference to material safety data sheets is essential
Corrosion	Using the wrong type and level of microbicide may cause corrosion or other damage to expensive items such as flexible endoscopes [61]
Improper mixing of chemicals	Incompatible chemicals may be mixed leading to either a violent chemical reaction, noxious gas production, or neutralization of microbicidal activity. Reference to material safety data sheets is essential
Improper monitoring of levels of chemicals in use	Dipsticks for monitoring microbicide levels are real-time based but may not give accurate information
Improper setting and monitoring of exposure	Sensors for detecting vapors of microbicides may not function properly or may not be used correctly [62]
Environmental	
Air quality	Use of gaseous or volatile microbicides may negatively affect indoor air quality [63]
Hormone disruption	Certain microbicidal chemicals or their breakdown products can disrupt hormone function in animals and possibly humans [64]
Water and food quality	Environmentally stable microbicides can contaminate food/water and groundwaters [5]

(continued)

TABLE 3.1 (continued)

Key Considerations in the Safe and Effective Use of Microbicides in Healthcare

Factors	Examples of Problems That May Arise
Product label	
Unclear directions	Unclear label directions may lead to improper use of microbicide
Unrealistically long contact time	In field use, contact times shorter than those in label directions may not achieve the desired level of disinfection
Purchasing	
Poor purchasing decisions	Decisions on purchasing microbicides may be made entirely based on cost, resulting in the stockpiling of inferior or inappropriate products
Training	
Insufficient or improper training	Personnel responsible for using microbicides may not receive any training or adequate instruction in the preparation, use, storage, and disposal of the formulation in use

Table 3.2 summarizes the major classes of widely used actives and lists some of their major disadvantages and limitations; it targets chemicals used with the aim of inactivating microorganisms, rather than those designed to protect from the impacts of microbial decay.

3.2.5   Spectrum of Activity

Even today, much of the microbicide manufacturing industry is coasting on the fundamental research and development of decades ago. Many available actives and products based on them are no more than remixes of the old, but perhaps with newer label claims. It is also noteworthy here that a great deal of the original innovation and testing of microbicides was carried out with bacterial pathogens as the primary targets [65]. While many bacterial diseases have become somewhat easier to deal with due to the availability of antibiotics, the relative importance of viral infections increased and still continues to the present. Moreover, many viral infections remain asymptomatic and unrecognized but can still affect the presentation and outcome of other infections [66]. That many bactericidal agents may be poor virucides attests to the fundamental differences in the structure and chemistry of bacteria and viruses. Microbicides that are simply bactericidal may be very useful to industry, but should those formulations approved for use in healthcare be required to have virucidal claims? This is certainly debatable but what is clear is that for most healthcare uses of microbicides the contaminant targets are unknown.

TABLE 3.2

Commonly Used Microbicides and their Disadvantages

Chemical Class	Major Uses	Major Disadvantages and Limits to Effectiveness in Healthcare Use
Acidified anionic detergents	Industrial surface sanitation, toilet bowl cleaners	Materials compatibility constrains formulations to prevent corrosion, intended only for industrial sanitizing, not suitable for general hospital hygiene
Alcohols	Topical antiseptic, disinfection of surfaces, additive in other formulations as potentiator	Volatile organic, flammability, potential for abuse, activity declines on dilution, does not clean and can fix material to surface, does not kill spores, may not be suitable for continuous exposure of some elastomers
Aldehydes	Medical instrument disinfection, area decontamination	Potential carcinogens, skin and respiratory sensitizers, toxic residues, fixative especially problematic on prolonged exposure, some stain proteins, sporicidal only on prolonged exposure, noxious gas produced when mixed with chlorine
Halogen-based products	Disinfection of drinking water and swimming pools, food processing, industrial processing, hygiene in hospitals, spill cleanup, domestic uses	Readily neutralized by organic materials, pungent irritating smell, toxic and mutagenic by-products, corrosivity and materials damage, bleaching inappropriately (chlorine), staining (iodine), allergic reactions (iodophors), efficacy and stability pH dependent
Peroxides and peracids	Reprocessing medical devices, surface disinfection, space decontamination as vapor, food processing and industrial processing	Corrosivity and materials damage, irritant to skin and mucous membranes at high concentrations, can be explosive at high concentrations, some may need generation on-site
Phenolics	Mainly used in healthcare for general disinfection	Toxic residues, not suitable for food contact surfaces, not recommended for use in pediatric settings, pungent smell, can cause skin burns at high concentrations, variable activity against viruses, not sporicidal
Quaternary ammonium compounds	Widely used for general disinfection in healthcare, industry, institutional and domestic settings	Readily neutralized by organic material, incompatible with anionic detergents, limited activity against viruses and mycobacteria, nonsporicidal, respiratory and skin sensitization

3.3   Innovative Products and Technologies

There is a vast market for microbicides in many different industries as well as significant demands in institutional and domestic settings, and it would be impossible to address all the different applications for them here. We will therefore provide only a general focus to the discussion and highlight some approaches to innovative technologies with actual or potential applications for areas with which we are most familiar, including healthcare, drinking water and food processing, and handling as well as domestic applications. It should be noted that it is not possible to predict the effectiveness of a product for a particular application, and so the following information should not be regarded as an endorsement for any particular technology or approach. The reader is also reminded that (1) the only way to validate effectiveness of a microbicide for a particular application is to test it under realistic conditions, and (2) formulations with apparently similar active ingredients may not behave equally in the field.

Interestingly, very few of the newer technologies accepted to date exploit new active ingredients. This is perhaps not surprising in view of the extensive toxicology packages that must be developed for registration of new actives. This comes in spite of the fact that there are vast arrays of natural products, some of which may have significant antimicrobial potential.

In general, the trend in microbicide innovation is to steer away from products that have significant recognized toxic residuals. This trend is less prominent in North America than in Europe where imminent regulation [67] will limit the types of chemicals used and discharged in many areas. Even in North America there is increasing recognition that it may be possible to use environmentally friendly microbicides. When this is combined with the well-recognized mantra that cleaning is an absolute prerequisite for effective disinfection of objects and surfaces, the trend toward oxidative products that are themselves relatively good cleaners is fairly obvious. There is a real upward, or one could say backward, swing to peroxygen compounds. Hydrogen peroxide is both a naturally occurring chemical and among the earliest recognized disinfectants. While peracids were not recognized early on, they are undoubtedly naturally generated in certain situations. Nowadays, peracids, especially peracetic acid, are a mainstay in the food industry and are becoming very well established in healthcare, especially in the reprocessing of medical devices. While hydrogen peroxide alone acts relatively slowly, modern formulations that incorporate improved and safer detergents allow these products to penetrate, act rapidly, and to be highly effective microbicides [68]. The lack of recognized toxic residues from peroxygen compounds does not necessarily mean that they are free of immunotoxic or genotoxic effects; as strong oxidizers that potential is still there, particularly when sensitive respiratory mucosa is exposed. Further, when these compounds are used at high concentrations, they may still carry the risk of corrosion.

The innovations for oxidizers are not limited to improved chemical formulations for hydrogen peroxide nor are they limited to peroxygen compounds. Even applications of halogen-based products are being rethought to improve their efficacy and safety. The general approaches focus on providing (1) conditions that improve the efficacy of existing microbicides, (2) innovations that deliver the active to target microorganisms in a more efficient manner, (3) unique combinations of actives, or (4) sequential application of technologies. Table 3.3 provides an incomplete summary of the approaches that have already been taken; the list is in no particular order and some applications use two or more innovations. Moreover, detailed understanding of mechanisms may show overlaps, but for the purposes of this summary we have taken a simplistic approach. We have deliberately tried to avoid providing trade names, though the reader should have no

TABLE 3.3

Innovative Approaches to Microbicide Generation and Application

Innovation	Examples (Applications)
Equilibrium-based chemistries that allow for demand release of actives	Iodophors (skin), hydrogen peroxides (skin, surfaces, medical devices)
pH modification	Bleach acidified with acetic acid (spills, remediation of contaminated sites, and high level disinfection), acidic hydrogen peroxide (skin, surfaces, medical devices)
Redox modification	Chlorine and mixed oxidants generated by the electrolysis of brine [69,70] (water disinfection and biofilm control), ozone [71,72] (medical devices, drinking water, and contained spaces with no incompatible material)
Electrical current to improve penetration of antimicrobials	Application in biofilm control [73,74]
Photocatalytic generation of oxidants	Titanium dioxide (water and wastewater disinfection, air decontamination) [75,76]
Nascent generation of actives through enzymes (haloperoxidases) or chemicals	Nascent oxidizer (oxygen, or mixed oxidants, or both) [77–80]
Improved access to targets through delivery of actives in micelles	Mainly applicable to topicals
Nanotechnologies	Being actively explored for topicals, general disinfectants, and drinking water disinfection
Safe on-site generation of actives that are unstable or require special transportation	Chlorine dioxide for water disinfection [81], peracetic acid for instrument reprocessing [82]
Unique or unusual combinations of actives to attack multiple targets	General disinfection and topical products
Sequential applications of different technologies to gain additive or synergistic effects	Advanced oxidation (drinking water disinfection)
Gas–plasma delivery system	Hydrogen peroxide (medical device reprocessing)

trouble tracking down manufacturers and suppliers on the internet or through peer-reviewed literature.

3.4   Concluding Remarks

Improvements in microbicide efficacy may not necessarily be accompanied by enhanced safety, unless it is clear that lower concentrations of actives can be used. Although our focus on oxidizers tends to show trends in the industry as a whole, other products have seen significant improvements as well. Newer generations of quaternary ammonium compounds may be more effective microbicides [83], but since these tend to have much higher concentrations of actives the same safety issues remain or even increase. There have even been attempts to improve phenolics by formulating them in nanoemulsions, for example. Nanotechnology is being targeted for inclusion in a variety of microbial control products. However, certain nanoparticulates incorporated into microbicides may present their own significant safety issues for the environment and user groups [84,85].

Fungi, particularly in domestic and damp environments, present their own unique control problems. Most domestic microbicides will be unlikely to eradicate these, for example in bathrooms. One unique solution has been to coat the surfaces with a mixture of inorganic salts and this appears to be effective for fungal control [86]. Increasingly we are coming to appreciate the role that free-living protozoa play in protecting certain pathogenic bacteria in protozoan cysts; with some exceptions [80], protozoa are rarely considered when evaluating microbicides. The conventional wisdom for a long period has been that low-level disinfection of environmental surfaces is adequate to prevent infection. Except for high-contact surfaces, it is questionable whether chemical disinfection of many environmental surfaces is needed at all because cleaning may be adequate. However, the recent emergence of *Clostridium difficile* as a prominent nosocomial pathogen [87] has raised the yardstick for disinfection in some circumstances. *C. difficile* spores are known to survive on environmental surfaces, and indeed it has been reported that sublethal exposure to environmental disinfectants encourages sporulation [88], though this needs corroboration. Finally, there are transmissible agents of spongiform encephalopathies (TSEs); there are intensive efforts underway to understand how microbicides and other technologies can be used to eliminate such transmissible agents, particularly on reusable medical devices. Although some success has been achieved [89], so far no microbicides have label claims against TSEs in North America.

Emerging and reemerging pathogens will always be with us, and provide continual challenges [90] for hospital epidemiologists [91] and infection control professionals to provide a safe environment for staff and patients. As awareness of risks from exposures to microbicides increases, the importance of assessing and minimizing such exposures in healthcare [92] and industry [93] becomes a priority. Since microbial control is a complex multifaceted field of study, there will always be innovators to produce new and improved microbicides. For the most part, the ability to measure tangible safety and efficacy improvements in the field is lacking. Moreover, regulators are not providing the necessary guidance to manufacturers regarding the need for safety as well as efficacy in registered products, and end users/purchasers may not be sufficiently informed to assess the relative risks and benefits of using the products they purchase.

It should be restated in closing that microbicides continue to be the backbone of infection control. Indeed, our reliance on them is increasing further in view of mounting antibiotic resistance and the ongoing assault from emerging and reemerging infectious agents. This fact has not been lost on those who manufacture and sell microbicides, and the intense competition in the marketplace is leading to a bewildering array of claims of effectiveness and safety. This is at a time when testing methods for microbicidal claims and regulations for their registration for sale in healthcare settings in particular remain essentially stagnant. The wide variations in national and regional requirements for testing and registration of microbicides, with the attendant investments in time and funds, are also a serious deterrent for innovation. In effect, the issues of safety and efficacy discussed here arise from a combination of factors [94]. Thus, any improvements for the future urgently require a comprehensive discussion of these issues among all relevant stakeholders.

References

1. Springthorpe, V.S. and Sattar, S.A., Carrier tests to assess microbicidal activities of chemical disinfectants for use on medical devices and environmental surfaces. *J. AOAC Int.*, 88, 182, 2005.
2. Sattar, S.A., Tetro, J.A., and Springthorpe, V.S., Effects of environmental chemicals and the host-pathogen relationship: Are there any negative consequences for human health? In *New Biocides Development: The Combined Approach of Chemistry and Microbiology*, Zhu, P. (ed.), American Chemical Society, Washington, DC, 2006.
3. Hayes, T.B., Case, P., Chui, S., Chung, D., Haeffele, C., Haston, K., Lee, M., Mai, V.P., Marjuoa, Y., Parker, J., and Tsui, M., Pesticide mixtures, endocrine disruption, and amphibian declines: Are we underestimating the impact? *Environ. Health Perspect.*, 114, Suppl. 1, 40, 2006.
4. Pomati, F., Castiglioni, S., Zuccato, E., Fanelli, R., Vigetti, D., Rossetti, C., and Calamari, D., Effects of a complex mixture of therapeutic drugs at environmental levels on human embryonic cells. *Environ. Sci. Technol.*, 40, 2442, 2006.
5. Adolfsson-Erici, M., Pettersson, M., Parkkonen, J., and Sturve, J., Triclosan, a commonly used bactericide found in human milk and in the aquatic environment in Sweden. *Chemosphere*, 46, 1485, 2002.

6. Brevard, T.A., Calvert, G.M., Blondell, J.M., and Mehler, L.N., Acute occupational disinfectant-related illness among youth, 1993–1998. *Environ. Health Perspect.*, 111, 1654, 2003.
7. Chan, T.Y. and Critchley, J.A., Pulmonary aspiration following Dettol poisoning: The scope for prevention. *Hum. Exp. Toxicol.*, 15, 843, 1996.
8. Preller, L., Doekes, G., Heederik, D., Vermeulen, R., Vogelzang, P.F., and Boleij, J.S., Microbicide use as a risk factor for atopic sensitization and symptoms consistent with asthma: An epidemiological study. *Eur. Respir. J.*, 9, 1407, 1996.
9. Di Stefano, F., Siriruttanapruk, S., McCoach, J.S., and Burge, P.S., Occupational asthma due to glutaraldehyde. *Monaldi Arch. Chest Dis.*, 53, 50, 1998.
10. Purohit, A., Kopferschmitt-Kubler, M.C., Moreau, C., Popin, E., Blaumeiser, M., and Pauli, G., Quaternary ammonium compounds and occupational asthma. *Int. Arch. Occup. Environ. Health*, 73, 423, 2000.
11. Waters, A., Beach, J., and Abramson, M., Symptoms and lung function in health care personnel exposed to glutaraldehyde. *Am. J. Indust. Med.*, 43, 196, 2003.
12. Kiec-Swierczynska, M. and Krecisz, B., Occupational skin diseases among the nurses in the region of Lodz. *Int. J. Occup. Med. Environ. Health*, 13, 179, 2000.
13. Shaffer, M.P. and Belsito, D.V., Allergic contact dermatitis from glutaraldehyde in health-care workers. *Contact Dermatitis*, 43, 150, 2000.
14. Dibo, M. and Brasch, J., Occupational allergic contact dermatitis from *N,N-bis*(3-aminopropyl) dodecylamine and dimethyldidecylammonium chloride in 2 hospital staff. *Contact Dermatitis*, 45, 40, 2001.
15. Kanerva, L., Estlander, T., and Jolanki, R., Occupational allergic contact dermatitis from alkylammonium amidobenzoate. *Eur. J. Dermatol.*, 11, 240, 2001.
16. Barbaud, A., Occupational dermatitis in health care personnel. *Rev. Prat.*, 52, 1425, 2002.
17. Reigart, J.R. and Roberts, J.R., *Recognition and Management of Pesticide Poisonings.* 5th edition. EPA #735-R-98-003. Washington, DC: US EPA, 1999.
18. Schnuch, A., Uter, W., Geier, J., Frosch, P.J., and Rustemeyer, T., Contact allergies in healthcare workers. *Result. IVDK*, 78, 358, 1998.
19. Thom, N.S. and Agg, A.R., The breakdown of synthetic organic compounds in biological processes. *Proc. R. Soc. Lond. B. Biol. Sci.*, 189, 347, 1975.
20. Battersby, N.S. and Wilson, V., Survey of the anaerobic biodegradation potential of organic chemicals in digesting sludge. *Appl. Environ. Microbiol.*, 55, 433, 1989.
21. Woodruff, N.W., Durant, J.L., Donhoffner, L.L., Penman, B.W., and Crespi, C.L., Human cell mutagenicity of chlorinated and unchlorinated water and the disinfection byproduct 3-chloro-4-(dichloromethyl)-5-hydroxy-2(5H)-furanone (MX). *Mutat. Res.*, 495, 157, 2001.
22. Mills, C.J., Bull, R.J., Cantor, K.P., Reif, J., Hrudey, S.E., and Huston, P., Health risks of drinking water chlorination by-products: Report of an expert working group. *Chronic Dis. Can.*, 19, 91, 1998.
23. Komulainen, H., Experimental cancer studies of chlorinated by-products. *Toxicology*, 198, 239, 2004.
24. Haddon, W.F., Binder, R.G., Wong, R.Y., Harden, L.A., Wilson, R.E., Benson, M., and Stevens, K.L., Potent bacterial mutagens produced by chlorination of simulated poultry chiller water. *J. Agric. Food Chem.*, 44, 256, 1996.
25. Richardson, S.D., Thruston, A.D., Caughran, T.V., Chen, P.H., Collette, T.W., Schenck, K.M., Lykins, B.W., Ray-Acha, C., and Clezer, V., Identification of new drinking water disinfection by-products from ozone, chlorine dioxide, chloramine, and chlorine. *Water, Air, Soil Pollut.*, 123, 95, 2000.

26. Richardson, S.D., New disinfection by-product issues: Emerging dbps and alternative routes of exposure. *Global NEST J.*, 7, 43, 2005.

27. Sattar, S.A., Tetro, J., and Springthorpe, V.S., Impact of changing societal trends on the spread of infectious diseases in American and Canadian homes. *Am. J. Infect. Cont.*, 27, S4, 1999.

28. Sakagami, Y., Yamazaki, H., Ogasawara, N., Yokoyama, H., Ose, Y., and Sato, T., The evaluation of genotoxic activities of disinfectants and their metabolites by umu test. *Mutat. Res.*, 209, 155, 1988.

29. Zeiger, E., Gollapudi, B., and Spencer, P., Genetic toxicity and carcinogenicity studies of glutaraldehyde—a review. *Mutat. Res.*, 589, 136, 2005.

30. Brusick, D., Analysis of genotoxicity and the carcinogenic mode of action for ortho-phenylphenol. *Environ. Mol. Mutagen.*, 45, 460, 2005.

31. Fang, H., Tong, W., Branham, W.S., Moland, C.L., Dial, S.L., Hong, H., Xie, Q., Perkins, R., Owens, W., and Sheehan, D.M., Study of 202 natural, synthetic, and environmental chemicals for binding to the androgen receptor. *Chem. Res. Toxicol.*, 16, 1338, 2003.

32. McGowan, J.E. Jr., Resistance in nonfermenting gram-negative bacteria: Multidrug resistance to the maximum. *Am. J. Med.*, 119, S29, discussion S62, 2006.

33. Piddock, L.J.V., Clinically relevant chromosomally encoded multidrug resistance efflux pumps in bacteria. *Clin. Microbiol. Rev.*, 19, 382, 2006.

34. De Rossi, E., Ainsa, J.A., and Riccardi, G., Role of mycobacterial efflux transporters in drug resistance: An unresolved question. *FEMS Microbiol. Rev.*, 30, 36, 2006.

35. Jana, S. and Deb, J.K., Molecular understanding of aminoglycoside action and resistance. *Appl. Microbiol. Biotechnol.*, 70, 140, 2006.

36. Cherepenko, Y. and Hovorun, D.M., Bacterial multidrug resistance unrelated to multidrug exporters: Cell biology insight. *Cell Biol. Int.*, 29, 3, 2005.

37. Aiello, A.E. and Larson, E., Antibacterial cleaning and hygiene products as an emerging risk factor for antibiotic resistance in the community. *Lancet Infect. Dis.*, 3, 501, 2003.

38. Aiello, A.E., Marshall, B., Levy, S.B., Della-Latta, P., and Larson, E., Relationship between triclosan and susceptibilities of bacteria isolated from hands in the community. *Antimicrob. Agent Chemother.*, 48, 2973, 2004.

39. Ledder, R.G., Gilbert, P., Willis, C., and McBain, A.J., Effects of chronic triclosan exposure upon the antimicrobial susceptibility of 40 ex-situ environmental and human isolates. *J. Appl. Microbiol.*, 100, 1132, 2006.

40. McMurry, L.M., Oethinger, M., and Levy, S.B., Triclosan targets lipid synthesis. *Nature*, 394, 531, 1998.

41. Escalada, M.G., Russell, A.D., Maillard, J.-Y., and Ochs, D., Triclosan–bacteria interactions: Single or multiple target sites? *Lett. Appl. Microbiol.*, 41, 476, 2005.

42. Gilbert, P., Allison, D.G., and McBain, A.J., Biofilms in vitro and in vivo: Do singular mechanisms imply cross-resistance? *Symp. Ser. Soc. Appl. Microbiol.*, 31, 98S, 2002.

43. Gaze, W.H., Abdouslam, N., Hawkey, P.M., and Wellington, E.M., Incidence of Class 1 integrons in a quaternary ammonium compound-polluted environment. *Antimicrob. Agent Chemother.*, 49, 1802, 2005.

44. Cole, E.C., Addison, R.M., Rubino, J.R., Leese, K.E., Dulaney, P.D., Newell, M.S., Wilkins, J., Gaber, D.J., Wineinger, T., and Criger, D.A., Investigation of antibiotic and antibacterial agent cross-resistance in target bacteria from homes of antibacterial product users and nonusers. *J. Appl. Microbiol.*, 95, 664, 2003.

45. Braoudaki, M. and Hilton, A.C., Low level of cross-resistance between triclosan and antibiotics in *Escherichia coli* K-12 and *E. coli* O55 compared to *E. coli* O157. *FEMS Microbiol. Lett.*, 235, 305, 2004.
46. Braoudaki, M. and Hilton, A.C., Adaptive resistance to biocides in *Salmonella enterica* and *Escherichia coli* O157 and cross-resistance to antimicrobial agents. *J. Clin. Microbiol.*, 42, 73, 2004.
47. McBain, A.J., Ledder, R.G., Moore, L.E., Catrenich, C.E., and Gilbert, P., Effects of quaternary-ammonium-based formulations on bacterial community dynamics and antimicrobial susceptibility. *Appl. Environ. Microbiol.*, 70, 3449, 2004.
48. Springthorpe, S., Nokhbeh, R., and Sattar, S.A., Acquisition of antibiotic resistance in bacteria exposed to disinfectant in drinking water distribution system biofilms. Presented at International Symposium on Microbial Ecology, Vienna, August 21–25, 2006.
49. Sniegowski, P.D., Gerrish, P.J., and Lenski, R.E., Evolution of high mutation rates in experimental populations of *E. coli, Nature*, 387, 703–705, 1997. Comment in: *Nature*, 387, 659, 661, 1997.
50. Kotewicz, M.L., Brown, E.W., LeClerc, J.E., and Cebula, T.A., Genomic variability among enteric pathogens: The case of the mutS-rpoS intergenic region. *Trends Microbiol.*, 11, 2, 2003.
51. Denamur, E., Bonacorsi, S., Giraud, A., Duriez, P., Hilali, F., Amorin, C., Bingen, E., Andremont, A., Picard, B., Taddei, F., and Matic, I., High frequency of mutator strains among human uropathogenic *Escherichia coli* isolates. *J. Bacteriol.*, 184, 605–609, 2002.
52. Springthorpe, V.S. and Sattar, S.A., Chemical disinfection of virus-contaminated surfaces. *Crit. Rev. Environ. Control*, 20, 169, 1990.
53. Springthorpe, S., Kibbee, R., and Sattar, S.A., The potential for peroxide-based products to clean milking equipment. Presented at The Milky Way Symposium on Canadian Dairy Research, Ottawa, January 25, 2001.
54. Ravis, S.M., Shaffer, M.P., Shaffer, C.L., Dehkhaghani, S., and Belsito, D.V., Glutaraldehyde-induced and formaldehyde-induced allergic contact dermatitis among dental hygienists and assistants. *J. Am. Dent. Assoc.*, 134, 1072, 2003.
55. Ryan, C.K. and Potter, G.D., Microbicide colitis. Rinse as well as you wash. *J. Clin. Gastroenterol.*, 21, 6, 1995.
56. Griffiths, P.A., Babb, J.R., Bradley, C.R., and Fraise, A.P., Glutaraldehyde-resistant *Mycobacterium chelonae* from endoscope washer disinfectors. *J. Appl. Microbiol.*, 82, 519, 1997.
57. Levy, S.B., Active efflux, a common mechanism for biocide and antibiotic resistance. *J. Appl. Microbiol.*, 92, 65S, 2002.
58. Anderson, R.L., Vess, R.W., Carr, J.H., Bond, W.W., Panlilio, A.L., and Favero, M.S., Investigations of intrinsic *Pseudomonas cepacia* contamination in commercially manufactured povidone-iodine. *Infect. Control Hosp. Epidemiol.*, 12, 297, 1991.
59. Wishart, M.M. and Riley, T.V., Infection with *Pseudomonas maltophilia*: Hospital outbreak due to contaminated disinfectant. *Med. J. Aust.*, 2, 710, 1976.
60. Dettenkofer, M., Wenzler, S., Amthor, S., Antes, G., Motschall, E., and Daschner, F.D., Does disinfection of environmental surfaces influence nosocomial infection rates? A systematic review. *Am. J. Infect. Control*, 32, 84, 2004.
61. Vizcaino-Alcaide, M.J., Herruzo-Cabrera, R., and Fernandez-Acenero, M.J., Comparison of the disinfectant efficacy of Perasafe and 2% glutaraldehyde in *in vitro* tests. *J. Hosp. Infect.*, 53, 124, 2003.

62. Cooke, R.P., Goddard, S.V., Whymant-Morris, A., Sherwood, J., and Chatterly, R., An evaluation of Cidex OPA (0.55% ortho-phthalaldehyde) as an alternative to 2% glutaraldehyde for high-level disinfection of endoscopes. *J. Hosp. Infect.*, 54, 226, 2003.

63. Wolkoff, P.H., Schneider, T., Kildeso, J., Degerth, R., Jaroszewski, M., and Schunk, H., Risk in cleaning: Chemical and physical exposure. *Sci. Total Environ.*, 23, 135, 1998.

64. Murr, A.S. and Goldman, J.M., Twenty-week exposures to the drinking water disinfection by-product dibromoacetic acid: Reproductive cyclicity and steroid concentrations in the female Sprague-Dawley rat. *Reprod. Toxicol.*, 20, 73, 2005.

65. Ayliffe, G.A.J. and English, M.P., *Hospital Infection: From Miasmas to MRSA.* Cambridge University Press, Cambridge, U.K., 2003.

66. Floret, D., Co-infections virus-bacteries. *Arch. Pediatr.*, 4, 1119, 1997.

67. Rasmussen, K., Chemin, P., and Haastrup, P., Regulatory requirements for biocides on the market in the European Union according to Directive 98/8/EC. *J. Hazard. Mater.*, 67, 237, 1999.

68. Omidbukhsh, N. and Sattar, S.A., Broad-spectrum microbicidal activity, toxicological assessment and materials compatibility of a new generation of accelerated hydrogen peroxide (ahp)-based environmental surface disinfectant. *Am. J. Infect. Control*, 34, 251, 2006.

69. Venczel, L.V., Arrowood, M., Hurd, M., and Sobsey, M.D., Inactivation of *Cryptosporidium parvum* oocysts and *Clostridium perfringens* spores by a mixed-oxidant disinfectant and by free chlorine. *Appl. Environ. Microbiol.*, 63, 1598, 1997.

70. Fenner, D.C., Burge, B., Kayser, H.P., and Wittenbrink, M.M., The anti-microbial activity of electrolyzed oxidizing water against microorganisms relevant in veterinary medicine. *J. Vet. Med.*, B53, 133, 2006.

71. Dufresne, S., Hewitt, A., and Robitaille, S., Ozone sterilization: Another option for healthcare in the 21st century. *Am. J. Infect. Control*, 32, E26, 2004.

72. Fan, L., Song, J., Hildebrand, P.D., and Forney, C.F., Interaction of ozone and negative air ions to control micro-organisms. *J. Appl. Microbiol.*, 93, 144, 2002.

73. Anpilov, A.M., Barkhudarov, E.M., Christofi, N., Kop'ev, V.A., Kossyi, I.A., Taktakishvili, M.I., and Zadiraka, Y., Pulsed high voltage electric discharge disinfection of microbially contaminated liquids. *Lett. Appl. Microbiol.*, 35, 90, 2002.

74. Blenkinsopp, S.A., Khoury, A.E., and Costerton, J.W., Electrical enhancement of biocide efficacy against *Pseudomonas aeruginosa* biofilms. *Appl. Environ. Microbiol.*, 58, 3770, 1992.

75. Ireland, J.C., Klostermann, P., Rice, E.W., and Clark, R.M., Inactivation of *Escherichia coli* by titanium dioxide photocatalytic oxidation. *Appl. Environ. Microbiol.*, 59, 1668, 1993.

76. Cho, M., Chung, H., Choi, W., and Yoon, J., Different inactivation behaviors of MS-2 phage and *Escherichia coli* in TiO_2 photocatalytic disinfection. *Appl. Environ. Microbiol.*, 71, 270, 2005.

77. Hansen, E.H., Albertsen, L., Schafer, T., Johansen, C., Frisvad, J.C., Molin, S., and Gram, L., *Curvularia* haloperoxidase: Antimicrobial activity and potential application as a surface disinfectant. *Appl. Environ. Microbiol.*, 69, 4611, 2003.

78. Winterton, N., Chlorine the only green element—Towards a wider acceptance of its role in natural cycles. *Green Chem.*, 2, 173, 2000.

79. Klebanoff, S.J., Myeloperoxidase: Friend or foe. *J. Leukocyte Biol.*, 77, 598, 2005.
80. Hughes, R., Andrew, P.W., and Klivington, S., Enhanced killing of Acanthamoeba cysts with a plant peroxidase-hydrogen peroxide-halide antimicrobial system. *Appl. Environ. Microbiol.*, 69, 2563, 2003.
81. Gagnon, G.A., Rand, J.L., O'leary, K.C., Rygel, A.C., Chauret, C., and Andrews, R.C., Disinfectant efficacy of chlorite and chlorine dioxide in drinking water biofilms. *Water Res.*, 39, 1809, 2005.
82. Sattar, S.A., Kibbee, R.J., Tetro, J.A., and Rook, T.A., Experimental evaluation of an automated endoscope reprocessor with *in situ* generation of peracetic acid for disinfection of semi-critical devices. *Infect. Control Hosp. Epidemiol.*, 27, 1193, 2006.
83. Kanazawa, A., Ikeda, T., and Endo, T., A novel approach to mode of action of cationic biocides: Morphological effect on antibacterial activity. *J. Appl. Bacteriol.*, 78, 55, 1995.
84. Donaldson, K., Stone, V., Tran, C.L., Kreyling, W., and Borm, P.J.A., Nanotoxicology. *Occup. Environ. Med.*, 61, 727, 2004.
85. Lam, C.W., James, J.T., McCluskey, R., Arepalli, S., and Hunter, R.L., A review of carbon nanotube toxicity and assessment of potential occupational and environmental health risks. *Crit. Rev. Toxicol.*, 36, 189, 2006.
86. Lea, P., Ding, S.-F., and Lemez, S.B., Ultrastructure changes induced by dry film formation of a trisodium phosphate blend, antimicrobial solution. *Scanning*, 25, 277, 2003.
87. Sunenshine, R.H. and McDonald, L.C., *Clostridium difficile*-associated disease: New challenges from an established pathogen. *Cleve. Clin. J. Med.*, 73, 187, 2006.
88. Wilcox, M.H. and Fawley, W.N., Hospital disinfectants and spore formation by *Clostridium difficile*. *Lancet*, 356, 1324, 2000.
89. McDonnell, G., Fichet, G., Antloga, K., Kaiser, K., Bernardo, M., Dehen, C., Duval, C., Comoy, E., and Deslys, J.-P., Cleaning investigations to reduce the risk of prion contamination on manufacturing surfaces and materials. *Eur. J. Parenteral Pharmaceut. Sci.*, 10, 67, 2005.
90. Fauci, A.S., Emerging and reemerging infectious diseases: The perpetual challenge. *Acad. Med.*, 80, 1079, 2005.
91. Daschner, F., The hospital and pollution: Role of the hospital epidemiologist in protecting the environment. In *Prevention and Control of Nosocomial Infections*, Wenzel, R.P. (ed.), Baltimore, MD: Williams & Wilkins, 1997.
92. Dettenkofer, M., Kuemmerer, K., Schuster, A., Mueller, W., Muehlich, M.S.M., and Daschner, F.D., Environmental auditing: Environmental auditing in hospitals: First results in a university hospital. *Environ. Manage.*, 25, 105, 2000.
93. Tielemans, E., Marquart, H., De Cock, J., Groenewold, M., and Van Hemmen, J., A proposal for evaluation of exposure data. *Ann. Occup. Hyg.*, 46, 287, 2002.
94. Sattar, S.A., The use of microbiocides in infection control: A critical look at safety, testing and applications. *J. Appl. Microbiol.*, 101, 743, 2006.

4

Microbiological Concerns in Non-Sterile Manufacturing

Joseph Kirby Farrington and Jim Polarine

CONTENTS

4.1 Introduction

Virtually all pharmaceutical manufacturing is performed under non-sterile conditions. Even for sterile products (aseptically filled or terminally sterilized), the majority of operations are carried out in non-sterile environments.

The finished product (either sterile or non-sterile) is stored under non-sterile conditions. What is sterility? The scientific definition is absence of life. In pharmaceuticals, the definition for products labeled sterile is based on the probability of a single unit having viable contaminants. The minimal acceptable standard is 1 in 10,000 for aseptic fills and 1 in 1,000,000 for terminally sterilized products. Risk and the actual contaminated product are not synonymous terms. However, current technology cannot support 100% sterility checks.

Products labeled as sterile or non-sterile are still required to be safe for their intended purposes. The microbial content and potential microbial content are major factors when performing a product safety assessment.

How does a product become contaminated with microorganisms? There is a very simple answer to this question and that is microorganisms entered or were incorporated into the product. What are the potential sources of contaminants?

1. People
2. Raw materials and components
3. Equipment
4. Environment

In non-sterile products the issue is not the potential for contaminated units but instead what are the acceptable levels and types of contaminants. The pharmaceutical microbiologist has the task of assessing the amount and types of contaminants based on actual data and not theoretical risk. Most non-sterile products should and would fail a standard sterility test. Non-sterile products whose base formulation will support microbial growth or survival usually contain antimicrobial preservatives. These preservatives are not ubiquitous in that there will be resistant organisms and capacities that prevent complete control. The various preservative evaluation methods available (including the compendial methods) are designed to predict certain aspects of preservative performance. These methods are not foolproof nor totally reliable. For example, the ability of laboratory methods to predict in-use efficacy of antimicrobial preservatives in an experimental cosmetic [1].

The single paramount microbiological issue for non-sterile manufacturers is how to produce a safe product and the secondary aspect is what a safe product is. For the purposes of this chapter, the in-use issues around microbial contaminants and non-sterile products will only be briefly addressed. Manufacturers have very little practical control over consumer-introduced contaminants. Products can be preserved, package design can limit contamination potential, and directions for use can include precautions and quantities of product limited to preclude extended use. However, consumers can be relied upon to do the unexpected. Unfortunately, manufacturers have been held liable on numerous occasions for product contaminated in-use by the consumer and causing some type of injury. Inadequate preservation has also been responsible for product recalls. This chapter concentrates on microbial contamination issues surrounding manufacturing operations when the manufacturer can have a reasonable opportunity to affect the finished product's microbial content.

It is not the intent or purpose of this chapter to quote regulations, guidelines, or standards. These are all in flux and such quotes would likely be inaccurate in a short period. In their place, the scientific approach will be used. That is to say, good science will be substituted for regulatory guidance and requirements.

4.2 What Is Good Science?

- Good science is data derived from methods of proven capability.
- Good science uses facts not opinions or paradigms.
- Good science questions before acceptance.
- Good science looks for the truth regardless of the potential outcome.
- Good science is data driven.

4.3 Origins of Microbial Contaminants

Microbial contaminants can have diverse origins. The paradigm that Grade A clean rooms are sterile contributes to the assumption that some significant error had to occur to result in a contaminated product. The several definitions of sterility add to the confusion and misunderstandings. The scientific definition of sterility is "without life." Processed foods use the term commercial sterility, which loosely means that the product does not contain potentially harmful microorganisms. For aseptically filled or terminally sterilized pharmaceutical products, sterility means the likelihood of a contaminated single unit. A sterility test can have different meanings based on the culture or regulations. For example, in the United States, a sterility test is a specified procedure used to detect the absence of specific types of microorganisms in a particular volume of product. In other countries, a sterility test may also simply mean the determination of microbial content.

There is nothing abnormal or strange about finding microorganisms in a non-sterile environment. Consequently, finding microbial contaminants in products not labeled or rendered sterile should be expected and anticipated. As discussed in Section 4.1, the possible sources of viable contaminants are not difficult to identify. The goal in non-sterile manufacturing should be to control and not eliminate potential contaminants.

Viable contaminants are unique in that they have the ability to increase in level and occasionally be replaced or supplemented by other types of viable contaminants. A chemical contaminant may increase in quantity with time only if it is a breakdown or degradation product of some formulation ingredient. That is to say there is a mass balance within the material in question. This approach does not apply to viable contaminants as the mass involved in significant levels of microbial contaminants is relatively small. What may happen is that microorganisms can metabolize many different materials and leave metabolic by-products that can have significant adverse consequences for the product. For example, pH shifts, color changes, odors, endo- and other toxins, various enzymes, etc., may all adversely affect products. A common mistake is to assume microorganisms have the ability to read and thus know what text books, reference books, study reports, and similar literature say that they can or cannot do.

The first response to a suspected contamination problem should not be to discount it or assume it is a lab error of some type. This same misunderstanding frequently occurs when evaluating the potential for microbial contamination or the possible consequences of such an event. Spontaneous generation—life arising from nonlife—was disproved in 1864 by Louis Pasteur. The fact that a contamination event occurred can only be refuted by another fact that it did not. The level of proof required for a scientific certainty is far greater than the legal requirement of "more likely than not."

4.4   Evaluating the Possible Effects of Microbial Content

The key issue concerning microbial contaminants and non-sterile products is the potential of these organisms to harm the user. If the product is protected from microbial contamination, degradation, or induced changes then the consumer will also be protected. Products can be involved in microbially related adverse events in several ways. The manufacturer can partially control only certain aspects and these are the microbial content of the finished product and, through antimicrobial preservatives, the in-use microbial content. Primary package design such as nonaspirating applicators, single-use containers, and single-use applicators also provide some protection. The manufacturer cannot prevent the consumer from intentionally or inadvertently contaminating the product by trying to dilute it, mixing it with other products, storing it under adverse conditions, sharing it with others, or using it contrary to label directions.

4.5   Microbial Growth and Characteristics

Before the potential effects of microbial contaminants can be evaluated, the presence of contaminants, their concentration, and their identity must be determined. Microorganisms do not exist in nature as pure cultures.

The use of pure cultures was developed to better study the characteristics of specific organisms. In nature, microorganisms many times exist in biofilms, which are composed of multiple species, each contributing to the evolution and survival of the biofilm structure. These biofilms can shed organisms into the surrounding environment at varying rates or be disrupted by mechanical or chemical means resulting in a massive release of organisms. A biofilm can contain many different environments, which are inhabited by organisms specific to that environment.

4.6 Detection Methods and Capabilities

The methods used to detect the presence and quantities of microorganisms are many and varied. However, none are totally reliable or capable of detecting all organisms. A method is composed of several components including sampling, sample holding, sample transport, sample processing, and sampling culturing. Each separate component or activity has multiple potential variables. The expectations and assumptions about method capabilities and its ability to provide accurate, reliable information have resulted in specifications and limits that are many times unrealistic. The United States Pharmacopoeia (USP) acknowledges at least 50% variability in the standard plate count procedure and states that a material or product with an aerobic plate count (APC) limit of 100 CFUs (colony-forming units) should not be considered out of specification unless the test shows 200 CFUs (USP 29 <1111> Microbiological Examination of Non-Sterile Products: Acceptance Criteria for Pharmaceutical Preparations and Substances for Pharmaceutical Use). The methods used to demonstrate method suitability acknowledge the inability of the methods to quantitatively recover organisms directly from the material or product tested. The suitability methods require that test organisms be quantitatively recovered from a dilution of the product or material and not the actual product or material. The antimicrobial preservative efficacy (APE) test (USP <51>) necessitates the recovery of organisms from the actual preserved product. However, it is not usually possible to recover the complete test inoculum immediately after inoculation in most products. Objective evaluations of commonly used methods have shown time after time that the expectations of that method as shown by specifications referencing the method are not within the capability of the method: for example, the ability to detect 1 CFU on a surface and 5 CFU on the same surface with confidence that the difference in numbers recovered is actually significant.

Microorganisms exist in many shapes, sizes, and configurations. Their nutritional state and growth conditions can affect all these characteristics unpredictably. Conventional detection methodology and attendant specifications assume that these factors are constant. In establishing specifications, the assumption is made that a CFU is synonymous with a single organism even though the term CFU was coined to acknowledge the fact that this was not the case for most species of microorganisms. In fact, a CFU can be one organism or a thousand and one or more organisms. This fact more than any other shows the fallacy of establishing specifications based on the number of recovered organisms, when using conventional test methodology. The term conventional refers to the classic culture techniques where the organisms must grow to be detected either by forming a visible colony, measurable change in their environment, or visible evidence of their presence. Newer, unconventional methodologies include gene probes, antigen detection, measuring ATP, ADP, DNA, or RNA levels and other such parameters. The basic science and first use of these unconventional methodologies actually goes back many years. Advances in instrumentation and the drive for more sophisticated and thus presumed better, faster methods has pushed the development and use of such alternative techniques.

4.7   Evaluating Recovered Organisms

When microorganisms are recovered from a product or material those organisms should be evaluated for significance. The presence of microbial contaminants in products purported to be sterile renders such products unfit for use. However, products not labeled as sterile require a different approach.

"Where warranted, a risk-based assessment of the relevant factors is conducted by personnel with specialized training in microbiology and the interpretation of microbiological data. For raw materials, the assessment takes account of processing to which the product is subjected, the current technology of testing and the availability of materials of the desired quality" (USP 29 <1111> Microbiological Examination of Non-Sterile Products: Acceptance Criteria for Pharmaceutical Preparations and Substances for Pharmaceutical Use).

4.8   How to Perform a Risk Assessment Using Compendial Guidance

There are a number of factors that comprise risk assessment for a non-sterile commercial product. Various guidance documents exist but there is no assurance that complying with guidance document requirements or suggestions will assure an absolutely risk-free product. A microbiological risk assessment is an informed judgment about the potential of any microbial contaminants to cause harm to a user. This assessment should be conducted by personnel with specialized training in microbiology and the interpretation of microbiological data.

In the simplest and most basic terms, the presence of viable (and sometimes non-viable) microorganisms poses a risk of these organisms causing an infection or some type of toxic reaction for the user. Even sterile products present a risk, particularly after first use. The environments in which we all live contain numerous different types and species of microorganisms. The fact that most of these organisms do not cause obvious or overt harm demonstrates that certain conditions must exist before the mere presence of viable microorganisms is a cause for concern. The expectations of most consumers is that products they buy and use should be safe for their intended purposes. Manufacturers do not produce and sell products that they know pose an unreasonable risk of causing harm. The purpose of risk assessment is to determine, from the facts available, if the product would pose an unreasonable risk if used as intended. The label directions should specify the intended method of use. A manufacturer may wish to consider the potential effect on safety if a consumer were to use the product in a foreseeable but contrary to the label directions manner: for example, drinking directly out of a bottle when the use of a spoon was specified on the label.

In making a risk assessment, the first evaluation should be the potential for the product to support microbial growth or survival. If microorganisms can proliferate within the product, then the potential for harm is greatly increased compared with a product that will only allow survival but not growth. Unless a product is naturally hostile to microbial growth or survival, i.e., pH extremes, anhydrous, low water activity, and antimicrobial components (excluding added preservatives), then the presence and efficacy of specifically added antimicrobial preservatives must be considered. Antimicrobial preservatives can have one of three effects on contaminating microorganisms:

1. No effect
2. Retard growth
3. Kill

The effect, if any, is very dependent on the type of organism, the quantity, storage conditions, other formulation components, and time. Whether a preservative system retains efficacy over time must be considered as well as what organisms will naturally challenge that system either during manufacturing and storage or during use. The manufacturer has far more control over conditions and potential microbial content during manufacturing than during use. If any particular type or species of organisms are known, safety issues with specific or similar products may arise and then these should be specifically evaluated for potential to cause harm in connection with the specific product being assessed. The various evaluation methods for preservative efficacy do not require the complete killing of challenge organisms even after 28 days.

Once the actual and potential microbial content of the product has been assessed, then this information can be used in further assessments based on the remaining compendial criteria. The genus and species of organisms in a product subsequent to manufacture but before distribution needs to be

identified. However, frequent name changes with time and improved identification methods make literature searches to the species level and many
times even to the genus level problematic. Consequently, assessments
need to be made using probabilities based on Gram reaction, colonial
morphology, microscopic appearance, etc., as well as various identification
methods. For example, only certain species of gram-positive spore-forming
rods capable of aerobic growth would normally be considered pathogenic
for humans. The majority of gram-negative organisms are opportunistic
pathogens and a large number are frank human pathogens.

An organism capable of causing a disease is referred to as a pathogen. An
organism that will cause a disease or set up an infection under certain
circumstances is called an opportunistic pathogen. The relative ability of
any organism to cause a disease or set up an infection is called virulence.
Those characteristics of the organism or conditions of the host that facilitate
an infection are called virulence factors. Certain microbial disease processes
are caused by the organism itself or by some metabolic by-product of that
organism. Humans as well as other animals harbor many different types
and species of microorganisms that are necessary for the body's proper
function. This is called a commensal or symbiotic relationship. When these
organisms migrate from their normal body location to other locations they
can and will cause an infection. Other organisms that would be considered
normal microflora can undergo genetic changes or be replaced by a different
strain of the same species and become pathogenic. A weed has been
described as a plant that is growing in the wrong place. This analogy can
be made for resident microflora. The point of this chapter is that any
microorganism is an unknown with regard to its ability to cause harm.
There are factors that can and should be considered when evaluating the
virulence of organisms in a product:

- The use of the product: hazard varies according to the route of
  administration (eye, nose, respiratory tract)
- The method of application
- The intended recipient: risk may differ for neonates, infants, and
  the debilitated
- Use of immunosuppressive agents, corticosteroids
- Presence of disease, wounds, organ damage

4.9   Use of the Product: Hazard Varies According to the Route of Administration (Eye, Nose, Respiratory Tract)

The ability of any microbial contaminant to cause harm is directly related
to its ability to colonize/grow on/infect the host. The body has certain
nonspecific defense mechanisms against infection including natural
barriers such as the skin, lysozyme in tears, low pH in the intestinal tract,

amebocytes and interferon in the circulating blood, mucous and hair to trap and filter organisms in the nose, etc. Consequently, the route of administration can and will affect the ability of organisms to infect the user. The eye provides a significant barrier unless the cornea is scratched. Such scratching can occur with contact lenses, fingernails, dirt, inadvertent injuries, etc. The eye has very little vascularization, which means that any type of defense depending on circulation will be hindered. The respiratory tract and intestinal tract provide routes where the skin barrier is bypassed. The low pH in the intestinal tract protects against many organisms. However, the respiratory tract provides only limited natural host defenses. Consequently, inhaled products such as nasal sprays can put the user at increased risk. An adverse reaction that is very unlikely to occur is called idiosyncratic. However, what would the individual affected call such a reaction?

4.10   Method of Application

Leave-on as opposed to apply-and-remove products provide longer contact times. Aerosols or sprays provide more surface area contact and also the possibility of inhalation. Reusable applicators are higher risk than one-time applicators. Diluting a product for use often dilutes the antimicrobial preservative system below useful levels. Any type of product designed to be injected or used on broken or abraded skin is a great risk because it bypasses the skin barrier.

4.11   Intended Recipient: Risk May Differ for Neonates, Infants, and the Debilitated

The common factor here is that natural host defenses may be compromised either because they are not fully developed (neonates/infants) or have deteriorated based on age or physical condition. Regardless of the root cause, the potential effect that being increased risk of infection is the same. Other than labeling, the manufacturer has very little if any real control over the users of their products. However, if it is conceivable that such individuals could use the product, the potential to cause risk for the user should be considered.

4.12   Use of Immunosuppressive Agents, Corticosteroids

Immunosuppressive drugs and corticosteroids will reduce the effectiveness of host defenses against infection. The ability of the manufacturers to restrict use of their products by such individuals is limited to labeling.

4.13   Presence of Disease, Wounds, Organ Damage

Disease, wounds, and organ damage will make an user more susceptible to infection through several mechanisms. Again, the manufacturer may intend their product for such individuals or may want to restrict use via labeling. A manufacturer who designs a product for use by diseased and debilitated people needs to consider the potential consequences of any microbial contaminants.

Taking the above considerations into account, it is still very difficult to make the judgment call to release a product containing viable organisms or which has a propensity to become contaminated during customary use. The manufacturer assumes the risk for potential harm to the user. The manufacturer must decide if the nonprescription product is unreasonably dangerous as unlike prescription products there is no one else such as a physician or pharmacist to inform or counsel the potential user.

4.14   Microbiological Control

In products labeled sterile, microbiological control is achieved in two ways. The first and most reliable is terminal sterilization. The second method is aseptic manufacture. The regulatory standards for each method reflect the differences in controllability. Regardless of method, the manufacturer must develop systems and procedures to control the bioburden of the product. In non-sterile products, the presence of viable organisms must be expected. A balance must be struck between the degree of control (including cost) and the needed level of control. The sources of potential contaminants are easily identified:

1. People
2. Raw materials and other components
3. Equipment
4. Environment

4.15   People

It is well known that the number one source of contamination in pharmaceutical manufacturing is people. It is equally well known that terminal sterilization would be an effective control measure but is not practical. Consequently, other measures must be used. Official guidelines and regulations specify minimum control measures, which primarily consist of "bagging" the workers and training them in various methods to minimize breaching the "bag" or coming in contact with external contaminants and exposing the product to such contaminants. Variations of this approach range from using clean uniforms to fully enclosed and pressurized suits that are externally sterilized.

Normal human skin is colonized with bacteria; different areas of the body have varied total aerobic bacterial counts (e.g., 1×10^6 CFUs/cm^2 on the scalp, 5×10^5 CFUs/cm^2 in the axilla, 4×10^4 CFUs/cm^2 on the abdomen, and 1×10^4 CFUs/cm^2 on the forearm). Total bacterial counts on the hands of medical personnel have ranged from 3.9×10^4 to 4.6×10^6. The issues with people are confining the inherent contaminants to prevent spread and controlling behavior so that fomites will not be a source of contamination [2].

4.16 Raw Materials and Other Components

Raw materials and other components represent a major control point for product contamination. Unless raw materials and primary packaging components are sterilized or the finished product sterilized, the bioburden of these items must be considered. A heat step in the process or a filtration step or some other part of the process can reduce the bioburden. However, reliability requires that such steps be clearly stated to be bioburden reducing and that appropriate controls and effectiveness evaluations must be included. Particularly, in raw materials and to a certain extent in bulk and finished products, the assumption cannot be made that microbial contamination is or will be uniform. Patterns of contamination are a topic unto themselves. It is sufficient at this point to state that contamination should be expected to be randomly distributed, nonuniform, and of variable levels in non-sterile materials that will support microbial growth or survival. Sampling plans should be designed with such variability in mind. A common mistake made in designing sampling plans and sampling methods to detect bioburden is to assume that the only concern is the bulk material and not areas of likely contamination. For example, sampling below the top surface or sampling the center of drums or bags only risk missing likely points of contamination. Materials that are hydroscopic will exhibit higher moisture levels on top or side surfaces. Airborne contaminants will more likely be on the top surface or bottom of bags and drums because containers are many times left open before filling and before closing after filling. Unless there is a policy of discarding the top and bottom material, the total contents will be included in the manufacture of the finished product.

4.17 Equipment

Equipment including primary contact and incidental contact can and will be a source of microbial contaminants. The ideal situation would be for all primary contact equipment to be sterilized. This is not possible in many situations. One alternative is to use disposable, sterile liners but these are not without cost. Even sterilized or sanitized equipment can be a problem based on storage conditions and the presence of low-level residuals. Condensate can be a major potential problem. A 500 L tank with 1 mL of condensate that contained 50,000,000 CFUs (5×10^{-7}) would start with 100 CFUs/mL of product.

As previously discussed, levels of microbial contaminants can and will increase under the proper conditions. Even the use of antimicrobial preservatives may not prevent or mitigate this problem. The best control measure is to sanitize the equipment immediately before use and to keep hold times to a minimum. Hot water rinse (60°C) has been shown to be a very effective control measure in numerous applications. Sanitizing conditions are generally acknowledged to be 60°C for 10 min with shorter exposure times using higher temperatures. Blocks of metal such as pump housings can act as heat sinks and require more exposure time to reach the required temperature.

4.18   Environment

Environmental conditions including airborne organisms (see Figure 4.1), general sanitation level of the facility, and similar concerns can have a real or presumed adverse effect on products manufactured in the area. Proving an effect or no-effect is not an activity that would be value-added in many situations. It is better to defer to the paradigm that says contaminated areas lead to contaminated products, i.e., the higher the level of ambient contamination, the greater the chance of a contaminated product. So every reasonable effort should be made to minimize area contamination and the opportunity to expose the product to contamination. Simple precautions include a cleaning sanitization program, physical barriers such as doors that stay closed and shields, use of gloves and appropriate uniforms, air flow patterns, etc. A monitoring program is useful but should be designed with the type and end use of the product in mind. For example, in manufacturing

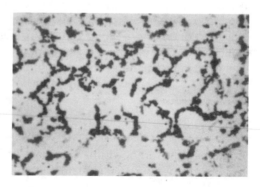

FIGURE 4.1   (See color insert following page 210)
Vegetative bacteria.

a non-sterile powder for topical use, the ability of microorganisms to survive or reproduce in such products is minimal when compared to water-based liquids. Extensive environmental controls and monitoring aimed at controlling potential microbial contamination would be of little value to product safety in such a situation.

References

1. Farrington, J.K., Martz, E.L., Wells, S.J., Ennis, C.C., Holder, J., Levchuk, J.W., Avis, K.E., Hoffman, P.S., Hitchins, A.D., and Madden, J.M., Ability of laboratory methods to predict in-use efficacy of antimicrobial preservatives in an experimental cosmetic. *Appl. Environ. Microbiol.*, 60: 4553–4558, 1994.
2. Selwyn, S., Microbiology and ecology of human skin. *Practitioner*, 224: 1059–1062, 1980.

5

New Technologies in Disinfection and Infection Control

José A. Ramirez

CONTENTS

5.1 Introduction

Considerable progress has been made in the field of disinfectants and infection control since the publication of Antoine van Leeuwenhoek's drawings of small living creatures in a variety of habitats in the *Philosophical Transactions of the Royal Society* in 1677. Surprisingly, disinfection technology continues to rely on essentially the same techniques discovered in the later nineteenth or early twentieth century. Hypochlorite solutions have been in use in hospital ward disinfection as early as 1854, hydrogen peroxide as a disinfectant since 1858, aqueous solutions of short-chain alcohols have been known to be effective since 1903, while phenolic compounds have been employed since 1877. Nonetheless, major breakthroughs occurred well into the twentieth century, most notably, the chlorinated phenols in 1933, long-chain quaternary ammonium salts in 1935, halogenated bisphenols in 1941, biguanides in 1950, amphoteric amine surfactants in 1954, and anionic surfactants in 1952 [1].

The present review of novel chemical disinfection technologies is deliberately not constrained by existing regulatory restrictions, as these vary greatly between different regions of the world. The technologies discussed are in most cases commercialized or close to commercialization, or in the worst case are highly feasible in practice as they meet many of the criteria of a superior disinfectant. The scope of application of these technologies is not only the health care environment, but they will also find application in other areas where hygiene is important such as food processing, food services, and institutional and household cleaning.

An optimal disinfectant will typically exhibit the following characteristics:

- Broad spectrum and rapid activity (effective against the widely differing scale of susceptibilities from hydrophobic viruses to bacterial endospores in practically realistic contact times)
- Favorable occupational and environmental safety profile
- Easy to use, which allows simplified training and minimizes user error
- Compatible with a wide variety of materials and surfaces
- Cost effectiveness

The importance of each criteria will depend on the intended use of the disinfectant. Considering the three types of disinfection objectives defined according to Spaulding should allow for a simplified discussion. Spaulding described disinfection objectives in terms of the intended use of the surface or material to be treated. He introduced the terms noncritical, semicritical, and critical surfaces to describe materials and surfaces that would only come in contact with intact skin, those that would regularly come in contact with mucous membranes or nonintact skin, and those that will come in contact with normally sterile areas of the body or the bloodstream, respectively [2]. Although the technologies discussed herein find application to all three types of surfaces, we focus on noncritical applications, as most recent reviews of novel disinfection technologies have mostly focused on semicritical or critical applications [3,4].

The basic strategies in development of new disinfection technologies can be summarized in Table 5.1. Significant incremental or breakthrough developments are typically achieved by combining two or more of these strategies.

The generation of compounds with biocidal activity at the point of use is a very effective way of circumventing problems associated with the stability of the biocidal compound or mixture in its active state. Packaging the components required, for example, in an anhydrous state or in separate containers avoids problems associated to stability or activity loss during transport or storage. It can also be used as a means of reducing or eliminating the hazard risk rating of the product, which brings considerable advantages in transportation and storage flexibility and improves the marketability of the product.

The drawbacks typically associated with in situ activation are mostly related to an increase in complexity during application. In situ activation can be simplified by employing sophisticated packaging or dispensing equipment, but this usually increases the cost in use and the training

TABLE 5.1

Summary of Strategies Available for Novel Disinfection Technology Development

Strategy	Advantages	Disadvantages
In situ generation of actives	• Overcome stability issues (longer shelf life) • Increase versatility in transport and storage • Reduced occupational and environmental risk during transport and storage	• May result in additional complexity/cost in packaging and delivery systems • Could increase difficulty of application leading to higher training requirements and more prone to user error
Synergistic mixtures	• Allows reliance of well-characterized molecules/ingredients • Allows lower active concentrations to be used, improving toxicology and environmental profiles • Easier to derive competitive benefits due to mixture complexity	• Many times result is complex mixture, prone to stability problems • Complex mixtures may also be difficult to manufacture • Increased complexity of registration process if relying on several actives
Synthesis of new molecules	• Patented molecule will yield singular competitive benefits	• Long and costly development and registration process • Comprehensive toxicological and environmental impact characterization usually required
Optimal actives delivery	• Allows reliance on well characterized molecules/ingredients • Allows lower active concentrations to be used, improving toxicology and environmental profiles • Easier to derive competitive benefits due to mixture or application process complexity	• May result in additional complexity/cost in packaging and delivery systems • Could result in increased difficulty in manufacturing • Potential for increased difficulty in registration • If tied to a device, increased capital and operating cost requirements or increased demand for operator training
Increasing internal energy of system	• Permits lower active concentrations to be used, improving toxicology and environmental profiles • Patented or trade secret process provides competitive advantage • Easier to achieve broader spectrum of activity with relatively simple chemistries	• Usually increases complexity (cost) of disinfection as an engineering device is typically required • Increased difficulty of application which includes training requirements for user, makes system more prone to operative errors • Registration process becomes more involved as it involves assurance of specific application conditions

requirements for the operating personnel. Some examples of in situ generation include the generation of peracetic acid on-site by perhydrolysis, mixing a peracid precursor such as tetraacetylethylenediamine (TAED) with a perhydroxyl anion from a percarbonate or perborate salt. Other examples are the on-site generation of chlorine dioxide by mixing a chlorite salt and a strong acid or the formation of diverse chlorinated biocidal species by mixing a monopersulfate salt and sodium chloride in water.

Synergism can be described as the additional biocidal effect obtained from a complex mixture, which is in excess of the additive effect observed from the individual components in the mixture. The principle has been explained for linear dose–effect relationships for mixtures by Zipf [5] as

$$\sum_{i=1}^{N} \left[\frac{C_i}{X_i} \right] \leq 1 \tag{5.1}$$

where $i = N \geq 2$ and C_i is the actual concentration of the combination of N components which exert the same biocidal effect as the minimum effective concentrations of the agents on their own (X_i). Lehmann [6] has described nonlinear dose–effect systems where, for example, the total effect may exceed the expected additive effect even when there are no apparent synergistic effects. Components for consideration range from traditionally accepted actives (e.g., quaternary ammonium chlorides, short-chain alcohols, phenolic compounds, peroxygen species, etc.) to nonbiocidal functional chemicals such as surface active and chelating agents.

The main advantage of synergistic formulations is that it results in products that are more effective at lower active concentrations, therefore improving their toxicological profiles. Furthermore, synergistic combinations of biocidal agents allow the formulator to address problems inherent to simple, single-ingredient-based products such as:

- Limited spectrum of activity: the combination of different types of biocidal actives allows one to exploit the varying activity of each biocidal active against each organism at relatively low levels instead of having to formulate with a high level of a particular active to overcome a less susceptible class of microorganism
- Limited activity by limited solubility: the saturation concentration of an active may be lower than the required biocidal concentration in which case the activity gap may be closed by adding a second active species
- Varying regulatory requirements: the different methodologies for biocide registration sometimes favor or disfavor a particular class of actives in which a synergistic combination allows for a more robust product for global or multiuse registration

The underlying mechanisms by which a combination of ingredients results in synergistic action can vary widely, but most likely involve improved kinetics of active uptake by the cell wall, more favorable partition coefficients of the active species between cell wall and bulk liquid, and multiple direct physiological effects on the cell wall. A predictive description of these mechanisms is difficult however; hence, most studies in this field have been limited to a phenomenological nature.

The synthesis of new molecules remains an approach of limited effectiveness in the field of disinfection and sterilization as this usually entails not only significant investment of upstream development resources but also a long and costly registration process. However, a strategy that merits consideration is the examination of chemistries, which have been registered or in use for some time in other regions of the world but which are not yet widely used. A couple of the more relevant examples are examined in more detail below.

Another strategy that can arguably be categorized on its own is that of optimizing the way in which the delivery of the actives is affected in order to maximize uptake rate and favor cell wall–bulk phase partitioning. This is perhaps the broadest and most difficult strategy to describe as it seldom is used on its own and the methods employed vary widely. This strategy has become more common in the last few years as advances in product packaging, dispensing engineering, and the evolution of the field of nanotechnology provide product developers new tools for optimizing how actives are delivered to a target site.

The activity of a biocidal product can also be greatly enhanced by increasing the internal energy of the thermodynamic control volume considered for disinfection or sterilization. This can be achieved by applying heat, mechanical work, or electromagnetic energy to the system. The advantage of this strategy is that it allows for lower minimum effective concentrations, which as mentioned before, can result in lower toxicity systems. The main drawback, however, is that controlling excitation to the system usually involves a sophisticated engineering setup, which reduces the flexibility of the system for widespread application, increases the cost and its complexity, and the demand for associated training requirements for use. Owing to its highly specialized nature and exclusive application to limited fields in medical instrument sterilization and large-scale food and water processing, we will not explore the field in detail herein. A good compilation on the topic has been presented elsewhere by Schneider [4].

5.2   In Situ Generation of Active Species

There has recently been renewed interest in old biocidal species with recognized activity but certain drawbacks concerning stability and ease of use. Two of these biocides are peracetic acid and hypochlorous acid. Because of their suspected mechanism of action and high reactivity based on their high oxidation potential, these biocides exhibit a relatively wide spectrum of activity.

Peracetic acid has long been described as having the beneficial properties of hydrogen peroxide (e.g., broad-spectrum activity, innocuous decomposition products, and infinite water solubility) but with greater potency, hydrophobicity, and resistance to inactivation by catalase and peroxidases [7]. Its main drawback has been the negative properties attached to its concentrated solutions, mainly its high reactivity and toxicity and strong odor. These characteristics have limited its use in nonindustrial applications although it has become widely popular in industrial environments (e.g., food-processing plants) where the handling of hazardous chemicals in bulk is part of the day-to-day operations.

The laundry detergents industry has long relied on the in situ generation of peracetic acid, but applications have been largely focused on bleaching. The main peracetic acid precursors in use today are tetraacetyl glycol urea, glucose pentaacetate, diacetyldioxohydrotriazine, sodium nonanoyloxybenzenesulfonate, and TAED. TAED was first used by Lever in France in 1978 [8], becoming one of the best known and most widely used activators worldwide. Recently, the largest manufacturer of TAED worldwide has been granted a U.S. Environmental Protection Agency (USEPA) registration for use in biocidal applications [9].

The biocidal properties of peracetic acid have been studied extensively in both the industrial and academic literature. Baldry [10] has reported extensively on the wide spectrum of activity of peracetic acid including confirmation of sporicidal activity against *Bacillus subtilis* using a carrier test method. Additional reviews have been compiled by Block [11], and Baldry and Fraser [12], among others. The basic reaction by which peracetic acid is formed is shown in Figure 5.1.

Antec International, which became a DuPont company in late 2003, has marketed a commercial TAED-based powdered product called PeraSafe®. The product consists of a blue-white powder packaged in sachets of predetermined weight for dissolution in prescribed amounts of water. The resulting solution is a mixture of peracetic acid and hydrogen peroxide at a concentration of roughly 0.26% peracetic acid (PAA) and a pH of around

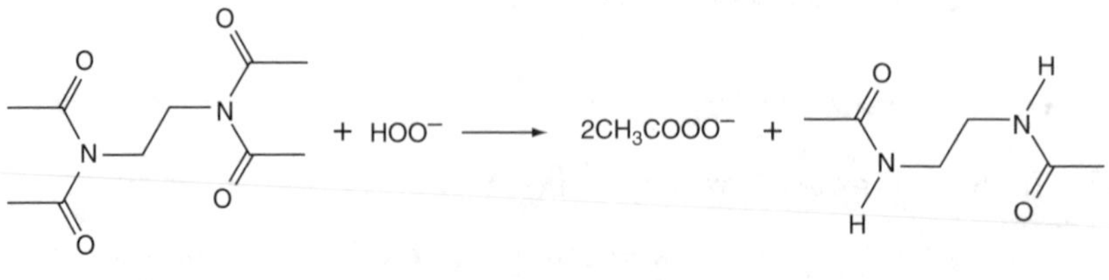

FIGURE 5.1
Formation of peracetic anion from the reaction of TAED with the peroxide anion.

8.0 [13]. Block [14] has shown in vitro work concerning the sporicidal activity of PeraSafe® versus *Clostridium difficile* spores and reported its activity was superior to 1000 ppm available chlorine as delivered by a sodium dichloroisocyanurate precursor.

Another old biocidal compound that has found increased application as an environmental disinfectant is hypochlorous acid. It has been well established that the biocidal activity of hypochlorite will depend greatly on the amount of undissociated hypochlorous acid in water solution [15,16]. The difference of activity between solutions of two different pH can be quite dramatic; for example, at pH 6, 15 ppm solution of hypochlorite will achieve the same sporicidal performance in 8.5 min as a solution at pH 8 in 42 min. This is due to the concentration of undissociated hypochlorous acid; in the first case, at pH 6, 96.5% of the hypochlorous anion is in the form of undissociated acid, while at pH 8, only 21% of the hypochlorous acid is undissociated [17]. The issue preventing widespread application of low pH solutions lies mainly in problems due to hydrolytic decomposition with storage and transportation.

There are two interesting commercial systems that have managed to circumvent this problem by generating the hypochlorous acid at the point of use. The first approach is based on the electrolysis of a dilute sodium chloride solution. An example of a commercial system is one manufactured and sold by PuriCore (formerly Sterilox). The system has been described in greater detail elsewhere [3] and has been tested in applications ranging from food processing to endoscope sterilization [18–21]. Although the microbiological and toxicological benefits have been widely demonstrated, there are inherent drawbacks concerning initial capital and operating costs related to the system as the one-time cost for the generating units and the maintenance costs for the electrodes can be significant.

A different yet simpler approach to the generation of hypochlorous acid in situ is by reacting a water-soluble inorganic halide (such as sodium chloride) with an oxidizing agent such as potassium monopersulfate, as described by Auchincloss [22]. The Auchincloss system is the basis for the commercialized version from DuPont (Virkon S)®, which is available in tablet or powder form. Extensive biocidal data have been published by DuPont [23] and the commercialized product in the U.S. has USEPA-approved bactericidal and virucidal disinfectant claims. The product has been marketed heavily toward the agricultural and veterinary sectors.

5.3 Synergistic Mixtures

The use of synergistic principles in biocidal mixture design has allowed for a resurgence of one of the oldest known antimicrobials, hydrogen peroxide. Hydrogen peroxide has failed to become widely popular as a disinfectant in spite of its broad spectrum of action and occupational and

TABLE 5.2

Summary of USEPA-Approved Claims for AHP Technology

Contact Time	Type of Germicidal Claim
30 s	Nonfood contact surface sanitizing

1 min	Bactericidal:
	    Wide spectrum across both gram (+) and gram (−) bacteria
	Virucidal:
	    Both hydrophilic and lipid viruses including Polio virus type 1,
	    Feline calicivirus, various influenza viruses among others

5 min	Mycobactericidal
10 min	Fungicidal

environmental safety benefits. The reason is that high concentrations and long contact times are often required if used on its own in an aqueous solution. For example, 3% solution of hydrogen peroxide requires approximately 15 min to reduce sixfold the viable numbers of *Staphylococcus aureus* in a suspension [24].

Attempts at improving the biocidal activity of hydrogen peroxide include formulation with low levels of polycarboxylic acids and low pH [25], with short-chain alcohols [26], and quaternary ammonium compounds [27]. However, the improvement in biocidal activity has not been dramatic enough, with concentrations above 1% hydrogen peroxide still required for practical disinfection.

Recently, the USEPA [28] has approved a hydrogen peroxide-based composition based on a mixture of surface active and chelating agents with exceptionally high biocidal activity at hydrogen peroxide concentrations below 0.5% [29]. The composition, referred to commercially as accelerated hydrogen peroxide (AHP)™, meets the USEPA requirements for hard-surface bactericide at a hydrogen peroxide active concentration of just under 700 ppm, as well as the European test for bactericidal activity (EN Norm 1276) under both clean and dirty conditions in 5 min at levels of about 300 and 1,800 ppm, respectively.

The AHP™ technology also meets the requirements of the USEPA for registration as a 5 min mycobactericide, as evidenced by the recent USEPA approval on a ready to use product with a hydrogen peroxide active level of 0.5% [30]. A summary of the biocidal activity for this AHP composition as recently approved by the USEPA is included in Table 5.2.

5.4   Novel Biocidal Actives

The biocidal activity of quaternary ammonium chlorides has been widely reported in the literature [31]. However, these compounds suffer from a

variety of drawbacks including low tolerance to water hardness, susceptibility to inactivation by surface active agents, and thermal stability problems
at very high active concentrations. Recently, it has been shown that quaternary ammonium carbonates can be used in lieu of quaternary ammonium
chlorides with benefits with respect to hard water and surfactant tolerance,
and corrosion protection as an additional benefit [32].

The quaternary ammonium carbonate is synthesized by taking a C_8–C_{12}
dialkyl quaternary ammonium chloride and reaching it with a metal
hydroxide to yield an intermediary quaternary ammonium hydroxide.
This intermediate is then reacted with carbon dioxide to yield the quaternary ammonium carbonate [32]. Lutz presents a comparison between the
biocidal efficacy of the carbonate and chloride versions and shows that
the carbonate quaternary ammonium is roughly five times more effective
when used against vegetative bacteria in a hard water suspension of
extreme hardness (1400 ppm) [32]. Although the difference at more reasonable hardness levels is less dramatic, it is still interesting. This is an area that
requires further research, but presumably one could be able to achieve
the same levels of efficacies at lower active quaternary ammonium levels
by using the more active carbonate quaternary ammonium molecules.
The compounds are currently registered with the USEPA for use as antimicrobial pesticides for public health claims [33].

The biocidal activity of peracid compounds has been discussed earlier
and has been widely documented in the literature. In recent years, considerable research has occurred in the field of peracid disinfectants with
the objective of improving or identifying peracids with similar disinfection power to peracetic acid but without the drawbacks associated to stability, materials compatibility, and strong odor. Research in Europe has
resulted in two very interesting peracid compounds which have been notified in the European Biocidal Products Directive (BPD) for a variety of
disinfectant uses.

The monoperoxy monoesters of dibasic acids (ester peracids) with the
general formula $ROOC-[CH_2]_n-COOH$, with $R = C_1-C_4$ and $n = 1-4$, are
reported to exhibit quite favorable disinfection, stability, and occupational
safety properties [34]. These compounds can be synthesized by reacting the
parent monoester with hydrogen peroxide in the presence of an acid catalyst. However, since these monoester acids are not readily available commercially as a raw material, dibasic esters are utilized in their place, and the
resulting products are an equilibrium mixture of various ester peracids in
hydrogen peroxide and water [34].

Carr [34] describes the mixture of peracids formed from the peracids of
monomethyl esters of adipic, glutaric, and succinic acids. Table 5.3 presents
comparative data on the biocidal performance of the mixture. Although
the effective concentrations for the mixture are about 40% higher against
vegetative bacteria as compared to peracetic acid and almost sixfold lower
against yeast, the odor stemming from the ester composition is cited as not
discernible, and the mixture exhibits improved compatibility to materials.

TABLE 5.3

Biocidal Activity of Peroxy Ester Acid Mixture as Measured by European
Standard Test Methods

Organism	Test Method	Peroxy Ester Acid Concentration	Peracetic Acid Concentration	Conditions
Staphylococcus aureus	EN1276	125 mg/L	75 mg/L	>5 Log_{10} reduction
Pseudomonas aeruginosa		125 mg/L	75 mg/L	Contact time = 5 min 3 g/L Bovine serum
Candida albicans	EN1650	2000 mg/L	300 mg/L	>4 Log_{10} reduction Contact time = 5 min 3 g/L Bovine serum

These two improvements alone in many applications may justify the
increased in-use active acid concentrations.

The mixture has been notified in the European Product Directive as a
biocidal substance for use in multiple product groups; however, no product
based on it has been reviewed by the USEPA or U.S. Food and Drug
Administration for use as a disinfectant or sterilant.

Another interesting new compound is the molecule ε-phthalimido-peroxy-
hexanoic-acid (PAP) (see Figure 5.2), first described by Venturello and
Cavallotti [35]. The compound is synthesized by reacting phthalic anhydride
with caprolactam to obtain phthalimido-hexanoic-acid. The peracid form is
obtained by reaction with hydrogen peroxide in the presence of an acid
catalyst. Recently, considerable research has concentrated on improving the
manufacturing process to obtain better yields and purities [36]. The result has
been that the current commercial versions of the product have improved
stability and comparable cost to other traditional in situ peracetic acid generation methods [37].

The biocidal activity of the compound is considerably weaker than peracetic acid. For example, 1000 mg/L of PAP is required in 5 min to reduce the
viable numbers of gram-positive and gram-negative bacteria in the presence
of a standard soil challenge, compared to 75 mg/L of PAA under identical
conditions. However, its materials compatibility and safety profile are such
that in many applications the increased in-use concentrations may be tolerated. The material has been notified to the European Product Directive for a

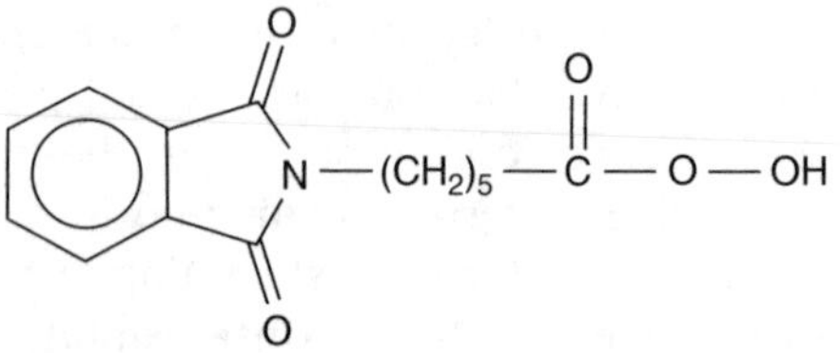

FIGURE 5.2
Molecular structure for ε-phthalimido-
peroxy-hexanoic-acid.

variety of application areas including human hygiene, food establishment, private and public area disinfection, veterinary disinfection, and cooling water treatment [37]. No biocidal product has currently been reviewed for public health claims in North America at the time of writing, however.

5.5 Optimized Actives Delivery

There has been considerable research in the past 10 years on the improvement of the biocidal efficacy of known antimicrobial compounds by facilitating cell absorption or membrane disruption through formulation in the form of nanoscale droplets or suspended particles. Hamouda and Baker [38] have investigated the use of surfactant–lipid emulsions for the inactivation of a broad spectrum of organisms including both gram-positive and gram-negative bacteria, bacterial endospores, and viruses [39,40]. One of the compositions investigated consists of a water-in-oil nanoemulsion made of soybean oil with an octyl phenol ethoxylate as a solubilizing agent. The composition has an average droplet size of between 400 and 800 nm and is said to be stable for at least 1 year. The composition shows activity against gram-positive bacteria and enveloped viruses at a dilution of 1% in water and is reported to exhibit a very favorable toxicological profile [41]. Another nanoemulsion described by Hamouda and others consists of a liposome made from soybean oil, cetylpyridinium chloride, refined soya sterols, glycerol monooleate, and a small amount of a polyethylene glycol sorbitan monostereate [39]. They also show that activity against gram-negative bacteria can be achieved or improved by the addition of a small amount of EDTA to the emulsion.

Another interesting approach is based on the use of nanoscale powders of nontoxic metal oxides such as magnesium oxide and calcium oxide for carrying active halogen species (i.e., $MgO \bullet Cl_2$ or $MgO \bullet Br_2$) as described by Koper and Klabunde [42]. The nanoscale powders can be used in either powder form or as dispersion in water. The particle sizes are typically less than 5 nm and exist as porous agglomerates of sizes between 1 and 2 μm. Reported data show that the nanoparticles reduce the viable numbers of gram-negative bacteria on a surface by more than 4 $\log_{10}$ at a contact time of 2 min. Promising data concerning the destruction of bacterial endospores in several minutes have also been published [42].

Although extensive data on efficacy have been developed for these compositions, it is necessary to further investigate their activity as measured by standard test methodology, required for registration as use for surface or instrument disinfection. Regulatory agencies may also require comprehensive toxicological characterization given that similar technologies have never been reviewed. One favorable aspect concerning registration is the fact that both technologies (cetylpyridinium chloride and chlorine, respectively) are based on known antimicrobial species; however, their efficacy is greatly enhanced by novel techniques for their delivery to the cell.

5.6   Conclusion

There continues to be a significant amount of research aimed at improving the efficacy, toxicology, reliability, and ease of use of disinfection technologies. Important trends have emerged in the past 5 years, of particular note is the renewed interest in oxygen-based chemistries. Interest in this type of chemistry has even spilled to the consumer arena, where many different types of oxygen-based cleaners occupy grocery store shelves worldwide. Another broad area of emerging importance is that of novel packaging and dispensing of the mixture or its components. As the globalization of economies proceeds at such a rapid pace, the cost of complex packaging and dispensing devices is reduced considerably by allowing high-quality design and manufacturing in emerging economic zones of eastern Europe, Asia, and Latin America. Finally, perhaps the field which holds the most promise is that of applied nanotechnology; we are already seeing the first manifestations in the form of nanoemulsions and nanodispersions for biocidal use and widespread adoption may only be a few years away.

References

1. Block, S.S., Historical review, *Disinfection, Sterilization and Preservation*, 4th Ed., Block, S.S. (ed.), Lea & Febirger, Philadelphia, 1991, chapter 1, pp. 3–17.
2. Spaulding, E.H., Chemical disinfection of medical and surgical materials, *Disinfection, Sterilization and Preservation*, Lawrence, C. and Block, S.S. (eds.), Lea & Febirger, Philadelphia, 1968, pp. 517–531.
3. Rutala, W.A. and Weber, D.J., New technologies for disinfection and sterilization, *Disinfection, Sterilization and Antisepsis: Principles and Practices in Healthcare Facilities*, Rutala, W.A. (ed.), Association for Professionals in Infection Control and Epidemiology Inc., Washington DC, 2001, chapter 11.
4. Schneider, P.M., New technologies for disinfection and sterilization, *Disinfection, Sterilization and Antisepsis: Principles, Practices, Challenges and New Research*, Rutala, W.A. (ed.), Association for Professionals in Infection Control and Epidemiology Inc., Washington DC, 2001, chapter 10, pp. 127–139.
5. Zipf, H.F., Practical points for experiments using combinations of two compounds, *Arzneimittelforsch.*, 3, 398, 1953.
6. Lehmann, R.H., Synergisms in disinfectant formulations, *Industrial Biocides— Critical Reports on Applied Chemistry*, vol. 23, Payne, K.R. (ed.), John Wiley & Sons, Chichester, 1988.
7. Block, S.S., Peroxygen compounds, *Disinfection, Sterilization and Preservation*, 4th Ed., Block, S.S. (ed.), Lea & Febirger, Philadelphia, 1991, chapter 9, pp. 167–181.
8. Ho, L.T.H., *Formulating Detergents and Personal Care Products: A Complete Guide to Product Development*, AOAC Press, Champaign, IL, 86, 2000.
9. *Laundry & Cleaning News*, December 21, 2005.
10. Baldry, M.G.C., The bactericidal, fungicidal and sporicidal properties of hydrogen peroxide and peracetic acid, *J. Appl. Bacteriol.*, 54, 417–423, 1983.

11. Block, S.S., Disinfection: Where are we headed. A discussion of hydrogen peroxide and peracetic acid, Proceedings of the 3rd Conference on Progress in Chemical Disinfection, Binghampton, NY, 1986.

12. Baldry, M.G.C. and Fraser, J.A.L., Disinfection with peroxygens, *Industrial Biocides*, in *Critical Reports on Applied Chemistry*, vol. 22, Payne, K.R. (ed.), John Wiley & Sons, Chichester, 1988, pp. 91–116.

13. Commercial literature for Perasafe, http://www.biosafetyusa.com/peraindex.html

14. Block, C., The effect of Perasafe and sodium dichloroisocyanurate (NaDCC) against spores of *Clostridium difficile* and *Bacillus atrophaeus* on stainless steel and polyvinyl chloride surfaces, *J. Hosp. Infect.*, 57(2), 144–148, 2004.

15. Rideal, E.K. and Evans, U.R., The effect of alkalinity on the use of hypochlorites, *J. Soc. Chem. Ind.*, 40, 64R–66R, 1921.

16. Johns, C.K., Germicidal power of sodium hypochlorite, *Ind. Eng. Chem.*, 26, 787–788, 1934.

17. Sheltmire, W.H., Chlorinated bleaches and sanitizing agents, *Chlorine: Its Manufacture, Uses and Properties*, Sconce, J.S. (ed.), Reinhold Publishing Corporation, New York, 1962, chapter 17, pp. 512–542.

18. Tanaka, T., Fujisawa, T., Daimon, T., Fujiwara, K., Yamamoto, M., and Abe, T., The cleaning and disinfecting of hemodialysis equipment using electrolyzed strong acid solution, *Artif. Organs*, 23(4), 303–309, 1999.

19. Kim, C., Hung, Y.C., and Brackett, R., Efficacy of electrolyzed oxidizing and chemically modified water on different types of foodborne pathogens, *Int. J. Food Microbiol.*, 61, 199–207, 2000.

20. Middleton, A.M., Chadwick, M.V., Sanderson, J.L., and Gaya, H., Comparison of a solution of super-oxidized water (Sterilox) with glutaraldehyde for the disinfection of bronchoscopes, contaminated in vitro with *Mycobacterium tuberculosis* and *Mycobacterium avium-intracellulare* in sputum, *J. Hosp. Infection*, 45, 278–282, 2000.

21. Fujita, J., Nanki, N., Negayama, K., Tsutsui, S., Taminato, T., and Ishida, T., Nosocomial contamination by *Mycobacterium gordonae* in hospital water supply and superoxidized water, *J. Hosp. Infection*, 51, 65–68, 2002.

22. Auchincloss, T.R., Biocidal, particularly virucidal compositions, US Patent 4,822,512.

23. DuPont company literature for Virkon S http://www.antecint.co.uk/main/virkons.htm

24. Block, S.S., Peroxygen compounds, *Disinfection, Sterilization and Preservation*, 4th Ed., Block, S.S. (ed.), Lea & Febirger, Philadelphia, 1991, p. 172.

25. Miner, N., Woller, W., Anderson, E., and Hobson, D.W., Quick acting chemical sterilant, US Patent 6,096,348.

26. Monticello, M.V. and Mayerhauser, R.G., Ready to use aqueous hard surface cleaning compositions containing hydrogen peroxide, US Patent 6,106,77.

27. Kramer, D. and Snow, P.A., Cleaning and disinfecting method, US Patent 5,620,527.

28. US EPA Notice of Pesticide Registration, EPA Reg. No. 70627–54, Date of Issuance: March 10, 2006.

29. Ramirez, J.A. and Rochon, M., Hydrogen peroxide disinfectant with increased activity, US Patent 6,803,057.

30. US EPA Notice of Pesticide Registration, EPA Reg. No. 70627–56, Date of Issuance: December 20, 2005.

31. Jungerman, E., Cationic Surfactants, Marcel Dekker Inc., New York, 1969, pp. 56–57.

32. Lutz, P.J., Disinfecting use of quaternary ammonium carbonates, US Patent 6,297,285.
33. US EPA Notice of Pesticide Registration, EPA Reg No. 6836-236, Date of Issuance: May 14, 2003.
34. Carr, G. and James, A.P., Peroxygen compositions, US Patent 6,207,108 B1.
35. Venturello, C. and Cavalotti, C., Imido-aromatic percarboxylic acids, European Patent No. 0325288 B1.
36. Cavallotti, C., Merenda, M., Zaro, A., Lagostina, A., and Bianchi, U.P., US Patent No. 5,208,340.
37. Solvay, Personal communication, 2005.
38. Hamouda, T.A. and Baker, J.R. Jr., Antimicrobial mechanism of action in negative gram negative bacilli, *J. Appl. Microbiol.*, 89, 397–403, 2000.
39. Hamouda, T., Hayes, M.M., and Cao, Z., A novel surfactant nanoemulsion with broad-spectrum sporicidal activity against Bacillus species, *J. Infect. Dis.*, 180, 1939–1949, 1999.
40. Donovan, B.W., Reuter, J.D., Cao, Z., Myc, A., Johnson, H.J., and Baker, J.R., Prevention of murine influenza A virus pneumonitis by surfactant-nanoemulsions, *Antivir. Chem. Chemother.*, 11, 41–49, 2000.
41. Hamouda, T., Myc, A., Donovan, B., Shih, A.Y., Reuter, J.D., and Baker, J.R. Jr., A novel surfactant nanoemulsion with a unique non-irritant topical antimicrobial activity against bacteria, enveloped viruses and fungi, *Microbiol. Res.*, 156(1), 1–7, 2000.
42. Koper, O. and Klabunde, K.J., Nanoparticles for the destructive sorption of biological and chemical contaminants, US Patent 6,057,488.

6

Biocides: Modes of Action and Mechanisms of Resistance

Gerald McDonnell

CONTENTS

6.1   Introduction

For the purpose of this chapter, an antimicrobial is considered as any physical or chemical agent that kills or inhibits the growth of microorganisms. These can be further classified as biocides and anti-infectives. Anti-infectives may be described as being rather specific drugs in their interaction and their use for the control of various microorganisms, including antibacterials (with the typical example being antibiotics), antifungals, antivirals, and antiprotozoal agents. Anti-infectives have a rather narrow spectrum of activity, in comparison to most biocides, but have been effective in the treatment of infections due to minimal associated toxicity in humans, animals, and plants, allowing for their systematic use. Examples include antibiotics that demonstrate specific mechanisms of action, narrow ranges of antimicrobial activity, and greater demonstrated risks in the development of bacterial resistance to their action [1]. If antibiotics are considered further, their mechanisms of action have been particularly well studied, and it is known that they specifically target important bacterial processes such as DNA replication (e.g., quinolones) and protein translation (e.g., aminoglycosides, tetracyclines). Their effects on the various types of bacteria can range, but they are generally much narrower than those described for biocides. For example, glycopeptides (e.g., vancomycin) are particularly effective against gram-positive bacteria (with the notable exception of mycobacteria), isoniazid is specifically antimycobacterial, and the macrolides and aminoglycosides are active against most gram-positive and gram-negative bacteria [1]. These limitations in the range of activity are linked to the presence of (or access to) specific targets in these bacteria, which are simply not found in other microorganisms or host cells. Therefore, the majority of antibiotics have no therapeutic uses against viruses and fungi. Others, especially specific antifungal agents such as the azoles and polyenes, target ergosterol synthesis and structure; ergosterol is a specific sterolic molecule found in fungal cell membranes [1]. Further, antiviral drugs such as amantadine inhibit viral (specifically influenza virus) penetration into host cells [2]. Their mechanisms of action [1,2] and resistance [1–3] have been particularly well described elsewhere and are not considered further in this chapter. In contrast, biocides are chemical or indeed physical agents that are used on inanimate surfaces or on the skin and mucous membranes. These antimicrobials are much more widely employed for various applications including preservation, sanitization, pasteurization, fumigation, antisepsis, disinfection, and sterilization. As these terms can vary in definition and use, some standardized definitions are provided in the Appendix for reference. These broad-spectrum chemical agents include chlorine, phenols, heavy metals, and quaternary ammonium compounds (QACs). Physical agents include heat and radiation. They demonstrate a wider range of antimicrobial activity but also have associated cellular toxicity, which has therefore limited their use to surface/air applications, with only a limited number being used on the skin, mucous membranes, or in wounds (e.g., iodophors, triclosan, and chlorhexidine (CHG)) [4]. The mechanisms of action of biocides are often difficult to study and determine as they have been shown to have multiple targets on surface components and interior structures in microorganisms [5,6]. Despite this, their principal modes of action may be considered under four general mechanisms including cross-linking or coagulation, oxidation, energy transfer, and other structure-disrupting agents (Table 6.1).

These are further considered in this chapter. Further, it is widely accepted that microorganisms can rapidly develop resistance to anti-infectives, presumably due to their specific mechanisms of action [1–3]. With biocides the opposite has been expected and observed; despite this, there are clear examples of microorganisms developing partial tolerance and extreme resistance to biocides and biocidal processes [5,6]. The mechanisms of resistance, in most cases and in contrast to anti-infectives, have not been fully investigated; however, the known mechanisms can be considered as being intrinsic and acquired. Intrinsic mechanisms are considered normal developments during microbial growth that allow microorganisms to survive biocide attack, some of which are dramatic developmental stages in their growth cycles. These include stress responses, biofilm development, and sporulation. Acquired resistance mechanisms have been shown to develop due to mutations or acquisition of genetic materials (such as

TABLE 6.1

General Biocidal Mechanisms of Action and Biocide Examples

Mechanism	Effects	Examples
Oxidation and oxidizing agents	Electron removal from macromolecules leading to loss of structure and function	Halogens, e.g., iodine, chlorine Peroxygens, e.g., hydrogen peroxide, peracetic acid
Cross-linking and coagulation	Cause interactions, cross-linking within/between macromolecules leading to clumping and coagulation	Aldehydes, e.g., formaldehyde, glutaraldehyde Alkylating agents, e.g., ethylene oxide Phenols Alcohols
Transfer of energy	Transfer of energy to various macromolecule bonds and structures leading to loss of structure and function	Heat Radiation, e.g., UV light, γ-radiation
Other structure-disrupting agents	Specific macromolecular structure disruption and loss of activity	Acridines Surfactants, e.g., quaternary ammonium compounds Metals, e.g., silver, copper

plasmids) that encode for specific resistance determinants. It is not the purpose of this chapter to explain all of these mechanisms in detail, but to highlight some of the more investigated and dramatic examples described in the literature. Other reviews provide further details and considerations of these mechanisms of resistance [5–7].

6.2   Modes of Action

6.2.1   General Considerations

Biocides and biocidal processes can have dramatic effects on the structure and function of the various macromolecules that make up microbial life, including carbohydrates, proteins, lipids, and nucleic acids, and the various essential structures that they form in combination, e.g., cell walls, cell membranes, and viral envelopes [5,6]. It is a culmination of these effects that leads to loss of microbial viability. Many of these primarily affect the various microbial surface structures, which are the first to experience contact with the biocide. These can include dramatic, discriminate effects due to the increase in heat, the oxidation of various components because of the presence of oxidizing agents, and the cross-linking of components in the presence of aldehydes. Other biocides may cause more subtle changes; for example, the disruption of bacterial/viral membrane (lipid bilayer) structures by intercalation/interaction of surfactants and QACs interferes with microbial viability and infectivity. Following various surface effects, further interactions with internal components can contribute to loss of viability. An understanding of the various mechanisms of action is important in understanding the risks associated with resistance development, in optimizing the efficiency of biocide use, in further biocide/anti-infective drug development, and in the consideration of efficacy against more atypical pathogens such as prions (that are considered to be composed exclusively of protein) or microbial populations (such as in biofilms).

An important consideration in the understanding of biocidal processes is that, in most cases, the biocide is not used on its own but in combination with other factors that can dramatically affect their efficacy and action. These include formulation effects in liquid biocidal products such as surface disinfectants and antiseptics. Formulation factors include the presence of buffers (to control pH), surfactants (to allow dispersion and even increased biocide penetration), and chelating agents (to prevent the interruption of the biocide action by the presence of metals or other interfering agents). The delivery process may also have an impact, for example, with increased temperature or under vacuum, the biocide in a liquid or gaseous application, or the distribution of temperature during heat disinfection–sterilization processes. These formulation and delivery processes can dramatically affect (both negatively and positively) the mode of action and the successful outcome of a biocidal process; therefore, although the following discussion on the mode of action of and resistance to biocides is based on a review of the literature, the overall activity and adequacy of specific biocide-containing products and processes can vary significantly. In addition, although various biocides and biocidal processes may be classified based on four basic mechanisms of action, it is clear that the effects of these may include a culmination of effects. For example, heat is clearly an effective mechanism of denaturing proteins and nucleic acid by the transfer of energy but will also lead to coagulation of these molecules. A further example may be given with ionizing radiation; the primary effects are also by the transfer of energy, which disrupts macromolecular structure causing the localized production of various oxidizing agents (in particular reactive oxygenated species like hydroxyl radicals).

6.2.2 Cross-Linking/Coagulation

Cross-linking or coagulating biocides cause specific interactions between and within various macromolecules to cause loss of structure and function. These biocides include aldehydes, alkylating agents, various phenolic compounds, and alcohols (Table 6.1). The primary mechanism with aldehydes (such as glutaraldehyde and formaldehyde) is to bind to and cause the cross-linking of various proteins, especially at microbial surfaces, and to a lesser extent with and between other macromolecules such as the bacterial cell-wall peptidoglycan layers [5,6]. The typical structure of some aldehydes used as biocides is given in Figure 6.1, with an example of a cross-link formed between two adjacent amino acids shown in Figure 6.2.

Glutaraldehyde is one of the most widely studied biocides and is known to be a potent cross-linking agent [5,8]. It can form cross-links between

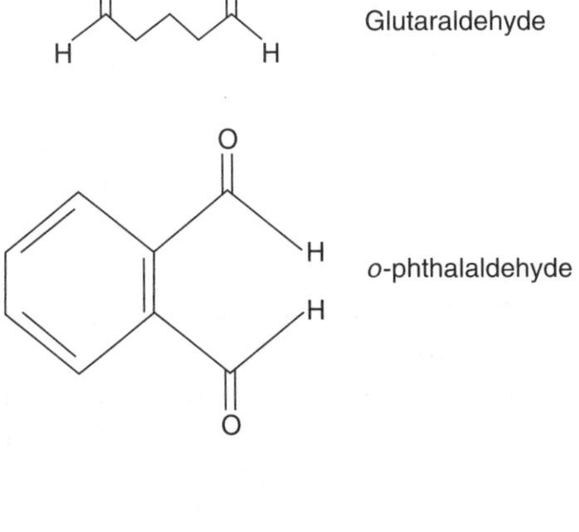

Formaldehyde

Glutaraldehyde

o-phthalaldehyde

FIGURE 6.1
Aldehydes widely used as biocides. Note, formaldehyde is a monoaldehyde (with a single aldehyde group). Glutaraldehyde and *o*-phthalaldehyde are both dialdehydes.

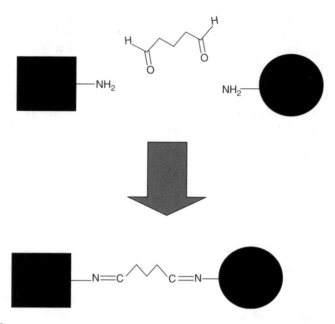

FIGURE 6.2

Typical cross-linking reaction with an aldehyde. Two amino acids are shown with free amine (NH$_2$) groups that react with the aldehyde groups of glutaraldehyde to form a cross-link (covalent bond). The formation of these bonds will disrupt the structure and function of the respective proteins.

amino acids both within the same protein as well as adjacent proteins (Figure 6.2). The amino acid free amine (NH$_2$) groups, in particular those of lysine or hydroxylysine, are particularly sensitive to form strong covalent bonds, which disrupt the essential structure and function of proteins and other macromolecules (e.g., nucleotides and polynucleotides). Similar reactions have been shown with other aldehydes [5]. Glutaraldehyde is primarily a surface-acting agent, binding strongly to various microbial surfaces, including bacterial/fungal cell walls and viral capsids/envelopes [8]. These interactions will disrupt the structure and function of key cell-surface roles including transport disruption, enzyme inhibition, and viral infectivity. Formaldehyde is also considered a surface-active biocide but is considered a less effective cross-linking molecule due to its structure (in contrast to the flexible glutaraldehyde and, to a certain extent *o*-phthalaldehyde (OPA), structures, Figure 6.2) [5]. More recent studies with OPA have shown that it demonstrates greater penetration of bacterial (in particular mycobacterial) cell surfaces and into the cell membrane and cytoplasm [9]. It has been proposed that this may be due to the ability of the biocide to adopt a less reactive (and more stable) conformational structure under hydrophobic in contrast to hydrophilic environmental conditions. For example, under hydrophilic conditions that are typical of the bacterial surface, it may adopt a 1,3-phthalandiol structure with unexposed/unreactive aldehyde groups, but as it passes through the cell surface (being exposed to the more hydrophobic conditions of the cell wall/membrane), it adopts a more reactive dialdehyde structure with potent cross-linking reactions with various proteins [10].

Ethylene oxide is a widely used alkylating biocide, particularly used in low-temperature chemical sterilization processes for medical devices and other materials. It has also been shown to form cross-links within proteins and nucleic acids. As an alkylating agent, ethylene oxide has been shown to particularly react with guanine nucleotide bases in DNA to introduce alkyl groups, which not only disrupt the structure and functions of DNA as an essential macromolecule but also lead to the formation of cross-links between adjacent nucleotides and protein hydroxyl groups [11]. Similar reactions with amino, carboxyl, sulfhydryl, and hydroxyl groups in protein amino acids will also lead to the formation of epoxide cross-linkages [11]. Certain amino acids such as cysteine, histidine, and valine appear to be particularly sensitive to epoxide reactions. It should be noted that, similar to the alkylating effects discussed, the reactions with other biocide types including oxidizing agents and heat-transfer mechanisms can also lead to the formation of reactive groups and cross-linkages; although these effects are considered secondary effects in the mode of actions of these agents (see below) they do also culminate in the overall biocidal effects [6].

Phenolics and alcohols are also considered to be coagulating agents as coagulation is the primary mechanism of their mode of action. Phenolics are widely used in disinfectant formulations and have been shown to particularly target surface and internal proteins by various interactions, including hydrogen bonding and hydrophobic interactions, which cause proteins and their associated structures to coagulate, cross-link, and precipitate. The associated hydroxyl (–OH) groups of phenolic structures are particularly reactive with macromolecules (Figure 6.3) [5,12]. Other associated structural groups may add directly to the antimicrobial activity or allow greater penetration of the phenolic group into target microorganisms. Alcohols, such as isopropanol and ethanol, are also widely used as disinfectants as well as antiseptics (so called hand-rubs or hand-rinses used for water-less applications), and also have reactive hydroxyl groups that form hydrogen bonding with macromolecules, causing similar coagulation and precipitation. In both groups of biocides, the precipitation effects will cause an overall loss of macromolecular structure and function, membrane/wall damage, leakage of cytoplasmic components, and loss of viability [5,6].

6.2.3   Oxidation and Oxidizing Agents

Oxidizing agents can be powerful antimicrobial agents owing to their ability to remove electrons from other substances such as proteins, lipids, and nucleic acids. The oxidation (or loss of an electron by a molecule, atom, or ion) and damage to these molecules will have dramatic effects on their structure and function, which culminate to give biocidal activity. Oxidizing agents are widely used for various antimicrobial applications in medicine,

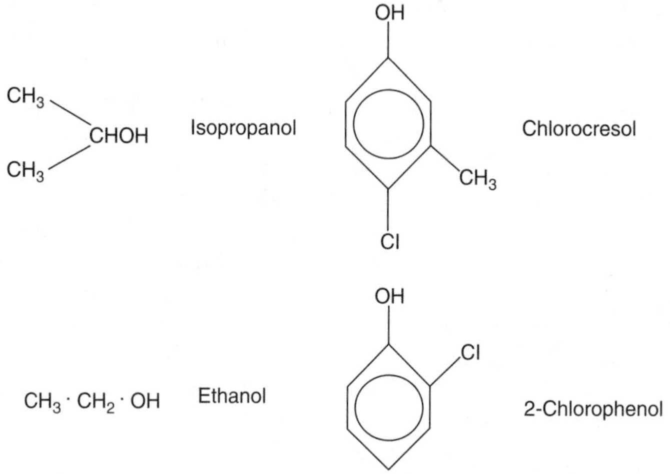

FIGURE 6.3
Examples of the structures of phenolics and alcohols. Note the major reactive hydroxyl (–OH) groups in each structure.

dentistry, industry, and agriculture [6]. These particularly include halogens and peroxygens [13]. Halogens are a group of highly reactive elements including chlorine (Cl_2), iodine (I_2), and bromine (Br_2). Simple solutions or delivery systems of these elements are rarely used as antiseptics or disinfectants, but compounds that contain or release the biocide are more widely used. These include the iodophors (iodine-releasing agents), hypochlorites (as sources of chlorine-active agents, in particular chlorine, hypochlorite ions, and hypochlorous acid), and bromine-releasing agents (e.g., sodium bromide and bronopol). The halogens are efficient oxidizing agents and will therefore cause dramatic effects on the structure and function of various external and internal microbial structures. Iodine, for example, has been shown to particularly target the amino acids arginine, lysine, cysteine, and histidine, but also various nucleic acids (such as DNA) and lipids (in particular unsaturated bonds in fatty acids) [5]. These effects are similar to the other halogens and oxidizing agents because of the overall susceptibility of these amino acids and unsaturated fatty acid bonds to oxidation. These primary effects are complemented by oxidation of other targets and secondary interactions because of the reactive nature of oxidized groups, culminating in irreversible loss of structure/function. Similar effects are seen with the other major class of oxidizing agents, the peroxygens or other forms of oxygen. These include the widely used biocides hydrogen peroxide, peracetic acid, chlorine dioxide, and ozone. The oxidation reactions with these biocides have been shown to cause macromolecular unfolding, fragmentation, and cross-reaction with oxidized groups [5,6]. Proteins, carbohydrates, and lipids on the surface of microorganisms are particularly accessible targets, followed by various intercellular components, including proteins and nucleic acids, as the structure of the microorganism disintegrates. Specific targets and effects with different peroxygens have been reported [5,14]. For example, chlorine dioxide has been shown to have particular effects against certain amino acids (tryptophan, cysteine, and tyrosine), while peracetic acid and hydrogen peroxide have been shown to disrupt sulfhydryl (–SH) groups and sulfur bonds (S–S) in proteins, and fatty acid unsaturated bonds. The effects of hydrogen peroxide on nucleic acid have been particularly well described as hydrogen peroxide and other reactive oxygenated species (including the short-lived but extremely reactive superoxide ions and hydroxyl radicals) are formed during cellular respiration and can have detrimental effects in eukaryotic cell components; indeed, these effects have been linked to cell aging and mutagenesis [15]. For example, hydrogen peroxide has a dramatic effect on DNA and RNA structures, attacking both the nucleotide bases and the sugar–phosphate backbone on these structures [15]. The oxidation of these groups will lead to strand breakage, and cross-reactions between converted bases/sugars, which will affect the replication, transcription, translation, and other roles of these essential structures. Differences have been observed in the effects of formulation and delivery processes with the mechanisms of action of oxidizing agents. For example, peracetic acid formulations at low pH can lead to protein clumping, thereby protecting susceptible molecules and microorganisms to the effects of the oxidizing agent; similarly, liquid (or condensed) hydrogen peroxide applications can have a similar effect in comparison to gaseous applications with the biocide [14,16,17].

6.2.4 Energy Transfer

Microorganisms are dependent on the correct structure and function of various macromolecules for growth and survival. These molecules can be particularly sensitive to changing environmental conditions and can be rapidly disrupted and inactivated by the transfer of energy in the form of heat or radiation. Heat (e.g., in the form of hot water, steam, or dry air) can be a simple, effective, and widely used method of microbial inactivation. The effects of heat on various macromolecules are summarized in Table 6.2.

Obviously, the culmination of these effects will lead to cell or viral death, due to their dependence on these structures. However, it should be noted that some active biological structures are particularly resistant to the inactivating effects of heat-based processes. These include, of notable mention, prions [18] and endotoxins [19]. Prions are unique infectious proteins (implicated in a group of diseases known as transmissible spongiform encephalopathies [20]), which present usual resistance profiles to heat inactivation, although hydration is believed to be important for effective inactivation [16]. Endotoxins are lipopolysaccharides that form a major component of the outer cell membrane of gram-negative bacteria and can lead to pyrogenic reactions when introduced directly in the bloodstream [19]; they are not significantly affected by steam sterilization and can only be

TABLE 6.2

Examples of the Various Effects of Heat on Macromolecules

Macromolecule	Effects of Heat
Nucleic acids (double stranded)	Denaturation (e.g., the hydrogen bonding between DNA strands is separated at ~85°C)
	Fragmentation
Proteins	Loss of secondary, tertiary, and quaternary protein structures (due to disruption of ionic and nonionic bonds)
	Fragmentation
	Coagulation
	Loss of protein hydration (as with dry heat)
	Loss of amino acid structure
Lipids	Oxidation of fatty acids
	Loss and disruption of phospholipid bilayer structure

inactivated by long exposure times to dry-heat sterilization [19]. In both cases, it is currently unknown why prions and endotoxins are particularly resistant to heat; of note, in the case of prions, the lack of any substantially associated nucleic acid with prions has been well cited as responsible for observed resistance to heat and other energy-transfer biocidal methods [21].

Energy transfer is also the mode of action of the various forms of radiation used for antimicrobial purposes. The various forms of radiation vary in energy levels and, therefore, their destructive effects on microorganisms [6,22]. Nonionizing radiation sources (such as infrared and ultraviolet [UV] light) are less penetrating and damaging than ionizing radiation (such as γ-rays and E-beams) because of the various degrees of associated damage to microbial macromolecules. In both cases, the destabilization of atomic structures is responsible for the overall loss of structure and function. Infrared and microwaves primarily cause a rise in temperature, in particular by direct effects on associated water molecules, leading to the variety of effects listed in Table 6.2 associated with heat inactivation. UV light, in contrast, has a greater associated energy level to allow for the penetration of microorganisms and direct disrupting effects on macromolecules, in particular targeting nucleic acids [23]. UV light has been shown to cause the formation of thymine dimers (or cross-links) between adjacent pyrimidine bases within a DNA molecule (Figure 6.4).

These effects on DNA structure cause a loss of structure and flexibility, as well as disrupting the essential replication and transcription functions for growth and survival. In addition to these well-cited effects, UV light is also known to affect the structures and functions of lipid membranes and proteins [6,23].

Ionizing radiation is particularly damaging to macromolecules, causing excitation and ionization (ejection) of electrons from atoms that make up the various macromolecules. They have, therefore, dramatic effects on the structure and functions of lipids, proteins, carbohydrates, and nucleic acids. The primary target for ionizing radiation also appears to be nucleic acids, in particular leading to fragmentation of single-stranded and double-stranded

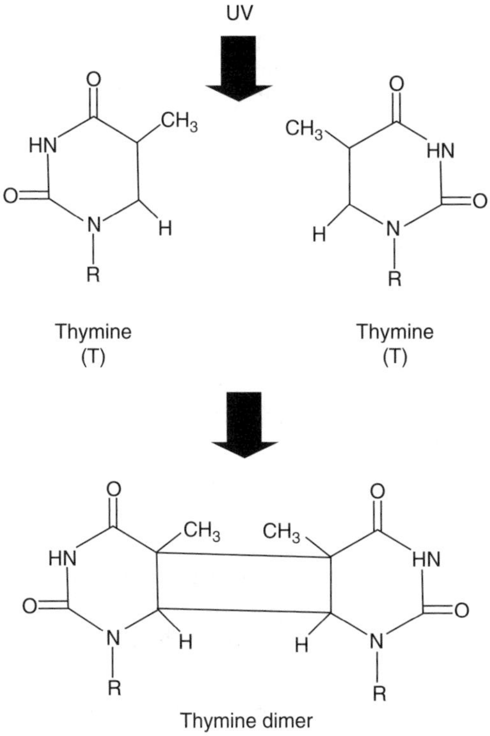

FIGURE 6.4
The formation of thymine dimers within DNA on exposure to UV light. The dimer formation is shown side by side in this representation, although spatially the two nucleotide residues are adjacent to each other in the DNA structure.

structures [6,22]. As for UV light, similar effects are expected for other molecules. These direct effects, due to the instability and resulting molecule reactivity, will also lead to further interactions between adjacent molecules, leading to coagulation and cross-linking. Further still, ionizing radiation (for example, in reaction with oxygen and water) will cause the production of reactive oxygenated species (such as ozone, hydroxyl radicals, and hydrogen peroxide) with their associated effects on proteins, nucleic acids, lipids, and proteins (as described earlier for oxidizing agents).

6.2.5 Other Structure-Disruption Mechanisms

The final group of biocides categorized based on their modes of action is described as having specific primary mechanisms against certain microbial structures, which in some cases may be responsible for their known limited spectrum of activity against microorganisms. Important examples of these include CHG, the QACs, the parabens, the acridines, and metals such as copper and silver. Examples of the various biocides and primary modes of action are given in Table 6.3.

The mechanisms of action of these biocides have been the focus of some reviews [5,6,24], but further discussion is given here to CHG, QACs, metals (in particular silver and copper), and triclosan.

TABLE 6.3

Examples of Biocides with Specific Structure-Disrupting Mechanisms of Action

Biocide	Mode of Action[a]
Chlorhexidine	Cell-membrane (lipid-bilayer) disruption
QACs[b]	Cell-membrane (lipid-bilayer) disruption
Parabens	Disruption of cell membrane and associated enzymes
Silver, copper	Interaction with microbial surfaces to disrupt cell membrane/wall-associated activities. Specific interaction with exposed sulfhydryl groups in proteins by silver
	Also affect nucleic acid structures
Acridines	Intercalates between double-stranded nucleic acid base pairs causing disruption of functions and structure
Triclosan	Inhibition of enoyl reductases and other proteins (at low concentrations)

[a] Principal mode of action listed but other effects may have been described and are further discussed.
[b] QACs: Quaternary ammonium compounds.

QACs are a class of surfactants classified as cationic, based on their overall charge [5,25]. They are widely used in antiseptic and disinfectant formulations. Their major target has been shown to be cell lipid-bilayer membranes, with particular studies focusing on the mode of action against bacterial cell membranes [5,26]. QACs have been shown to penetrate through the bacterial cell wall and interact with lipids/proteins in the membrane to cause disruption of its structure and associated functions. The main effects appear to be due to direct interaction with or between the phospholipids or both of these mechanisms, which disrupts the structure and leads to leaking of cytoplasmic materials from the cell [5]. Membrane disruption interferes with many associated functions of the cell membrane including transport and energy generation mechanisms (including ATP production and the proton motive force). Although the inner cell membrane is the primary target, the outer membrane of the gram-negative cell-wall structure may also be disrupted, and other effects include interference with cytoplasmic proteins and nucleic acids that cumulate to cause loss of cell viability [26]. Similar primary mechanisms of action on membranes have been shown in yeast studies and on effects against enveloped (lipid) viruses [27]. As may be expected from the mechanism of action, QACs have limited activity against mycobacteria, presumably because of the lack of penetration into the mycobacterial cell wall, and have little or no activity against nonenveloped viruses (because of the lack of any external lipid membrane, in contrast to the enveloped viruses).

Overall, a similar mode of activity has been shown with CHG. Chlorhexidine is a widely used biocide in antiseptic applications and its mechanisms of action have been particularly studied in bacteria [5,6]. It also shows rapid penetration through bacterial and fungal cell walls, and interacts with the cell membrane to disrupt structure and functions. Studies have suggested that the CHG molecule interacts with and integrates into the membrane, with similar effects to those described for QACs [5]. Further similarities are shown at higher concentrations of the biocide in various cytoplasmic effects, including protein and nucleic acid loss of function, and coagulation.

Copper and silver, unlike other heavy metals such as mercury, have seen continued use as biocides, having potent antimicrobial activity and broader spectrum of activity in comparison to QACs and CHG. Examples of metal compounds used as biocides include copper sulfate, silver nitrate, and silver sulfadiazine [5,6]. Silver has also been increasingly used as integrated or coated onto the surface of medical devices such as in-dwelling catheters [28]. The heavy metals have similar mechanisms of action, particularly targeting and binding to proteins to disrupt their secondary and tertiary structures required for structural and enzymatic activities [5,6]. Protein binding can initially be reversible (causing growth stasis), but can cumulate to cause protein denaturation and coagulation. With both copper and silver, proteins with thiol (sulfhydryl) groups are particularly sensitive and targeted. As the cell wall/membrane proteins are initially exposed to the effects of the metals, their various associated functions such as respiration, transport, and energy generation are also disrupted. Further effects of the biocides in the cytoplasm may also be expected and observed, not only because of protein interactions but also binding to and interfering with DNA functions. The more generalized mechanism of action is thought to be responsible for the broader spectrum of activity of the heavy metals including bacteria, fungi, some viruses, and protozoa.

Finally, the bisphenols, in particular triclosan, deserve particular discussion owing to a number of excellent studies on the modes of action and resistance to this biocide [5,29,30]. Triclosan is one of the most widely used biocides in antiseptic products, including mouthwashes, antimicrobial soaps, toothpastes, and impregnated surfaces. The mechanisms of action of triclosan had previously been considered to be similar to other phenolics, targeting macromolecules, in particular proteins, leading to coagulation and precipitation [5]. These effects clearly would include the cell surface (cell wall and cell membrane). However, more recent studies, especially those based on *Escherichia coli* mutants with increased tolerance (or minimum inhibitory concentrations, MICs) to triclosan, have shown that triclosan specifically binds to enoyl reductases, enzymes involved in fatty acid biosynthetic pathways in bacteria [29,30]. It was surprising to find that triclosan and an another bisphenol hexachlorophene specifically bind to the substrate binding site of the enzyme to reversibly inhibit (as is the case with hexachlorophene) and cause a conformational change (as in the case of triclosan) in the protein structure to form an irreversible complex which precipitates [31]. This finding was remarkable and yet familiar, as previous studies had shown a similar mode of action for the antibiotic isoniazid, an antimycobacterial antibiotic. However, unlike isoniazid, higher concentrations of triclosan (even below the minimum bactericidal concentration of the biocide) were found to be effective against these mutants, and some bacteria were found to have enoyl reductases that were unaffected by triclosan [6,30], suggesting further mechanisms of biocidal activity. Subsequently, triclosan has actually been shown to have multiple intracellular protein targets at low concentrations including some transferases, as well as targeting other macromolecules and cell-membrane/cell-wall interactions that appear to culminate in the biocidal activity [6,29,30]. Despite this, the overall spectrum of activity of triclosan is rather narrow in comparison to other biocides, presumably because of the lack of appreciable or slow penetration into cells and viruses.

Overall, despite the demonstration of specific primary targets for this group of biocides in disrupting macromolecular structures, it is clear that in each case further targets are affected that culminate in the overall biocidal activity of the active.

6.3 Modes of Resistance

6.3.1 Introduction

Microorganisms have been shown to have a variety of intrinsic and acquired mechanisms of resistance to biocidal chemistries and processes. Examples of these are given in Table 6.4.

Overall, the microbial world can be generally classified based on their known or estimated resistance to physical and chemical processes (Figure 6.5). This classification system, however, can only be used as a general guide as the microbial resistance profiles will vary depending on the biocide or antimicrobial process. Further, there are many exceptions to this profile, including various extremophiles with unique resistance profiles to environmental growth conditions (e.g., *Pyrolobus* and *Thermococcus* bacterial growth at >80°C [32]), unique strains of mycobacteria demonstrating unique resistance to

TABLE 6.4

Mechanisms of Biocidal Resistance

Resistance Mechanisms	Examples
Intrinsic	
Microbial structure	Bacterial cell-wall structure, e.g., gram-positive, gram-negative, and mycobacterial cell-wall structures. Virus structure, i.e., enveloped viruses, nonenveloped viruses
Growth phase responses	Clumping stationary phase growth stress responses—efflux
Biofilm development	Carbohydrate/protein production—growth phase
Dormancy	Bacterial or fungal spore production—protozoal cyst production
Revival	Repair mechanisms—viral reactivation
Extremophiles	*Deinococcus* resistance to radiation—thermophilic resistance to heat
Acquired	
Mutations	Efflux enzyme mutations—cell-surface structure
Transmissible elements	Efflux—biocide inactivation

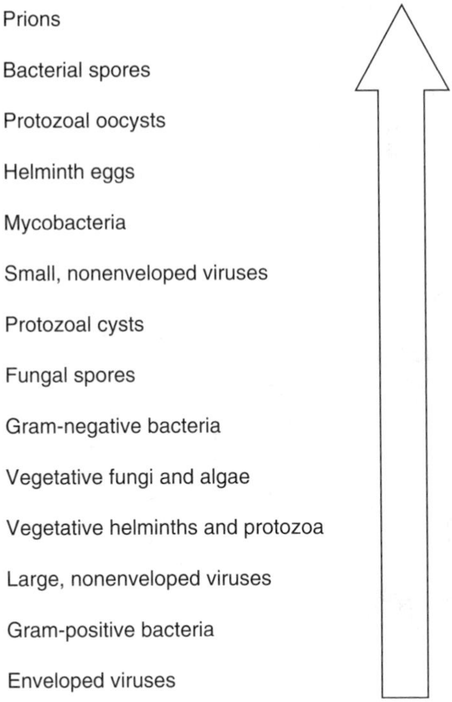

Prions

Bacterial spores

Protozoal oocysts

Helminth eggs

Mycobacteria

Small, nonenveloped viruses

Protozoal cysts

Fungal spores

Gram-negative bacteria

Vegetative fungi and algae

Vegetative helminths and protozoa

Large, nonenveloped viruses

Gram-positive bacteria

Enveloped viruses

FIGURE 6.5
Microbial resistance to biocides. This list is a guide displaying those organisms that are generally less resistant to biocides (from the bottom) and those that show the greatest resistance (at the top). This profile can vary depending on the biocide or biocidal process under consideration. (From McDonnell, G., *Antisepsis, Disinfection and Sterilization: Types, Action and Resistance*, ASM Press, Washington DC, 2006.)

glutaraldehyde [33], and prion inactivation by simple cleaning chemistries [16]. In addition, this profile does not consider the various environmental growth and survival characteristics displayed by these microorganisms in their various habitats, which are important practical considerations in their control.

The basic intrinsic mechanisms of resistance are due to various structures, in particular surface structures that can protect the microorganism from the biocide. In the case of vegetative bacteria, the cell wall and associated structures such as capsules or slime layers are important considerations [34]. Bacteria can be generally categorized based on their cell-wall structure in four groups: cell-wall free, gram-positive, gram-negative, and mycobacterial cell-wall groups (Figure 6.6).

The simplest structures described are present in the cell-wall free bacteria, which consist of a simple cell-wall structure consisting of a cell membrane [35]. Although these bacteria have a typical lipid-bilayer cell membrane as described for prokaryotes and eukaryotes, the overall structures are unique as they contain various lipids (in particular sterols) and carbohydrates that appear to add greater rigidity to their structure and resistance to biocides. Overall, the true resistance patterns of mycoplasma have not been studied in detail; the overall resistance profiles are not considered greater than that of gram-positive bacteria [36]. The Gram stain has been used for over 120 years to differentiate bacteria into two main groups, gram-positive and gram-negative, based on their cell-wall structure (Figure 6.6). In general, gram-positive bacteria (such as *Bacillus, Clostridium, Staphylococcus,*

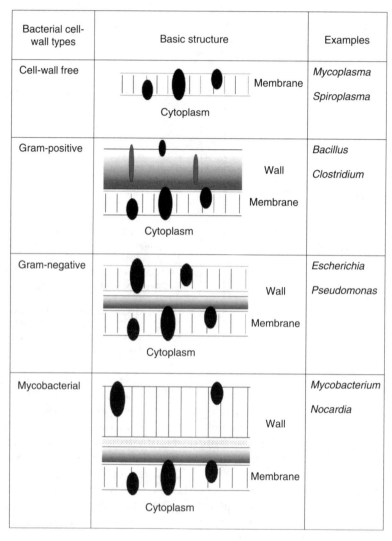

Bacterial cell-wall types	Basic structure	Examples
Cell-wall free	Membrane / Cytoplasm	*Mycoplasma* *Spiroplasma*
Gram-positive	Wall / Membrane / Cytoplasm	*Bacillus* *Clostridium*
Gram-negative	Wall / Membrane / Cytoplasm	*Escherichia* *Pseudomonas*
Mycobacterial	Wall / Membrane / Cytoplasm	*Mycobacterium* *Nocardia*

FIGURE 6.6
The basic structures of bacterial cell walls.

and *Enterococcus*) have a typical internal cell membrane surrounded by an external peptidoglycan layer with associated polysaccharides and proteins. The gram-negative cell wall also contains a (albeit thinner) peptidoglycan layer that is linked to an outer membrane structure. These cell-wall structures do vary in intrinsic resistance to biocides, with gram-negative bacteria generally shown to be more resistant than gram-positive bacteria [5]. In addition to the cell wall, various bacteria produce other surface structures such as capsules that can further protect the cell from the environment (e.g., *Bacillus, Staphylococcus, Pseudomonas* species), S-layers (e.g., *Bacillus* species), and slime layers (e.g., *Streptococcus, Myxococcus*). These layers are primarily composed of polysaccharides and proteins [34]. The protection mechanisms include reaction with the biocide (leading to neutralization), adsorption of the biocide, or prevention of penetration, which can overall lead to a significantly lower dose of chemical biocide reaching the cell surface to impact its antibacterial activity. The expression of these external layers is an important consideration in the surface attachment/interaction and development of biofilm that is discussed later as an important mechanism of intrinsic resistance to biocides. Another group of bacteria, based on their cell-wall structure, have been well described in the literature as being an effective barrier to biocide and antibiotic penetration: those containing a mycobacterial cell-wall structure. These bacteria, including *Mycobacterium, Nocardia,* and *Corynebacterium,* also demonstrate an internal cell membrane not unlike those described for other bacteria, but have an external tripartite cell-wall structure [37]. This cell-wall structure type consists of an internal peptidoglycan layer, a polysaccharide layer (arabinogalactan), and an external layer of long-chained lipids (mycolic acids). Overall, this structure presents a unique barrier to biocides, with mycobacteria generally considered the most resistant vegetative bacteria to antiseptics and disinfectants [5,6].

Similarly, the various surface structures presented by yeasts, fungi, protozoa, algae, and viruses are responsible for the first line of protection to the attacking biocide [6]. Vegetative fungi and yeasts are known to have greater intrinsic resistance to biocides in comparison to vegetative bacteria [5]. In general, molds present a greater resistance profile to yeast, presumably because of their cell-wall structure. The fungal cell wall is composed of fibrils of chitin or cellulose within a polysaccharide matrix [6,34]; however, the various structures present in the many various forms of yeasts and molds and their resistance profiles remain to be studied in detail. Viruses and their resistance to biocides have been traditionally considered under two groups: enveloped and nonenveloped, depending on the presence of an external lipid-bilayer envelope, which is derived from the host cell on release [27,38]. The presence of a viral envelope (e.g., as present in HIV, influenza viruses, and HBV [39]) offers an obvious target to the biocide to render the virus noninfectious. These viruses are found to be significantly more sensitive to biocides. In contrast, nonenveloped viruses, in particular, small nonenveloped viruses such as parvoviruses and poliovirus [39] are known to demonstrate the highest resistance to biocides [5,27,40].

6.3.2   Intrinsic Mechanisms

6.3.2.1   Introduction

In addition to the basic vegetative structures of microorganisms that present an innate resistance to biocides, the following sections discuss some of the major mechanisms of intrinsic resistance. Most of the research in this area has focussed on bacteria, and to a lesser extent on fungi, but references are made to similar resistance mechanisms in protozoa and viruses.

6.3.2.2 Growth Phase Responses

Bacteria and fungi can grow freely as planktonic (or free-growing) cells, but generally demonstrate the same overall growth phase patterns. These growth phases are simply demonstrated under laboratory conditions where the test organism initially grows slowly in a lag phase, entering a rapid exponential phase over time, but eventually demonstrating a slower stationary phase of growth which, in the case of lack of nutrients or unfavorable growth conditions, can be followed by a death phase [34]. Bacteria, which have been studied in more detail, have been shown to be particularly sensitive to biocides during the exponential phase of growth and equally more resistant to biocides during stationary phase. During the stationary phase, cells demonstrate various responses to allow them to survive more toxic or competitive growth conditions (Figure 6.7).

A very simple example is clumping, which can be a particularly effective process in some bacteria (e.g., mycobacteria), fungi (e.g., mold hyphal growth), and viruses. Overall, this poses a penetration challenge to the biocide and increases the chances of microbial survival due to inefficient disinfection. Further, the microorganisms are known to be metabolizing at a slower rate, making them overall less sensitive than in a typical exponential phase [6]. Finally, other specific stationary phase responses described can contribute to biocide resistance including

1. Motility, allowing microorganisms to move to a less harsh environment
2. Production of extracellular matrix (e.g., capsule or slime) and enzymes

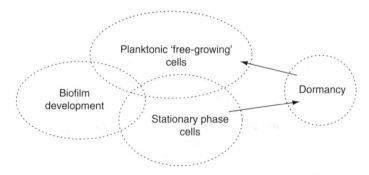

FIGURE 6.7
Bacteria and fungi are the most sensitive to biocides in their planktonic or exponential phases of growth. However, as growth conditions become unfavorable (e.g., in the presence of biocides), they demonstrate greater resistance due to a number of mechanisms including slower growth rate, production of protective mechanisms, and stress responses. Similarly, as they develop biofilms under various conditions, they demonstrate greater resistance as a community. The ultimate stress response in some bacteria and most fungi, as well as for protozoa, is to develop dormant forms of themselves (such as endospores, exospores, cysts, and eggs), which generally have greater biocidal resistance profiles than their vegetative progenitors.

3. Heat-shock response
4. Oxidative stress response

For example, the heat-shock response can allow for the survival of a cell under the stress of increased temperature conditions and is also triggered in the presence of biocides [6,41]. This response includes the production of a variety of proteins, which aid the cell to protect and repair itself from damage due to adverse conditions. Another response is triggered due to oxidative stress. This is indeed a natural response to increasing levels of reactive oxygen species (including the presence of hydrogen peroxide and hydroxyl radicals) that are produced during normal aerobic/facultative anaerobic cell metabolism [41]. A key component in this response is the production of various enzymatic and nonenzymatic mechanisms to neutralize these actives. Enzymes include catalase, which degrades hydrogen peroxide, and chemical antioxidants such as glutathione [6,41]. These mechanisms will also play a role in protecting the cell against oxidizing agents that are used as biocides, in particular when produced at high levels in the extracellular matrix. Although many of these stress responses have been investigated in some detail in bacteria (in particular in *E. coli* and *Bacillus subtilis*), they have also been described in at least some fungi, including *Streptomyces* [42]. Overall, these growth phase responses can play a significant role in the initial intrinsic resistance to biocides and particularly in cases where lower concentration of biocides are used (e.g., as preservatives) or when sublethal concentrations have been applied.

Another mechanism often associated with stationary phase is the efflux of the biocide out of the cell by various cell membrane-associated transport systems. Efflux has been shown to be an important mechanism of resistance to antibiotics and has only recently been found to contribute as an intrinsic resistance mechanism to certain biocides including QACs, antimicrobial dyes, antimicrobial metals, and certain phenolic compounds [43]. These efflux mechanisms have been described in both gram-positive and gram-negative bacteria. For example, the metal-efflux system CzrAB-OpmD has been particularly well studied in *Pseudomonas aeruginosa* [44]. CzrAB-OpmD is an example of a three component system consisting of a cytoplasmic membrane-associated transporter protein, a periplasmic-associated protein, and an outer membrane-associated protein (the outer membrane factor, OMF), which uses energy from the cell surface-associated proton motive force to pump metals like cadmium and zinc out of the cell. Other examples in *P. aeruginosa* include the MexAB-OprM and MexCD-OprJ efflux systems that express antibiotic (β-lactams and fluoroquinolones) and biocide (including antimicrobial dyes, triclosan, and toluene) resistance characteristics [43,44]. Examples of gram-positive bacteria efflux system include the NorA system in *Staphylococcus aureus* and the BmrR system in *Bacillus atrophaeus* that allows for the efflux of fluoroquinolone antibiotics and biocides such as QACs and antimicrobial dyes [43]. Overall, although these systems can only provide minimal protection against some biocides (and particularly under preservative or microstatic conditions), their importance in the promoted expression of antibiotic resistance may be regarded as a concern [45].

6.3.2.3 Biofilm Development

Biofilms have been defined in many ways but can be considered as communities of microorganisms developed on or associated with surfaces. They can consist of single or multiple microbial species and while they have been particularly studied with certain bacterial (e.g., *Pseudomonas* and *Staphylococcus*) and fungal (e.g., *Candida*) species, they are often found to have other associated microorganisms such as viruses and protozoa [46]. Biofilms have been shown to develop on a variety of surfaces including solid/inanimate surfaces (in particular those associated with water or other liquids), foods, tissues (e.g., the skin and the lower intestine are both notable examples of active biofilms), and any liquid–air or liquid–liquid interfaces (as found in formulated products).

Biofilms develop through a series of stages (Figure 6.8). The initial stage is attachment to a surface or association with an interface. Bacteria and fungi

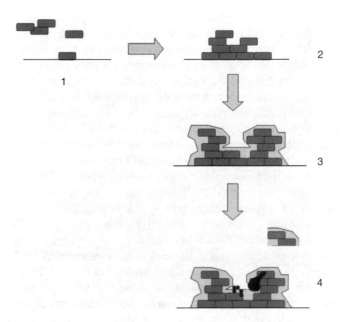

FIGURE 6.8

Stages in surface biofilm development. Stage (1) is microbial attachment to a surface. The microorganism divides to produce microcolonies (2) but as they change growth phase (3) they produce an extracellular matrix consisting of carbohydrates as the biofilm develops. If the biofilm is allowed to further mature (4), other microorganisms can interact and survive within the community and fragments of the biofilm can dislodge to allow further attachment.

have been shown to express various surface structures that can aid in this interaction, including slime layers and capsules (described earlier). The microorganism will then initiate a cycle of multiplication on or at the surface. As development continues, individual cells will vary in their stage of growth depending on their location within the biofilm as well as various environmental factors. Generally, the cells at the outer layers of the biofilm are faster growing than the lower or inner layers. These inner cells can be considered in stationary phase with the various protective responses described earlier, including the production of extracellular protein–carbohydrate matrices. As the biofilm community further matures, growth will continue and can include the interaction and survival of various other microorganisms, both actively growing/multiplying (e.g., bacteria, fungi, and protozoa) or by association (viruses). Further, sections of the biofilm can slough off and recolonize other surfaces.

Biofilm formation can have many negative consequences including damage to surfaces, contamination of products and liquids (e.g., in-dwelling medical devices, water lines, and preserved products), and, of particular pertinence to this chapter, increased resistance patterns to biocides [5,47]. An important example of product contamination was shown with biofilm formation in an iodophor (iodine-releasing) antiseptic preparation, which was unfortunately contaminated with *Pseudomonas* isolated from pipe work in a manufacturing facility [48]. As in most cases, when the individual cells from the biofilm were tested for intrinsic resistance under laboratory conditions, they were found to be rapidly sensitive to the antiseptic preparation, suggesting that biofilm resistance is afforded by an accumulation of protective mechanisms within a matrix rather than an intrinsic resistance mechanism in individual cells. Certain medical devices are particularly prone to biofilm development owing to their design or use, and have been associated with infections; these medical devices include in-dwelling catheters, contact lens, flexible endoscopes, and washer disinfectors [49,50].

There are many individual mechanisms of intrinsic resistance to biocides that combine to provide a unique environment within biofilms, protecting microorganisms against the normal effects of antiseptics, disinfectants, and sterilants [46,47]. These include

1. Limited access of the biocide (and indeed antibiotics or other infective drugs) to the cells
2. Chemical interaction or neutralization of the biocide with various components of the biofilm structure
3. Presence of degradative enzymes and other neutralizing chemicals
4. Expression of stationary phase phenomena, including efflux mechanisms, slower growth phase, and stress responses

In addition to the various intrinsic mechanisms of resistance described, it is also important to note that biofilms can provide a unique environment for the subsequent development of acquired resistance by mutation or genetic exchange between cells in a biofilm. As an important example, it has been suggested that this has occurred in the development of mycobacterial resistance to glutaraldehyde in washer disinfectors, which is further discussed as an acquired mechanism of resistance with significant therapeutic consequences [50,51]. Further, it is easy to speculate that biofilms can provide a unique opportunity for the transfer of genetic material between microorganisms, which can include genes that promote biocide and anti-infective resistance mechanisms (see Section 6.3.3).

6.3.2.4 Dormancy

The impact of a decrease in metabolism during stationary phase of growth and the overall resistance to biocides has been discussed in Section 6.3.2.2. During this stage of growth, microorganisms can show dramatic changes in their responses to various environmental stresses that challenge their survival, including the presence of biocides. The most dramatic changes described are the development of dormant (or low metabolizing) forms of the microorganisms themselves (Table 6.5).

Probably the most studied of these developmental changes is bacterial sporulation, which is only found in a restricted number of bacterial species, notably *Geobacillus*, *Bacillus*, and *Clostridium* [5,6]. These bacterial types have the ability to develop endospores that are considered the most resistant forms of life to various physical and chemical biocidal processes. For

TABLE 6.5

Examples of the Development of Dormant Forms of Microorganisms

Dormant Forms	Description	Examples
Endospores	Development of a dormant spore within a mother cell. Bacterial endospores are some of the most resistant forms to biocides.	*Geobacillus*, *Bacillus*, *Clostridium*
Exospores	Spores developed by specialized fruiting bodies of other structures external to the cell. Resistance can vary significantly depending on the microbial species and spore type (e.g., fungal asexual and sexual spores).	*Streptomyces*, *Aspergillus*, *Penicillium*
Cysts	Dormant cysts are produced during the life cycles of various protozoa. Cysts are produced asexually.	*Giardia*, *Entamoeba*, *Acanthamoeba*
Oocysts	Oocysts are produced during the normal life cycles of some protozoa and helminths. Oocysts are produced sexually.	*Cryptosporidium*, *Toxoplasma*
Eggs	Dormant eggs (or ova) are produced by a diverse group of multicellular eukaryotes including various types of worms. These eggs will vary considerably in size, shape, components, and resistance to biocides.	*Ascaris*, *Enterobius*

example, *Geobacillus stearothermophilus* endospores are regarded as the most resistant microorganisms to steam-based sterilization processes and are widely used as indicators for monitoring and studying the effectiveness of steam sterilization. Another example is *Bacillus atrophaeus* (previously known as *Bacillus subtilis*) spores, one of the most resistant forms of microorganisms to ethylene oxide sterilization cycles. The development of *Bacillus atrophaeus* spores has been the subject of particularly detailed studies and may be summarized as a series of genetically coordinated phases, which shift the normal vegetative functions of a cell to the development, maturation, and release of an endospore from a single mother cell (Figure 6.9) [52].

During the various stages of spore development (sporulation) and subsequent germination/outgrowth of the endospore, biocides show various degrees of resistance profiles [5,6]. For example, the development of resistance to biocides such as formaldehyde and CHG has been described earlier, during the sporulation cascade, and resistance to moist heat and glutaraldehyde occurs at a later stage in endospore development. The endospores may be particularly resistant to the presence of biocides (depending on the biocidal type and concentration that lacks sporicidal activity); however, biocides remain sporistatic or being capable of inhibiting germination and subsequent growth. Similarly, on germination, the spores become readily sensitive to heat and aldehydes, with sensitivity to other biocides (including QACs and CHG) observed during the subsequent outgrowth of the spore into a true vegetative cell [6].

An important factor in the resistance of endospores to biocides is the various protective layers that surround the internal spore core structure, which contains the various components necessary for spore survival

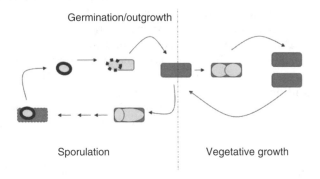

Germination/outgrowth

Sporulation   Vegetative growth

FIGURE 6.9

The typical development of bacterial endospores. Sporulating bacteria normally grow vegetatively by binary fusion (to the right), but under restrictive growth conditions the cell will asymmetrically divide and dedicate itself to the development of an endospore in a series of phases known as sporulation (to the left). The endospore will eventually be released from the expired cell and can remain dormant until suitable growth conditions are present. The endospore will then proceed though a series of further developmental stages (germination and outgrowth) to produce a vegetative cell that can reenter the normal growth cycle.

(including DNA, RNA, and proteins). A typical bacterial endospore consists of various, tightly associated surface layers, depending on the bacterial genus or species. These layers can consist of an inner coat of lipid/protein surrounding the spore core, overlaid by a cell wall and cortex (consisting of peptidoglycan or modified peptidoglycan), an outer membrane of lipid/ protein, various inner and outer spore coats (mainly consisting of protein), and finally an external protein-based exosporium. The main function of these various layers is to provide an effective barrier to the penetration of various physical and chemical effects [5,6,53]. In addition, the internal spore core is known to have further resistance factors including low water content, high calcium levels, either limited or no metabolism in comparison to vegetative cells, and other protective chemicals (such as dipicolinic acid) and proteins (small acid-soluble proteins). For example, the complex of calcium and dipicolinic acids has been described as playing an important role in spore resistance to both heat and chemicals [53]. The small acid-soluble proteins also appear to function by tight association to the bacterial DNA as a protective mechanism, as well as serving as a subsequent efficient source for the germinating spore [53].

A limited number of other bacteria, notably some actinomycetes such as *Streptomyces*, are also known to produce different forms of spores known as exospores. Morphological differentiation into exospores has also been studied in some detail and follow a similar series of development phases [6,42]. Exospore development in bacteria is similar to that described in fungi, with the development of aerial hyphae, separation of a pre-spore compartment that develops a thicker cell wall, and subsequent spore desiccation. Despite these changes and notable survival under dry conditions, bacterial exospores have not been described to be particularly resistant to various biocidal processes in comparison to their vegetative forms or bacterial endospores [6,53].

The growth of various filamentous fungi can enter into a stationary phase, which can trigger the development of spores. The various types and structures of fungal spores have not been studied in great detail but they are known to be developed both sexually and asexually, depending on the fungal species. The various types of spores or spore-containing bodies have been traditionally used for the classification of fungi, based on morphology [54]. Examples of asexual spores include sporangiospores and conidiospores, and sexual spores include ascospores and basidiospores. In the case of *Aspergillus*, which is widely considered as one of the more resistant fungi to biocides, both conidiospores (asexual spores) and ascospores (sexual spores) are produced. Asexual sporulation is triggered by various environmental stresses including nutrient limitation. The resulting spores are generally simple structures developed from fungal cells, consisting of a low metabolizing structure, with a nutrient source and surrounded by a specialized spore wall. Sexual spores are generally more structurally distinct and are usually developed within specialized structures known as fruiting bodies; for example, ascospores develop within an ascus. Sexual spores develop and mature external wall or walls around a central core.

The spore walls generally consist of polysaccharides, including various proteins, pigments, and lipids, and can vary in thickness. In addition to these protective layers, the spores can have various degrees of desication and also have the added protection from the inclusion within a defined fruiting body structure. Overall, the protective mechanisms are considered similar to those described for bacterial endospores, although in general most fungal spores are more sensitive to thermal and chemical biocides and biocidal processes.

Other dormant forms include helminth eggs and protozoal cysts and oocysts. The helminths are a group of multicellular eukaryotes and are generally known as worms. They include various human and animal pathogens including roundworms (e.g., *Wuchereria* and *Onchocerca*) and flukes (e.g., *Schistosoma*). Overall, the effects of biocides on this group have not been well studied. It is known that typical life cycles display adult worms reproducing sexually to produce eggs (ova) and in some cases cysts. Helminth eggs have been associated with outbreaks due to survival on environmental surfaces, despite disinfection precautions. The egg structure varies in size and shape [34,55]. They consist of a central, viable core with external layers of protein, and carbohydrate or lipid or even both, depending on the species, and of various thicknesses. Various levels of resistance to water treatment methods including UV radiation, oxidizing agents (including chlorine and ozone), and even heat up to 70°C have shown various resistance patterns [6,56]. The main mechanism of resistance would appear to be reduced penetration to the central core of the egg, although the actual mechanisms have not been studied. Similarly, protozoa (that include pathogens such as *Giardia*, *Acanthamoeba*, and *Cryptosporidium*) produce cysts or oocysts during their normal life cycles. These dormant forms have also been shown to demonstrate greater resistance profiles than other structures, during their life cycles. Survival of these dormant forms in various chemical biocide formulations or water has been linked to various protozoal-infection outbreaks [6,57]. Of particular note, *Cryptosporidium parvum* oocysts are known to be particularly resistant to chemical biocides [58], although they are readily inactivated at 50°C. Resistance is also considered to be due to the various thicknesses of the outer oocyst wall, with particular consideration given to the protein component of the wall [6].

6.3.2.5 Revival

The concept of revival as a true mechanism of intrinsic resistance has been debated [5,6]. When a given microorganism is exposed to a biocide or biocidal process, the extent of damage and subsequent cell death will determine the overall effectiveness. It has been shown, in particular in bactericidal and fungicidal studies, that damage (even to internal structures) can be repaired over time and, although the microorganism may not initially demonstrate growth by normal laboratory methods, they can be recovered (or revived) over time. Examples of various revival mechanisms include

The various stress responses described in Section 6.3.2.2, including heat-shock and SOS responses

Cell wall and membrane regeneration and repair

DNA repair mechanisms (or particular note those described for the bacteria *Deinococcus* [59])

Viral reassociation [60]

The latter example in viruses is unique because of their ability to remain potentially infectious despite significant damage or even physical breakdown of viral structure. This can be simply due to the ability of the viral nucleic acid to infect cells or indeed reassociate with other viral components to allow for active cell infection [5,27]. The debate regarding revival as a true mechanism of resistance revolves around the fact that many of the observations may be simply explained by the inadequate penetration or activity of the biocide to the target microorganisms. Examples include the penetration of biofilm growth or fixation by cross-linking agents (leaving certain cells protected from attack and remaining viable but uncultivable under normal growth conditions).

6.3.2.6 Extreme Mechanism Examples

In addition to dormancy, unique resistance or protective mechanisms are displayed by some notable microorganisms that allow them to survive and grow under what we may consider as extreme environments. Examples of these are given in Table 6.6.

These organisms include thermophiles (or hyperthermophiles) that grow at high temperatures, psychrophiles at cold/freezing conditions, and acidophiles under high acid conditions, often in combination with high growth temperatures [34]. The growth characteristics may seem remarkable but it should be noted that many of these bacteria or archaea actually require these growth conditions to survive and do not grow under so-called normal or ambient conditions. Therefore, "extreme" is certainly only from

TABLE 6.6

Examples of Extremophiles or Microorganisms That Grow and Survive under Extreme Conditions

Microorganisms	Description	Mechanisms of Resistance
Pyrococcus, Pyrolobus	Heat tolerance	Cell-wall structure, protein structure, protective and repair mechanisms
Thiobacillus, Thermoplasma, Helicobacter	Acid tolerance	Cell-wall structure, exclusion mechanisms, efflux mechanisms
Cyanobacteria	Cold or freezing tolerance	Protein structure, cell-wall structure, secretion of protective chemicals/exopolysaccharide
Deinococcus, Deinobacter	Radiation resistance	Cell-wall structure, tetrad growth, DNA repair mechanisms

our perspective and not that of the microorganisms. Despite this, it is of particular interest to understand the mechanisms of resistance to allow for their growth given that their genetic and biochemical natures are not vastly distinct from other prokaryotic or eukaryotic cells. In brief, there are surprisingly many similarities that include [6,34]

Exclusion mechanisms, predominantly due to cell-wall/membrane structures

Protein/enzyme structure (e.g., in psychrophiles proteins show an increase in α-helical structure, or in thermophiles proteins show increased ionic bonding within their structures)

Lipid membrane structures (in psychrophiles the cell membrane can have a high proportion of unsaturated fatty acids, while in thermophiles increased saturated fatty acids and even lipid mono-layer cell membranes have been identified)

Presence of protective cytoplasmic proteins and sugars

Repair mechanisms

Active efflux systems, particularly in various acidophilic bacteria to maintain cytoplasmic pH levels in the neutral range

Most extremophiles are found under unique environmental conditions and are currently not considered significant pathogens or environmental concerns (with the notable exception of *Helicobacter* and their association with peptic ulcers and survival within the low acid environment of the stomach [61]). In *Helicobacter*, two important mechanisms of resistance appear to be active efflux and a urease-mediated production of neutralizing chemicals [6,61]. However, there are other examples of extreme resistance (in addition to dormant forms like endospores discussed previously) that have been linked to product spoilage. Radiation, particularly ionizing radiation, processes based on gamma (γ)-rays or electron beams are considered some of the most effective sterilization methods. It was therefore of some surprise to identify nonsporulating gram-positive bacteria that survived an irradiation process in canned products. These bacteria were identified as *Deinococcus radiodurans*, and other gram-negative bacteria of the genus *Deinobacter* have also been described with similar resistance patterns [6,34]. These bacteria can be frequently isolated as environmental contaminants (of interesting note, including in nuclear waste) and demonstrate dramatic tolerance to radiation, some oxidizing agents and drying. Similar to other extremophiles, these bacteria have been shown to have multiple mechanisms of intrinsic resistance that cumulate to allow for the observed radiation tolerance. These include [6]

Cell-wall structure (in comparison to other bacteria, *Deinococcus* has been particularly shown to have a thicker cell wall, a distinct peptidoglycan structure, and an unique cell-wall lipid profile)

Production of protective enzymes, in particular, to neutralize oxygenated species such as catalase

Production of protective chemical pigments, which can neutralize the
localized production of hydroxyl radicals

Efficient stress response mechanisms, including cell-wall metabolism

Efficient repair mechanisms for DNA damage (note: DNA is a key
target for radiation-based processes)

Tetrad growth, where cells grow in groups of four and can contain
multiple copies of their DNA to allow for more efficient recombi-
national repair mechanisms

Overall, these extreme examples of resistance to chemical and physical biocidal
processes are instructive in the constant development of mechanisms of
bacterial tolerance. It is not hard to speculate that similar mechanisms
may be identified in the future in the development of resistance in more
significant pathogenic or spoilage-associated microorganisms.

6.3.3 Acquired Mechanisms

6.3.3.1 Introduction

While intrinsic mechanisms of resistance are considered as a normal property
of given microorganisms, acquired mechanisms are distinct owing to single or
multiple genetic mutations (defined as inherited changes in the sequence of
nucleic acids which can result in altered structures and functions) or
the acquisition of transposable genetic elements such as plasmids. Acquired
mechanisms of antibiotic resistance have been particularly well described in
the literature and have dramatic effects as far as therapeutic applications are
considered [1]. In contrast, acquired biocide resistance mechanisms have been
less described or studied [5,6]. It is important to note that antibiotics (or indeed
other anti-infectives such as antiviral and antifungal drugs) are known to have
specific mechanisms of action, which despite the therapeutic advantages of
these chemicals have allowed for the development of acquired resistance
mechanisms and loss of any practical effect. Considering the broad-spectrum
activity of many biocides, their modes of action and particularly the in-use
concentrations typically used in antiseptic, disinfectant, and sterilization applications, it has been considered unlikely that practical resistance to biocides would develop by acquired mechanisms. However, research in the
last 10 years has shown that this can be a significant concern. First,
low-level resistance (or tolerance) to certain biocides may have little practical
impact in their safe use but have also been linked to the co-development of
antibiotic resistance of much greater clinical significance. Further, bacterial
strains have been identified, presumably due to mutations, demonstrating
clinically significant resistance patterns to widely used biocides at actual use-
concentrations. Examples of mutational and transmissible-element associated
resistance mechanisms are further discussed in this section.

6.3.3.2 *Mutation*

The development of resistance to antibiotics and other anti-infectives because of mutations is a subject of much research [1,3]. In contrast, little has been studied of similar or indeed distinct mechanisms of resistance with biocides [5,6]. Despite this, there have been some interesting reports and investigations on the development of resistance that has been shown to be due to mutations; examples of some of these are given in Table 6.7.

In a number of cases, although many of these reports remain to be confirmed (particularly at the genetic level), it would seem that changes in the cell-wall/membrane structure are responsible [5,6]; the expectation that this could indeed occur is strengthened by the observation of opposite effects, where mutations affecting the structure or functions of bacterial or fungal cell-wall structure have lead to increased sensitivity to biocides [6]. Probably the most dramatic reports of mutational resistance were the isolation and identification of atypical *Mycobacterium* isolates (in particular *Mycobacterium chelonae* strains) that were resistant to the activity of ~1%– 3% glutaraldehyde solutions, which are widely used for the low-temperature disinfection of thermosensitive medical devices such as flexible endoscopes [51]. Little to no activity was observed with multiple glutaraldehyde-based formulations, despite isolation and cultivation under laboratory conditions. Further, these strains have been shown to be less sensitive to other aldehydes, such as OPA solutions [62]; it should be noted that these strains remain sensitive to OPA (over-extended exposure times) and equally sensitive to other chemical disinfectants (including oxidizing agent-based formulations [63]). Although the exact mechanisms of resistance have not been firmly identified, initial investigations of these strains showed phenotypic differences in their colony morphology (appearing drier and waxier), and cell-wall analysis found differences in the cell-wall polysaccharides arabinogalactan and arabinomannan [64]. The proposal that the cell wall in these mutants

TABLE 6.7

Examples of Mutation-Associated Resistance in Bacteria

Mutation/Mechanism	Bacteria	Biocides
Unconfirmed, but considered altered lipid content in cell wall/membrane	*Serratia marcescens* *Escherichia coli*	QACs, triclosan
Modified target protein (e.g., enoyl reductase)	*E. coli*	Triclosan, hexachlorophene
Mutations involving the expression of efflux-associated proteins	*E. coli, Pseudomonas*	Triclosan, some detergents, QACs, CHG
Cell-wall polysaccharide expression/changes (?)	*Mycobacterium chelonae*	Glutaraldehyde, *o*–phthaldehyde

Note: In most cases, resistance refers to an increase in minimum inhibitory concentration (MIC) of the biocide, unless otherwise indicated.

excludes the penetration of glutaraldehyde is supported by recent investigations into the mode of action of glutaraldehyde and OPA, suggesting that OPA demonstrates greater penetration into the cell [10].

Two other mechanisms of mutational resistance to biocides in bacteria of notable interest, in particular with respect to cross-resistance to antibiotics, include the investigations of triclosan resistance in *E. coli* and other bacteria [66] and efflux-associated mechanisms in *E. coli* [43,44]. Mutations associated with the expression and control of efflux proteins have been shown (under laboratory conditions) to have cross-resistance to antibiotics and some biocides, although biocide tolerance in these investigations has only been found to have small changes in the MIC of the biocide against the mutant [43,44]. These include mutations at the *acr* locus, leading to increased QAC and antimicrobial dye MICs, and overexpression of the *E. coli* RobA protein, leading to increased heavy metal resistance. Triclosan is a bisphenol biocide with particular effectiveness against gram-positive bacteria, including *Staphylococcus* [5,65]. It is also somewhat effective against gram-negative bacteria, demonstrating low MICs but requiring significantly higher concentrations (and the assistance of chelating agents) to inactivate bacteria such as *E. coli*. A number of laboratory investigations have shown the development of stable *E. coli* mutants with increased MICs to triclosan [6,65]. Investigations of these strains have identified a number of effects on the bacteria, including some specific cellular targets:

- Overproduction of efflux proteins
- Exclusion, due to loss of surface proteins or changes in fatty acid structure
- Changes in active site structure of the enoyl-[acyl carrier protein]-reductase, which is involved in fatty acid synthesis
- Overproduction of enoyl-[ACP]-reductase or glucomamine-6-phosphate aminotransferase (involved in amino sugar biosynthesis)

In the case of enoyl-[ACP]-reductase mutants in *E. coli*, cross-resistance to another biphenol (hexachlorophene) and the antimycobacterial antibiotic isoniazid has been shown [66,67]. Further investigations have confirmed that, similar to isoniazid, triclosan binds to certain enoyl reductases in some gram-positive and gram-negative bacteria (including *S. aureus*, *Mycobacterium tuberculosis*, *E. coli*, and *Pseudomonas aeruginosa*) but not other bacteria such as *Streptococcus pneumoniae*. Specific mutations that lead to changes in the amino acid structure in the enzyme active site have been shown to cause a decreased affinity for the biocide. These results were quite dramatic showing that the particular mode of action of this biocide was in many ways similar to an antibiotic; however, it should be noted that the mutations only lead to an increased triclosan MIC and that the strains remained equally sensitive to the minimum bactericidal concentration of the biocide as well as to various triclosan-containing products and formulations [68].

These results, and the lack of sensitivity of enoyl reductases in bacteria such as *S. pneumoniae* despite being very sensitive to the biocide, would confirm that triclosan has multiple effects on bacteria, both at the cell surface and within the cytoplasm.

Considering the mechanisms of resistance that have been described so far in bacteria, it is more than likely that mutation-related tolerance mechanisms have yet to be confirmed in fungi as well as other microorganisms.

6.3.3.3 Transmissible Elements

There have been some notable, confirmed examples of plasmid-associated resistance mechanisms, particularly in bacteria (Table 6.8). Plasmids are extrachromosomal DNA molecules that replicate independently of the chromosomal DNA. In addition to plasmids, transposons are also mobile DNA sequences, but are not capable of autonomous replication. These transmissible elements are found to express many functions including biocide and antibiotic resistance mechanisms. In consideration of Table 6.8, it is clear that many of these resistance mechanisms have previously been described as being similar to known intrinsic mechanisms.

Plasmid expression of degradative enzymes is generally specific to their respective target biocides or chemicals [5,6]. An example is the expression of mercuric reductases as a mechanism of resistance against mercury or organomercurials. Mercury is not a widely used biocide (although was used in the past), but organomercurials are used as preservatives; in addition, mercury can be a significant concern as a wastewater contaminant, and the use of bacteria (based on the expression of mercuric reductases) for bioremediation is of environmental interest. Mercuric reductases cause the conversion of the mercury ion into a vapor, which is removed by vaporization.

TABLE 6.8

Examples of Plasmid-Acquired Resistance in Bacteria

Plasmid Examples	Bacteria	Biocides	Mechanisms
TOL plasmids	*Pseudomonas*	Toluene, phenol	Expression of degradative enzymes
pLVPK	*Klebsiella*	Silver, copper	Sequestration
pMGH100	*Salmonella*	Silver	Efflux
pMER327	*Pseudomonas*	Mercury and some organomercurial compounds	Expression of mercuric reductases
pR773	*Escherichia*	Arsenic	Efflux
pR124	*Escherichia*	Cetrimide (a QAC)	Alteration of the outer membrane of the cell wall (decreased penetration)
pSK1	*Staphylococcus*	QACs, CHG, antimicrobial dyes	Efflux

Mercuric reductase genes have been identified as plasmid based on a wide variety of bacteria, including *Staphylococcus aureus* [69]. Other enzymes (e.g., lysases) can also be present on the same plasmid that allows for the conversion of various organomercuric compounds into mercury ions as subsequent substrates for reductase activity. Further examples of plasmid-associated degradative enzyme expression include formaldehyde dehydro-genase (as a mechanism of formaldehyde tolerance [70]) and enzymes involved in the degradation of toluene/phenols (e.g., as expressed on the TOL or TOM plasmids in various pseudomonads [6,71]). Plasmid-encoded factors have been shown to cause changes in the structure, and therefore the biocide penetration capabilities of the cell wall/membrane, with associated resistance to some QACs (e.g., in *E. coli*) and aldehydes (in some gram-negative bacteria species). Sequestration, by binding of a plasmid-encoded molecule, can also be an effective protective mechanism, as shown in heavy-metal (silver and copper) resistance in *Klebsiella pneumoniae* [5,72]. It is interesting to note that many plasmid-associated mechanisms of silver resistance have been described [73,74]. Although silver and silver compounds have become less used as biocides, recent advances in medical device technology have seen the use of impregnated or associated silver to prevent device contamination and biofilm development, particularly with in-dwelling catheters. The mechanisms of silver resistance have been previously reviewed [72,73] and include:

> Sequestration or metal ion binding (including cell surface and cytoplasmic-associated mechanisms)
>
> Efflux
>
> Cell-wall structure changes (exclusion)
>
> Other speculated mechanisms that remain to be fully elucidated

Similar mechanisms have been identified for other toxic metals, including copper, arsenic, and cadmium.

Efflux has previously been discussed as an important mechanism of intrinsic resistance associated with stationary phase growth in bacteria. This mechanism has often been shown (and increasingly so) to also be plasmid encoded. Plasmid-associated efflux mechanisms have been descri-bed for a wide variety of biocides including heavy metals, CHG, QACs, acridines, diamidines, antimicrobial dyes, and triclosan [43,45]. One of the most studied mechanisms is associated with the pSK41 and pSK1 families of plasmids found in staphylococci, including *S. aureus* and *S. epidermidis* [5,43]. Strains that contain these plasmids are found to have increased MICs to cationic biocides such as QACs, diamidines, dyes, and CHG. These plasmids have been found to express various different *qac* genes that encode for efflux proteins (e.g., QacA-J); in addition, these plasmids may also carry antibiotic resistance factors such as β-lactamases, apart from antibiotic (e.g., tetracyc-line) efflux characteristics. Some of these plasmids have been shown to contain transposons, with other associated antibiotic resistance determinants (as observed with aminoglycoside resistance expressed from pSK1 [43,45]). These are clear examples of the development of co-resistance to both biocides and anti-infective drugs, despite the fact that the actual biocide-tolerant phenotype is marginal and from a biocide application concern is limited (e.g., in the case of cationic biocides used at preservative concentrations). As for mutational resistance, other possible cases of plasmid-associated resistance have been speculated, in particular in environmental isolates, but remain to be confirmed or further investigated [5].

6.4 Conclusions and Perspectives

Biocides and biocidal processes are widely used for antisepsis, disinfection, and sterilization in medical, pharmaceutical, production, industrial, and household applications. They may be considered as causing generalized toxic effects leading to inactivation of microorganisms as, in most cases, they are associated with multiple mechanisms of action that cumulate to cause loss of viability. Their mechanisms of action may be classified based on their primary effects on macromolecules, including oxidizing agents, cross-linking/coagulating agents, energy-transfer mechanisms, and specific structure-disrupting mechanisms; however, this classification system is clearly an oversimplification considering other interactions and effects that have been observed and the variable impact of the biocide depending on its specific use and formulation. A greater understanding of these effects may allow for their more efficient and optimal uses in various surface and air applications. The widespread use of biocides, in particular chemical biocides, is of some concern, not only when considering their optimal and prudent use but also when considering their safety [75]. Clearly, the environmental impact alone is of some concern, including bioaccumulation and the risks associated with the development of biocide resistance. Microorganisms display an array of intrinsic or innate tolerance and resistance mechanisms to environmental stresses, including the presence of biocides. These mechanisms should be considered in the optimization of biocidal methods for given applications. An example is the case of water-line disinfection, which should optimally consider biofilm penetration and removal in addition to the antimicrobial activity of the biocidal process. Over the last few years, investigations of acquired biocide resistance mechanisms, not only to individual biocides but also cross-resistance to antibiotics, have shown some dramatic results. Similar to anti-infectives, these mechanisms may be due to mutational changes in the genetic material of the microorganism or the acquisition of plasmids/transposons. Although in some of these cases the impact of low-level resistance in the use of biocides is debatable [7], an increasing number of investigations have shown the development of even high-level resistance and the failure of biocidal products under typical in-use conditions.

The identification of resistant strains and an understanding of the mechanisms of resistance are clearly of further research interest, not the least of which to understand how resistance may be overcome by various formulation or process (e.g., temperature, concentration, etc.) effects to allow greater activity and penetration of the biocide. Further, a greater understanding of the modes of action and resistance to biocides may give some important strategies or targets for next-generation anti-infective developments.

Appendix

The terms and definitions associated with biocidal applications can vary in both general and scientific use. The following references can be used as a guide to understanding the basic definitions for each of these terms, with consideration of those harmonized according to international standards [5,6]. It should be noted that the use of these terms can vary in various industrial/medical uses, and from a regulatory perspective depending on the country, region, or specific application.

Antisepsis. Destruction or inhibition of microorganisms in or on living tissue, e.g., on the skin or mucous membranes.

Disinfection. The antimicrobial reduction of the number of viable microorganisms to a level previously specified as appropriate for its intended further handling or use.

Fumigation. Disinfection with a liquid or gas indirectly within an enclosed area.

Pasteurization. Disinfection, usually by heat, of microorganisms that can be harmful or cause product spoilage.

Preservation. The prevention of the multiplication of microorganisms in products.

Resistance. The inability of an anti-infective or biocide to be effective against a target microorganism.

Sanitization. Disinfection of microorganisms that pose a threat to public health.

Sterilization. Defined process used to render a surface or product free from viable organisms, including bacterial spores.

Tolerance. Decreased effect of an active against a target microorganism and requiring increased concentration or other effects to be effective.

References

1. Bryskier, A., *Antimicrobial Agents: Antibacterials and Antifungals*, ASM Press, Washington DC, 2005.
2. Driscoll, J.S., *Antiviral Drugs*, John Wiley & Sons, Hoboken, NJ, 2005.

3. Coffin, S., Antiviral agents, *Pediatric Infectious Diseases*, Shah, S. (ed.), Blackwell, Oxford, 2004.
4. Crabtree, T.D., Pelletier, S.J, and Pruett, T.L., Surgical antisepsis, *Disinfection, Sterilization, and Preservation*, Block, S.S. (ed.), Lippincott Williams & Wilkins, Philadelphia, 2001, chapter 44.
5. McDonnell, G. and Russell, A.D., Antiseptics and disinfectants: Activity, action, and resistance. *Clin. Microbiol. Rev.*, 12, 147–179, 1999.
6. McDonnell, G., *Antisepsis, Disinfection and Sterilization: Types, Action and Resistance.* ASM Press, Washington DC, 2006.
7. Gilbert, P. and McBain, A.J., Potential impact of increased use of biocides in consumer products on prevalence of antibiotic resistance, *Clin. Microbiol. Rev.*, 16, 189–208, 2003.
8. Scott, E.M. and Gorman, S.P., Glutaraldehyde, *Disinfection, Sterilization, and Preservation*, Block, S.S. (ed.), Lippincott Williams & Wilkins, Philadelphia, 2001, chapter 17, pp. 361–382.
9. Fraud, S., Hann, A.C., Maillard, J.Y., and Russell, A.D., Effects of ortho-phthalaldehyde, glutaraldehyde and chlorhexidine diacetate on *Mycobacterium chelonae* and *Mycobacterium abscessus* strains with modified permeability, *J. Antimicrob. Chemother.*, 51, 575–584, 2003.
10. Zhu, P.C., Roberts, C.G., and Favero, M.S., Solvent or matrix-mediated "molecular switches," the lipophilic dialdehyde (OPA) and the amphiphilic 1,3-phthalandiol and OPA disinfection mechanism, *Curr. Org. Chem.*, 9, 1155–1166, 2005.
11. Parisi, A.N. and Yound, W.E., Sterilization with ethylene oxide and other gases, *Disinfection, Sterilization, and Preservation*, Block, S.S. (ed.), Lea & Febirger, Philadelphia, 1991, chapter 33, pp. 580–595.
12. Hugo, W.B., Disinfection mechanisms, *Principles and Practice of Disinfection, Preservation and Sterilization*, 2nd edition, Russell, A.D., Hugo, W.B., and Ayliffe G.A.J. (eds.), Blackwell Science, Oxford, 1992, chapter 9, pp. 158–185.
13. Hugo, W.B. and Russell, A.D., Types of antimicrobial agents, *Principles and Practice of Disinfection, Preservation and Sterilization*, 2nd edition, Russell, A.D., Hugo, W.B., and Ayliffe, G.A.J. (eds.), Blackwell Science, Oxford, 1992, chapter 2, pp. 8–106.
14. McDonnell, G., Peroxygens and other forms of oxygen: Their use for effective cleaning, disinfection and sterilization, *New Biocides Development: The Combined Approach of Chemistry and Microbiology*, Zhu, P. (ed.), ACS Symposium Series, American Chemical Society, New York, 2006.
15. Cabiscol, E., Tamarit, J., and Ros, J., Oxidative stress in bacteria and protein damage by reactive oxygen species. *Int. Microbiol.*, 3, 3–8, 2000.
16. Fichet, G., Comoy, E., Duval, C., Antloga, K., Dehen, C., Charbonnier, A., McDonnell, G., Brown, P., Lazmezas, C.I., and Deslys, J.P., Novel methods for disinfection of prion-contaminated medical devices, *Lancet*, 364(9433), 521–526, 2004.
17. Kampf, G., Bloss, R., and Martiny, H., Surface fixation of dried blood by glutaraldehyde and peracetic acid, *J. Hosp. Infect.*, 57, 139–143, 2004.
18. Brown, P., Liberski, P.P., Wolff, A., and Gajdusek, D.C., Resistance of scrapie infectivity to steam autoclaving after formaldehyde fixation and limited survival after ashing at 360°C: Practical and theoretical implications, *J. Infect. Dis.*, 161, 467–472, 1990.
19. Ludwig, J.D. and Avis, K.E., Dry heat inactivation of endotoxin on the surface of glass, *J. Parenter. Sci. Technol.*, 44, 4–12, 1990.
20. Silveira, J.R., Caughey, B., and Baron, G.S., Prion protein and the molecular features of transmissible spongiform encephalopathy agents, *Curr. Top. Microbiol. Immunol.*, 284, 1–50, 2004.

21. Antloga, K., Meszaros, J., Malchesky, P.S., and McDonnell, G.E., Prion disease and medical devices, *ASAIO J.*, 46, S69–72, 2000.
22. Hansen, J.M. and Shaffer, H.L., Sterilization and preservation by radiation sterilization, *Disinfection, Sterilization, and Preservation*, Block, S.S. (ed.), Lippincott Williams & Wilkins, Philadelphia, 2001, chapter 37, pp. 729–746.
23. Blatchley, E.R. and Peel, M.M., Disinfection by ultraviolet irradiation, *Disinfection, Sterilization, and Preservation*, Block, S.S. (ed.), Lippincott Williams & Wilkins, Philadelphia, 2001, chapter 41.
24. Russsell, A.D., Antibiotic and biocide resistance in bacteria: Introduction, *Symp. Ser. Soc. Appl. Microbiol.*, 31, 1S–3S, 2002.
25. Merianos, J.J., Surface active agents, *Disinfection, Sterilization, and Preservation*, Block, S.S. (ed.), Lippincott Williams & Wilkins, Philadelphia, 2001, chapter 14, pp. 283–320.
26. Salton, M.R.J., Lytic agents, cell permeability and monolayer penetrability, *J. Gen. Physiol.*, 52, 277 (Suppl.) 1968.
27. Maillard, J.-Y. and Russell, A.D., Viricidal activity and mechanisms of action of biocides, *Sci. Progr.*, 80, 287–315, 1997.
28. Borschel, D.M., Chenoweth, C.E., Kaufman, S.R., Hyde, K.V., Vanderelzen, K.A., Raghunathan, T.E., Collins, C.D., and Saint, S., Are antiseptic-coated central venous catheters effective in a real-world setting?, *Am. J. Infect. Control*, 34, 388–393, 2006.
29. Gilbert, P. and McBain, A.J., Literature-based evaluation of the potential risks associated with impregnation of medical devices and implants with triclosan, *Surg. Infect.*, 3.1, S55–S63, 2002.
30. Schweizer, H.P., Triclosan: A widely used biocide and its link to antibiotics, *FEMS Microbiol. Lett.*, 202, 1–7, 2001.
31. Heath, R.J. and Rock, C.O., Fatty acid biosynthesis as a target for novel antibacterials, *Curr. Opin. Investig. Drugs*, 5, 146–153, 2004.
32. Tehei, M. and Zaccai, G., Adaptation to extreme environments: Macromolecular dynamics in complex systems, *Biochim. Biophys. Acta.*, 1724, 404–410, 2005.
33. Manzoor, S.E., Lambert, P.A., Griffiths, P.A., Gill, M.J., and Fraise, A.P., Reduced glutaraldehyde susceptibility in *Mycobacterium chelonae* associated with altered cell wall polysaccharides, *J. Antimicrob. Chemother.*, 43, 759–765, 1999.
34. Madigan, M.T., Martinko, J.M., and Park, J., *Brock Biology of Microorganisms*, 10th edition, Pearson Education, Upper Saddle River, NJ, 2003.
35. Trachtenberg, S., Mollicutes-wall-less bacteria with internal cytoskeletons, *J. Struct. Biol.*, 124, 244–256, 1998.
36. Boddie, R.L., Owens, W.E., Ray, C.H., Nickerson, S.C., and Boddie, N.T., Germicidal activities of representatives of five different teat dip classes against three bovine mycoplasma species using a modified excised teat model, *J. Dairy Sci.*, 85, 1909–1912, 2002.
37. Dover, L.G., Cerdeno-Tarraga, A.M., Pallen, M.J., Parkhill, J., and Besra, G.S., Comparative cell wall core biosynthesis in the mycolated pathogens, *Mycobacterium tuberculosis* and *Corynebacterium diphtheriae*, *FEMS Microbiol. Rev.*, 28, 225–250, 2004.
38. Klein, M. and Deforest, A., Principles of viral inactivation, *Disinfection, sterilization and preservation*, 3rd edition, Block, S.S., (ed.), Lea & Febirger, Philadelphia, 1983, pp. 422–434.
39. Collier, L. and Oxford, J., *Human Virology*, 2nd edition, Oxford University Press, New York, 2000.
40. Springthorpe, V.S. and Satter, S.A., Chemical disinfection of virus-contaminated surfaces, *CRC Crit. Rev. Environ. Cont.*, 20, 169–229, 1990.

41. Foster, P.L., Stress responses and genetic variation in bacteria, *Mutat. Res.*, 569, 3–11, 2005.

42. Sevcikova, B., Benada, O., Kofronova, O., and Kormanec, J., Stress-response sigma factor sigma(H) is essential for morphological differentiation of *Streptomyces coelicolor* A3(2), *Arch. Microbiol.*, 177, 98–106, 2001.

43. Poole, K., Efflux-mediated antimicrobial resistance, *J. Antimicrob. Chemother.*, 56, 20–51, 2005.

44. Poole, K., Efflux-mediated multiresistance in gram-negative bacteria, *Clin. Microbiol. Infect.*, 10, 12–26, 2004.

45. Levy, S.B., Active efflux, a common mechanism for biocide and antibiotic resistance, *J. Appl. Microbiol.*, 92, 65S–71S, 2002.

46. Lappin-Scott, H.M. and Costerton, J.W., *Microbial Biofilms*, Cambridge University Press, Cambridge, 2003.

47. Gilbert, P. and McBain, A.J., Biofilms: Their impact on health and their recalcitrance toward biocides, *Am. J. Infect. Control*, 29, 252–255, 2001.

48. Anderson, R.L., Vess, R.W., Carr, J.H., Bond, W.W., Panlilio, A.L., and Favero, M.S., Investigations of intrinsic *Pseudomonas cepacia* contamination in commercially manufactured povidone–iodine, *Infect. Control Hosp. Epidemiol.*, 12, 297–302, 1991.

49. Mack, D., Rohd, H., Harris, L.G., Davies, A.P., Horstkotte, M.A., and Knobloch, J.K., Biofilm formation in medical device-related infection, *Int. J. Artif. Organs*, 29, 343–359, 2006.

50. Kressel, A.B. and Kidd, F., Pseudo-outbreak of *Mycobacterium chelonae* and *Methylobacterium mesophilicum* caused by contamination of an automated endoscopy washer, *Infect. Control Hosp. Epidemiol.*, 22, 414–418, 2001.

51. Griffiths, P.A., Babb, J.R., Bradley, C.R., and Fraise, A.P., Glutaraldehyde-resistant *Mycobacterium chelonae* from endoscope washer disinfectors, *J. Appl. Microbiol.*, 82, 519–526, 1997.

52. Errington, J., Regulation of endospore formation in *Bacillus subtilis*, *Nat. Rev. Microbiol.*, 1, 117–126, 2003.

53. Russell, A.D., Chemical sporicidal and sporostatic agents, *Disinfection, Sterilization, and Preservation*, Block, S.S. (ed.), Lippincott Williams & Wilkins, Philadelphia, 2001, chapter 27, pp. 529–542.

54. Guarro, J., Gene, J., and Stchigel, A.M., Developments in fungal taxonomy, *Clin. Micro. Rev.*, 12, 454–500, 1999.

55. Bernard, E., *Introduction to Parasitology*, Mellon's Books, East Sussex, United Kingdom, 1998.

56. Brownell, S.A. and Nelson, K.L., Inactivation of single-celled *Ascaris suum* eggs by low-pressure UV radiation, *Appl. Environ. Microbiol.*, 72, 2178–2184, 2006.

57. Backer, H., Water disinfection strategies for *Cryptosporidium*, *Wilderness Environ. Med.*, 8, 75–77, 1997.

58. Barbee, S.L., Weber, D.J., Sobsey, M.D., and Rutala, W.A., Inactivation of *Cryptosporidium parvum* oocyst infectivity by disinfection and sterilization processes, *Gastrointest. Endosc.*, 49, 605–611, 1999.

59. Cox, M.M. and Battista, J.R., *Deinococcus radiodurans*—the consummate survivor, *Nat. Rev. Microbiol.*, 3, 882–892, 2005.

60. Young, D.C. and Sharp, D.C., Virion conformational forms and the complex inactivation kinetics of echovirus by chlorine in water, *Appl. Environ. Microbiol.*, 49, 359–364, 1985.

61. van Amsterdam, K., van Vliet, A.H., Kusters, J.G., and van der Ende, A., Of microbe and man: Determinants of *Helicobacter pylori*-related diseases, *FEMS Microbiol. Rev.*, 30, 131–156, 2006.
62. Fraud, S., Hann, A.C., Maillard, J.Y., and Russell, A.D., Effects of ortho-phthalaldehyde, glutaraldehyde and chlorhexidine diacetate on *Mycobacterium chelonae* and *Mycobacterium abscessus* strains with modified permeability, *J. Antimicrob. Chemother.*, 51, 575–584, 2003.
63. Stanley, P.M., Efficacy of peroxygen compounds against glutaraldehyde-resistant mycobacteria, *Am. J. Infect. Control*, 27, 339–343, 1999.
64. Manzoor, S.E., Lambert, P.A., Griffiths, P.A., Gill, M.J., and Fraise, A.P., Reduced glutaraldehyde susceptibility in *Mycobacterium chelonae* associated with altered cell wall polysaccharides, *J. Antimicrob. Chemother.*, 43, 759–765, 1999.
65. Yazdankhah, S.P., Scheic, A.A., Hoiby, E.A., Lunestad, B.T., Heir, E., Fotland, T.O., Naterstad, K., and Kruse, H., Triclosan and antimicrobial resistance in bacteria: An overview, *Microb. Drug Resist.*, 12, 83–90, 2006.
66. Russell, A.D., Whither triclosan?, *J. Antimicrob. Chemother.*, 53, 693–695, 2004.
67. McMurry, L.M., McDermott, P.F., and Levy, S.B., Genetic evidence that InhA of *Mycobacterium smegmatis* is a target for triclosan, *Antimicrob. Agents Chemother.*, 43, 711–713, 1999.
68. McDonnell, G. and Pretzer, D., Action and targets of triclosan, *ASM News*, 64, 670–671, 1998.
69. Bogdanova, E.S. and Mindlin, S.Z., Occurrence of two structural types of mercury reductases among gram-positive bacteria, *FEMS Microbiol. Lett.*, 62, 277–280, 1991.
70. Dorsey, C.W. and Actis, L.A., Analysis of pVU3695, a plasmid encoding glutathione-dependent formaldehyde dehydrogenase activity and formaldehyde resistance in the *Escherichia coli* VU3695 clinical strain, *Plasmid*, 51, 116–126, 2004.
71. Shields, M.S., Reagin, M.J., Gerger, R.R., Campbell, R., and Somerville, C., TOM, a new aromatic degradative plasmid from *Burkholderia* (*Pseudomonas*) *cepacia* G4, *Appl. Environ. Microbiol.*, 61, 1352–1356, 1995.
72. Sikka, R., Sabherwal, U., and Arora, D.R., Association of R plasmids of *Klebsiella pneumoniae* with resistance to heavy metals, klebocin and beta lactamase production and lactose fermentation, *Indian J. Pathol. Microbiol.*, 32, 16–21, 1989.
73. Russell, A.D. and Chopra, I., *Understanding Antibacterial Action and Resistance*, 2nd edition, Ellis Horwood, Chichester, United Kingdom, 1996.
74. Silver, S., Bacterial silver resistance: Molecular biology and uses and misuses of silver compounds, *FEMS Microbiol. Rev.*, 27, 341–353, 2003.
75. Sattar, S.A., Tetro, J.A., and Springthorpe, V.S., Effects of environmental chemicals and the host–pathogen relationship: Are there any negative consequences for human health?, *Biocides Old and New: Where Chemistry and Microbiology Meet*, Zhu, P.C. (ed.), ACS Symposium Series, Oxford University Press, New York, 2006.

7

A Pragmatic Approach to Judicious Selection and Proper Use of Disinfectant and Antiseptic Agents in Healthcare Settings

Vesna Šuljagić

CONTENTS

7.1 Introduction

In today's world we are becoming increasingly aware of the problem of nosocomial infections. At any time about 1.4 million people worldwide are suffering from some type (urinary tract infection, surgical site infections, lung infection, bloodstream infection, etc.) of nosocomial infection. About 5%–10% of hospitalized patients in developed countries and much more in developing countries (over 25% patients in some countries) acquire one or more infection during hospitalization [1]. Also, we are aware that approximately one-third of nosocomial infections are preventable and infection control can be very cost-effective. Because of this, infection prevention is important in healthcare settings. We have to protect the patients from infectious agents likely to be acquired during different medical procedures (surgical, invasive, etc.). Also, we have to protect medical staff from infectious agents, because they maintain high level of close interpersonal contact with patients and their blood, body fluids, tissues, mucous membranes, nonintact skin, as an essential part of delivery of care.

Presently, the cornerstone of an effective program of prevention and control of nosocomial infections is asepsis—a concept that includes cleaning, disinfection, sterilization, and aseptic techniques—hand hygiene, surveillance, epidemiologic methods, and patient isolation [2]. Spread of multiresistant pathogens, possible acquisition and transmission of H5N1 among humans, potential bioterrorist events and spread of bioterror agents, healthcare-associated transmission of Creutzfeldt–Jakob disease agent, all constitute challenges for today's infection control professionals and major tasks of most infection control program in healthcare institutions.

The primary goal of this chapter is to provide a framework for the selection and use of disinfectant and antiseptic agents in healthcare settings.

7.2 Cleaning

Cleaning and disinfection or sterilization are the basic activities in the decontamination process of medical equipment and hospital environments. There are some differences in the meaning of the word decontamination in the United States and Europe. In Europe, decontamination basically means cleaning. In the United States, this term describes all steps required to eliminate risks of infection to the medical staff, including cleaning. However, cleaning is always part of the decontamination process in both continents [3].

Cleaning is the removal of soil, organic matter, chemical residues, etc. Organic material left on medical equipment or in the environment may shield microorganisms and protect them from antimicrobial agents, and may also bind and inactivate the chemical activity of antimicrobial agents. Any contamination should be cleaned as soon as possible. If the soiled materials become dry, the cleaning process becomes more difficult and disinfection and sterilization less effective or ineffective. Cleaning is done manually or by automatic cleaners, with use of water, detergents, enzymatic cleaners, or elevated temperatures. There is a wide choice of cleaning agents available—alkalis, both mineral and organic acids and cationic, anionic, or nonionic surfactants [4]. The best choice is a neutral detergent solution with enzymes [5], but new hydrogen peroxide-based formulations are as effective as enzymatic detergents in removing protein, blood, carbohydrate, and toxin from surface test carriers and are also very effective in reducing microbial loads present on contaminated equipment [6]. Finally, rinsing, ideally with soft water, is necessary to remove cleaning agents and soil suspended in water. Studies in endoscopes have shown that cleaning is extremely effective in removing microorganisms with a reduction of >99.99% or 4-logs for viruses and [7,8] vegetative bacteria.

All items must be dried before the next step, disinfection or sterilization; when they are moist, they may become coated with biofilms, population of microbial cells, enveloped with extracellular polymeric matrices, which firmly bind to the colonized surface, and further entrap excreted products such as enzymes and virulent factors [9,10]. Biofilms may have serious implications for resistance of microorganisms to a wide variety of antimicrobial agents.

Staff responsible for decontamination must be properly trained and standard precautions with appropriate personnel protective equipment should always be used when dealing with contaminated items, during cleaning and all other decontamination procedures.

7.3 Disinfection

Disinfection is a process that reduces the number of pathogenic microorganisms on an inanimate object to a level which is not harmful to health. It does not mean that all microorganisms are killed, and in particular, bacterial endospores are frequently not killed. In healthcare settings, disinfection can be carried out by wet pasteurization or by using liquid chemicals. If an item is able to withstand the process of heat and moisture and is not required to be sterile, pasteurization is appropriate. Using heat and water at temperatures and times to destroy microbial agents is a very efficient method of disinfection. In most developed countries in Europe, pasteurization is preferred whenever possible. It is generally more reliable than chemical processes, leaves no residues, is nontoxic, shows lack of emergence of resistance, and the process is automated and validated, similar to the process of sterilization [3].

The effectiveness of the disinfection process depends on several factors [5,11,12]. We discussed earlier how important the proper cleaning of the object is. The type and level of microbial contamination is also important, because different microorganisms have different degrees of resistance to disinfectants. Vegetative bacteria and enveloped viruses are usually the most sensitive, and prions, bacterial spores, and protozoan cysts are the most resistant (Figure 7.1).

Each disinfectant has a minimum concentration for potency against a particular pathogen in suspension, so manufacturer recommendations have to be followed. The disinfectant should be applied for at least the time specified on the product label as this has been validated by the manufacturer (called exposure or contact time). Efficacy of disinfection generally increases with temperature, and therefore it is important to observe the minimum temperatures given on label instructions. Also, surface compatibility with certain types of disinfectants have to be checked. The disinfectant must be able to contact all areas of the object including lumens and cervices (scrubbing with brush and fluids under pressure).

Certain products and processes may provide different levels of disinfection. After 35 years, Spaulding's suggestion of three levels of disinfection is still "low level," "intermediate level," and "high level" [13,14]. Low-level disinfection inactivates vegetative bacteria, fungi, enveloped viruses (e.g., human immunodeficiency virus, influenza virus), and some non-enveloped viruses (e.g., adenoviruses). Quaternary ammonium compounds (QACs), some phenolics, and some iodophors are low-level disinfectants [15]. Intermediate-level disinfection kills *Mycobacterium tuberculosis* var. *bovis*, vegetative bacteria, most fungi, and medium to small viruses (with

Prions

Bacterial spores (*Bacillus subtilis*)

Protoza with cystis (*Giardia lamblia, Cryptosporidium parvum*)

Decreasing resistance Mycobacteria (*Mycobacterium tuberculosis*)

Nonlipid or small viruses (*Polio* virus)

Fungi (*Trichophyton* spp.)

Vegetative bacteria (*Pseudomonas aruginosa, Staphylococcus aureus*)

Lipid or medium-sized viruses (Hepatitis B virus, Herpes simplex virus, Human immunodeficiency virus)

FIGURE 7.1
Relative resistance of microorganisms to disinfection process.

or without lipid envelopes), but not necessarily bacterial spores. Alcohols, chlorine-containing compounds (e.g., sodium hypochlorite), some phenolics, and some iodophors are disinfectants with sufficient potency to achieve intermediate-level disinfection. High-level disinfection inactivates all microorganisms (vegetative bacteria, mycobacteria, viruses, fungi) with exception of high numbers of bacterial spores. Only powerful chemicals like glutaraldehyde, *O*-phthalaldehyde (OPA), hydrogen peroxide, peracetic acid, chlorine dioxide, and succinaldehyde are high-level disinfectants. At similar concentration, but with prolonged exposure times, these agents are sporicidal (kill spores) and are called chemical sterilants [11,15,16].

Sterilization is the closely monitored, validated process used to render a product free of all forms of viable microorganisms, including all bacterial endospores. It is an absolute state. Sterilization can be achieved by either high-temperature or low-temperature methods. High-temperature methods are steam under pressure, which is the most efficient and reliable method of sterilization, and dry-heat sterilization. There are few low-temperature methods and a lot of dilemma as to the method of choice. ETO gas, hydrogen peroxide, gas–plasma, low temperature, and liquid chemicals are the standard sterilizing agents used in healthcare facilities in the United States. In addition to these methods, in Europe's healthcare settings, low-temperature steam formaldehyde process is in use [17].

In 1972, Spaulding identified a classification scheme for use when assessing disinfection and sterilization procedures of certain items. This classification scheme is still applicable for use today. His plan divided equipment, instruments, and other device-related surfaces into three categories taking into account the degree of infection risk involved in the use of each one: noncritical items, semicritical items, and critical items [5,11,13,14]. Environmental surfaces in healthcare settings have also been implicated for infection risk and were added to the original Spaulding's classification by Catheter-related infections (CDC) in 1991 [18]. Very similarly, Ayliffe and colleagues categorized risk to patients and treatment of equipment and environments into four: minimal risk, low risk, intermediate risk, and high risk [19]. Anyway, professionals in infection control all over the world have to make right decisions when assessing disinfection and sterilization.

Environmental surfaces include medical equipment surfaces, such as knobs or handles of hemodialysis machines, x-ray machines, and housekeeping surfaces, such as floors, walls, and table tops [18]. They are not in close contact with patients and infection risk is minimal. Cleaning with warm water and detergent will achieve a safe level of decontamination of items listed in this category. Disinfection may be required in some special circumstances [11,15,19]. Although cleaning is preferred for spillage of blood, urine, or sputum, disinfection with sodium hypochlorite before cleaning is recommended. Also, surface disinfection is important during the outbreaks caused by organisms such as enterococci [20], *Acinetobacter baumannii* [21,22], and *Clostridium difficile* [23], which survive well in inanimate environments.

Noncritical items are those that come in contact with intact skin, which is an effective barrier to most microorganisms, or those items that do not make contact with the patient. So, there is a low risk of transmitting micro-organisms through these items. However, these items could potentially contribute to secondary transmission by contaminating hands of healthcare workers. Examples in this category are stethoscope, bedpans, blood pressure cuffs, ECG cables and electrodes, water glasses, etc. Detergent and water and low-level disinfectants may be used to process noncritical items [11,15,19].

Semicritical items do not penetrate the skin or enter sterile areas of the body but are in contact with intact mucous membranes and nonintact skin. Intact mucous membranes are generally resistant to infection by common bacterial spores, but susceptible to viruses and tubercule bacilli, and use of semicritical items is considered as intermediate infection risk. Instruments for gastrointestinal endoscopy, respiratory and anesthetic equipment, thermometers, vaginal instruments, and hydrotherapy units used on nonintact skin are included in this category, which usually requires cleaning followed by, at minimum, high-level liquid chemical disinfection or pasteurization [11,15,19].

Critical items penetrate sterile tissues, including sterile cavities and vascular system. This is high-infection risk contact. So these items must be sterile at the time of use. There are some exceptions. Heat-label endoscopes require high-level disinfection, usually by chemical agents. Examples of critical devices include surgical instruments, needles, syringes, implants, catheters (urinary, angiography), intravenous fluids, and all other equipment used in surgical operations or antiseptic techniques [11,15,19].

7.3.1 Commonly Used Disinfectants in Healthcare Settings

7.3.1.1 *Quaternary Ammonium Compounds*

Some QACs are widely used in infection control in hospitals. Older generation compounds like alkyl dimethyl benzyl ammonium chloride, alkyl didecyl dimethyl ammonium chloride, dialkyl dimethyl ammonium chloride and newer generation compounds like didecyl dimethyl ammonium bromide and dioctyl dimethyl ammonium bromide are the most popular in human medicine. Their biocidal actions appear to result from the disruption of the cell membrane, inactivation of enzymes, and denaturation of cell proteins [11]. QACs are cationic surfactants with weak detergent and strong activity, primarily reacting against gram-positive bacteria and lipophilic viruses. Also, they are not active against mycobacteria [24] and hydrophilic viruses [11,25]. Gram-negative bacteria have often been reported as contaminants of QACs, which were sources of infection in outbreaks [26,27]. Because of this CDC recommends to eliminate the use of these solutions as antiseptics [28]. Narrow microbicidal spectrum limits the use of QAC solutions as low-level disinfectants for cleaning walls, floors, and furnishings in healthcare settings. They bind to organic material including soaps, so the area to be disinfected must be cleaned and rinsed free of soap. Also, they show limited effectiveness in hard-water salts.

7.3.1.2 Phenolics

Lister was the first to use carbolic acid in antiseptic methods (for instruments, ligatures, surgeon's hands, etc.) [29]. Today, *ortho*-phenylphenol and *ortho*-benzyl-*para*-chlorophenol are derivates of phenol that are used as constituents of hospital disinfectants [5,11]. Different studies have shown different results about the antimicrobial efficacy of these disinfectants. Studies by manufacturers demonstrated that commercial phenolics were bactericidal, fungicidal, virucidal, and tuberculocidal, but were not sporicidal [5,11]. On the other side, some studies from independent laboratories demonstrated their antimicrobial ineffectiveness [30,31]. Anyway, they may be used for low- and intermediate-level of disinfection. They are commercially available with added detergents to provide one-step cleaning and disinfection.

Phenolics leave residual films on disinfected surfaces that can cause irritation of skin or tissues. This is one of the reasons why phenolics are not recommended for semicritical items and are recommended for hard surfaces and equipment that do not contact mucous membranes (e.g., i.v. poles, wheelchairs). Also, they are not recommended for use in nurseries because hyperbilirubinemia occurs in phenol-exposed infants [32].

7.3.1.3 Iodophors

Iodophors are combinations of iodine and a solubilizing agent or carrier which provides a sustained-release of iodine. Carriers usually are polyvinyl-pyrrolidone (povidone) or ethoxylated derivates (poloxamers) [3]. Different concentrations of iodophors are used for antisepsis and disinfection. However, antiseptic iodophors are not suitable for use on hard surfaces. For disinfection in healthcare settings, iodophors are generally used at concentrations that will provide 75 ppm of available iodine. Broad spectrum microbicidal activity depends on the concentration of free iodine. It is interesting that they are effective at low concentrations. This effect is called a paradoxical effect, as bactericidal activity increases if the dilution of povidone–iodine solution increases [33]. So, they must be used at a dilution stated by manufacturers. These compounds are bactericidal, virucidal, and fungicidal, but have limited activities against spores (require a prolonged contact time). They have shown to be most effective in acid solutions, and have limited activity in the presence of organic matter [11].

As disinfectants, iodophors have been used for the disinfection of medical equipment such as hydrotherapy tanks, thermometers, and blood culture bottles [5].

7.3.1.4 Chlorine and Chlorine Compounds

Chlorine and chlorine compounds are among the oldest disinfectants used in healthcare settings. The discovery of chlorine was reported in 1774 by the Swedish chemist Schelle. A few years later saw the use of calcium hypochlorite for general sanitation and for treatment of wounds with dressings containing a diluted aqueous solution of hypochlorite.

All chlorine-based disinfectants work by releasing chlorine ions. Effectiveness increases with the concentration of available chlorine. Liquid form of hypochlorite, such as sodium hypochlorite, or solid form of hypochlorite, such as calcium hypochlorite, are the most common bases for chlorine disinfectants [11]. Chlorine dioxide, chloramine-T, dichloroisocyanuric acid, and sodium and potassium dichloroisocyanurates are alternative compounds that retain chlorine longer and show prolonged microbicidal effect than hypochlorite [11,25].

Chlorine-based disinfectants have broad antimicrobial spectra that include *Mycobacterium tuberculosis* and bacterial spores at higher concentration. Also they are fast-acting and inexpensive, but their use in healthcare settings is limited because of their relative instability, corrosiveness to metals, inactivation by organic materials, and their ability to irritate mucous membranes, eyes, and skin [34]. It is important to know that activities of these disinfectants are greatly affected by pH and water hardness. At high pH, available chlorine or hypochlorous acid, the germicidal component of hypochlorites, is present in very low concentrations, resulting in lower levels of disinfection and in prolonged disinfection times [35].

Chlorine compounds can be used as low-level disinfectants for noncritical environmental surfaces. However, they can be used as intermediate-level and high-level disinfectants, only on selected items whose structural materials are not altered by chlorine, such as hydrotherapy tanks used for patients who have damaged skin, dialysis equipment, dental equipment, tonometers, and cardiopulmonary training manikins. Also, they are very useful for decontamination of blood spillage before cleaning [11].

7.3.1.5 Alcohols

Alcohols with antimicrobial properties have been important in infection control for centuries. Ethyl alcohol and isopropyl alcohol are very widely used for antisepsis and disinfection in healthcare settings today. They are effective against gram-positive and gram-negative bacteria, enveloped viruses, fungi, and mycobacteria, but not effective against bacterial spores and non-enveloped viruses [25]. The most effective concentration is 60%–90% solution in water. They evaporate fast, and therefore have limited exposure time. Also, alcohols are flammable and consequently must be kept away from heat sources, electrical equipment, flames, and hot surfaces.

Alcohols are intermediate- and low-level disinfectants. They are used for disinfection of external surfaces of some equipment (e.g., stethoscopes, thermometers, and ventilators), or small surfaces such as rubber stoppers of multiple-dose medication vials, or small and clean environments such as medication preparation areas, tabletops, etc. [5]. However, they are too expensive for general use as a surface disinfectant.

7.3.1.6 Peracetic Acid

Peracetic acid is a mixture of acetic and hydrogen peroxide in a water solution. Also, it can be produced by oxidation of acetaldehyde. It oxidizes the outer cell membranes of microorganisms, including bacteria and their spores, mycobacterium, fungi, algae, and viruses, causing the microorganism to be deactivated rapidly. It is active in the presence of organic materials, with rapid action at low temperature. Products of degradation of peracetic acid, water and oxygen, are nontoxic and can easily dissolve in water. It is corrosive to copper, brass, bronze, galvanized iron, and plain steel, but its effect can be reduced using additives and pH modification [3].

This acid is a high-level disinfectant or sterilant for heat-sensitive equipment. Higher concentrations of peracetic acid may be used as a chemosterilant in specially designed machines for decontamination of heat-sensitive medical devices (especially all kinds of endoscopes) [5,11].

7.3.1.7 Hydrogen Peroxide

Hydrogen peroxide was recognized as a very potent antimicrobial agent early, about 1818 [25]. Today, we know of their disinfectant, antiseptic, and deodorant properties; it is one of the most commonly available oxidizing agents. Solutions of hydrogen peroxide are unstable, and different stabilizers (benzoic acid) are added to stabilize solutions. Also, it is blended with peracetic acid in high concentration to achieve sporicidal effects [35]. Hydrogen peroxide can be corrosive to zinc, brass, copper, and aluminum.

Stabilized hydrogen peroxide is effective against a broad range of pathogens, including both enveloped and non-enveloped viruses, vegetative bacteria, fungi, mycobacteria, and bacterial spores [5,25]. This strong oxidant is a generator of free hydroxyl radicals, which attack essential cell components [36]. Hydrogen peroxide breaks down into water and oxygen, which are environment friendly decomposition products.

A 3% solution of hydrogen peroxide is a low-level disinfectant and is used on inanimate surfaces, whereas 6% solutions are effective for high-level disinfection (soft contact lenses, ventilators, endoscopes, and tonometer biprisms). Higher concentrations are used as chemosterilants in specially designed machines for decontamination of heat-sensitive medical devices, very similar to peracetic acid.

During the severe acute respiratory syndrome (SARS) outbreak in Canada in 2003, hydrogen peroxide-based disinfectants were recommended for use in decontamination of equipment and vehicles [37].

Vaporized hydrogen peroxide generally shows good penetration and is also used for low-temperature sterilization.

7.3.1.8 Aldehydes

Few aldehydes are currently of considerable importance as disinfectants: formaldehyde, glutaraldehyde, OPA, and succinaldehyde.

Formaldehyde, one of the oldest aldehydes used in healthcare settings, in aqueous solution or as a gas, is used as a disinfectant and sterilant. These substances are hazardous to health of employers. They are highly toxic, carcinogenic, strongly irritating, and have pungent odor. Because of this, its use in healthcare settings was limited in the past. U.S. Federal Government agencies, such as OSHA, and the Health and Safety Executive of the United Kingdom started legislation to ban formaldehyde for disinfection and sterilization process [3].

It has broad spectrum of microbicidal activity. The aqueous solutions are bactericidal, fungicidal, virucidal, and sporicidal. Formaldehyde inactivates microorganisms by alkylating the amino and sulfhydryl groups of proteins and nitrogen-ring atoms of purine bases [18].

In the past, formaldehyde was mixed with ethanol and used for sterilizing surgical instruments. Today, it is used to prepare viral vaccines, to preserve anatomic specimens, as an embalming agent, and for reprocessing hemodialyzers. Also, as vaporized paraformaldehyde, it is used in biomedical laboratories for the decontamination of biological safety cabinets [5,11].

Glutaraldehyde has been the high-level disinfectant of choice for more than 30 years. It is commonly used as 2% aqueous solution. This solution is acidic and must be activated by an alkalizing agent before use. The activated solution loses activity by polymerization of glutaraldehyde molecules at alkaline pH within 14–28 days, depending on formulation. This time is the shelf life of the disinfectant.

Glutaraldehydes inactivate microorganisms by alkylating sulfhydryl, hydroxyl, carboxyl, and amino groups of a wide spectra of microorganisms, including bacteria, mycobacteria, viruses, fungi, and spores [38]. They have moderate residual activity and are effective in the presence of limited amounts of organic material. So, monitoring of residual concentration in reusable solutions is recommended. Compatibility with various materials is excellent. However, they coagulate blood and fix tissue. Also, they are extremely irritating to skin and mucous membranes, and this chemical must be used under very strict controlled conditions and in a safe working environment.

Glutaraldehyde-based solutions are used most commonly as a high-level disinfectant for endoscopes, respiratory therapy equipment, anesthesia equipment, transducers, etc. Sterilization by such solutions may be accomplished in 6–10 h [11].

Today, a lot of hospitals in developed countries try to eliminate or reduce dependence on glutaraldehyde by investing in new enclosed equipment technologies or by using different disinfectants instead of glutaraldehyde, such as OPA, peracetic acid, or hydrogen peroxide.

OPA, the youngest generation of aldehydes, received clearance by the Food and Drug Administration (FDA) in 1999. OPA solution contains 0.55% of OPA (1,2-benzenedicarboxaldehyde) and has very similar mode of action, antimicrobial activity, and material compatibility to glutaraldehyde [5,39]. But OPA has several potential advantages compared with glutaraldehyde. It requires no activation, because of superior stability over a wide pH range (pH 3–9), and requires no exposure monitoring because it is not known to be an irritant to skin and mucous membrane. However, it stains proteins gray (including unprotected skin). Also, it is observed that repeated exposure to OPA, during the reprocessing of urologic instruments, may have anaphylaxis-like reactions in some patients with history of bladder cancer undergoing repeated cystoscopy [40].

In different countries, different test methodologies are required for licensure of the disinfectant. So, the high-level label claims for OPA solution at 20°C vary worldwide, e.g., 5 min in Europe, Asia, and Latin America; 10 min in Canada and Australia; and 12 min in the United States [39].

7.3.2 Regulatory Framework for Disinfectants

In the United States, liquid chemical disinfectants are regulated by the U.S. Environmental Protection Agency (EPA) and U.S. FDA. EPA regulates disinfectants used in environmental surfaces in healthcare settings. EPA requires manufacturers to test formulations by using accepted methods for microbicidal activity, stability, and toxicity to animals and humans. To be labeled as an EPA hospital disinfectant, the product must pass the Association of Official Analytical Chemists' (AOAC) effectiveness test against three target organisms: *Salmonella choleraesuis* for effectiveness against gram-negative bacteria, *Staphylococcus aureus* for effectiveness against gram-positive bacteria, and *Pseudomonas aeruginosa* for effectiveness against a primarily nosocomial pathogen. Manufacturers might also test specifically against organisms of known concern in healthcare practice (e.g., HIV, HBV, hepatitis C virus, and herpes). However, potency against *Mycobacterium tuberculosis* is used only as a bench mark to measure germicidal potency. Because mycobacteria have the highest intrinsic levels of resistance among the vegetative bacteria, viruses, and fungi, any germicide with a tuberculocidal claim on the label is considered capable of inactivating a broad spectrum of pathogens, including less-resistant organisms as blood-borne pathogens (e.g., HBV, HCV, and HIV). So, EPA registers environmental surface disinfectants as hospital disinfectants with or without tuberculocidal claim. In accordance with a practical approach and Spaulding classification, CDC recommends any EPA-registered hospital disinfectant without tuberculocidal claim as a low-level disinfectant and any EPA-registered hospital disinfectant with tuberculocidal claim as an intermediate-level disinfectant [41].

FDA regulates liquid chemical sterilants/high-level disinfectants (e.g., peracetic acid, hydrogen peroxide, glutaraldehyde, OPA) used on critical and semicritical items [42].

In Europe, disinfectants are regulated by the European Committee for Standardization. Disinfectants are registered based on agreed standard suspension, and by surface and practical tests of bactericidal, viricidal, fungicidal, mycobactericidal, and sporicidal activities. These tests are defined by CEN/TC 216—Chemical Disinfectant and Antiseptic standards. In phase 1 tests, basic efficacy of a product is demonstrated in suspension assays using bacteria [43] and fungi [44] as test organisms. In phase 2/step 1, specific suspension assays are carried out to examine the antimicrobial impact under representative conditions using interfering substances [45,46]. Phase 2/step 2 tests comprise assays mimicking real conditions. Phase 3 tests are field trials. Within this test hierarchy, the phase 2/step 2 tests are of greatest importance. If some products do not meet the requirements after phase 1 tests or phase 2/step 1 tests, they are still acceptable if they pass the phase 2/step 2 tests.

7.4 Antiseptic Agents

In contrast to disinfectants that are used on inanimate objects, antiseptic agents are compounds applied to living tissue to reduce the number of microbial flora, without damaging these tissues. They not only remove or kill microorganisms, but may also prevent the growth and development of some types of microorganisms.

Normal human skin is colonized with microorganisms, which Price, an American surgeon, divided into two types: resident and transient flora [47]. Resident flora live and grow on normal skin. Predominant flora is normally harmless and consists of coagulase-negative staphylococci, mainly *Staphylococcus epidermidis*, micrococci, and a smaller number of diphtheroids. In some areas of the body, such as moist areas (e.g., axillae), gram-negative bacteria are more common (e.g., *Acinetobacter*, *Klebsiella*) [48]. However, composition of skin flora depends on location, sex, age, health condition, hospitalization, season of the year, and frequency of hand washing [49]. Resident flora are protective to skin because they prevent invasion by other harmful species. They are difficult to remove, because they are attached to deeper layers of the skin. These microorganisms can cause infections during surgery and other invasive procedures. Also, skin bacteria can create problems when transferred from healthcare workers to immunocompromised patients [50].

Transient flora does not normally colonize skin. They are often acquired accidentally, from the environment and colonize superficial layers of the skin. So, transient flora consists of microorganisms with high pathogenic potential that are most frequently associated with healthcare-associated infections. These microorganisms are easily removed by mechanical means such as handwashing [50].

Infectious flora is also very important. In this group, the etiologic agents of infections such as abscesses, panaritium, paronychia, or infected eczema on the hand are bacteria like *Staphylococcus aureus* and β-hemolytic strepto-cocci. They are of proven pathogenicity [49].

The purpose of reduction of total microorganism counts of both resident and transient flora is threefold [51]:

1. To protect vulnerable sites (e.g., surgical sites) from microorganisms, which can be transferred from the hands of healthcare workers

2. To protect patients' tissues against endogenous flora during inva-sive procedures (e.g., surgery, placement of vascular or urinary catheters, etc.)

3. Occasionally, to treat carriers and dispersers of multiresistant strains of microorganisms (MRSA [methicillin-resistant *Staphylo-coccus aureus*], VRE, and multidrug-resistant *Acinetobacter*)

7.4.1 Commonly Used Antiseptic Agents in Healthcare Settings

In the past, ordinary foodstuffs such as wine, vinegar, and honey were used as dressing and cleansing agents for wounds. But, in 1750, John Pringle first used the term "antiseptic," derived from Greek, meaning "against putre-faction." He described substances that prevent putrefaction as antiseptic. But antisepsis was not generally accepted until Pasteur's publication in 1863 of the microbial origin of putrefaction [52].

Today we have a more scientific approach to this topic. Activity of an antiseptic agent varies depending on the formulation in which it is con-tained and the nature of contamination that is being used against. We choose appropriate agents for appropriate situations. The most commonly used antiseptic agents are alcohols, chlorhexidine, chlorine, hexachloro-phene, iodine and iodophors, QACs, triclosan, and octenidine dihy-drochloride [48,49].

7.4.1.1 Alcohols

As mentioned before alcohols are used in the disinfection of inanimate items in healthcare settings often. Also, they are among the most important and safest known antiseptic agents in healthcare settings. Alcohols provide the most rapid and greatest reduction in microbial counts on skin [53,54]. When applied to the skin they are rapidly germicidal, but without residual activ-ity. Alcohols kill by dehydration, and they must be allowed to evaporate to have any effect.

There are three main alcohols used in antiseptic solutions: isopropyl alcohol, ethyl alcohol, and *n*-propanol. They are effective against gram-positive and gram-negative bacteria, enveloped viruses, fungi, and mycobacteria and are not effective against bacterial spores and non-enveloped viruses [25]. Bactericidal activity decreases in the following order: *n*-propanol > isopropyl alcohol > ethanol [49]. *n*-Propanol has been used in alcohol-based hand rubs in Europe for years, but it is not registered as an antiseptic agent for hand hygiene in the United States [48].

Some alcohol-based hand antiseptic preparations contain some of these products, or a combination of two of them, or alcohol solutions containing limited amounts of chlorhexidine gluconate, povidone–iodine, QACs, triclosan, hexachlorophene, octenidine, or hydrogen peroxide [55–59]. Since alcohol alone has no lasting effect, another compound with antiseptic activity may be added to the disinfection solution to prolong the effect. Most alcohol-based hand rubs contain emollients, such as glycerol, volatile silicone oils, refattening agents, and probably most importantly, rehydrating agents, to reduce the drying effect on the skin. They are available as low-viscosity rinses, gels, and foams for use in healthcare settings [49]. Alcohol does not require water and is easy and convenient to use, which makes it ideal for use in hospital wards.

7.4.1.2 *Iodine and Iodophors*

The first specific reference to the use of iodine for the treatment of infected wounds was in 1839. Iodine was officially recognized by the pharmacopoeia of the United States in 1830. In studies till 1874, iodine was found to be one of the most efficacious antiseptics, a notion still valid today [60].

Elemental iodine is normally used in an aqueous or alcoholic solution, but it often causes irritation and stains skin. Because of better acceptability, it has been replaced by iodophors, iodine complex with carriers, polyvinylpyrrolidone (i.e., povidone) and ethoxylated nonionic detergents (i.e., poloxamers). The amount of molecular, free iodine present determines the level of antimicrobial activity of iodophors. Typical 10% povidone–iodine formulations contain 1% available iodine and yield free iodine concentrations of 1 ppm [61]. The concentration of iodine in antiseptic agents is much lower than in disinfectants.

Iodine and iodophors have a wide range of activity against gram-positive and gram-negative bacteria, tubercle bacillus, fungi, and viruses. Their activity is based on the oxidizing potential of iodine. Some new studies have shown that povidone–iodine is effective against MRSA, *Chlamydia*, *Herpes simplex*, adenoviruses, and enteroviruses [62].

Today, iodophors are widely used for antisepsis of skin, mucous membranes, and wounds. Iodine-based creams may also be used to treat nasal colonization with MRSA [63]. As a result of poor manufacturing processes, iodophor antiseptics have become contaminated and have caused outbreaks or pseudo-outbreaks of infections [64,65].

7.4.1.3   Chlorhexidine

One of the best and most widely used cationic antiseptics is chlorhexidine. It is used in the form of acetate, gluconate, or hydrochloride, either alone or in combination with other antiseptic agents such as cetrimide, alcohol, etc. Chlorhexidine gluconate was introduced as an antiseptic in the United States during 1970s, several decades after approval for use in Europe and Canada [51]. Preparations of chlorhexidine are available as aqueous and alcoholic solutions, detergents, or powders [49].

Chlorhexidine attains its level of antimicrobial action by causing destruction of microbial cytoplasmic membranes, resulting in precipitation and coagulation of cell contents. It has a wide spectrum of antibacterial activity against both gram-positive and gram-negative bacteria [25]. However, some gram-negative bacilli, such as *Pseudomonas aeruginosa*, *Pseudomonas cepacia*, *Proteus mirabilis*, and *Serratia marcescens*, may be highly resistant to chlorhexidine [66]. Similar to the contamination of the preparation of iodophors, contamination of chlorhexidine solutions may cause outbreaks of nosocomial infections [67]. Chlorhexidine has little activity against acid-fast organisms, and bacterial and fungal spores. Also, its activity in preventing transmission of viruses to healthcare workers is unknown [51]. Chlorhexidine gluconate has greater residual activity than other available antiseptics.

Long-term experience with chlorhexidine use has shown that incidence of skin irritation is minimal. Preparations with >1% chlorhexidine can cause damage to conjunctiva and cornea. Also, ototoxicity can occur if chlorhexidine is instilled directly into the middle ear. Direct contact of chlorhexidine with brain tissue and the meninges should be avoided [48].

Chlorhexidine is widely employed in healthcare settings for hand hygiene [48], preoperative antiseptic showering, patient skin preparation in the operating room [68], and catheterization procedures [69]. It is also widely used in the field of dentistry [25] for treating sore gums and mouth ulcers, preventing plaque formation on teeth, and as a potential endodontic irrigant in combination with sodium hypochlorite [70].

7.4.1.4   Hexachlorophene

Hexachlorophene is a chlorinated bisphenol. It is almost insoluble in water, but soluble in ethanol, ether, acetone, and alkaline solutions. It is much more active against gram-positive bacteria than gram-negative bacteria, fungi, and mycobacteria. Hexachlorophene acts slowly, but binds strongly to the skin [51]. Absorption through the skin can cause damage to the central nervous system, particularly in infants [71] and patients with burn injuries [51]. After overdose incidents in infants in France in the 1970s, in many countries use of hexachlorophene was restricted. In the United States, hexachlorophene is now available only by prescription in 3% formulation [51]. In the United Kingdom, hexachlorophene is used in cosmetic products and the maximum concentration allowed is 0.1% with the stipulation that it is not to be used in products for children or personal hygiene products [25]. It has been largely replaced by the related compound triclosan.

7.4.1.5 Triclosan

Triclosan is a chlorinated bisphenol, which dissolves well in some detergents such as anionic soaps, and alcohol. It is only sparingly soluble in water. Concentration of 0.2%–2% are effective against gram-positive bacteria and gram-negative bacteria but with variable and poor activity against *Pseudomonas aeruginosa*. This agent is also active against mycobacteria and *Candida* spp., but it has limited activity against filamentous fungi such as *Aspergillus* spp. [49,72]. Because of acceptable antimicrobial activity, use of triclosan has risen dramatically in the past few years. Today it is used as an additive to hand soaps, dish-washing products, cosmetics, and toothpaste. It is also used as an additive to plastics, polymers, and textile and implant devices to confer these materials with antibacterial properties.

Because of its fat solubility, triclosan penetrates the bacterial cell wall and attacks enzymes that are used to produce fatty acids vital to cell function [73,74]. This mode of action could ultimately lead to development of antibiotic resistance in bacteria. So, scientists are concerned that overuse and misuse of triclosan may lead to increased resistance of bacteria.

7.4.1.6 Quaternary Ammonium Compounds

Of this large group of compounds, alkyl benzalkonium chlorides are the most widely used antiseptics. Other compounds that have been used as antiseptics include benzethonium chloride, cetrimide, and cetylpyridium chloride. As discussed above, QACs are primarily bacteriostatic and fungistatic, although they are microbicidal against certain organisms at high concentrations; they are more active against gram-positive than against gram-negative bacilli. QACs have relatively weak activity against mycobacteria and fungi and have greater activity against lipophilic viruses. Initially, they were used as an adjunct in surgery, such as preoperative patient skin treatment, cleaning of surgeon's hands, and disinfection of surgical instruments. During 1980–2000, in the United States, they are seldom used alone for skin and hand disinfection, but rather in combination with alcohols to produce sustained residual activity [49]. Weak activity against gram-negative bacilli influenced the occurrence of outbreaks of infection or pseudoinfection with *Serratia marcescens* [75] and *Pseudomonas* spp. [76], traced to contaminated quaternary ammonium preparations.

7.4.2 Regulatory Framework for Antiseptic Agents

In different regions of the world investigators use different methods to regulate the use of antiseptic agents. In the United States, antiseptic agents are regulated by the FDA. Requirements for in vitro and in vivo testing of products are outlined in the 1994 FDA Tentative Final Monograph (TFM) for Antiseptic Drug Products. According to the TFM, healthcare antiseptic drug products are divided into three categories: patient preoperative skin preparations, preparations for antiseptic hand wash or healthcare workers hand wash, and preparation for surgical hand scrubs. The FDA TFM classified 60%–95% ethanol and 5%–10% povidone–iodine as a category I agent (i.e., generally safe and effective for use in antiseptic hand wash or healthcare workers' hand-wash products). Although TFM placed 70%–91.3% isopropanol, benzalkonium chloride, benzethonium chloride, and ≤1.0% triclosan as category III agents (i.e., insufficient data exist to classify them as safe and effective for use as antiseptic hand wash). Hexachlorophene is classified as not safe and effective for use as an antiseptic hand wash [77].

In Europe, disinfectants and antiseptics are regulated by the European Committee for Standardization. They are registered based on agreed standard suspension, surface and practical tests of bactericidal, virucidal, fungicidal, mycobactericidal, and sporicidal activities as defined by CEN/TC 216—Chemical Disinfectant and Antiseptic standards. It is important to outline that the final evaluation of the benefits of an antiseptic treatment for specific indications requires clinical studies (phase 3 tests).

So, strategies to prevent infections in healthcare settings by using antiseptic agents are hand hygiene, antiseptic showers, skin preparation of surgical sites, and preparation of skin or mucosa during placement of vascular and urinary catheters.

7.4.3   Hand Hygiene

Hands of medical personnel are the most common source of cross infection. So, hand hygiene is the first principle of infection control. It is the simplest, most effective measure for preventing healthcare-associated infections and spread of antimicrobial resistance [78,79]. However, healthcare workers' adherence to good practice is unacceptably low. Average compliance with hand hygiene recommendations varies between hospital wards and among professional categories of healthcare workers, and is usually estimated as <50% [80].

The term hand hygiene includes several actions intended to decrease colonization with microorganisms: handwashing, antiseptic hand wash, antiseptic hand rubs, and surgical hand antisepsis [48]. The preferred method for hand hygiene depends on the type of procedure, the degree of contamination, and the desired persistence of antimicrobial action on the skin.

Handwashing refers to washing hands with non-antimicrobial soap and water [48]. Handwashing suspends microorganisms and allows them to be rinsed off. This action is often referred to as mechanical removal of microorganisms and dirt when hands are visibly soiled. Although adequate in many instances, its efficacy is not sufficient in some, especially when dealing with certain antibiotic-resistant pathogens or in high-risk areas. Antiseptic hand wash refers to the same procedure, when an antiseptic agent is added to soap (antimicrobial soap). This action is often referred to as chemical removal of microorganisms, because antiseptic agents kill or inhibit the growth of microorganisms. However, frequent handwashing is potentially damaging to skin, time consuming, and expensive [81]. So, antiseptic hand rub is considered to be the most effective way of breaking the chain of transmission of microorganisms [48]. That is rubbing of all surfaces of the hands with small amount (3–5 ml) of, usually, an alcohol-containing preparation until dry or for a preset duration recommended by the manufacturer. The purpose of this action is to eliminate, as quickly as possible, microorganisms that survive on hands after contact with a contaminated object or colonized/infected patient. This method ensures that the concerned persons do not develop an infection and that contamination will not be transmitted from their hands to other patients. Surgical hand antisepsis refers to antiseptic hand wash or antiseptic hand rubs performed preoperatively by members of surgical team who have direct contact with the sterile operating field, or sterile instruments, or supplies used in the field. The purpose of these actions is to prevent the transfer of microorganisms from members of the surgical team to the operation site, and to ensure that even if the surgical gloves become perforated during the operation, only few microorganisms from the hands will be implanted in to the surgical site. This means that transient flora as well as resident flora should be reduced, and that this reduction should last during the operation, for several hours. Selecting the most appropriate antiseptic agents for surgical hand antisepsis requires consideration of certain characteristics, especially efficacy and acceptability by the surgical team after repeated use [48,68]. Alcohol-containing products are the gold standard in several European countries [54,82,83]. On the other side povidone–iodine and chlorhexidine gluconate are agents of choice in the United States [84]. However, studies have demonstrated that products of alcohol (60%–70%) combined with chlorhexidine gluconate (0.5%) compared with products of povidone–iodine (7.5%) or chlorhexidine gluconate (4%) alone have greater residual antimicrobial activity [85,86]. This topic is not clear yet, because most studies evaluated the reduction of bacterial counts on hands of surgical personnel. Hence, there is an urgent need for randomized controlled trials that evaluate the impact of antiseptic agents for surgical hand antisepsis on surgical site infection rate [68].

7.4.4 Antiseptic Shower

Different categories of patients showering with antiseptic agents have been tested for studying the effect on nosocomial infection rates. In some studies with surgical patients, antiseptic preoperative bath or showers with different antiseptic agents (chlorhexidine gluconate and povidone–iodine) have not been associated with reduced postoperative infections rates, but in others significant differences were observed [87–89]. Anyway, the CDC strongly recommends (category IB) that healthcare facilities "require patients to shower or bathe with an antiseptic agent on at least the night before the operation" [68]. Other studies with surgical patients have showed that whole-body washing with triclosan-containing products substantially reduced rates of acquisition of methicillin-resistant *Staphylococcus aureus* and helped to successfully control the outbreak [90,91]. Also, whole-body washing of neonates with chlorhexidine-containing formulations has been associated with reduction in the rates of infections [92].

7.4.5   Skin Preparation of Surgical Sites

Like surgeons' hands, the surgical sites of patients require skin preparation. This action is directed against resident as well as transient flora. Sometimes, for example in emergency surgery, it requires maximal effect in a single treatment without the progressive effect of repeated applications. FDA defines a patient preoperative skin preparation as a "fast-acting, broad-spectrum, and persistent antiseptic-containing preparation that significantly reduces the number of microorganisms on intact skin" [77]. Several antiseptic agents are used for skin preparation of surgical sites in the operating room. The iodophors (i.e., povidone–iodine), alcohol-containing products (FDA-required concentrations for ethyl alcohol are 60%–95%, by volume in aqueous solution, or 50%–91.3% for isopropyl alcohol, by volume in an aqueous solution), and chlorhexidine gluconate are the most commonly used agents. Data on their use are insufficient, because no randomized controlled trials which evaluate the impact of antiseptic agents for skin preparation of surgical sites for surgical site infection rates have been conducted yet [68].

Spores are usually not of concern, but the skin of buttocks and upper leg often has transient contamination with spores, especially that of *Clostridium perfringens* from feces, and operations with poor arterial supply (e.g., amputation of a leg for diabetic gangrene) carry special risk of endogenous gas gangrene. Bacterial spores are resistant to alcohol, chlorhexidine, and QACs, and only halogens, like povidone–iodine, have some activity, but require prolonged contact time [93]. This should be considered before high-risk operations. FDA categorized only iodine products with alcohol as category 1 agents (generally recognized as safe and effective) [77].

7.4.6   Preparation of Skin or Mucosa during Placement of Invasive Devices

Urinary tract infections (UTIs) are the most common nosocomial infections. They account for about 40% of all hospital-acquired infections and constitute a major source of nosocomial septicemia and related mortality [94,95]. Nearly all such infections are associated with urinary tract instrumentation, such as temporary indwelling bladder catheterization more frequently, and urologic procedures less frequently. According to a recent National

Nosocomial Infections Surveillance (NNIS) system report, nosocomial UTI rates ranged from 3.0 to 6.7 per 1000 urinary catheter-days in intensive care patients [96]. The most important efforts in prevention or at least delay of these infections are elimination of unnecessary urethra catheterization, reduction of duration of catheterization, and promotion of aseptic care of closed drainage systems [97]. One of the universally recommended practices for aseptic care of closed drainage systems is insertion of the urinary catheter using an aseptic technique. It should be inserted by trained healthcare professionals, using aseptic technique (sterile gloves, sterile drape). Meatus urethrae should be cleaned with an effective antiseptic, such as 10% povidone–iodine or 1%–2% aqueous chlorhexidine. Also, the catheter and drainage systems should be manipulated as little as possible [98]. Studies have shown that regular daily meatal cleaning with iodophore and daily application of iodophore antiseptic ointment to meatus were not associated with decreased incidence of catheter-associated UTIs [99,100]. Only one small study, conducted about 20 years ago among patients after single or intermittent urethral catheterization, found that povidone–iodine bladder irrigation after each episode of catheterization decreased the incidence of bacteriuria [101]. The effect of antiseptic agents in patients with indwelling catheters is not clear, and these findings need confirmation in other studies, by other investigators.

Thus, the most important role of antiseptic agents in prevention of UTIs is in preparation of meatus urethra during placement of a urinary catheter.

Intravascular catheters are essential to the practice of modern medicine, but they are the major cause of morbidity and mortality. Catheter-related bacteremia is the most frequent serious complication related to the use of central venous catheters (CVCs) [102]. The serious complications in patients and the difficult and expensive management of vascular CDC [103] have prompted a major interest in preventing these infections [104]. CDC can be reduced by scrupulous adherence to evidence-based techniques for catheter insertion and maintenance. According to the CDC Guidelines for the Prevention of Intravascular Catheter-Related Infections, released on August 9, 2002, the most important efforts in prevention of these infections are healthcare education and training of healthcare providers, surveillance, hand hygiene, aseptic technique during catheter insertion and care, catheter site care, catheter site dressing regimens, selection and replacement of intravascular catheters, replacement of administration sets, needleless systems and parenteral fluids, cleaning of IV-injection ports, etc. [105]. Some of them may be achieved only by using appropriate antiseptic agents. Antiseptic agents for hand hygiene were discussed above. A lot of studies were conducted to compare efficacy of different antiseptic agents for preparing the site before catheter insertion and with each dressing change. Many of them showed that chlorhexidine-based solutions were associated with lower colonization of vascular catheters and lower bloodstream infection rates than were povidone–iodine or alcohol [106,107]. Thus, the Guidelines for the Prevention of Intravascular Catheter-Related Infections (CDC) for cutaneous antisepsis state, as a category IA recommendation: "Disinfect clean skin with an appropriate antiseptic before catheter insertion and during dressing changes. Although a 2% chlorhexidine-based preparation is preferred, tincture of iodine, an iodophore, or 70% alcohol can be used." A 1% tincture of chlorhexidine preparation is available in Canada and Australia, but not yet in the United States [105].

7.5   Special Considerations in Twenty-First Century

At the onset of the twenty-first century, SARS and avian flu, two infectious diseases, have significantly aroused the attention of both healthcare workers and the public likewise.

7.5.1   Severe Acute Respiratory Syndrome

SARS is a newly recognized acute viral respiratory syndrome caused by a novel SARS coronavirus (SARS-CoV) [108]. The first case was retrospectively recognized as having occurred in Guandong province, China, in November 2002. By July 2003, the international spread of SARS-CoV resulted in 8098 SARS cases in 26 countries, with 774 deaths [109]. The disturbing characteristic of this illness was the apparent ease of transmission in healthcare facilities. The novel coronavirus can be found in sputum, tears, blood, urine, and feces of an infected person. It can be shed in the feces for 30 days and has been shown to survive on hard surfaces for more than 24 h. This raises the possibility of transmission by fomites and the amplification of its communicability within hospital settings, becoming an almost "perfect pathogen" [110]. The primary mechanism by which SARS-CoV appears to spread is through droplets and through contact either directly (person-to-person) or via contaminated environmental surfaces or equipment [111,112]. Although, not proven, it is possible that SARS-CoV can also be spread by the airborne route, particularly in healthcare settings during aerosol-generating medical procedures [113]. It was calculated, by mathematical modeling of data from studies conducted in Hong Kong and Singapore, that without control measures, each SARS case will lead to 2.2–3.7 secondary cases in susceptible populations [114,115]. Thus, prevention of transmission of SARS-CoV within healthcare settings will play a crucial role in prevention and control of future SARS outbreaks. The first line of defense for prevention and control of SARS is to practice standard precautions, and all additional precautions like preventing droplet, contact, and airborne transmissions. A combination of these precautions will give an appropriate infection control [116].

Role of disinfectants and antiseptic agents in circumstances of occurrence of SARS in healthcare settings is great. Studies have shown that SARS-CoV loses infectivity after exposure to commonly used disinfectants, which are also effective against other enveloped viruses [117].

Hand hygiene, by hand washing or use of an alcohol-based hand rub, as part of standard precautions is critical to prevent possible transmission of viruses. It should be performed before and after each contact with patients or their care environment, and after removing gloves [118]. For cleaning and disinfection, and sterilization of reusable equipment used for care of patients infected with SARS, CDC recommends EPA-registered disinfectants that provide low- or intermediate-level disinfection, because these products are known to inactivate related viruses with physical and biochemical properties similar to suspected SARS agents. Disinfectants such as fresh bleach solutions should be widely available at appropriate concentrations [118].

Cleaning and disinfection of environmental surfaces are important components of preventing nosocomial infections in healthcare settings. Although little is known about the extent of environmental contamination in SARS patients' rooms, epidemiologic and laboratory evidence suggests that the environment could play a role in transmission [113]. Therefore, cleaning and disinfection are critical to the control of SARS-CoV transmission too. Environmental cleaning and disinfection for SARS-CoV follow the same principles generally used in healthcare settings. CDC recommends using EPA-registered hospital detergents or disinfectants and following manufacturer's recommendations for use-dilution (concentration), contact time, and care in handling. Special attention should be given to frequently touched surfaces and horizontal surfaces around the patient after aerosol-generating procedures [116].

Currently, there is no known SARS transmission anywhere in the world. The most recent human cases of SARS-CoV infection were reported in China in April 2004 in an outbreak resulting from laboratory-acquired infections [119].

7.5.2 Avian Influenza

During 2003, the public has become keenly aware of the global spread of SARS. At the same time, in southeast Asia, highly pathogenic strains of avian influenza virus including the H5N1 again crossed from birds to humans and caused fatal disease. Since 2003, avian H5N1 influenza virus has infected 217 persons and killed 123, in 10 countries all over the world [120]. Normally, avian influenza viruses do not infect humans because of host barrier and direct avian-to-human influenza transmission was unknown before 1997. Direct transmission of H5N1 viruses of purely avian origin from birds to humans was first described during an outbreak among poultry in Hong Kong in 1997. In this outbreak, 6 of 18 confirmed human H5N1 patients died [121]. And serologic evidence was found for asymptomatic infection in humans after exposure to infected poultry [122]. To date, the risk of human-to-human transmission of influenza H5N1 has been suggested in several household clusters [123], and in one case of apparent child-to-mother transmission [124], but patient-to-healthcare workers transmission has not been suggested [125,126].

However, enhanced infection control precautions for patients with suspected or confirmed avian influenza infection appear to be warranted because of the uncertainty about modes of human-to-human transmission, the high lethality of human avian influenza A (H5N1) infection to date, and the possibility that the virus could mutate or resort at any time into a strain capable of efficient human-to-human transmission [127].

Human influenza viruses are among the most communicable viruses infecting humans and their transmission occurs through multiple routes including large droplets, direct and indirect contact, and droplet nuclei [128]. So, minimal requirements for caring for avian-influenza infected patients should include standard contact and droplet precautions [127]. Role of disinfectants and antiseptic agents in these precautions are great.

Avian influenza virus is inactivated by standard hospital disinfectants with tuberculocidal claim on the label, including phenolic disinfectants, QACs, peroxygen compounds, sodium hypochlorite (household bleach), and alcohol [15,129]. Environmental cleaning and disinfecting healthcare settings that care for avian-influenza infected patients should be done with noted germicides, with special attention to frequently touched surfaces and horizontal surfaces around the patient after aerosol-generating procedures. For healthcare settings with limited resources, which do not have access to standard hospital disinfectants, alcohol and bleach containing 5% sodium hypochlorite are acceptable alternatives [127].

Hand hygiene, by hand washing or use of an alcohol-based hand rub, as part of standard precautions is critical to prevent possible self-inoculation, and transfer of viruses to the environment or other patients by contaminated hands. It should be performed before and after each contact with the patient or their environment of care, and after removing gloves as precaution for avian influenza [127].

We do not know whether H5N1 influenza will spread or not. But we must be well prepared for any future surge of either small or large outbreaks in healthcare settings. Infection control measures with use of appropriate disinfectant and antiseptic agents will be critical for prevention and control of avian influenza in healthcare settings.

7.6 Conclusion

We live in times where risk of bioterrorist attacks with known or unknown agents is extremely high. Avian flu epidemic is a constant possible threat; SARS epidemic represents a reality, and resistance of microorganisms to antibiotics is increasing. In such times, correct use of disinfectant and antiseptic agents in healthcare settings is of critical importance for the prevention and control of spread of infections.

References

1. World Health Organization, World Alliance for Patient Safety, Global Patient Safety Challenge: 2005–2006, Clean Care Is Safer Care, Geneva, Switzerland, 2005, p. 1.
2. Van Den Broek, J.P., Historical perspectives for the new millennium, *Prevention and Control of Nosocomial Infections*, 4th edition, Wenzel, R.P. (ed.), Lippincott Williams & Wilkins, Philadelphia, 2003, p. 3.
3. Widmer, A.F. and Frei, R., Decontamination, disinfection, and sterilization, *Manual of Clinical Microbiology*, 7th edition, Murrey, P.R., editor-in-chief, Baron, J.E., Pfaller, M.A., Tenover, F.C., and Yolken, R.H. (eds.), ASM Press, Washington, 1999, p. 138.
4. Underwood, E., Good manufacturing practice, *Russel, Hugo & Ayliffe's Principles and Practice of Disinfection, Preservation & Sterilization*, 4th edition, Fraise, A.P., Lambert, P.A. and Maillard, J.Y. (eds.), Blackwell Publishing, Oxford, 2004, p. 622.
5. Rutala, W.A. and Weber, D.J., Selection and use of disinfectant in healthcare, *Hospital Epidemiology and Infection Control*, 3rd edition, Mayhall, C.G. (ed.), Lippincott Williams & Wilkins, Philadelphia, 2004, p. 1473.
6. Alfa, M.J. and Jackson, M., A new hydrogen peroxide-based medical-device detergent with germicidal properties: Comparison with enzymatic cleaners, *Am. J. Infect. Control*, 29, 168, 2001.
7. Hanson, P.J., et al., Enteroviruses, endoscopy and infection control: An applied study, *J. Hosp. Infect.*, 27, 61, 1994.
8. Hanson, P.J., et al., Viral transmission and fiberoptic endoscopy, *J. Hosp. Infect.*, 18, 136, 1991.
9. Costerton, J.W., et al., Bacterial biofilms, in nature and disease, *Annu. Rev. Microbiol.*, 41, 435, 1987.
10. Costerton, J.W., et al., Biofilms, the customized microniche, *J. Bacteriol.*, 176, 2137, 1994.
11. Rutala, W.A., 1994, 1995, and 1996 APIC Guidelines Committee, APIC guideline for selection and use of disinfectants, Association for Professionals in Infection Control and Epidemiology, *Am. J. Infect. Control*, 24, 313, 1996.
12. Russell, A.D., Factors influencing the efficacy of antimicrobial agents in *Russel, Hugo & Ayliffe's Principles and Practice of Disinfection, Preservation & Sterilization*, 4th edition, Fraise, A.P., Lambert, P.A., and Maillard, J.Y. (eds.), Blackwell Publishing, Oxford, 2004, p. 98.
13. Spaulding, E.H., Role of chemical disinfection in the prevention of nosocomial infections, *Proceeding of the International Conference of Nosocomial Infections*, Brachman, P.S. and Eickhoff, T.C. (eds.), American Hospital Association, Chicago, 1070, 247, 1971.
14. Spaulding, E.H., Chemical disinfection and antisepsis in hospital, *J. Hosp. Res.*, 9, 5, 1972.
15. CDC, Guidelines for environmental infection control in health-care facilities: Recommendations of CDC and Healthcare Infection Control Practices Advisory Committee (HICPAC), *MMWR*, 52, 1, 2003.
16. WHO, *Prevention of Hospital-acquired Infections. A Practical Guide*, 2nd edition, World Health Organization, 2002.

17. Keene, J.H., Sterilization and pasteurization, *Hospital Epidemiology and Infection Control*, 3rd edition, Mayhall, C.G. (ed.), Lippincott Williams & Wilkins, Philadelphia, 2004, p. 1523.
18. Favero, M.S. and Bond, W.W., Chemical disinfection of medical and surgical materials, *Disinfection, Sterilization and Preservation*, 5th edition, Block, S.S. (ed.), Lippincott Williams & Wilkins, Philadelphia, 2001, p. 881.
19. Fraise, A.P., Decontamination of the environment and medical equipment in hospitals, *Russel, Hugo & Ayliffe's Principles and Practice of Disinfection, Preservation & Sterilization*, 4th edition, Fraise, A.P., Lambert P.A., and Maillard, J.Y., (eds.), Blackwell Publishing, Oxford, 2004, p. 563.
20. Notskin, G.A., et al., Recovery of vancomycin-resistant enterococci fingertips and environmental surfaces, *Infect. Control Hosp. Epidemiol.*, 16, 577, 1995.
21. Wang, S.H., et al., Healthcare associated outbreak due to pan-drug resistant *Acinetobacter baumannii* in surgical intensive care unit, *J. Hosp. Infect.*, 53, 97, 2003.
22. El Shafiee, S.S., Alishaq, M., and Garcia, L.M., Investigation of an outbreak of multidrug resistant *Acinetobacter baumannii* in trauma intensive care units, *J. Hosp. Infect.*, 56, 101, 2004.
23. Mayfield, J.L., et al., Environmental control to reduce transmission of *Clostridium difficile*, *Clin. Infect. Dis.*, 31, 995, 2000.
24. Rutala, W.A., et al., Inactivation of *Mycobacterium tuberculosis* and *Mycobacterium bovis* by 14 hospital disinfectants, *Am. J. Med.*, 91, 267S, 1991.
25. Moore, S.L. and Payne, D.N., Types of antimicrobial agents, *Russel, Hugo & Ayliffe's Principles and Practice of Disinfection, Preservation & Sterilization*, 4th edition, Fraise, A.P., Lambert, P.A., and Maillard, J.Y. (eds.), Blackwell Publishing, Oxford, 2004, p. 8.
26. Ehrenkranz, N.J., Bolyard, E.A., and Wiener, M., Antibiotic sensitive *Serratia marcescens* infections complicating cardiopulmonary operations: Contaminated disinfectant as a reservoir, *Lancet*, ii, 1289, 1980.
27. Frank, M.J. and Schaffner, W., Contaminated aqueous benzalkonium chloride, An unnecessary hospital infection hazard, *JAMA*, 236, 2418, 1976.
28. Simmons, B.P., CDC guidelines for the prevention and control of nosocomial infections, Guideline for hospital environmental control, *Am. J. Infect. Control*, 11, 97, 1983.
29. Newsom, S.W.B., Pioneers in infection control—Joseph Lister, *J. Hosp. Infect.*, 55, 246, 2003.
30. Rutala, W.A. and Cole, E.C., Ineffectiveness of hospital disinfectant against bacteria: A collaborative study, *Infect. Control*, 8, 501, 1987.
31. Narang, H.K. and Cod, A.A., Action of commonly used disinfectants against enteroviruses, *J. Hosp. Infect.*, 4, 209, 1983.
32. Wysowski, D.K. et al., Epidemic neonatal hyperbilirubinemia and use of a phenolic disinfectant detergent, *Pediatrics*, 61, 165, 1978.
33. Gottardi, W., Iodine and iodine compounds, *Disinfection, Sterilization and Preservation*, Block, S.S. (ed.), Lea & Febirger, Philadelphia, 1991, p. 152.
34. Rutala, W.A. and Weber, D.J., Uses of inorganic hypochlorite (bleach) in healthcare facilities, *Clin. Microbiol. Rev.*, 10, 597, 1997.
35. Leaper, S., Influence of temperature on the synergistic sporicidal effect of peracetic acid plus hydrogen peroxide in *Bacillus subtilis* SA22 (NCA 72-52), *Food Microbiol.*, 1, 1999, 1984.
36. Block, S.S., Hydrogen peroxide, *Disinfection, Sterilization and Preservation*, Block, S.S. (ed.), Lea & Febirger, Philadelphia, 1991, p. 167.

37. Ontario Ministry of Health and Long-Term Care–Emergency Health Services Branch, Severe Acute Respiratory Syndrome (SARS), Training Bulletin, Version 1.1 MOHLTC, 2003, p. 3.
38. Scott, E.M. and Gorman, S.P., Glutaraldehyde, *Disinfection, Sterilization and Preservation*, 4th edition, Block, S.S. (ed.), Lea & Febirger, Philadelphia, 1991, p. 596.
39. Rutala, W.A. and Webere, D.J., New disinfection and sterilization methods, *Emerg. Infect. Dis.*, 7, 348, 2001.
40. Rutala, W.A., Disinfection and sterilization: New HICPAC guidelines, Available at: http://www.unc.edu/depts/spice/dis/APIC-disinfection-sterilization 2005.pdf
41. Antimicrobial Chemical/Registration Number Index, Available at: http://www.epa.gov/oppad001/chemregindex.htm
42. Food and Drug Administration, Guidance for industry and FDA reviewers: Content and format of premarket notification [510(k)] submissions for liquid chemical sterilants/high level disinfectants, Rockville, MD: US Department of Health and Human services, Food and Drug Administration, 2000, Available at http://www.fda.gov/cdrh/ode/397.pdf
43. European Committee for Standardization (CEN), EN 1040 Chemical disinfectant and antiseptics, Basic bactericidal activity, Test method and requirements (phase 1), Brussels: CEN, 1997.
44. European Committee for Standardization (CEN), EN 1275 Chemical disinfectants and antiseptics, Basic fungicidal activity, Test method and requirements (phase 1), Brussels: CEN, 1997.
45. European Committee for Standardization (CEN), EN 13713 Chemical disinfectants and antiseptics, Surface disinfectants used in human medicine, bactericidal activity, Test method and requirements (phase 2/step 1), Brussels: CEN, 1999.
46. European Committee for Standardization (CEN), EN 13727 Chemical disinfectants, Quantitative suspension test for the evaluation of bactericidal activity for instruments used in medical area, Test method and requirements (phase 2/step 1), Brussels: CEN, 2000.
47. Price, P.B., The bacteriology of normal skin: A new quantitative test applied to a study of the bacterial flora and the disinfectant action of mechanical cleansing, *J. Infect. Dis.*, 63, 301, 1938.
48. Boyce, J.M. and Pitet, D., Guideline for hand hygiene in health-care settings, Recommendations of the healthcare infection control practices advisory committee and the HICPAC/SHEA/APIC/IDSA hand hygiene task force, Centers for Disease Control and Prevention, *MMWR*, 51(RR-16), 1.
49. Rotter, M.L., Hand washing and hand disinfection, *Hospital Epidemiology and Infection Control*, 3rd edition, Mayhall, C.G. (ed.), Lippincott Williams & Wilkins, Philadelphia, 2004, p. 1727.
50. Garner, J.S. and Favero, M.S., CDC guideline for handwashing and hospital environmental control, 1985, *Infect. Control*, 7, 231, 1986.
51. Larson, E., APIC guideline for handwashing and hand antisepsis in health care settings, *Am. J. Infect. Control*, 23, 251, 1995.
52. Fraise, A.P., Historical introduction, *Russel, Hugo & Ayliffe's Principles and Practice of Disinfection, Preservation & Sterilization*, 4th edition, Fraise, A.P., Lambert, P.A., and Maillard, J.Y. (eds.), Blackwell Publishing, Oxford, 2004, p. 3.

53. Lowbury, E.J.L., Lilly, H.A., and Ayliffe, G.A.J., Preoperative disinfection of surgeons' hands: Use of alcoholic solutions and effects of gloves on skin flora, *Br. Med. J.*, 4, 369, 1974.
54. Ayliffe, G.A.J., Surgical scrub and hand disinfection, *Infect. Control*, 5, 23, 1984.
55. Ayliffe, G.A.J., et al., Hand disinfection: A comparison of various agents in laboratory and ward studies, *J. Hosp. Infect.*, 11, 226, 1988.
56. Babb, J.R., Davies, J.G., and Ayliffe, G.A.J., A test procedure for evaluating surgical hand disinfection, *J. Hosp. Infect.*, 18, 41, 1991.
57. Kjrlen, H. and Andersen, B.M., Handwashing and disinfection of heavily contaminated hands—effective or ineffective? *J. Hosp. Infect.*, 21, 61, 1992.
58. Namura, S., Nishijima, S., and Asada, Y., An evaluation of the residual activity of antiseptic handrub lotions: An 'in use' setting study, *J. Dermatol.*, 21, 481, 1994.
59. Renner, P., Unger, G., and Peters, J., Efficacy of hygienic hand disinfectants in the presence of blood, *Hyg. Med.*, 18, 153, 1993.
60. The Iodine Source, The History of Iodine, Available at: http://www.iodinesource. com/HistoryOfIodine.asp
61. Gottardi, W., Iodine and iodine compounds, *Disinfection, Sterilization and Preservation*, 4th edition, Block, S.S. (ed.), Lea & Febirger, Philadelphia, 1991, p. 152.
62. Reimer, K., et al., Antimicrobial effectiveness of povidone iodine and consequences for new application areas, *Dermatology*, 204, 114, 2002.
63. Boyce, J.M., MRSA patients: Proven methods to treat colonization and infection, *J. Hosp. Infect.*, 48, S9, 2001.
64. Anderson, R.L., Iodophor antiseptics: Intrinsic contamination with resistant bacteria. *Infect. Control Hosp. Epidemiol.*, 10, 443, 1989.
65. Anderson, R.L., et al., Prolonged survival of *Pseudomonas cepacia* in commercially manufactured povidone–iodine. *Appl. Environ. Microbiol.*, 56, 3598, 1990.
66. Stickler, D.J. and Thomas, B., Antiseptic and antibiotic resistance in gram-negative bacteria causing urinary tract infection, *J. Clin. Pathol.*, 33, 288, 1980.
67. Vigeant, P., et al., An outbreak of *Serratia marcescens* infection related to contaminated chlorhexidine, *Infect. Control. Hosp. Epidemiol.*, 19, 791, 1998.
68. Mangram, A.J., et al., Guidelines for prevention of surgical site infection, *Infect. Control Hosp. Epidemiol.*, 20, 247, 1999.
69. Centers for Disease Control and Prevention, Guidelines for prevention of intravascular catheter related infections, *MMWR*, 51, 1, 2002.
70. Kuruvilla, J.R. and Kamath, M.P., Antimicrobial activity of 2.5% sodium hypochlorite and 0.2% chlorhexidine gluconate separately and combined, as endodontic irrigants, *JOE*, 24, 472, 1998.
71. Shuman, R.M., Leech, R.W., and Allvord, E.C. Jr., Neurotoxicity of hexachlorophene in humans: II, A clinicopathological study of 46 premature infants, *Arch. Neurol.*, 32, 320, 1975.
72. Jones, R.D., et al., Triclosan: A review of effectiveness and safety in health care settings, *Am. J. Infect. Control*, 28, 184, 2000.
73. McMurray, L.M., Oethhinger, M., and Levy, S.B., Triclosan targets lipid synthesis, *Nature*, 394, 531, 1998.
74. Levy, C.W., et al., Molecular basis of triclosan activity, *Nature*, 398, 383, 1999.
75. Sautter, R.L., Mattman, L.H., and Legaspi, R.C., *Serratia marcescens* meningitis associated with contaminated benzalkonium chloride solution, *Infect. Control*, 5, 223, 1984.
76. Oie, S. and Kamiya, A., Microbial contamination of antiseptics and disinfectants, *Am. J. Infect. Control*, 24, 389, 1996.

77. Food and Drug Administration, Tentative final monograph for healthcare anti-septic drug products; proposed rule, Federal register, 59, 31441, 1994.
78. WHO Guidelines on Hand Hygiene in Health Care (Advanced draft): A Summary, http://www.who.int/patientsafety/information_centre/who_ghhhcad/en/
79. Global consensus conference: Final recommendations, *Am. J. Infect. Control*, 27, 503, 1999.
80. Pitett, D., Improving adherence to hand hygiene practice: A multidisciplinary approach, *Emerg. Infect. Dis.*, 7, 234, 2001.
81. Voss, A. and Widamer, A.F., No time for handwashing? Handwashing versus alcoholic rub: Can we afford 100% compliance? *Infect. Control Hosp. Epidemiol.*, 28, 205, 1997.
82. Rotter, M.L., Hygienic hand disinfection, *Infect. Control*, 5, 18, 1984.
83. Lowbury, E.J., Lilly, H.A., and Ayliffe, G.A.J., Preoperative disinfection of surgeons' hands: Use of alcoholic solutions and effects of gloves on skin flora, *Br. Med. J.*, 4, 369, 1974.
84. Hardin, W.D. and Nichols, R.L., Handwashing and patient skin preparation, *Critical Issues in Operating Room Management*, Malangoni, MA (ed.), Lippincott-Raven, Philadelphia, 1997, p. 133.
85. Nichols, R.L., et al., Current practices of preoperative bowel preparation among North American colorectal surgeons, *Clin. Infect. Dis.*, 24, 609, 1997.
86. Wade, J.J. and Casewell, M.W., The evaluation of residual antimicrobial activity on hands and its clinical relevance. *J. Hosp. Infect.*, 18, 41, 1991.
87. Mackenzie, I., Preoperative skin preparation and surgical outcome, *J. Hosp. Infect.*, 11, 27, 1988.
88. Rotter, M.L., et al., A comparison of the effects of preoperative whole-body bathing with detergent alone and with detergent containing chlorhexi-dine gluconate on the frequency of wound infections after clean surgery. The European Working Party on Control of Hospital Infections, *J. Hosp. Infect.*, 11, 310, 1988.
89. Ayliffe, G.A., et al., A comparison of pre-operative bathing with chlorhexidine-detergent and non-medicated soap in prevention of wound infection, *J. Hosp. Infect.*, 4, 237, 1983.
90. Tuffnell, D.J., et al., Methicillin resistant *Staphylococcus aureus*; the role of anti-sepsis in the control of an outbreak, *J. Hosp. Infect.*, 10, 255, 1987.
91. Bartzokas, C.A., et al., Control and eradication of methicillin-resistant *Staphylococcus aureus* on a surgical unit, *N. Engl. J. Med.*, 311, 1422, 1984.
92. Meberg, A. and Schoyen, R., Bacterial colonization and neonatal infections. Effects of skin and umbilical disinfection in the nursery, *Acta. Pediatr. Scand.*, 74, 366, 1985.
93. Rotter, M.L., Special problems in hospital antisepsis, *Russel, Hugo & Ayliffe's Principles and Practice of Disinfection, Preservation & Sterilization*, 4th edition, Fraise, A.P., Lambert, P.A., and Maillard, J.Y. (eds.), Blackwell Publishing, Oxford, 2004, p. 540.
94. Centers for Disease Control and Prevention, Public health focus: Surveillance, prevention and control of nosocomial infections, *MMWR*, 41, 783, 1992.
95. National Nosocomial Infections Surveillance (NNIS) report, data summary from October 1986–April 1996, issued May 1996, A report from National Nosocomial Infections Surveillance (NNIS) System, *Am. J. Infect. Control*, 24, 380, 1996.

96. National Nosocomial Infections Surveillance (NNIS) System Report, data summary from January 1992 through June 2004, issued October 2004, *Am. J. Infect. Control*, 32, 470, 2004.

97. Burke, P.B. and Yeo, T.W., Nosocomial urinary tract infections, *Hospital Epidemiology and Infection Control*, 3rd edition, Mayhall, C.G. (ed.), Lippincott Williams & Wilkins, Philadelphia, 2004, p. 267.

98. Maki, D.G. and Tambyah, P.A., Engineering out the risk of infection with urinary catheters, *Emerg. Infect. Dis.*, 7, 1, 2001.

99. Burke, J.P., et al., Efficacy of daily meatal care regimens, *Am. J. Med.*, 70, 655, 1981.

100. Classen, D.C., et al., Prevention of catheter-associated bacteriuria: Clinical trial of methods to block three known pathways of infection, *Am. J. Infect. Control*, 19, 136, 1991.

101. van den Broek, P.J., Daha, T.J., and Mouton, R.P., Bladder irrigation with povidone–iodine in prevention of urinary tract infections associated with intermittent urethral catheterization, *Lancet*, 1(8428), 563, 1985.

102. Richards, M.J., et al., Nosocomial infections in medical intensive care units in the United States, *Crit. Care Med.*, 27, 887, 1999.

103. Raad, I., Management of intravascular catheter-related infections, *J. Antimicrob. Chemother.*, 45, 267, 2000.

104. Mermel, L.A., Prevention of intravascular catheter-related infections, *Ann. Intern. Med.*, 132, 391, 2000.

105. Centers for Disease Control and Prevention, Guideline for the prevention of intravascular catheter-related infections, *MMWR*, 51(RR10), 1, 2002.

106. Maki, D.G., Ringer, M., and Alvarado, C.J., Prospective randomized trial of povidone–iodine, alcohol, and chlorhexidine for prevention of infection associated with central venous and arterial catheters, *Lancet*, 338(8763), 339, 1991.

107. Mimoz, O., et al., Prospective, randomized trial of two antiseptic solutions for prevention of central venous or arterial catheter colonization and infection in intensive care unit patients, *Crit. Care Med.*, 24, 1818, 1996.

108. Kuiken, T., et al., Newly discovered coronavirus as the primary cause of severe acute respiratory syndrome, *Lancet*, 362, 263, 2003.

109. World Health Organization, Summary of probable SARS cases with onset of illness from 1 November 2002 to 31 July 2003, based on data as of the 31 December 2003, Available at: http://www.who.int/csr/sars/country/table2004_04_21/en/

110. Wenzel, R.P. and Edmond, M.B., Listening to SARS: Lessons for infection control, *Ann. Intern. Med.*, 139, 592, 2003.

111. Seto, W.H., et al., Effectiveness of precautions against droplets and contact in prevention of nosocomial transmission of severe acute respiratory syndrome (SARS), *Lancet*, 361, 1519, 2003.

112. Varia, M., et al., Investigation of a nosocomial outbreak of severe acute respiratory syndrome (SARS) in Toronto, Canada, *CMAJ*, 169, 285, 2003.

113. Butler, J.C. and Jernigan, J.A., Severe acute respiratory syndrome, *Hospital Epidemiology and Infection Control*, 3rd edition, Mayhall, C.G. (ed.), Lippincott Williams & Wilkins, Philadelphia, 2004, p. 1979.

114. Riley, S., et al., Transmission dynamics of the etiological agent of SARS in Hong Kong: Impact on public health interventions, *Science*, 300, 1961, 2003.

115. Lipstich, M., et al., Transmission dynamics and control of severe acute respiratory syndrome, *Science*, 300, 1966, 2003.

116. Centers for Disease Control and Prevention, Public health guidance for community level preparedness to severe acute respiratory syndrome, 2004, January 8, Available at: http://www.cdc.gov/ncidod/sars/guidance/I/healthcare.htm
117. World Health Organization, First data on stability and resistance of SARS corona virus compiled by members of WHO laboratory network, 2003, Available at: http://www.who.int/csr/sars/survival_2003_05_04/en/index.html
118. World Health Organization, Hospital infection control guidance for severe acute respiratory syndrome (SARS), Available at: htpp://www.who.int/csr/sars/infectioncontrol/en/
119. World Health Organization, Western Pacific Region, Investigation into China's recent outbreak yields important lessons for global public health, July 2004, Available at: http://www.wpro.who.int/sars/docs/update/update_07022004.asp
120. World Health Organization, Cumulative number of confirmed human cases of avian influenza A/ (H5N1) reported to WHO, 19 May 2006, Available on: http://www.who.int/csr/disease/avian_influenza/country/cases_table_2006_05_19/en/print.html
121. Yuen, K.Y., et al., Clinical features and rapid viral diagnosis of human disease associated with avian influenza A H5N1 virus, *Lancet*, 351, 467, 1998.
122. Bridges, B.C., et al., Risk of influenza A (H5N1) infection among poultry workers, Hong Kong, 1997–1998, *J. Infect. Dis.*, 185, 1005, 2002.
123. Tran, T.H., et al., Avian influenza A (H5N1) in 10 patients in Vietnam, *N. Engl. J. Med.*, 350, 1179, 2004.
124. Ungchusak, K., Probable person-to-person transmission of avian influenza A (H5N1), *N. Engl. J. Med.*, 352, 333, 2005.
125. Liem, N.T., World Health Organization International Avian Influenza Investigation Team, Vietnam, Lim, W, Lack of H5N1 avian influenza transmission to hospital employees, Hanoi, 2004, *Emerg. Infect. Dis.*, 11, 210, 2005.
126. Schultsz, C., et al., Avian influenza H5N1 and healthcare workers (letter), *Emerg. Infect. Dis.*, 11, 1158, 2005.
127. WHO, Avian influenza, including influenza A (H5N1), in humans: WHO interim infection control guideline for health care facilities, 9 February 2006, www.wpro.who.int/NR/rdonlyres/177DBE1C-4E3C-4C7E-B784-4124E6C600CF/0/InfectionControl.pdf
128. Valenti, W.M., Influenza viruses, *Hospital Epidemiology and Infection Control*, 3rd edition, Mayhall, C.G. (ed.), Lippincott Williams & Wilkins, Philadelphia, 2004, p. 1979.
129. Suarez, D.L., et al., The effect of various disinfectants on detection of avian influenza virus by real time RT-PCR, *Avian Dis.*, 47, 1091, 2003.

8

Extended Activity of Healthcare Antiseptic Products

Jerry L. Newman and Nancy E. Kaiser

CONTENTS

8.1 Introduction

The extended activity of antiseptic agents has been studied for many years. The literature describing extended activity uses many terms and these terms have been used interchangeably and with little discrimination, based on mechanisms or effects. Examples of this term proliferation in the literature can often be found within the same paper. As well as being confusing, this practice has served to potentially diminish the perception of the benefit of such effects. Persistent, residual, and cumulative are the terms most commonly used to describe extended activity. These three terms have been applied somewhat indiscriminately to different active agents, which have different modes of action and different use classifications. There is therefore confusion as to whether these words describe differentiated characteristics of extended activity or modes of action. However, some clarity can be achieved by looking at the terms with a systematic approach based on standards for English terminology and most common use patterns.

The use of terms to describe extended antiseptic activity goes back to at least the early 1960s when studies of hexachlorophene attempted to describe

TABLE 8.1

List of the Terms That Have Been Used to Describe Various Aspects of Extended Antiseptic Activity

Term	References
Absorbed	[1]
Accumulation	[13]
Continuing effect	[5]
Continuous control (or release)	[5,14]
Cumulative	[2,5,7,9,15–21]
Delayed (effects, drying time)	[14,22]
Deposited or deposition (of antiseptic)	[4]
Long term (action)	[5,12,15]
Long lasting (or acting)	[6,11,13,15,23]
Maintained or maintenance (of lower bacterial populations)	[18,24]
Persistent (or persistence)	[3,8,9,11,19,20,23–31]
Progressive reduction	[2,9,23,32]
Prolonged (effect, activity, release)	[5,13,14,23,29]
Remnant	[13,21,23]
Reservoir (effect)	[8,11,19]
Residue (of antiseptic agents)	[8,22,32,33]
Residual (effect or action)	[5,6,10,13,16–18,20,22,26,27,31,33–37]
Substantive (or substantivity)	[8,11,13,14,20,21,25,28,29,35,36,38]
Suppressive effect/suppression of regrowth	[3,39]
Sustained (efficacy or activity)	[9,23,35,36,40]

how the agent is absorbed in the skin [1] and how it produces progressive or cumulative reductions [2]. In fact, these hexachlorophene effects and mechanisms were well studied and were published throughout the 1960s and 1970s [3–10]. Iodophors are another active ingredient that have been studied to determine properties of extended antiseptic activity as early as the 1960s [3,6,10–12]. Table 8.1 contains a list of the terms that have been used to describe various aspects of extended antiseptic activity in the course of evaluating these and other antiseptic products.

A lack of standard terminology is demonstrated by the list of 20 terms for prolonged activity, contained in Table 8.1. This long list also demonstrates why there is a lack of clear understanding of the benefits of prolonged activity. Understanding the differences between effects and modes of action indicates the difference between what it does and how it does it. We start our systematic approach with separate analyses based on effects and mechanisms (or modes) of action.

8.2　Effects (or Activity)

From the viewpoint of users of an antiseptic product, the effect (or activity) exerted by that product is its most important aspect. The U.S. Food and Drug

Administration (FDA) has defined use classifications for types of products and included persistence requirements in the definitions [41,42]. The FDA has asserted that surgical hand scrubs and patient preoperative skin preparations should exert persistent activity. FDA states that, the healthcare personnel handwash products should be "broad spectrum, fast-acting and, if possible, persistent." While there is no standard definition for persistent effect, it seems that the following is appropriate based on customary practice:

> Persistent antiseptic effect: Suppression of regrowth of resident flora or antimicrobial activity (inhibition or biocidal) against transient flora maintained on skin over an extended period of time after initial use of an antiseptic product.

To exert a persistent antiseptic effect on transient bacteria (as is the case with healthcare personnel handwash products) it is logical that the antiseptic product would have to exhibit substantivity, that is, antimicrobial components need to remain in or on the epidermis for an extended period of time and be present to act as a reservoir. Substantivity may or may not be involved in the persistent effects demonstrated on resident flora, as seen with surgical hand scrubs and patient preoperative skin preparations. Inhibition of regrowth of resident flora may be accomplished through the use of non-substantive products such as alcohols and iodophors. In the case of alcohols, which evaporate and therefore are clearly not substantive, a persistent effect is demonstrated through continued death of damaged organisms [20,30]. This has been well documented and helps explain a mechanism which allows alcohol-based products to meet the persistence requirement for use as a surgical hand scrub. Persistent effects can also be contributed through the inclusion of an additional substantive active ingredient such as chlorhexidine gluconate [28,35,37,43] or preservative elements in the formulation, as suggested by the FDA [42].

Progressive microbial reductions after multiple uses have sometimes been confused with persistence. A definition of cumulative effects based on prevalent usage and English norms would be as follows:

> Cumulative antiseptic effect: A progressive increase in reductions in microbial counts (from baseline) on skin after multiple uses of an antiseptic product.

While these cumulative effects may be imparted by the same means—such as substantivity—the effects are a truly different phenomena from that of the persistent antiseptic effect. For instance, triclosan-based products, which are well known to provide persistent effects, do not appear to produce significant cumulative activity comparable to chlorhexidine gluconate-based products [17]. While the benefit of cumulative activity can be debated, FDA requirements include higher log reductions after multiple uses for both healthcare personnel handwashes, as well as surgical hand scrubs [42]. It should be

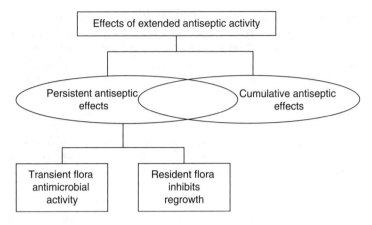

FIGURE 8.1
Effects of extended antiseptic activity.

noted that these efficacy requirements can also be met with products that have high initial activity and maintain that activity through multiple applications, without exhibiting significant cumulative effects. Products with lower initial activity can meet the FDA requirements based on outstanding cumulative activity.

In understanding the effects of antiseptic products, it is important to differentiate between persistent and cumulative effects. A clear distinction exists between persistent effects demonstrated on transient flora versus those demonstrated against resident flora. Persistent effects against transient organisms are based on antimicrobial activity or a reduction of organisms by an antimicrobial agent. This is opposed to suppression of regrowth of organisms as is the case of persistent effects on resident flora. The above graphic (Figure 8.1) illustrates the relationship between these distinct effects that in practicality often overlap.

8.3   Modes of Action

The most obvious mechanism for extended antiseptic action is that of substantivity. A working definition for this mode is as follows:

> Antiseptic substantivity: Extended antiseptic activity provided by an agent that remains in the stratum corneum for prolonged period of time after use (including rinsing, if appropriate), providing a reservoir of the agent on the skin.

Notable examples of this mode of action include chlorhexidine and triclosan. While these actives are both substantive, they illustrate two different ways in which active agents can adhere to the skin. In the case of chlorhexidine, the strong affinity to skin is due to electrostatic interactions—chlorhexidine gluconate is cationic and binds to anionic sites in the stratum corneum [35]. Triclosan, on the other hand, is hydrophobic and partitions into oily parts of the skin and is bound due to intermolecular forces (i.e., van der Waals) or perhaps lipophilic affinity for nonpolar environments.

While substantivity is clearly a well-known mechanism for extended activity, there are other ways in which antiseptics provide prolongation of effect. In fact, these modes will generally be referred to as antiseptic effect prolongation. Antiseptic effect prolongation pertains to the delayed activity observed for alcohol-based products described above [20,22,26,30,33]. An example is the previously discussed case of persistent effects observed with alcohol. These effects are clearly not because of substantivity since alcohol evaporates from skin rapidly. These effects were first noted as effectiveness under gloves for up to 3 h [26,33]. Delayed or continued death of sublethally damaged organisms is the recognized source of this phenomenon [22,26,33].

Another form of antiseptic effect prolongation is demonstrated through retardation of evaporation resulting in prolonged contact with the skin [43]. This phenomenon is one of several formulation techniques that is used with alcohol-based antiseptic products to help comply with the worldwide needs for improved hand hygiene products. Thus, a definition for this mechanism is as follows:

> Antiseptic effect prolongation: Extended activity due to factors not related to substantivity.

The modes of action of antiseptic activity are depicted graphically below (Figure 8.2). While actives such as alcohol clearly do not exhibit substantivity, some agents that act predominantly through substantivity also likely exhibit effects through the means described as prolongation. Thus, the overlap of these mechanisms is illustrated here.

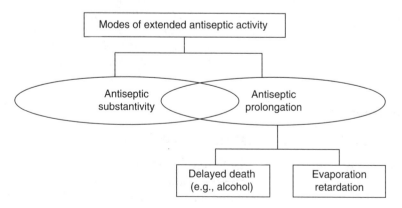

FIGURE 8.2
Modes of extended antiseptic activity.

8.4 Surgical Hand Scrubs

Surgical hand scrubs are used before surgeries that last for a number of hours, during which time surgeons would be expected to wear gloves on their hands—an ideal environment for regrowth of organisms on the skin. Thus, the benefit of a persistent antiseptic effect is readily apparent. The FDA performance criteria require that resident flora should not regrow above baseline or pre-use levels for 6 h under gloves. The FDA surgical hand scrub performance criteria also incorporate a requirement for progressively higher reductions throughout 11 uses over 5 days. This scheme allows products with lower immediate activity to meet the requirement by producing a cumulative effect.

Chlorhexidine-based products are ideally suited to meet this requirement. The value of the persistent effect for surgical hand scrubs (preventing regrowth under gloves) is well documented but that of the cumulative antiseptic effect is less clear. Although an argument can be made for the value of increasing reductions in resident bacteria over multiple applications, the ultimate result is that a product with lower initial efficacy can meet the requirement by providing a build in efficacy over time. It seems reasonable that cumulative effect is desirable but by virtue of the FDA criteria, cumulative effect is in effect mandatory in those cases where a product has relatively low initial efficacy. It is this build in efficacy with time that allows such products to be marketed.

8.5 Patient Surgical Skin Preparations

Patient preoperative skin preparations are applied to the patients' skin before surgery to reduce flora at the surgical site during the course of the surgical procedure. Since these procedures can last for a number of hours the value of a persistent antimicrobial effect on the skin at the surgical incision site is clear-cut. This is the apparent rationale behind the FDA requirement that the reduction in flora below baseline be maintained for a minimum of 6 h. A cumulative antiseptic effect is of little value since multiple uses at the surgical site is highly unlikely and is not anticipated in the FDA test methods.

8.6 Healthcare Personnel Handwash

While the health care personnel handwash is primarily intended for reducing transient organisms on the skin versus resident organisms, the way these products are used is similar to that of surgical hand scrubs in that their use involves multiple applications to the hands of health care workers (HCW). The FDA definition for this product type indicates that a persistent antiseptic effect is desirable, but not required. However, the test methods required by FDA do not measure a persistent effect. Conversely, the methods actually can serve to measure a cumulative antiseptic effect. In fact, as in the case of the surgical hand scrub, the FDA performance requirements allow a product that has lower initial activity to be marketed if it exhibits a cumulative antiseptic effect. Again a question may exist about the value of products that exhibit cumulative effects versus products that have higher initial activity and maintain that activity over multiple applications. One possible reason for measuring cumulative effects is that they could be viewed as an indirect indicator of persistent antiseptic effects. However, as noted above, these are indeed separate effects, with a less than perfect correlation. The confusion of these separate effects has contributed to some of the uncertainty about the value of the persistent antiseptic effect for this class of products. The benefit of a persistent antiseptic effect against transient organisms has been suggested to be a solution to lapses in compliance with hand hygiene [3], a well-known problem that continues to plague the healthcare system [20].

8.7 Summary

The following graph (Figure 8.3) illustrates the interaction of effects (on the x axis) and modes of action (on the y axis). It is clear that antiseptic

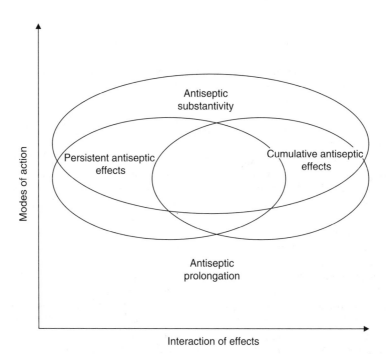

FIGURE 8.3
Interaction of various antiseptic effects versus the modes of antiseptic action.

substantivity accounts for a significant amount of persistent and cumulative effects. Those that are not due to substantivity are explained by antiseptic effect prolongation. While products that exhibit substantivity are also likely to produce other antiseptic effect prolongation, it is virtually impossible to determine the degree of overlap.

References

1. Miller, J.M., Jackson, D.A., and Collier, C.S., The microbicidal property of Phisohex, *Mil. Med.*, 127, 576, 1962.
2. Lowbury, E.J., Lilly, H.A., and Bull, J.P., Disinfection of hands: Removal of resident bacteria, *Br. Med. J.*, 1, 1251, 1963.
3. Lowbury, E.J., Lilly, H.A., and Bull, J.P., Disinfection of hands: Removal of transient organisms, *Br. Med. J.*, 5403, 230, 1964.
4. Stoughton, R.B., Hexachlorophene deposition in human stratum corneum. Enhancement by dimethylacetamide, dimethylsulfoxide, and methylethylether, *Arch. Dermatol.*, 94 (5), 646, 1966.
5. Van der Hoeven, E. and Hinton, N.A., An assessment of the prolonged effect of antiseptic scrubs on the bacterial flora of the hands, *Can. Med. Assoc. J.*, 99 (9), 402, 1968.
6. Kundsin, R.B. and Walter, C.W., The surgical scrub—practical consideration, *Arch. Surg.*, 107 (1), 75, 1973.
7. Lowbury, E.J. and Lilly, H.A., The effect of blood on disinfection of surgeons' hands, *Br. J. Surg.*, 61 (1), 19, 1974.
8. Marples, R.R. and Kligman, A.M., Methods for evaluating topical antibacterial agents on human skin, *Antimicrob. Agents. Chemother.*, 5 (3), 323, 1974.
9. Michaud, R.N., McGrath, M.B., and Goss, W.A., Application of a gloved-hand model for multiparameter measurements of skin-degerming activity, *J. Clin. Microbiol.*, 3 (4), 406, 1976.
10. Eitzen, H.E., Ritter, M.A., French, M.L., and Goe, T.J., A microbiological in-use comparison of surgical hand-washing agents, *J. Bone Joint Surg. Am.*, 61 (3), 403, 1979.
11. Leyden, J.J., Stewart, R., and Kligman, A.M., Updated in vivo methods for evaluating topical antimicrobial agents on human skin, *J. Invest. Dermatol.*, 72 (4), 165, 1979.
12. Hartmann, A.A., A comparison of the effect of povidone–iodine and 60% *n*-propanol on the resident flora using a new test method, *J. Hosp. Infect.*, 6 (Suppl. A), 73, 1985.
13. Bartzokas, C.A., Corkill, J.E., Makin, T., and Pinder, D.C., Assessment of the remnant antibacterial effect of a 2% triclosan-detergent preparation on the skin, *J. Hyg.* (Lond), 91 (3), 521, 1983.
14. Larson, E.L., Eke, P.I., and Laughon, B.E., Efficacy of alcohol-based hand rinses under frequent-use conditions, *Antimicrob. Agents Chemother.*, 30 (4), 542, 1986.
15. Hartmann, A.A., Pietzsch, C., Elsner, P., Lange, T., et al., Antibacterial efficacy of Fabry's tinctura on the resident flora of the skin at the forehead. Study of bacterial population dynamics in stratum corneum and infundibulum after single and repeated applications, *Zentralblatt fur Bakteriologie, Mikrobiologie und Hygiene. Serie B, Umwelthygiene, Krankenhaushygiene, Arbeitshygiene, praventive Medizin*, 182 (5–6), 499, 1986.

16. Lilly, H.A. and Lowbury, E.J., Disinfection of the skin with detergent preparations of Irgasan DP 300 and other antiseptics, *Br. Med. J.*, 4 (5941), 372, 1974.

17. Faoagali, J., Fong, J., George, N., Mahoney, P., and O'Rourke, V., Comparison of the immediate, residual, and cumulative antibacterial effects of Novaderm R,* Novascrub R,* Betadine Surgical Scrub, Hibiclens, and liquid soap, *Am. J. Infect. Control*, 23 (6), 337, 1995.

18. Smylie, H.G., Logie, J.R., and Smith, G., From Phisohex to Hibiscrub, *Br. Med. J.*, 4 (5892), 586, 1973.

19. Bartzokas, C.A., Corkill, J.E., and Makin, T., Evaluation of the skin disinfecting activity and cumulative effect of chlorhexidine and triclosan handwash preparations on hands artificially contaminated with *Serratia marcescens*, *Infect. Control—IC*, 8 (4), 163, 1987.

20. Boyce, J.M. and Pittet, D., Guideline for Hand Hygiene in Health-Care Settings, Recommendations of the Healthcare Infection Control Practices Advisory Committee and the HIPAC/SHEA/APIC/IDSA Hand Hygiene Task Force, *Am. J. Infect. Control*, 30 (8), S1, 2002.

21. Jones, R.D., Jampani, H.B., Newman, J.L., and Lee, A.S., Triclosan: A review of effectiveness and safety in health care settings, *Am. J. Infect. Control*, 28 (2), 184, 2000.

22. Lilly, H.A., Lowbury, E.J., Wilkins, M.D., and Zaggy, A., Delayed antimicrobial effects of skin disinfection by alcohol, *J. Hyg.* (Lond), 82 (3), 497, 1979.

23. Bartzokas, C.A., Corkill, J.E., Makin, T., and Parry, E., Comparative evaluation of the immediate and sustained antibacterial action of two regimens, based on triclosan- and chlorhexidine-containing handwash preparations, on volunteers, *Epidemiol. Infect.*, 98 (3), 337, 1987.

24. McBride, M.E., Montes, L.F., and Knox, J.M., The persistence and penetration of antiseptic activity, *Surg. Gynecol. Obstet.*, 127 (2), 270, 1968.

25. Bruch, M., Newer germicides: What they offer, *Skin Microbiology: Relevance to Clinical Infection*, Maibach, H.I. (ed.), Springer-Verlag, New York, 1981, p. 103.

26. Ayliffe, G.A., Surgical scrub and skin disinfection, *Infect. Control*, 5 (1), 23, 1984.

27. Rosenberg, A., Alatary, S.D., and Peterson, A.F., Safety and efficacy of the antiseptic chlorhexidine gluconate, *Surg. Gynecol. Obstet.*, 143 (5), 789, 1976.

28. Aly, R. and Maibach, H.I., Comparative study on the antimicrobial effect of 0.5% chlorhexidine gluconate and 70% isopropyl alcohol on the normal flora of hands, *Appl. Environ. Microbiol.*, 37 (3), 610, 1979.

29. Larson, E., Mayur, K., and Laughon, B.A., Influence of two handwashing frequencies on reduction in colonizing flora with three handwashing products used by health care personnel, *Am. J. Infect. Control*, 17, 83, 1989.

30. Larson, E.L., APIC guideline for handwashing and hand antisepsis in health care settings, *Am. J. Infect. Control*, 23 (4), 251, 1995.

31. Manivannan, G., Brady, M.J., Cahalan, P.T., et al., Immediate, persistent and residual antimicrobial efficiency of Surfacine (TM) hand sanitizer, *Infect. Control Hosp. Epidemiol.*—The official journal of the Society of Hospital Epidemiologists of America, 21, 105, 2000.

32. Lilly, H.A., Lowbury, E.J., and Wilkins, M.D., Limits to progressive reduction of resident skin bacteria by disinfection, *J. Clin. Pathol.*, 32 (4), 382, 1979.

33. Lowbury, E.J., Lilly, H.A., and Ayliffe, G.A., Preoperative disinfection of surgeons' hands: Use of alcoholic solutions and effects of gloves on skin flora, *Br. Med. J.*, 4 (5941), 369, 1974.

34. Werner, H.P., Rotter, M., and Flamm, H., The immediate and residual effectiveness of surgical scrubbing with ethyl alcohol and hexachlorophene, *Zentralblatt*

fur Bakteriologie, Parasitenkunde, Infektionskrankheiten und Hygiene. Erste Abteilung Originale. Reihe B-Hygiene, praventive Medizin, 155 (5), 512, 1972.

35. Rotter, M., Hand washing and hand disinfection (Chapter 87), *Hospital Epidemiology and Infection Control*, 2nd edition, Mayhall, C.G. (ed.), Lippincott Williams & Wilkins, Philadelphia, 1999.

36. Bhargava, H.N. and Leonard, P.A., Triclosan: Applications and safety, *Am. J. Infect. Control*, 24 (3), 209, 1996.

37. Larson, E.L., Butz, A.M., Gullette, D.L., and Loughon, B.A., Alcohol for surgical scrubbing? *Infect. Control Hosp. Epidemiol.*, 11 (3), 139, 1990.

38. O'Connor, D.O. and Rubino, J.R., Phenolic compounds, *Disinfection, Sterilization and Preservation*, 4th edition, Block, S.S. (ed.), Lea & Febirger, Philadelphia, 1991, p. 204.

39. Peterson, A.F., Rosenberg, A., and Alatary, S.D., Comparative evaluation of surgical scrub preparations, *Surg. Gynecol. Obstet.*, 146 (1), 63, 1978.

40. Kampf, G. and Ostermeyer, C., Efficacy of two distinct ethanol-based hand rubs for surgical hand disinfection—a controlled trial according to prEN 12791, *BMC Infect. Dis. Electronic Resource*, 5 (1), 17, 2005.

41. US. FDA, OTC topical antimicrobial products: Over-the-counter drugs generally recognized as safe, effective, and not misbranded, *Fed. Regist.*, 1978, p. 1210.

42. US. FDA, Topical antimicrobial drug products for over-the-counter human use: Tentative final monograph for health-care antiseptic drug products-proposed rule (Title 21 CFR Parts 333 and 369), *Fed. Regist.*, 1994, p. 31441.

43. Larson, E.L., Eke, P.I., and Laughon, B.E., Efficacy of alcohol-based hand rinses under frequent-use conditions, *Antimicrob. Agents. Chemother.*, 30 (4), 542, 1986.

9

Disinfectant Rotation in a Cleaning/ Disinfection Program for Clean Rooms and Controlled Environments

Scott V. W. Sutton

CONTENTS

9.1 Introduction

An effective cleaning and disinfection program is an essential component for the successful manufacture of products in the pharmaceutical, personal products, and medical device industries. This cleaning and disinfection program is designed to minimize the cross contamination of chemicals between manufacturing runs and to maintain the microbial control of the facility. A major concern in this effort is the potential for microbial development of resistance to the agents used in this program. This concern has led to a regulatory expectation in some circles that manufacturers employ a program of disinfectant rotation in the cleaning and disinfection program to prevent the development of resistant strains of bacteria in their facilities. We examine the scientific underpinnings of this concern in this chapter.

There are three popular positions to assume when discussing disinfectant rotation:

1. Rotate a low pH phenol and high pH phenolic disinfectant
2. Rotate disinfectants with different active chemicals
3. Rotate regular use of a disinfectant with periodic use of a sporicide

It is the position of this author that the third option is the only scientifically justifiable approach to a reasoned disinfection program. Both the first and second options assume the development of relevant levels of resistance to biocides under the conditions of use in pharmaceutical manufacturing facilities.

A recent article by Sartain [1] explored these issues in some depth where it was argued that the appropriate approach was to "rotate" a disinfectant with a sporicide. This issue developed a life of its own on the internet discussion group "PMFList," an e-mail discussion list dedicated to topics of interest to microbiologists in the pharmaceutical and personal care industries.* A great deal of discussion revolved around the concept that a microorganism could become resistant to a disinfectant. Here is where the first bit of clarification is required. A microorganism will not become resistant to much of anything. It either is or it is not affected by the compound. Within a very large population of microorganisms, there is a chance (normally a chance of approximately 10^{-6}) that a cell within the population will have a mutation at a specific gene. This might provide some competitive advantage under some environments, perhaps survival in the presence of an elevated level of a chemical. It is also possible (and usually more likely) that the chance mutation will have no competitive advantage, be somewhat deleterious, or be lethal. The likelihood of a beneficial mutation occurring in the correct environment to confer a competitive advantage in that environment is extremely low. However, should it happen, the mutation that promoted survival under these conditions could, over the course of a few generations, become the dominant genotype in that population. This is, of course, assuming that the challenge is not too great. For example, it does not matter how acidotolerant a particular mutant is, growth in 6 N hydrochloric acid is unlikely—the conditions are too inhospitable to life. Similarly, biocide resistance to slightly elevated levels of the active compound does not confer resistance to use-levels of the biocide several thousand times more concentrated.

Before going further into this discussion, it will be useful to examine the basic mechanisms of biocidal activity. This topic has been the subject of several excellent reviews, particularly those of Denyer and Russell [2] and Lambert [3] who stress that chemical biocides have, as their target, chemical structures (see Table 9.1). In direct contrast to antibiotics, which have specific cellular targets (e.g., the ribosome), chemical biocides may target all proteins as a class.

---

*The PMFList is run by the author, continuously since 1996 (see http://www.microbiol.org/pmflist.htm).

TABLE 9.1
Microbial Targets for Biocides

Biocide	Cell Wall	Cytoplasmic Membrane	Cytoplasm	Effect
Alcohols		Disruption of membrane		Denaturation of proteins
Bronopol		Proteins with thiol groups	Proteins with thiol groups	Oxidation of thiol groups to disulfides leading to structural changes in proteins
Chlorhexidine (*bis*-biguanides)		Disruption of membrane		Binding to phosphate head of fatty acids—disruption of membrane by the reduction in essential membrane fluidity
Ethylene oxide		Proteins with thiol groups	Thiol and amino groups	Alkylation of amino, carboxyl, sulfhydryl, and hydroxyl groups on proteins resulting in structural damage
Aldehydes (formaldehyde, glutaraldehyde)	Fixative (cross-linking)	Fixative	Fixative	Alkylation of amino, carboxyl, sulfhydryl, and hydroxyl groups on proteins resulting in structural damage
Hexachlorophene		Disruption of membrane		Disruption of membrane resulting in cellular lysis
Hydrogen peroxide		Oxidative damage	Oxidative damage	Oxidation of thiol groups to disulfides, leading to structural changes in proteins
Hypochlorites, chlorine releasers	Thiol and amino groups	Thiol and amino groups	Thiol and amino groups	Halogenation of aromatic amino acids
Iodine		Proteins with thiol groups	Thiol groups	Modification of protein structure
Mercurials	Active	Proteins with thiol groups	Thiol groups	Binding to thiol groups causing structural damage to proteins
Phenols		Disruption of membrane	Denaturation of proteins	Denaturation of proteins
Quaternary ammonium compounds		Disruption of membrane		Binding to phosphate head of fatty acids—disruption of membrane by the reduction in essential membrane fluidity

Source: Based on Denyer, S. and Russel, A.D., in S.P. Denyer, M.A. Hodges, and S.P. Gorman (eds.), *Hugo & Russell's Pharaceutical Microbiology*, Blackwell Science, Inc., Malden, MA, 2004, 306–322 and Lambert, P.A., in A.P. Fraise, P.A. Lambert, and J.-Y. Maillard (eds.), *Principles and Practices of Disinfection, Preservation and Sterilization*, Blackwell Science, Inc., Malden, MA, 2004, 139–153.

With this in mind, let us examine the phenomenon of biocide resistance, which is well known and has been extensively reviewed in the literature [4–7]. Extensive work on the mechanism of action allowing previously susceptible microorganisms to survive elevated levels of the biocide have highlighted the participation of efflux pumps as a major contributor to biocide resistance [8,9] as well as physiological adaptation [10].

9.2 The Development of Resistance in Bacterial Populations

9.2.1 Genetic Resistance

The literature provides many examples of microorganisms able to survive in disinfectants. This can be either in laboratory experiments using an increasing level of biocide to select variants in the population or by examination of biocidal solutions for the presence of resistant microorganisms. The gram-negative bacilli are the most frequent isolates from this type of evaluation [11–17]. This may be due to a combination of activities including alterations in outer membrane permeability because of changes in porin diameters [18–21]. In addition, it is not clear that the outer membrane-mediated resistance is in fact due to the selection of a mutant genotype from the population or rather phenotypic adaptation, as this trait has been reported to be lost rapidly once the selective pressure is removed.

The other type of experiment performed to demonstrate development of genetic resistance is one in which a laboratory researcher takes a microorganism in culture and exposes it to increasing levels of biocides, selecting resistance variants at each stage. There is one such set of experiments of particular relevance to the pharmaceutical microbiologist that we will examine in some detail as an example.

In 1992, Conner and Eckman published a study purporting to show the need for rotation of an alkaline phenolic and an acidic phenolic disinfectant to prevent the generation of resistant *Pseudomonas aeruginosa* [22]. This study was a repetitive zone of inhibition design, where an alkaline (pH 10.4) and an acidic (pH 2.6) phenolic disinfectant were placed in paper discs at the center of a bacterial lawn. As the lawn grew to visible turbidity, the disinfectant diffusing out from the disc created a concentration gradient. Conner and Eckman then picked colonies that grew closest to the disc (and so in the highest concentration of disinfectant from the diffusion gradient) for the next cycle. Four treatments were used: water, high pH phenolic, low pH phenolic, and alternating (rotating) the two phenolics for the challenge over 40 cycles of picking a resistant colony, creating a lawn using this new isolate, and then looking for another resistant colony from the first variant.

What the authors found was that the alkaline phenolic treatment eventually resulted in a variant of *P. aeruginosa* that produced no zone of inhibition around the alkaline phenolic-impregnated disc, while the acidic phenolic and the rotated phenolics had measurable zones to the end of the 40 cycles. One plausible explanation for these results is that the alkaline phenolic is a poor disinfectant. However, Conner and Eckman concluded that these data demonstrated the need for rotating disinfectants. They provided no explanation for the fact that the acidic phenolic seemed to be as effective as the "rotated" phenolics at preventing resistance from developing.

They continued this work in another study published in 1993 [23] and 1994 [24] where an adherent *P. aeruginosa* biofilm was formed on a stainless steel coupon and then subjected to repeated cycles of disinfection (by dipping the coupon in the use-dilution) and reinoculation. Survivors were determined by sampling with RODAC plates. The authors showed that none of the treatments eliminated the established biofilm, but the rotation provided a statistically significant decrease in the numbers of CFU than in either the high or the low pH treatments. Unfortunately, the difficulty with using biofilms in this type of study is that the biofilm itself can adapt to different environments [25,26]. This well-studied phenomenon confounds interpretation of the data. In addition, there was no attempt by the authors to determine the sampling efficacy of RODAC plates on biofilm generated under the various conditions. Given the well-documented inefficiency of this sampling method [27,28], and the variations in biofilm structure [29–33], interpretation of the study becomes a bit more difficult.

Unfortunately, there is a dearth of articles in the literature demonstrating the need to rotate disinfectants. Given what we know about population variability, the infrequent rate of favorable genetic mutations, and the mechanism of action of many of these biocides, it seems that the probable scenario for selection of a resistant variant would require exposure of an extremely large number of cells (in excess of 1,000,000 CFU) to a low level of the toxic chemical to result in the selection of resistant microorganisms (variants) from the population. It is not surprising that this has not been reported in the literature from a pharmaceutical manufacturing clean room facility. In fact, this has not been reported even for hospital situations where both the level of microorganisms are much higher and the potential for recognition of the event is more likely [34,35]. Put simply, selection of mutants that are resistant to in-use levels of disinfectants has not been shown to happen in clean room settings. Literature reports of resistance to in-use levels are restricted to descriptions of survival of specific microorganisms in contaminated solutions [9,14,36]. The most common mechanism for this seems to be a modification of the porin structure in the outer membrane of gram-negative cell [37] or an enhancement of the naturally occurring efflux mechanism [38].

9.2.2 Physiological Adaptation

While we have discussed genetic adaptation, there is another mechanism for resistance. In the previous section we touched on biofilms. A biofilm is a complex community of microorganisms suspended in a polysaccharide glycocalyx. The extracellular structure provides a foundation, nutrients for some members of the community, and also physical protection from chemical treatment as it impedes the diffusion of chemicals to the cells in the interior. The ability of the biofilm to withstand large levels of disinfectants is well established [39–41]. This does not, however, speak of the need to rotate disinfectants as the biofilm will provide protection against whatever chemical treatment you attempt, as shown earlier by Conner and Eckman.

There is one other aspect of this discussion we should explore. The disinfection of vegetative cells is quite different from the need to provide sporicidal activity to eliminate spores of bacteria and fungi. Although there are clear differences among different bacterial species in their sensitivities to disinfectants, and these differences can result in persistent bacterial load under some conditions, these differences pale beside the resistance of spores to environmental insult. The spore form is naturally more resistant to chemical treatments, and harsher agents must be used to combat these organisms. This is the basis of the common industrial practice of alternating the daily use of a disinfectant with the periodic use of a sporicide in a manufacturing facility's sanitization program. The obvious approach to this problem might seem to be the exclusive use of the sporicide. However, as Sartain points out, the consistent use of common sporicides (frequently strong oxidizing agents) will result in corrosion of equipment in a relatively short period, and pose safety issues for the technicians applying them to the clean room. It is far preferable to use the gentler disinfectant for as much of the time as possible, reserving the sporicide for periodic cleaning, response to an event (power failure or catastrophic excess of the established action levels for environmental monitoring), or bringing a facility back online after a shutdown.

9.2.3 Relevance

The next question is one of significance. Is this phenomenon of any consequence to the pharmaceutical clean room sanitization program? Arguing logically, it is difficult to justify the concern that a chemical compound that acts by destroying protein, by shredding the microbial membrane, or by any of the other physical mechanisms commonly employed by chemically toxic active agents could select for significant mutational resistance. After all, the assumption is that there is a selective pressure to shift the proportion of resistant phenotypes in the development of widespread resistance—an overwhelming assault leaving none alive selects for nothing.

Sheldon argues the demonstration of antibiotic resistance can be directly correlated with clinical outcome, but that there are no such correlations available for biocide resistance models [42]. Although his analysis focuses on the potential for cross-selection to antibiotic resistance, he notes little evidence of generation of significant levels of biocide resistance in the literature.

Lambert conducted a study using bacteria from a culture collection to perform a survey on biocide resistance over a 10 year period. The bacteria were tested by MIC* to various biocides and antibiotics. The author concludes that there is insufficient evidence to argue that biocide resistance is increasing in the natural state.

McBain et al. evaluated the effect of sublethal dosing of triclosan in a sink drain biofilm [43]. The biofilm present in a household drain was followed before, during, and after exposure to two levels of triclosan. While the relative levels of microorganisms in the biofilm shifted (favoring the naturally more resistant species), the study found no evidence of mutational changes within species toward more resistant strains.

The food industry is concerned about the topic as well and in 2002 a thorough review of the literature was provided by Davidson and Harrison [44]. This excellent review examined many different aspects of the issue, but one in particular bears examination. The "resistance" reported for biocides is an academic definition. That is to say that the difference in wild-type susceptibility and resistance is statistically significant when measured by an MIC. For example, one study reviewed the described resistance of *Listeria monocytogenes* to benzalkonium chloride at 16 μg/mL (see note below about reservations to this testing method, let alone concern over resistance to this level of biocide even with the test used).

Lear et al. examined microbial isolates from biocide-producing plants in two European locations [45]. These facilities manufactured *para*-chloro-*meta*-xylenol (PCMX—a phenolic) and triclosan, and environmental samples were collected from areas likely to have exposure to the biocidal agents. After evaluating 100 isolates by MIC, the authors failed to demonstrate significant resistance (by suspension testing), and concluded

> Attempts to select or generate increased tolerance in the standard strains were unsuccessful. High tolerances in terms of MIC were not reflected in terms of lethal effects. This study did not produce any evidence suggesting that the presence of residual biocide concentrations in the industrial environment promotes the emergence of bacterial tolerance for them.

In short, there is little evidence in the literature that the undisputed resistance seen to biocides in some situations is relevant.

9.2.4 On the Need for Clarity in Terms

A final aspect of this discussion is the semantics of the subject matter. A real problem is the use of the term resistance. This term originated in the clinical microbiology arena where the inhibition of bacterial growth provides a true

---

*MIC testing involves determining the minimum inhibitory concentration of bacterial strains. This test involves making a growth media with varying levels of the biocide, and then inoculating the mixture with the challenge microorganism at low levels. It is a measure of the ability of the biocide to inhibit growth of low numbers of cells in the background of the growth media. The author of this chapter is uncertain how relevant this design might be in evaluating biocides designed to kill (not inhibit growth) of bacteria.

measure of the efficacy of a particular chemical agent against that bacterial species. This measure has little meaning in disinfection where the true test is the ability of the agent to kill bacteria, not prevent them from growing. So, we start off with the wrong test. From this poor beginning we then argue that a slight but measurable increase in the ability of the organism to grow in low levels of the chemical agent is proof of resistance, ignoring the fact that the use-dilution of the agent may be thousands of times more concentrated than the concentration used in the study [46]. Finally, we ignore the basic mechanism of action of disinfectants, incorrectly applying the model of antibiotic resistance (where a mutant develops resistance to a "magic bullet" by altering the specific target of the antibiotic) with the mode of action of disinfectants (most of which act at a basic chemistry level—completely disrupting the cell membrane, or fundamentally altering entire classes of cell components). A resistance that can be measured only in the lab, not in the field, is of little practical concern.

A second problem is the term used to describe the pharmaceutical practice. Given our current understanding, describing what we do as disinfectant rotation is grossly inaccurate. We in fact are not discussing rotating disinfectants at all. Rather we are urging the routine use of an effective disinfectant with the periodic use of a sporicide [47,48]. Block [49] defines a disinfectant as "...an agent that frees from infection, ...that destroys disease or other harmful microorganisms but may not kill bacterial spores. It refers to substances applied to inanimate objects." He goes on to define a sporicide as "...an agent that destroys microbial spores, especially a chemical substance that kills bacterial spores."

What we are discussing is a practice more accurately described as a sanitization program, but certainly not disinfectant rotation—a term that continues to confuse practitioners and regulators alike.

9.3 Summary

The need for rotation of disinfectants in a pharmaceutical clean room sanitization program is unsupportable from a scientific basis. The assumptions that proponents of the practice assert as facts (generation of resistant organisms, greater efficacy of alternating agents) are not supported in the literature. However, even when using a validated disinfectant as part of a well-managed clean room sanitization program, periodic use of a sporicide is a prudent, even an essential component of the sanitization program. It is needed to address the occasional appearance of spore-forming organisms in the environmental monitoring program, and hence ensure the cleanest possible environment for manufacturing. We need to clearly describe our practices, and leave behind the inaccurate phrase "disinfectant rotation" as it does not describe current practice of an effective clean room sanitization program and only confuses discussion of the issues involved.

References

1. Sartain, E.K. 2005. Regulatory update: Rotating disinfectants in cleanrooms: Avoid going in circles. *A2C2*. 8(3):32–33.
2. Denyer, S. and A.D. Russel. 2004. "Non-antibiotic antibacterial agents: Mode of action and resistance" in *Hugo & Russell's Pharmaceutical Microbiology.* S.P. Denyer, M.A. Hodges, and S.P. Gorman (eds.), Blackwell Science, Inc., Malden, MA, pp. 306–322.
3. Lambert, P.A. 2004. "Mechanism of action of biocides" in *Principles and Practices of Disinfection, Preservation & Sterilization*. A.P. Fraise, P.A. Lambert, and J.-Y. Maillard (eds.), Blackwell Science, Inc., Malden, MA, pp. 139–153.
4. Russell, A.D. 1986. Mechanisms of resistance to antiseptics, disinfectants and preservatives. *Pharmacy Int.*, 7:305–308.
5. Poole, K. 2002. Mechanisms of bacterial biocide and antibiotic resistance. *J. Appl. Microbiol. Symp. Suppl.* 92:55S–64S.
6. White, D.G. and P.F. McDermott. 2001. Biocides, drug resistance and microbial evolution. *Curr. Opin. Microbiol.* 4:313–317.
7. Russell, A.D. et al. 1997. Microbial susceptibility and resistance to biocides. *ASM News.* 9:481–487.
8. Levy, S.B. et al. 2002. Active efflux, a common mechanism for biocide and antibiotic resistance. *J. Appl. Microbiol. Symp. Suppl.* 92:65S–71S.
9. Nies, D. 2003. Efflux mediated heavy metal resistance in prokaryotes. *FEMS.* 27:313–319.
10. Liu, X. et al. 2000. Global adaptations resulting from high population densities in *Escherichia coli* cultures. *J. Bacteriol.* 182(15):4158–4164.
11. Ayliffe, G.A.J. et al. 1969. Contamination of disinfectants. *Br. Med. J.* 1:505–511.
12. Geftic, S.G. et al. 1979. Fourteen-year survival of *Pseudomonas cepacia* in a salts solution preserved with benzalkonium chloride. *Appl. Environ. Microbiol.* 37(3):505–510.
13. Berkleman, R. 1981. Pseudobacteremia attributed to contamination of povidone–iodine with *Pseudomonas cepacia*. *Ann. Int. Med.* 32–36.
14. Marrie, T. and J.W. Costerton. 1981. Prolonged survival of *Serratia marcescens* in chlorhexidine. *Appl. Environ. Microbiol.* 42(6):1093–1102.
15. Sautter, R.L. et al. 1984. *Serratia marcescens* meningitis associated with a contaminated benzalkonium chloride solution. *Infect. Control.* 5:223–225.
16. Anderson, R.L. et al. 1984. Investigations into the survival of *Pseudomonas aeruginosa* in poloxamer-iodine. *Appl. Environ. Microbiol.* 47(4):757–762.
17. Anderson, R.L. et al. 1990. Prolonged survival of a strain of *Pseudomonas cepacia* in a commercially manufactured povidone–iodine. *Appl. Environ. Microbiol.* 56:3598.
18. Hancock, R.E.W. 1984. Alterations in outer membrane permeability. *Ann. Rev. Microbiol.* 38:237–264.
19. Nikaido, H. and M. Vaara. 1985. Molecular basis of outer membrane permeability. *Microbiol. Rev.* 49:1–32.
20. Nikaido, H. 1989. Outer membrane barrier as a mechanism of antimicrobial resistance. *Antimicrob. Agents Chemother.* 33(11):1831–1836.
21. Anderson, R.L. 1991. Investigations of intrinsic *Pseudomonas cepacia* contamination in commercially manufactured povidone-iodine. *Infect. Control Hosp. Epidemiol.* 12(5):297–302.

22. Conner, D.E. and M.K. Eckman. 1992. Rotation of phenolic disinfectants. *Pharm. Technol.* (Sept.):148–160.
23. Conner, D.E. and M.K. Eckman. 1993. Rotation of phenolic disinfectants enhances efficacy against adherent *Pseudomonas aeruginosa. Pharm. Technol.* 6(1):94–104.
24. Conner, D.E. and M.K. Eckman. 1994. Rotation of phenolic disinfectants enhances efficacy against adherent *Pseudomonas aeruginosa. Pharm. Tech. Eur.* 6(1):44–47.
25. Lewis, K. 2001. Riddle of biofilm resistance. *Antimicrob. Agents Chemother.* 45(4):999–1007.
26. Rachid, S. et al. 2000. Effect of subinhibitory antibiotic concentrations on polysaccharide intercellular adhesin expression in biofilm-forming *Staphylococcus epidermidis. Antimicrob. Agents Chemother.* 44(12):3357–3363.
27. Favero, M.S. et al. 1968. Microbiological sampling of surfaces. *J. Appl. Bacteriol.* 31:336–343.
28. Whyte, W. 1989. Methods for calculating the efficiency of bacterial surface sampling techniques. *J. Hosp. Infect.* 13:33–41.
29. Stoodley, P. et al. 2001. Growth and detachment of cell clusters from mature mixed-species biofilms. *Appl. Environ. Microbiol.* 67(12):5608–5613.
30. Tolker-Nielsen, T. et al. 2000. Development and dynamics of *Pseudomonas* sp. biofilms. *J. Bacteriol.* 182(22):6482–6489.
31. Sauer, K. et al. 2002. *Pseudomonas aeruginosa* displays multiple phenotypes during development as a biofilm. *J. Bacteriol.* 184(4):1140–1154.
32. Sauer, K. et al. 2004. Characterization of nutrient-induced dispersion in *Pseudomonas aeruginosa* PAO1 biofilm. *J. Bacteriol.* 186(21):7312–7326.
33. Wagner, V.E. et al. 2003. Microarray analysis of *Pseudomonas aeruginosa* quorum-sensing regulons: Effects of growth phase and environment. *J. Bacteriol.* 185(7):2080–2095.
34. Suller, M.T.E. and A.D. Russell. 1999. Antibiotic and biocide resistance in methicillin-resistant *Staphylococcus aureus* and vancomycin-resistant *Enterococcus. J. Hosp. Infect.* 43:281–291.
35. Suller, M.T. and A.D. Russell. 2000. Triclosan and antibiotic resistance in *Staphylococcus aureus. J. Antimicrob. Chemother.* 46:11–18.
36. Langsrud, S. et al. 2003. Characterization of *Serratia marcescens* surviving in disinfecting footbaths. *J. Appl. Microbiol.* 95:186–195.
37. De Spiegeleer, P. et al. 2005. Role of porins in sensitivity of *Escherichia coli* to antibacterial activity of the lactoperoxidase enzyme system. *Appl. Environ. Microbiol.* 71(7):3512–3518.
38. Poole, K. 2005. Efflux-mediated antimicrobial resistance. *J. Antimicrob. Chemother.* 56:20–51.
39. Stickler, D.J. 2002. Susceptibility of antibiotic-resistant gram-negative bacteria to biocides: A perspective from the study of catheter biofilms. *J. Appl. Microbiol. Symp.* (Suppl. 92):163S–170S.
40. Ntsama-Essomba and C. Ntsama. 1997. Resistance of *Escherichia coli* growing as biofilms to disinfectants. *Vet. Res.* 353–363.
41. Gilbert, P. et al. 1990. Influence of growth rate on susceptibility to antimicrobial agents: Biofilms, cell cycle, dormancy and stringent response. *Antimicrob. Agents Chemother.* 34(10):1865–1868.
42. Sheldon, A. 2005. Antiseptic "resistance": Real or perceived threat? *CID.* 40(1):1650–1656.

43. McBain, A.J. et al. 2003. Exposure of sink drain microcosms to triclosan: population dynamics and antimicrobial susceptibility. *Appl. Environ. Microbiol.* 69(9):5433–5442.
44. Davidson, P.M. and M.A. Harrison. 2002. Resistance and adaptation to food antimicrobials, sanitizers, and other process controls. *Food Technol.* 56(11):69–78.
45. Lear, J.C. et al. 2002. Chloroxylenol and triclosan-tolerant bacteria from industrial sources. *J. Ind. Micro. Biotech.* 29:238–242.
46. Gilbert, P. et al. 2002. Biocide abuse and antimicrobial resistance: Being clear about the issues. *J. Antimicrob. Chemother.* 50:137–148.
47. USP. 2004. <1072> Disinfectants and antiseptics. *Pharm. Forum.* 30(6):2108–2118.
48. USP. 2005. <1116> Microbiological control and monitoring environments used for the manufacture of healthcare products. *Pharm. Forum.* 31(2):524–549.
49. Block, S.S. 1991. Definition of terms, *Disinfection, Sterilization and Preservation*, 4th edition. S.S. Block (ed.), Lea & Febirger, Philadelphia. pp. 18–25.

10

Reprocessing Flexible Endoscopes: Origin of Standards, Overview of Structure/Function, and Review of Recent Outbreaks

Terri A. Antonucci, Nicole Williams, and Nancy A. Robinson

CONTENTS

10.1   Introduction

The use of flexible endoscopes in minimally invasive procedures has increased dramatically since the early 1960s. As with most fields of practice, the development, use, and reprocessing standards for flexible endoscopes have evolved through time. Some key highlights of this evolution include

- The first clinical report of flexible fiberoptic endoscopy was published in 1961 [1]. At first, the endoscope was prepared for the next usage by a variety of methods, typically flushing the channels with fluid (sometimes just water) and wiping the exterior.

- In the mid-1960s, as the potential for transmission of disease was recognized, a flush with a disinfectant such as quaternary ammonium compounds or alcohol was added.

- The first of the sterilant/high-level disinfectant chemical germicides, buffered 2% glutaraldehyde, was introduced in the mid-1970s.

- In 1978, the Society of Gastroenterology Nurses and Associates (SGNA) first published guidelines for care of endoscopic equipment.

- It was not until after 1985 that endoscopes were fully immersible for cleaning and disinfection, and the suction channel was first made brush-accessible. In this same period, the endoscopes became more complex, for example, the addition of the "elevator" channel.

- In 1988, the American Society for Gastrointestinal Endoscopy, the American Gastroenterological Association, and the American College of Gastroenterology jointly published guidelines for cleaning and disinfection of gastrointestinal (GI) endoscopes which stressed:

  - Cleaning of all channels
  - Use of an approved high-level disinfectant
  - A water rinse to remove residual germicide
  - Forced air drying after rinsing

The objectives of this chapter on flexible endoscope reprocessing are to

1. present an understanding of flexible endoscope reprocessing practices,
2. provide a systematic explanation of flexible endoscope structure and function as it relates to effective reprocessing, and
3. review flexible endoscope true and pseudo-outbreaks published from 1995 to 2005 and summarize the suspected causes of these more recent reprocessing failures.

10.2 The Principles of Practice: Where Do They Come From?

10.2.1 Introduction

As healthcare professionals, we are dedicated to delivering the best possible care to our patients. But how do we know what to do and how to do it?

For years, healthcare practices have evolved based on historical data, research, expert opinions, and consensus. Different professional organizations have studied practice needs and have systematically developed models to guide the activities of patient care. Many organizations have collaborated with others to provide interrelated and consistent guidelines and recommendations that are attainable, manageable, up to date, and relevant for medical and nursing practices.

International healthcare arenas and organizations often look to the United States for assistance in developing their own standards of practice. For example, the American National Standards Institute Z136.3 *American National Standard for Safe Use of Lasers in Health Care Facilities* is being used throughout the world to govern the practice of laser surgery and treatments. This standard implies a consensus of those significantly concerned with the scope of laser safety and application.

Endoscopy is another area where U.S. standards tend to be relied upon to guide international organizations. The popularity and demand for endoscopy has grown significantly in the past 20 years. Because of this rapid growth, dependable practices have been introduced and accepted in most areas. So, what are these resources that guide the practices of endoscopy?

10.2.2 Standards, Recommended Practices, Guidelines, and Statements

Practice resources can be divided into four major categories which include standards, recommended practices (RPs), guidelines, and statements. Each has its own suggested definition and meaning as characterized by the Association of periOperative Registered Nurses (AORN) in the *Standards, Recommended Practices, and Guidelines*, 2006 edition [2].

10.2.2.1   Standards

"Authoritative statements that describe the responsibilities for which . . . nurses are accountable. Standards reflect the values and priorities of the profession. They are a means to direct and evaluate professional nursing practice." [2]

Standards are categorized into standards of care and standards of professional performance. Standards of care reflect the nursing process of assessment, planning, implementation, and evaluation. They address structure (the administrative aspect of nursing), process (nursing competency), and outcome (the patient's observable and measurable responses). Standards of professional performance deal with areas such as performance appraisal, ethics, collegiality, research, leadership, etc. Standards are reviewed approximately every 5 years to make sure they are consistent with the current healthcare environment.

10.2.2.2   Recommended Practice

"Achievable recommendations representing what is believed to be an optimal level of . . . nursing practice. RPs are broad statements to be used to guide policy and procedure development in specific work environments. Although they are considered to represent the optimal level of practice, variations in practice settings and clinical situations may limit the degree to which each recommendation can be implemented." [2]

AORN's RPs, as stated, are based on an optimal level of practice. Compliance is then voluntary. On the other hand, some professional organizations' RPs reflect minimal performance, meaning that this is the least that a healthcare professional is expected to do. All RPs are based on research, a review of scientific literature, and expert opinion.

10.2.2.3   Guidelines

"Addresses a specific medical diagnosis or clinical condition and is based on empirical data. A guideline should assist practitioners in clinical decision making, be used to assess and assure the quality of care, and to guide clinical practice." [2]

A guideline reflects the current state of knowledge. Some examples of guidelines address latex allergies or the reuse of single-use devices.

10.2.2.4   Statements

There are three different types of statements:

- Guidance statement "provides suggested strategies to assist practitioners in developing facility-specific processes related to clinical and administrative issues"
- Competency statement "represents expected, measurable, . . . nursing behaviors. Defines the knowledge, skills, and abilities necessary to fulfill the professional role functions"

- Position statement "articulates... official position or belief about certain... topics" [2]

The review and revision processes for statements are often undefined but many suggest a 5-year review cycle. Some organizations will sunset a position statement meaning that the statement will be reaffirmed, archived, becoming a policy, or dropped after 5 years.

10.2.3 Development and Use of Recommended Practices

Since healthcare is a standards-seeking environment, resources that lead to consistent and safe practices are vital and needed. The resources used must reflect agreement and consensus so that they will be followed and valued.

The three organizations that nursing professionals turn to for best practices in flexible endoscopy are the Society of Gastroenterology Nurses and Associates (SGNA), Association of periOperative Registered Nurses (AORN), and Association of Practitioners in Infection Control and Epidemiology (APIC). These organizations have collaborated and continually share and update their recommendations for care of the endoscopy patient and reprocessing of the instrumentation.

Since most healthcare professionals look at RPs to develop their own facility policies and procedures, an understanding of how these RPs are developed is helpful. The process of development is outlined below:

1. Assessment—A needs assessment is performed to determine if an RP is considered necessary.
2. Planning—A team is formed to determine how to address the documented need. Group meetings are held to validate the need though consensus.
3. Implementation—A draft of the RP is written by a committee of experts and then reviewed by the organization's board. Changes are made as needed.
4. Evaluation—A peer review of the RP is conducted. Suggestions are addressed and changes made as needed.
5. RP finalization—The final version of the RP is published for healthcare providers to reference.
6. Review—An RP is usually reviewed every 3 years unless drastic changes in technology or practices warrant changes before the 3 year period has passed.

Standards, RPs, guidelines, and statements have many uses in the healthcare arena today. They not only guide practice as policies and procedures but also are used for competency reviews, lawsuits, industry instruction manuals, and even as references for the media and legislators. Organizations will usually preface their standards, RPs, guidelines, and statements with disclaimers that state that the organization will assume no liability or responsibility for the result of following one of these resources.

10.2.4   Flexible Endoscope Reprocessing Practices

Healthcare professionals can become confused with the number of resources available today. If the practice of reprocessing flexible endoscopes were examined, SGNA, AORN, and APIC recommendations would be referenced. SGNA's standards and guidelines represent a minimal level of expected practice and address personnel training, quality assurance, reprocessing, cleaning, disinfection, sterilants, and endoscope storage. AORN's RPs represent the optimal level of expected practice and address care, cleaning, decontamination, maintenance, storage, sterilization, and disinfection of flexible and rigid endoscopes and accessories. APIC's guideline for endoscope care reflects a minimum standard of care and addresses infections, flexible endoscope design, cleaning, sterilization, disinfection, storage, quality control, outbreak management, and endoscopy personnel.

Each organization (SGNA, AORN, and APIC) consistently recommends that healthcare professionals should follow manufacturers' written instructions on the care and handling of endoscopes. Manual cleaning should be performed immediately after use followed by thorough rinsing. Spaulding's classifications should be used to determine the type of reprocessing needed (sterilization, disinfection, or cleaning).

These guidelines can also be contrasted. For example, SGNA recommends that high-level disinfection is recognized as the standard of care while AORN suggests that high-level disinfection of endoscopes reflects a minimum standard of care. Drying the endoscope between patients following liquid sterilization is not recommended by AORN and APIC but is recommended by SGNA. Drying the flexible endoscope before storage and after liquid sterilization is recommended by SGNA and APIC but not by AORN, which believes that endoscopes should be processed immediately before use if not packaged and sterilized.

The SGNA has noted that there can be wide practice variations with poor outcomes as a result of inconsistencies in guideline recommendations. More uniformity and consistency are needed in the care and handling of flexible endoscopes.

A survey conducted by SGNA noted that compliance with existing guidelines ranged from 67% to 93% when 2030 surveys were tallied [3]. New guidelines would not be needed today if existing guidelines were strictly adhered to by healthcare professionals. But because of the inconsistencies of the different recommendations, a consensus guideline was deemed necessary.

In 2003, a multi-society meeting was scheduled by the American Society for Gastrointestinal Endoscopy (ASGE) and the Society for Healthcare Epidemiology of America (SHEA). They invited physician and nursing organizations, infection control organizations, federal and state agencies, and industry leaders to develop a consistent document for the care and handling of flexible endoscopes. The purpose was to discuss inconsistencies in the various guidelines and develop a consensus document that could be published.

The format for the "Multi-society guideline for reprocessing flexible gastrointestinal endoscopes" follows the CDC listing for categorizing recommendations. Each specific statement is rated as to category or weight:

Category IA: Strongly recommended for implementation and strongly supported by well-designed experimental, clinical, or epidemiologic studies.

Category IB: Strongly recommended for implementation and supported by some experimental, clinical, or epidemiologic studies, and a strong theoretical rationale.

Category IC: Required by state or federal regulations. Because of state differences, readers should not assume that the absence of an IC recommendation implies the absence of state regulations.

Category II: Recommended for implementation and supported by suggestive clinical or epidemiologic studies or theoretical rationale.

No recommendation: Unresolved issue. Practices for which insufficient evidence or no consensus regarding efficacy exists.

Having these categories helps the healthcare professional understand the importance of mandating some of the recommendations versus not needing to be as compliant with some of the others. This allows facilities to determine the best practices for their individual settings.

For each category, examples of recommendations are noted: [4]

Category IA: Strongly recommended, strongly supported

- "Reusable endoscopic accessories that break the mucosal barrier should be cleaned and sterilized."
- High-level disinfection: "Studies and organizations support the efficacy of >2% glutaraldehyde at 20 min at 20°C."
- "Temporary personnel should not be allowed to reprocess endoscopes until competency has been established."

Category IB: Strongly recommended, some support

- "Nonimmersible GI endoscopes should be phased out immediately."
- "High-level disinfect or sterilize the water bottle and its connecting tube at least daily. Sterile water should be used to fill the water bottle."
- "Discard enzymatic detergents after each use because these products are not microbicidal and will not retard microbial growth."

Category IC: Required by state or federal regulations

- "All personnel who use chemicals should be educated about the biologic and chemical hazards present while performing procedures that use disinfectants."

- "Personal protective equipment should be readily available and should be used . . . to prevent workers from exposure to chemicals, blood, or other potentially infectious material."

Category II: Supported by suggestive studies

- "Cleaning items (i.e., brushes) should be disposable or thoroughly cleaned and disinfected/sterilized between uses."
- "Endoscopes should be stored in a manner that will protect the endoscope from contamination."
- "Healthcare facilities should develop protocols to ensure that users can readily identify whether an endoscope is contaminated or is ready for patient use."

No recommendation: Unresolved issue

- "The utility of routine environmental microbiologic testing of endoscopes for quality assurance has not been established."

Compliance with these recommendations that address the reprocessing of flexible endoscopes between patient use can prevent the transmission of pathogens. Efforts to encourage healthcare professionals to understand and employ these recommendations should be increased for the safety and success of flexible endoscopic procedures. The organizations that participated in the development of this multi-society guideline are committed to addressing critical device reprocessing issues for flexible endoscopes and accessories.

10.3 Flexible Endoscope Structure/Function

10.3.1 Introduction

Endoscopes are medical instruments used for minimally invasive evaluation and surgical procedures. There are many kinds of endoscopes; those used for diagnostic purposes are used for viewing only, while others are used for surgical procedures, and have channels that allow introduction of surgical instruments. The three basic types of endoscopes are rigid, semirigid, and flexible. This chapter focuses on the flexible endoscopes that are complex, and therefore challenging to reprocess.

Flexible endoscopes can be broken into two groups: simple (for example bronchoscopes) and GI (for example, colonoscopes) endoscopes. Table 10.1 lists various types of simple and GI flexible endoscopes and the body area they are used to examine, and also provides a brief descriptive summary of the associated medical procedure. From the information presented in Table 10.1, it is clear that flexible endoscopes are diverse in their application; however,

TABLE 10.1

Endoscope and Procedure Summary

Flexible Endoscope Type	Body Area Examined	Procedure Summary
Simple		
Bronchoscope	Trachea, bronchi, and lungs	Procedure is called bronchoscopy. It is usually not performed in an operating room. Endoscope is inserted through the mouth or nose. Procedure can last from 30 min–2 h [5].
Cystoscope	Urinary bladder	Procedure is called cystoscopy. It is sometimes performed in an operating room. Endoscope is inserted through the urethra. Procedure can last 15–30 min [5].
Ureteroscope	Ureters and bladder	Procedure is called ureteroscopy. It is usually performed in an operating room. Endoscope is inserted through the urethra. Procedure can last 15–45 min [6].
Hysteroscope	Vagina and uterus	Procedure is called hysteroscopy. It is performed in an operating room. Endoscope is inserted through the vagina. Procedure can last 30 min [7].
Gastrointestinal (GI)		
Colonoscope	Colon—large intestine	Procedure is called a colonoscopy or lower endoscopy and is usually preformed in a GI clinic or treatment room. Endoscope is inserted through the anus. Procedure lasts 30–60 min [5].
Gastroscope	Stomach	Procedure is called gastroscopy or upper endoscopy and is usually performed in a GI clinic or treatment room. Endoscope is inserted through the mouth. Procedure lasts 15–30 min [5].
Duodenoscope	Duodenum (first part small intestine)	Procedure is called endoscopic retrograde cholangiopancreatography (ERCP) and is usually performed in GI clinics and treatment rooms. The scope is inserted through the mouth. Procedure lasts 30 min–2 h [8].
Sigmoidoscope	Rectum and sigmoid colon (lower part of large intestine)	Procedure is called sigmoidoscopy and is usually preformed in a doctor's office. Endoscope is inserted through the anus. Procedure lasts 15–30 min [5].
Small Bowel Scope	Small intestine	Procedure is called push or double balloon enteroscopy and is usually performed in a GI clinic or treatment room. Endoscope is inserted through the mouth. Procedure lasts 15–45 min [9].

the structural elements are surprisingly similar. A detailed understanding of the internal and external structural features is imperative for safe and effective cleaning and disinfection of flexible endoscopes between patient uses. The following sections provide an overview of the structure and function of flexible endoscope features. The features described have been generalized. It is extremely important to refer to the flexible endoscope manufacturer's instruction manual for a detailed understanding of a specific endoscope.

10.3.2 External Features

Flexible endoscopes have many similarities in their designs. Each manufacturer may use different terminology for similar parts of the endoscopes. Table 10.2 lists the terms that will be used in this section to identify the main external components and ports of a flexible endoscope. Also provided is a listing of alternative nomenclature used for equivalent features.

While all of the components listed in Table 10.2 will never exist on the same endoscope model, for demonstration purposes, all of the parts are shown on the same generic endoscope in Figure 10.1.

10.3.3 Removable Components

Table 10.3 lists the common removable components used with flexible endoscopes. Again, because each endoscope manufacturer may use different terminology for these components, alternative names are also provided.

TABLE 10.2

External Components: Summary

Terms Used in Chapter	Can Also Be Referred to as [10]
Air pipe	Air supply connector
Air/water port	Water container connector, water supply connector, feed water connector
Air/water valve port	Air/water button, air/water feeding valve
Auxiliary water inlet/channel	Auxiliary water supply inlet, water jet inlet, forward water jet, water jet channel
Biopsy port	Forceps inlet, channel inlet, instrument channel port
CO_2 valve port	Same
Control handle	Control portion, control section, control body
Distal tip	Distal end
Elevator guide wire (EGW)	Forceps elevator wire washing inlet, elevator wire cleaning inlet, wire port, elevator channel plug
Insertion tube	Insertion portion, flexible portion
Light guide cable	Umbilical cable, umbilical, universal cord, LG flexible portion
Light guide connector	Endoscope connector, EVE connector, PVE connector, connector
Suction barb	Suction connector, suction nipple
Suction valve port	Suction button, suction control valve

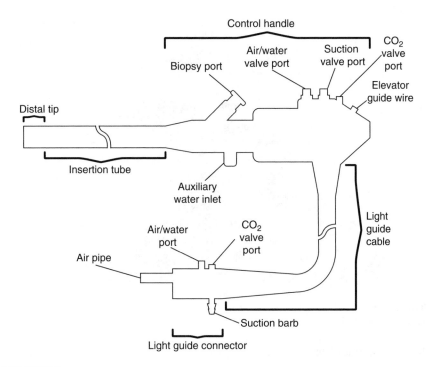

FIGURE 10.1
Generie flexible endoscope.

Figure 10.2 shows examples of some common removable parts. Before cleaning and disinfecting an endoscope, all components are removed to allow access to the internal channels and areas that may otherwise be masked, and therefore hinder reprocessing. For this reason, it is extremely important to identify all removable components of a specific flexible endoscope. The endoscope manufacturer's instruction manual will list all components for the endoscope and the instructions for the proper removal.

TABLE 10.3

Common Removable Parts

Terms Used in Chapter	Can Also Be Referred to as [10]
Air/water valve	Air/water button, air/water feeding valve
Biopsy cap	Biopsy valve, rubber inlet seal, forceps inlet cap
Distal hood	Same
Channel selector	Suction switch-over lever, instrument channel selector, suction selector knob
CO_2 valve	Same
Irrigation tube	Auxiliary water tube, water jet irrigation tube
Suction valve	Suction control valve

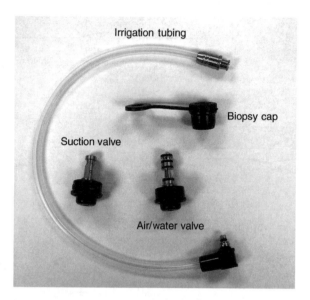

FIGURE 10.2
Common removable accesories.

10.3.4 Channel Configurations

While some flexible endoscopes are used solely for visualization, and there-fore do not contain internal channels, most flexible endoscopes contain one to three internal channels. In all, there are five types of internal channel systems: suction/biopsy, air/water, auxiliary water, CO_2, and elevator guide wire. Every channel of the endoscope must be cleaned and disinfected after each use regardless of whether or not the channel was used during the procedure. An understanding of the internal features of the channels and the various entry points will help ensure that all portions of every channel are properly reprocessed. With this goal in mind, each of the four channel systems and their various design configurations are described below.

10.3.4.1 Suction/Biopsy System

Channeled flexible endoscopes have, at a minimum, a suction/biopsy system. The purpose of the suction/biopsy channel is twofold:

1. To allow access for endoscopic accessories, such as biopsy forceps or snares, for sampling (biopsy portion of the system)
2. To remove fluids from the patient (suction portion of the system)

The suction/biopsy channel may be accessed via three ports: the suction barb (which is located on the control handle for simple endoscopes and on the light guide connector for GI endoscopes), the suction valve port (located on the control handle), and the biopsy port (which is located on the control

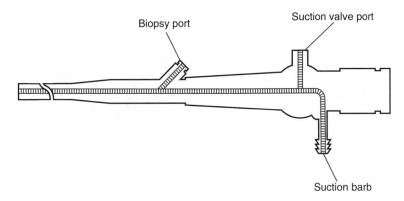

FIGURE 10.3
Suction/biopsy channel of a simple flexible endoscope.

handle). The suction/biopsy channel exits the endoscope at the distal tip. Figure 10.3 shows a schematic of the suction/biopsy channel for a simple flexible endoscope such as a bronchoscope, and Figure 10.4 shows that for a GI flexible endoscope such as a colonoscope.

Some GI endoscopes have dual suction/biopsy channels. This type of endoscope is considered a specialty scope. It is used during procedures where there may be excess fluids or where many samples will be taken. With two suction/biopsy channels, one can be used for sampling and the

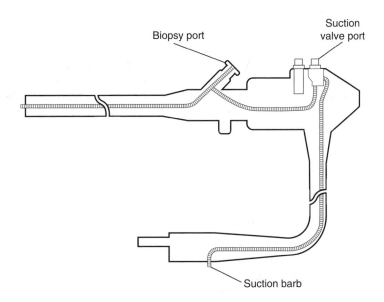

FIGURE 10.4
Suction/biopsy channel of a gastrointestinal flexible endoscope.

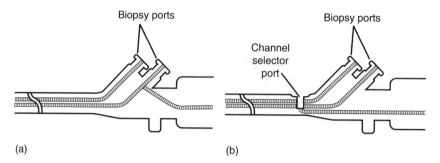

FIGURE 10.5
Dual suction channel configurations.

other used for the removal of fluids. These two channels may be separate as in Figure 10.5a or joined together with the use of a channel selector as in Figure 10.5b. The channel selector allows the user to choose which channel is used during the procedure channel: a, b, or both.

10.3.4.2 Air/Water System

Simple flexible endoscopes do not have air/water systems; however, flexible GI endoscopes always have an air/water system in addition to the biopsy/suction system. There are two purposes of the air/water system:

1. To provide insufflation to the area being examined
2. To clear the lens of debris

The air/water system consists of two channels that begin at the air/water port at the light guide connector. The channels extend through the light guide cable and join at the air/water valve port on the control handle. The two channels exit the air/water valve port separately and continue down the insertion tube, where they may connect before exiting at the distal tip. Some flexible GI endoscopes have an air pipe, which is also connected to the air/water port. Figures 10.6 and 10.7 show air/water systems where the air and water lines do and do not join in the insertion tube, respectively.

10.3.4.3 Auxiliary Water System

Some GI endoscopes have an auxiliary water channel. The auxiliary water channel is used to irrigate the area in front of the endoscope's distal tip to improve visibility. The auxiliary water system may be an independent channel that begins at one of two areas on the control handle (Figures 10.8a and b) or at the light guide connector (Figure 10.9). The auxiliary water channel may also join the water channel in the control handle as shown in Figure 10.10.

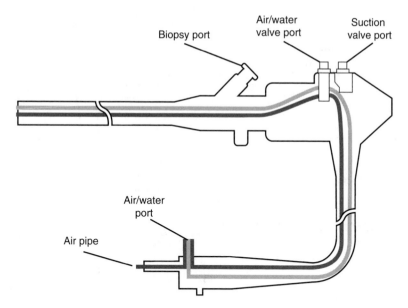

FIGURE 10.6
Air/water system not joined.

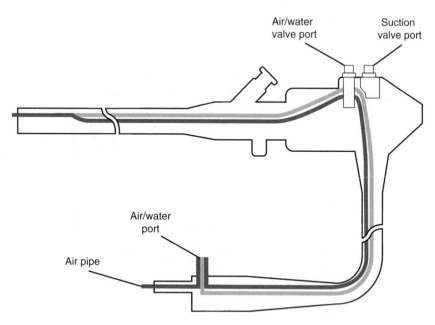

FIGURE 10.7
Air/water system joined.

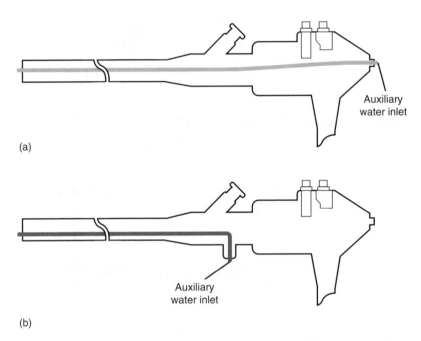

(a)

(b)

Auxiliary
water inlet

Auxiliary
water inlet

FIGURE 10.8
Independent auxillary water channel with inlet on the control handle.

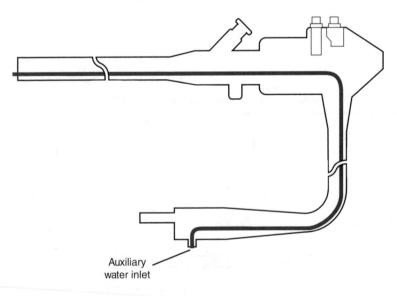

Auxiliary
water inlet

FIGURE 10.9
Independent auxillary water channel with inlet on the light guide.

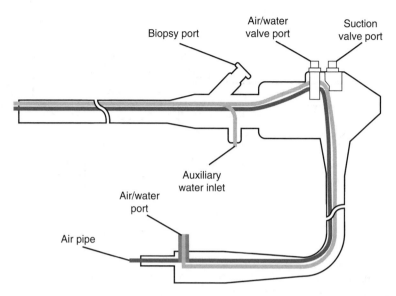

FIGURE 10.10
Auxillary water channel inlet on the control handle, channel joins the air/water system.

10.3.4.4 *CO₂ System*

Some older colonoscopes and sigmoidoscopes may have a CO_2 lumen. The CO_2 lumen joins the air line in various spots, usually by the air pipe on the light guide connector or in the air/water valve port on the control handle. The CO_2 lumen was used to deliver CO_2 to the patient for insufflation instead of air. This practice is no longer followed and not many devices with CO_2 lumens are in use today; because of this a figure will not be shown. When using an endoscope with the CO_2 feature, be sure to refer to the manufacturer's instruction manual for a complete understanding of the device.

10.3.4.5 *Elevator Guide Wire System*

The elevator guide wire is found on duodenoscopes (also referred to as ERCP endoscopes) and some specialty endoscopes such as dual-channel endoscopes. It is used during endoscopic retrograde cholangiopancreato-graphy (ERCP) to lift and move instruments to a 40°–90° angle for visualization and placement. The elevator guide wire is operated by an extra angulation lever located on the control handle. This channel can be a closed system that is not exposed to any patient soil during use, and therefore does not require cleaning and decontamination, or it can be an open system, where it is an additional channel that needs special attention during reprocessing because of its small size. Figure 10.11 shows an open elevator guide wire system.

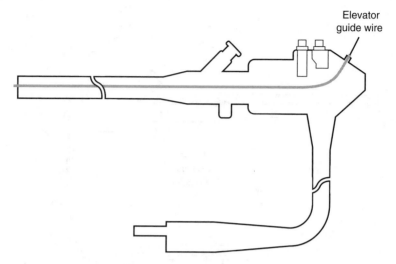

FIGURE 10.11
Open elevator guide wire system.

10.3.5   Importance of Understanding Flexible Endoscope Design

Flexible endoscopes can range from being very simple in design, such as endoscopes that do not contain internal channels, to very complex colono-scopes that contain dual biopsy channels in addition to an air/water and auxiliary water system. In order to safely and effectively reprocess these medical devices, review the manufacturer's instruction manual specific to the device model being reprocessed to

- Identify all removable components
- Determine the configuration of internal channels
- Identify the access ports for each channel

By understanding these critical design elements, the personnel responsible for reprocessing will be sure to remove all removable components and clean and disinfect all portions of all internal channels.

10.4   Endoscopy-Related Contamination Outbreaks

10.4.1   Introduction

The risk of exogenous infection caused by flexible endoscopes (patient-to-patient or environment-to-patient transmission of microorganisms during endoscopy) is considered to be low. In general, breaches in reprocessing protocol have been the assigned cause when evidence for contamination has been traced to the flexible endoscope. The intent of this section is to provide an overview of flexible endoscope-associated contamination outbreaks with emphasis on reports from the past 10 years.

Many of the early documented reports of endoscopy-related contamination outbreaks were a result of ineffective decontamination directly attributable to the challenges of adopting a new and revolutionary technique and the difficulties in properly reprocessing the early endoscope designs. These cases were first reviewed by Spach et al. [11] in a 1993 publication. This review concluded that the three major reasons for transmission were

- Improper cleaning and disinfection procedures
- Contamination of endoscopes by automatic washers
- Inability to decontaminate the endoscope because of complex channel and valve systems

Since the Spach et al. article, several other excellent reviews have reported infections associated with use of flexible endoscopes. In 2001, Weber and Rutala [12] reviewed outbreaks associated with bronchoscopy and in 2003, Nelson [13] published a review which discussed patient-to-patient transmission of pathogens during GI endoscopy. Using the articles referenced in these reviews, as well as literature searches to identify more recent articles, case reports from 1995 to 2005 were analyzed to determine the attributed causes of true and pseudo-outbreaks in the last 10 years. The reports associated with GI endoscopy procedures are listed in Table 10.4 while those that occurred during bronchoscopy procedures are presented in Tables 10.5 and 10.6 for true and pseudo-outbreaks, respectively.

10.4.2 GI Endoscopy–Related Outbreaks

10.4.2.1 Case Review

Ten reports of GI endoscopy-related outbreaks were identified from 1995 to 2005. These reports are listed chronologically (based upon publication date) in Table 10.4.

Case 4.1: The first case report listed in Table 10.4 was published in 1995 by Miyaji et al. [14]. This incident is unusual in that it was one of the authors who became ill after being subjected to a phenol-red-dye spraying endoscopy procedure under development for diagnosis of *Helicobacter pylori* infection. The subject had no history of gastroduodenal disease before the procedure and the phenol-red-dye spraying as well as biopsy specimens taken during the procedure were negative for *H. pylori*. Three days post-procedure, he began to display symptoms. After 7 days, the procedure was repeated and this time *H. pylori* was detected in an erosive

TABLE 10.4

GI Endoscopy–Associated Outbreaks

Case	Reference	Publication Year	Country	Organism	Suspected Cause of Outbreak
4.1	[14]	1995	Japan	*Helicobacter pylori*	Inadequate cleaning and disinfectant
4.2	[15]	1996	Taiwan	*H. pylori*	Inadequate disinfectant
4.3	[16]	2001	Italy	*Trichosporon asahii*	Forceps disinfected, not sterilized
4.4	[17]	1996	Australia	Hepatitis B virus (HBV), Hepatitis C virus (HCV)	Unknown—colonoscopy with biopsy risk factor suggested. Evidence that the endoscopic procedure caused transmission is lacking
4.5	[19]	1997	France	HCV	Inadequate cleaning—biopsy channel was not brushed. Forceps disinfected, not sterilized
4.6	[20]	1999	France	HCV	Contaminated syringe or multiuse vial
4.7	[21]	1999	United States	HBV	Unknown—evidence that the endoscopic procedure caused transmission is lacking
4.8	[22]	1999	Spain	*Aeromonas hydrophila*	Inadequate disinfectant—quaternary ammonia and glutaraldehyde phenate
4.9	[23]	2004	United States	*Pseudomonas aeruginosa*	Use of a particular endoscope (loaner). Cause unknown—endoscope culture was negative
4.10	[24]	2005	India	*P. aeruginosa*	Inadequate cleaning, possible inadequate disinfectant exposure time

gastritis lesion. By day 14, the lesion was healed and *H. pylori* was not detected. Investigation of the incident revealed that the endoscope had been used on a patient who was infected with *H. pylori* just before the subject's procedure. The endoscope had been washed manually by wiping the sheath with a hyamine-soaked paper towel and 0.2% hyamine was

TABLE 10.5

Bronchoscopy-Related True Outbreaks

Case	Reference	Publication Year	Country	Organism	Suspected Cause of Outbreak
5.1	[25]	1997	United States	Multidrug-resistant *Mycobacterium tuberculosis*	Inadequate cleaning, no leak test, no potency testing of glutaraldehyde, bronchoscope not totally immersed, tap water for final rinse without use of alcohol flush
5.2	[26]	1997	United States	*M. tuberculosis*	Did not use enzymatic cleaner, scope not fully immersed, biopsy forceps not sterilized
5.3	[27]	2000	United Kingdom	*Pseudomonas aeruginosa*	Lack of reprocessing system maintenance
5.4	[28]	2001	Germany	*P. aeruginosa*	Disinfectant concentration inadvertently set too low (0.04% rather than 3.0%)
5.5	[29]	2001	United States	*P. aeruginosa*	Incorrect channel connectors used
5.6	[30]	2002	United States	*M. tuberculosis*	Leak tests were not performed. Hole in sheath provided access to a space that was difficult to clean and disinfect.
5.7	[31]	2003	United States	*P. aeruginosa*	Bronchoscope design issue loose port on biopsy channel-recalled
5.8	[32]	2003	United States	*P. aeruginosa* and *Serratia marcescens*	Bronchoscope design issue loose port on biopsy channel-recalled
5.9	[33]	2005	France	*Klebsiella pneumoniae, Proteus vulgaris, Morganella morganii, Proteus mirabilis*	Bronchoscope design issue loose port on biopsy channel-recalled
5.10	[34]	2005	France	*P. aeruginosa*	Damaged biopsy channel caused by defective biopsy forceps

TABLE 10.6

Bronchoscopy-Related Pseudo-Outbreaks

Case	Reference	Publication Year	Country	Organism	Suspected Cause of Outbreak
6.1	[35]	1995	United States	*Rhodotorula rubra*	Inadequate drying of suction channel
6.2	[36]	1995	Taiwan	*M. chelonae*	Contaminated suction channel, inadequate drying
6.3	[37]	1997	Australia	*Legionella pneumophila*	Contaminated rinse water
6.4	[38]	1999	United States	*M. tuberculosis*	Failure to follow reprocessor connection instructions
6.5	[38]	1999	United States	*M. avium-intracellular*	Failure to use the reprocessor developed connectors
6.6	[39]	1999	United States	Acid-fast bacilli	Failure to use the reprocessor developed connectors
6.7	[40]	2000	United States	*Aureobasidium* species	Reuse of single-use stopcocks
6.8	[41]	2000	India	Nonviable *Mycobacterium* species	Difficulties in cleaning distal tip area
6.9	[42]	2001	United States	*M. tuberculosis*	Errors in cleaning, incompatible reprocessor
6.10	[43]	2001	United States	*M. chelonae, Methylobacterium mesophilicum*	Inadequate disinfection of reprocessor which lead to biofilm contamination
6.11	[44]	2002	Italy	*M. gordonae*	Failure to properly maintain and change rinse water filter
6.12	[45]	2003	United States	*Trichosporon mucoides*	Bronchoscope design issue, loose port on biopsy channel-recall
6.13	[46]	2003	Brazil	*P. aeruginosa* and *S. marcescens*	Inappropriate bronchoscope disinfection practices

suctioned through the suction–biopsy channel followed by rinsing the endoscope with water. The authors report that wash-out samples from the biopsy suction channel of the endoscope after cleaning were positive for *H. pylori*. This case provides very strong evidence that inadequate cleaning—the device was not immersed in a cleaning solution and the suction–biopsy channel was not brushed—and use of a solution not cleared for reprocessing endoscopes lead to patient-to-patient transmission of *H. pylori*.

Case 4.2: In the second report, Wu et al. [15] describes a study in which stored serum samples were tested for IgG antibodies against *H. pylori*. The participants were required to have at least two serum samples available for analysis, one before endoscopic examination and another 6–12 months after the examination. The subjects were split into two groups. The first group was composed of individuals who underwent endoscopy using endoscopes that were cleaned manually followed by use of 70% ethyl alcohol in a manual disinfection step. For the second group the endoscopes were processed in an automated machine using 2.3% glutaraldehyde as the disinfectant. Five of 60 initially seronegative patients became seropositive when the endoscopes were manually reprocessed while none of the 72 individuals in the mechanically reprocessed group seroconverted. These results provide indirect evidence that use of an inadequate disinfectant can lead to endoscopic transmission of *H. pylori*.

Case 4.3: In the report by Lo Passo et al. [16], although the evidence for patient-to-patient transmission of *Trichosporon asahii* is convincing, the cause of the transmission is not clear. The two patients involved underwent gastroesophageal endoscopy examination on the same day. The endoscopic examination revealed that the first patient was infected with multiple strains of *T. asahii*. No pathological findings were noted for the second patient. However, 1 week following the endoscopic procedure, the second patient complained of mild chest pain and dysphagia. During a second endoscopic evaluation, 2 weeks after the first, the second patient was confirmed to be infected with one of the *T. asahii* strains isolated from the first patient. The authors suggest that the mode of transmission was the biopsy forceps. This suggestion is based upon the observation that the biopsy forceps were positive for glutaraldehyde-resistant *T. asahii* and the forceps had been disinfected with glutaraldehyde rather than sterilized. Although this appears to be the likely source, it was not clear if the same biopsy forceps was used on both patients. Nor was it stated if the endoscope or endoscopes used in the procedures were similarly cultured to determine if they were likewise contaminated.

Case 4.4: The fourth article by Davis et al. [17] suggests, but does not present definitive evidence, of exposure to Hepatitis B and C viruses (HBV and HCV) during colonoscopy. The patient was positive for hepatitis B and C antibodies but denied a history of the common significant risk factors. The patient reported having an annual colonoscopy for the past 15 years. In a previous study conducted in France [18], it was reported that for those over 45 years of age, a history of endoscopic biopsy was a significant independent risk factor for exposure to HCV. The cause of the purported transmission was not investigated.

Case 4.5: The report by Bronowicki et al. [19] describes a patient-to-patient transmission of HCV. Three patients sequentially underwent colonoscopy using the same colonoscope. The first procedure was conducted on a patient who was known to be HCV positive. The next two patients (husband and wife) were both confirmed HCV negative when tested after blood donation ~2–3 months earlier. The husband and wife developed symptoms ~3 months after the colonoscopy. All three patients were infected with the same isolate, which showed 100% nucleotide homology. Investigation into the reprocessing procedure revealed that the biopsy channel was not brushed during the cleaning procedure and that the endoscopic accessories were disinfected rather than sterilized. The authors propose that these lapses in recommended procedures resulted in ineffective disinfection and were the cause of the HCV transmission.

Case 4.6: For the sixth case reported by Le Pogam et al. [20], the cause of the transmission of HCV either could not or was not fully investigated. The evidence for patient-to-patient transmission is however relatively strong. The two patients underwent colonoscopy on the same day, one after the other. The viral clones isolated from the two individuals were shown to be virtually identical, differing by only three nucleotides. The authors did not believe that the endoscope was the transmission agent because it seemed unlikely that the two consecutive colonoscopies were performed with the same endoscope. The other suggested cause was use of a multidose anesthesia vial or use of the same syringe for both patients.

Case 4.7: Similar to case 4.4, the report by Federman and Kirsner [21] concerns infection with HBV in the absence of significant risk factors. The patient had undergone colonoscopic decompression 4 months before the onset of infectious symptoms. The authors were unable to contact the two patients that underwent colonoscopy before the patient in question. No further investigation of reprocessing routine was conducted. In this case, evidence that ties the infection and the endoscopic procedure is lacking.

Case 4.8: Esteban et al. [22] describe a pseudo-outbreak of *Aeromonas hydrophila* in 1998. An increase in the frequency of colon biopsy samples positive for *A. hydrophila* was noted from January to March without evidence of patient infection. The endoscope and biopsy forceps pincers were sampled during the outbreak investigation, and both were positive for *A. hydrophila*. Analysis of the isolates (biotyping, antibiotyping, plasmid analysis, and SDS-PAGE analysis) indicated that they were identical to that isolated in the biopsy samples. During the outbreak, the endoscopic instruments were at first being disinfected using quaternary ammonia. Later in the outbreak, the disinfection solution was switched to glutaraldehyde-phenate. Finally, the disinfection solution was switched to 2% glutaraldehyde, after which the incidence of *A. hydrophila* in biopsy specimens dramatically decreased.

Case 4.9: The article by Fraser et al. [23] describes an apparent patient-to-patient transmission of *Pseudomonas aeruginosa* during ERCP. Samples (bile, blood, or peripancreatic fluid) from five patients cultured positive for *P. aeruginosa* after ERCP during July 2002. The same endoscope was used during four of the five patients' procedures; however, the serial number of the endoscope used for the fifth patient was not recorded. The endoscope in question was on loan from the manufacturer and had been sampled on June 18, 2002 during the routine quarterly surveillance program. The culture sample was negative. Twelve patients underwent ERCP with this endoscope in the period from after the negative surveillance culture until the endoscope was removed from service in response to the outbreak investigation. The purported index patient was the second case to use the endoscope after the negative surveillance culture. Three of the other 10 patients who underwent ERCP with this endoscope after the index patient presented positive *P. aeruginosa* cultures. The antimicrobial susceptibility patterns for the isolates were similar, and molecular typing of available isolates revealed that the strains were identical. The suspect endoscope was cultured and again was found to be negative. Inspection of the endoscope revealed no mechanical damage. Once the endoscope was removed from service no further episodes of *Pseudomonas* infection were recorded. The cause of the transmission is unknown.

Case 4.10: The final article by Ranjan et al. [24] describes another outbreak of *P. aeruginosa* after ERCP. The antibiotic sensitivity of the cultures from the three patients involved was similar. In the ensuing investigation, cultures taken from the duodenoscope, water bottle, and endoscope accessories all cultured positive for *P. aeruginosa*. When the cleaning and disinfection procedure used for the instruments was evaluated, it was found that the cleaning was inadequate (blood and bile stains remained on instruments post-cleaning) and that the water bottle had not been cleaned for 4 weeks. In addition, it was not possible to confirm that the glutaraldehyde exposure time was appropriate. Post-outbreak the center instituted supervision of the manual cleaning procedure and ethylene oxide sterilization rather than glutaraldehyde disinfection of the accessories. After making these changes, all instruments cultured sterile.

10.4.2.2 Suspected Causes: Summary

From 1995 to 2005, 10 published articles were identified that described suspected outbreaks related to GI endoscopy. Of these 10 reports, the root cause for seven of the cases (70%) appear to be lapses in the recommended endoscope reprocessing practices or use of multiple dose anesthesia vials/reuse of syringe.

For two of the remaining three cases, evidence that the endoscopic procedure was the cause of the viral infections is not conclusive. The remaining case is very unusual in that the evidence that the loaner endoscope was the source of the transmission is strong; however, the root cause was not identified. There is not enough information to postulate why the loaner endoscope would harbor contamination. It is however tempting to speculate that because the staff was unfamiliar with the endoscope, it was not effectively cleaned or it was not properly connected to the automated reprocessing equipment. However, there is no evidence to support this theory.

10.4.3 Bronchoscopy-Associated Outbreaks

10.4.3.1 True Outbreaks

Table 10.5 lists true outbreaks of contamination associated with broncho-scopes. These are cases in which patient infection accompanied the outbreak.

Case 5.1: Agerton et al. [25] report the transmission of multidrug-resistant *Mycobacterium tuberculosis* strain W1. The only connection between the four patients involved in the outbreak was that they all underwent a bronchos-copy procedure in May at the same hospital. All four procedures were performed by the same pulmonologist using the same bronchoscope. Three of the cases had their bronchoscopies performed 1, 12, and 17 days after the index patient, respectively. Only two other bronchoscopy proced-ures were performed in May at this hospital, neither of which resulted in infection. The authors identified several breaches to the written hospital procedure for reprocessing flexible endoscopes.

- Cleaning was substandard lasting 2.5–3 min with the brush passed through the channel only one time.
- A leak test was not performed.
- The potency of the glutaraldehyde solution was not monitored.
- Only the tips of the scopes were placed in the glutaraldehyde solution.
- The disinfectant contact period was not timed.
- The scope was rinsed with tap water but not flushed with alcohol afterwards.

These breaches were suggested as the cause of the outbreak.

Case 5.2: Michele et al. [26] report a patient-to-patient transmission of an identical strain of *M. tuberculosis* by a bronchoscope. The link uncovered in the investigation was care in the same hospital. The index patient under-went bronchoscopy 2 days before the second patient. The two cases used the same endoscope without an intervening procedure. Investigation of the reprocessing practices revealed that the cleaning procedure used 4.5% chlorhexidine gluconate rather than an enzymatic cleaner and the broncho-scope was only partially immersed in the 2.4% glutaraldehyde solution. The biopsy forceps were decontaminated rather than sterilized.

Case 5.3: Schelenz and French [27] describe the outbreak of two multi-drug-resistant *P. aeruginosa* strains over a 2 month period in 1998. Eleven patients were involved and all isolates came from respiratory specimens. Eight of the patients had undergone bronchoscopy previous to the infection. The outbreak investigation began with environmental sampling. All sam-ples were negative for *P. aeruginosa* or other gram-negative pathogens. The possibility that contaminated bronchoscopes were the contamination source was also investigated. Two of three disinfected bronchoscopes that were sampled cultured positive for antibiotic-susceptible *P. aeruginosa*.

Although not the same strain as the outbreak organism, this finding pointed to a failure to disinfect the bronchoscopes. Next the bronchoscope washer was disassembled and it was found that the inside of the circulatory plastic tubing was covered with a dark brown film. Various samples were taken from the washer–disinfector. All but one yielded various *Pseudomonas* species and one from the pump water grew *P. aeruginosa*. The organism was similar to that isolated from the two bronchoscopes. It appeared that the washer–disinfector had not received maintenance for the year or so for which it had been in service. The washer–disinfector was thoroughly disinfected, including replacing the pump housing, pump filters, and all of the circulatory tubing. A rigorous maintenance schedule was put in place for the washer–disinfector and intermittent sampling confirmed that the instruments remained free of bacterial contamination for 12 months post-outbreak. Although neither of the two epidemic organisms were isolated from the bronchoscopes or disinfector, the authors presumed that this equipment was the most likely source of the outbreak.

Case 5.4: In this case [28], six patients were diagnosed with endotracheal colonization with *P. aeruginosa*. All six had undergone bronchoscopy before the diagnosis. Investigation of the fiberoptic bronchoscope, the endoscope reprocessing equipment, the disinfectant, and the automated disinfectant dispenser unit detected a *P. aeruginosa* strain that matched the strain identified for two of the patients on all equipment. Further investigation of the reprocessing equipment revealed that the disinfectant concentrate had mistakenly been preset to 0.04% (an insufficient level) instead of 3.0%.

Case 5.5: Sorin et al. [29] describe an outbreak of imipenem-resistant *P. aeruginosa* after bronchoscopy. Of the 18 patients who had this organism isolated from bronchoscopic or post-bronchoscopic specimens, 3 patients demonstrated clinical evidence of infection. Investigation of the bronchial-wash isolates (18 isolates) revealed all strains but one was >95% related. The outbreak coincided with introduction of an automated reprocessor. Further investigation revealed that the bronchoscope was connected to the reprocessor using incorrect connectors that obstructed the flow of the germicide through the bronchoscope lumen. After the appropriate connectors were implemented, no further imipenem-resistant *P. aeruginosa* washings were obtained.

Case 5.6: An outbreak of *M. tuberculosis* was identified after numerous positive bronchial washings were noted for patients who underwent bronchoscopy in July 1999 [30]. The historical incidence of *M. tuberculosis* positive washing for this hospital was two per year during 1995 to 1998, prompting an investigation. Of the 19 bronchoscopy procedures performed in July 1999, 18 had bronchial washings and 10 of these were positive for *M. tuberculosis*. Two of the patients had symptoms of active tuberculosis before bronchoscopy. *M. tuberculosis* infections developed in two other patients while the remaining six patients had no evidence of infection. All of the isolates were shown to be identical by restriction fragment length polymorphism (RFLP) analysis. Nine of the ten procedures were performed with the same bronchoscope. The bronchoscope was found to have a small leak located in the external sheath of the flexible tip, very close to the end. The reprocessing procedure used at the hospital did not routinely employ leak testing, which would have detected the leak. The bronchoscope was negative when cultured; however, it had been processed in an ethylene oxide gas sterilizer before the culturing in response to the outbreak. The authors, Ramsey et al. [30], suggest that the contamination occurred during the index patient's procedure and that the hole in the sheath provided an area that was difficult to clean and disinfect.

Cases 5.7–5.9: The cases identified as 5.7–5.9 in Table 10.5 are grouped together because the suspected outbreak cause was the same for all three: a defective biopsy post cap which could become loose and harbor contamination. The design was ultimately subjected to a recall by the manufacturer; however, not all of the hospitals were aware of the action and continued to use the defective bronchoscopes.

Case 5.7: In the report by Srinivasan et al. [31], an increase in occurrence of *P. aeruginosa* recovery from bronchoalveolar lavage specimens from 10.4% to 31.0% prompted the investigation. Of the 414 patients who underwent bronchoscopy during the outbreak (June 2001 to January 2002), 39 patients were identified who had respiratory tract or bloodstream infections or both (48 infections identified). *P. aeruginosa* was confirmed as a potential causative organism in 66.7% of the infections. Cultures were taken of the automatic endoscope reprocessors, treatment rooms, cleaning rooms, ancillary equipment, GI endoscopes, and the endoscopy suite bronchoscopes. Only 3 of the 10 bronchoscopes cultured positive for *P. aeruginosa*. All three bronchoscopes had been recalled by the manufacturer in November 2001. All three bronchoscopes were confirmed to have had loose biopsy ports. The three bronchoscopes were removed from service in early February 2002 after which the rate of recovery of *P. aeruginosa* from bronchial washings returned to the baseline level.

Case 5.8: Kirschke et al. [32] report similar findings in which the same design defect was implicated in an outbreak of *P. aeruginosa* and *Serratia marcescens* after bronchoscopy. From July to October 2001, 60 procedures were performed and 43 specimens obtained for bacterial culture. Twenty of the specimens were positive for *P. aeruginosa* and 6 of the 20 were also positive for *S. marcescens*. One patient was hospitalized with *P. aeruginosa* pneumonia 11 days post-bronchoscopy. All of the positive samples were associated with three new bronchoscopes.

The outbreak coincided with the introduction into service of two new bronchoscopes (8 days before the collection of the first positive sample). When these two bronchoscopes were removed from service, no *P. aeruginosa* positive cultures were collected for the next 5 weeks. Investigation of the bronchoscopes revealed that the biopsy ports were loose. The threads and the inside of the cap were swabbed and a dark green film was noted. Of the 12 samples taken of this area, nine were positive for *P. aeruginosa* and *S. marcescens* and the other three were positive for *P. aeruginosa* alone. Two new bronchoscopes (same model numbers as the ones removed from service due to the outbreak) were obtained and put into service. It was noted after the second use that one of the biopsy port was loose. Upon investigation it was found that the second bronchoscope also had a loose biopsy port cap. Three *P. aeruginosa* positive cultures were identified for patients that underwent a procedure with this endoscope. The biopsy port cap was swabbed and all five swabs were *P. aeruginosa* positive. Both bronchoscopes were removed from service.

The strain of *P. aeruginosa* isolated from 10 patient samples that were available for analysis and the three bronchoscope cultures were indistinguishable when analyzed by pulse field gel electrophoresis (PFGE). The authors concluded that the loose biopsy cap was the most likely contaminating source for the following reasons:

1. No breach in cleaning or disinfection protocol was identified.
2. The bronchoscope contamination persisted through three cycles of cleaning and disinfection.
3. No other bronchoscopes in use during the outbreak period were associated with positive cultures.

Case 5.9: For the three outbreaks described by Cetre et al. [33], a similar pattern emerged. The first outbreak was identified after it was noted that there was an unusual increase in the occurrence of specific pairs of bacteria in the bronchoalveolar lavage samples (*Klebsiella pneumoniae* and *Proteus vulgaris* or *Morganella morganii* and *Proteus mirabilis*). Although no patients developed pneumonia, the sputum from two patients yielded *M. morganii* some days after the bronchoscopy procedure. Analysis of the records for all procedures performed during the outbreak showed that of 342 samples taken, 26% (89) were contaminated with the suspect organisms. All of the procedures which yielded the contaminated cultures were performed with two bronchoscopes. Cultures from one of the identified endoscopes confirmed the presence of *K. pneumoniae* and *P. vulgaris*. Patient isolates and those from the bronchoscope were indistinguishable by PFGE. The two bronchoscopes were removed from service and returned to the manufacturer for evaluation. The outbreak end coincided with removal of the bronchoscopes from service.

After being received back from the manufacturer, the two bronchoscopes were returned to service. Second and third outbreaks were subsequently identified. When the two bronchoscopes were again removed from service, the outbreaks ended. The disinfection procedure was audited shortly after each of the three outbreaks and each time was found to be in compliance with local guidelines. The models of the bronchoscopes were of the same design as those that the biopsy cap would become loose after use. It was therefore hypothesized that the loose cap allowed for sequestering of the bacteria.

Case 5.10: In this last article by Corne et al. [34], two outbreaks were detected because of an increase in respiratory tract specimens positive for *P. aeruginosa*. Investigation of the outbreaks revealed that four patients had developed pneumonia as a result of undergoing a bronchoscopy procedure with the two bronchoscopes used by the facility. PFGE was used to link the transmission of two *P. aeruginosa* strains. When the bronchoscopes were flushed with saline, the two *P. aeruginosa* strains identified in the outbreaks were recovered. Review of the reprocessing procedure revealed no significant breaches. The two bronchoscopes were removed from service and returned to the manufacturer for inspection. This inspection revealed large breaches in the internal channel of both bronchoscopes. The defects were not detected by leak tests. The cause of the breach was attributed to use of defective biopsy forceps. It was stated that "the forceps had defective closing that was not visually detectable, probably due to repeated use and decontamination." It was suggested that the damaged inner channels sheltered organisms and led to improper cleaning and disinfection of the bronchoscopes despite adherence to reprocessing standards.

10.4.3.2 *Pseudo-Outbreaks*

Pseudo-outbreaks or pseudo-epidemics are clusters of false infections in which patients are not infected and so do not exhibit symptoms of disease. In these cases, however, the patients may be at risk for complications because of unnecessary therapeutic intervention. Table 10.6 compiles bronchoscopy related pseudo-outbreaks identified from 1995 to 2005.

Case 6.1: Hagan et al. [35] describe the pseudo-outbreak of *Rhodotorula rubra* recognized when four bronchoalveolar lavage samples cultured this unusual fungus. In all, 11 different patient specimens yielded growth of *R. rubra*. None of the patients had a clinical course consistent with pulmonary infection with a fungal pathogen nor were any subsequent cultures from the patients positive for *R. rubra*. Investigation of the bronchoscope in use during the pseudo-outbreak revealed that the suction channel was contaminated with *R. rubra* as well as *P. aeruginosa* and a *Flavobacterium* species. The reprocessing procedure was reviewed and the following changes were made. The bronchoscopes were sterilized with ethylene oxide and the cleaning procedure was changed to include a step in which alcohol was flushed through the suction channel followed by air to dry the bronchoscope thoroughly between patient use. Once these changes were instituted, there were no further isolates of *R. rubra* from bronchoscopy specimens. The authors suggested that the probable cause of the pseudo-outbreak was inadequate drying of the bronchoscope. Although not explicitly stated, it appears that the bronchoscope was sterilized using ethylene oxide at the end of each day and was disinfected, including the alcohol flush and air dry step, between patients. It is confusing that the authors used the term "cleaning" and drying between patient uses rather than cleaning, disinfection, and drying; however, the article does not provide additional details concerning the disinfection procedure used.

Case 6.2: The root cause of the pseudo-epidemic described by Wang et al. [36] was not determined. The outbreak occurred during September to December of 1992 and was identified by a cluster of *M. chelonae* positive bronchial washings. Cultures from four bronchoscope suction channels were taken and all four samples were positive for *M. chelonae*. The broncho-scopes had been processed in an automated reprocessor, which had been put into service in June 1992. The authors thought that the most likely contamination source was the water used to rinse the endoscopes; however, *M. chelonae* was not isolated from either the tap water or the reprocessor. The facility instituted 70% alcohol rinsing followed by extensive suctioning through the channel after disinfection. Once this procedure was instituted, no further isolates of *M. chelonae* were identified.

Case 6.3: The article by Mitchell et al. [37] describes a pseudo-outbreak of *Legionella pneumophila*, which was traced to the water used to rinse the bronchoscope after disinfection. Several upgrades to the water preparation procedure were put in place including placing a 5 μm filter in the cold-water line. After introduction of the changes, *Legionella* species were not isolated from water samples for 6 months, after which the microbiological sampling was discontinued. However, 6–12 months afterward, *L. pneumo-phila* was again isolated from two patients' bronchoalveolar lavage speci-mens. It was determined that inadequate maintenance of the water filter was the cause. After introduction of a regular maintenance schedule and reintroduction of routine microbiological monitoring, no further contamin-ation of bronchoalveolar lavage samples with *Legionella* species occurred.

Cases 6.4 and 6.5: A CDC report [38] describes two clusters in which failure to connect the bronchoscopes to the automated reprocessor was the suspected cause of the pseudo-outbreaks. In the first, which occurred between November and December of 1996, a common strain of *M. tuberculosis* was identified in bronchial specimens from five patients. The index patient was identified and the four subsequent patients did not manifest clinical symptoms of infection, although one patient had a positive tuberculin skin test 6 weeks post-procedure. The procedures had been performed with three different but identical model bronchoscopes. All three cultured nega-tive 5 weeks after the last case. The cleaning brushes also cultured negative. Investigation of the reprocessing procedure revealed that the bronchoscopes had not been properly connected to the automated reprocessing system. This lapse in reprocessing procedure is presumably the cause of the pseudo-outbreak.

In the second cluster, which occurred during March and April in 1998, seven incidents of *M. avium-intracellulare* positive bronchial specimens were identified. All of the patients had undergone bronchoscopy with the same bronchoscope. The bronchoscope cultured negative when sampled 12 days after diagnosis of the last case. Investigation of the reprocessing methods revealed that the bronchoscope was connected to the reprocessor using

adapters supply by the bronchoscope manufacturer rather than the connectors developed for that bronchoscope by the reprocessor manufacturer. It was
this breach in reprocessing procedure that was implicated in the outbreak.

Case 6.6: In this case reported by Strelczyk [39], 10 bronchial washing
specimens over a 19 month period grew acid-fast bacilli in the absence of
identifiable disease. Cultures from the bronchoscope were negative. Investigation of the reprocessing procedure revealed that the automated reprocessing system was not being appropriately maintained and that, once
again, the connectors used to interface the bronchoscope with the reprocessor were not those developed by the reprocessing system's manufacturer.
This breach was suggested as the cause of the contamination.

Case 6.7: The article by Wilson et al. [40] described the outbreak of an
unusual mold contamination (*Aureobasidium*) in bronchoscopy specimens.
Ten specimens, taken from nine patients, were involved; however, none of
the patients had a true infection either before or after the bronchoscopy
procedures. The contaminated samples were not associated with a particular bronchoscope. Investigation of the bronchoscopy procedure revealed
that single-use plastic stopcocks were being reused on different patients.
After use the stopcocks were placed in an automated reprocessor designed
for bronchoscopes. When the stopcocks were cultured after the disinfection
procedure a heavy growth of *Aureobasidium* species was identified; however, the fluid from the reprocessor was negative. Reuse of the stopcocks
was halted and for 6 months *Aureobasidium* was not recovered from any
specimens.

Case 6.8: Chaturvedi and Narang [41] describe a situation in which non-
viable *Mycobacterium* species were responsible for smear-positive results in
patient aspirates leading to false-positive diagnosis and unnecessary medication in six of eight patients. Investigation of the endoscopic instruments
revealed that the distal tip was smear-positive for *Mycobacterium*, which
failed to grow when subcultured. It was suggested that cleaning of this
area was difficult due to the delicate nature of the distal tip.

Case 6.9: Inadequate cleaning and disinfection of a bronchoscope was the
suspected cause of a pseudo-outbreak in October 2000. The investigation
described by Larson et al. [42] was prompted by the occurrence of three
patient cultures that were positive for *M. tuberculosis*. The index patient was
identified and the two subsequent patients (4 and 8 days after the index
patient) did not present symptoms of *M. tuberculosis* infection. Laboratory
analysis indicated a common source of the contamination. There was no
evidence of laboratory cross contamination. The three specimens had
been collected with the same bronchoscope. Investigation of the suspect
bronchoscope by the manufacturer revealed residual patient debris on the
suction connector, suction valve, suction cylinder, and distal tip. The reprocessing procedure was reviewed and found to be in compliance with the
following exceptions: the concentration of the cleaning solution was not
accurately measured, the bronchoscope channels were brushed only once

rather than several times to remove debris, and the entire endoscope was not fully submerged in the detergent for 2–3 min. In addition, the broncho-scope was not compatible with the automated reprocessor used at the institution.

Case 6.10: Kressel and Kidd [43] describe a pseudo-outbreak attributed to contamination of the automated reprocessors with a biofilm that was resist-ant to subsequent decontamination. The outbreak took place from July 21 to October 2, 1998. Early in July 1998, the endoscopy department began disin-fecting bronchoscopes in an automated reprocessor that was used to wash and disinfect GI endoscopes. In August, the microbiology lab reported an unusual number of *M. chelonae* and *M. mesophilicum* positive samples taken during bronchoscopy. Investigation of the source determined that *M. chelo-nae* grew from samples of endoscopes, the automated reprocessors, and the glutaraldehyde from the washers. However, samples of the tap water and unopened glutaraldehyde were negative for this organism. It was discovered that the recommended routine maintenance of the reprocessors, including a monthly 8 h disinfection cycle of the unit, was often omitted. Once the reprocessor was contaminated, the disinfection cycle did not eliminate contamination. The hospital purchased new endoscopes and a different reprocessing system and, after correction of the adapters used for reprocessing, surveillance cultures of the endoscopes were negative.

Case 6.11: A pseudo-outbreak of *M. gordonae* was described by Rossetti et al. [44] after 16 of 267 bronchial aspirate cultures in a 10-month period were positive for this organism. Historically, only one such contamination was noted in 1368 samples over a 7-year period. Investigation of the fluids involved in the disinfection procedure identified *M. gordonae* in two samples of tap water feeding the reprocessor and from the filtering unit, indicating that the filter was not effectively trapping bacteria in the water. A failure in filter replacement and maintenance was uncovered. After restoration of the correct procedures, no further isolation of *M. gordonae* was noted.

Case 6.12: Singh et al. [45] report a pseudo-outbreak of *Trichosporon mucoides* in September 2000. The same bronchoscope had been used for all six procedures where the *T. mucoides* positive bronchoalveolar lavage sam-ples were obtained. Cultures of the bronchoscope lumen yielded growth of this organism while those taken from the external body of the endoscope, the reprocessor, the 2% glutaraldehyde disinfecting solution, and water filters/supply were negative for *T. mucoides*. The bronchoscope was returned to the manufacturer. This bronchoscope model along with 14 others was recalled by the manufacturer in November 2001. According to the manufacturer, loosening of the biopsy channel port housing during use could lead to contamination of the port.

Case 6.13: Silva et al. [46] report a bronchoscope-related pseudo-outbreak of *P. aeruginosa* and *S. marcescens* in July through September 1995. Obser-vation of the reprocessing procedure revealed inadequacies which were addressed with the following updated procedures:

- Prompt scrupulous cleaning using brushes and sinks dedicated to bronchoscopes (before the changes the same area and tools were used for colonoscopes and bronchoscopes)
- Strict control of disinfectant immersion time
- Rinsing with filtered water followed by an alcohol flush and forced air drying
- Use of chemical monitoring strips
- Cleaning the disinfectant immersion tubing each time the disinfectant was changed
- Separation of procedure area from cleaning area

After implementation of these measures the incident rate fell from 12.6% to 0.6%.

10.4.3.3   Suspected Causes: Summary

Regarding bronchoscope-associated true and pseudo-outbreaks, a total of 23 cases were reviewed and the suspected cause was assigned as follows:

- Nine instances of deviations from standard cleaning and disinfection practices
- Five cases of not following the automated reprocessor or device (single-use stopcock) instructions
- Five instances of device design defect, four of which were limited to one specific bronchoscope design series
- Four examples of failing to maintain automated reprocessor or water treatment system

10.4.4   Summary of Suspected Causes of Outbreaks

The suspected cause of the 33 outbreaks published between 1995 and 2005 that were discussed above are grouped into five categories and graphed in Figure 10.12. In total, 49% of the outbreaks were attributed to errors in the reprocessing procedure, 15% to not following either the reprocessor or the medical device manufacturer's instructions for use, 15% to a device defect, 12% to failure to properly maintain the reprocessing equipment, and 9% the cause was not identified. Thus, a majority of the outbreaks could have been avoided if the reprocessing personnel had faithfully followed standard reprocessing procedures and endoscope reprocessors' instruction for use and maintenance. However, this outcome also stresses the manufacturers' responsibility to provide users with clear and definitive reprocessing and maintenance instructions as well as to design instruments that can be repetitively reprocessed without difficulty in detecting defects that impair reprocessing efficacy.

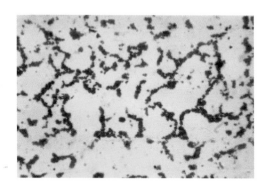

FIGURE 4.1
Vegetative bacteria.

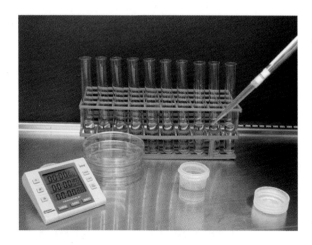

FIGURE 11.1
Typical time-kill setup.

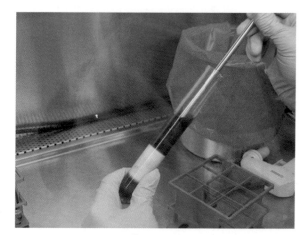

FIGURE 11.2
Homogenization of fungal culture.

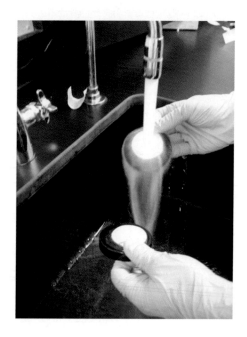

FIGURE 11.3
Washing of pigskins.

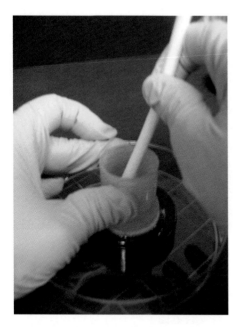

FIGURE 11.4
Debriding with Teflon policeman.

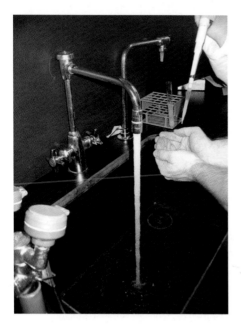

FIGURE 11.5
Inoculum application in a HCPHW.

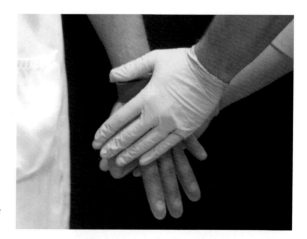

FIGURE 11.6
Sampling using a glove juice procedure.

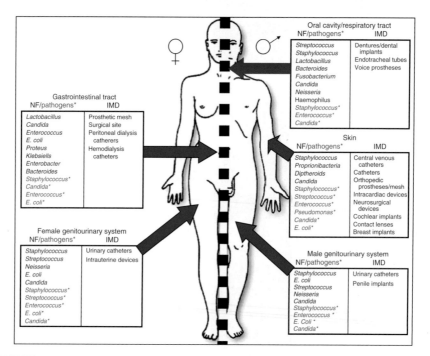

Oral cavity/respiratory tract

NF/pathogens*	IMD
Streptococcus	Dentures/dental implants
Staphylococcus	Endotracheal tubes
Lactobacillus	Voice prostheses
Bacteroides	
Fusobacterium	
Candida	
Neisseria	
Haemophilus	
Staphylococcus*	
Enterococcus*	
Candida*	

Gastrointestinal tract

NF/pathogens*	IMD
Lactobacillus	Prosthetic mesh
Candida	Surgical site
Enterococcus	Peritoneal dialysis catherers
E. coli	Hemodialysis catheters
Proteus	
Klebsiells	
Enterobacter	
Bacteroides	
Staphylococcus*	
Candida*	
Enterococcus*	
E. coli*	

Skin

NF/pathogens*	IMD
Staphylococcus	Central venous catheters
Proprionibacteria	Catheters
Diptheroids	Orthopedic prostheses/mesh
Candida	Intracardiac devices
Staphylococcus*	Neurosurgical devices
Streptococcus*	Cochlear implants
Enterococcus*	Contact lenses
Pseudomonas*	Breast implants
Candida*	
E. coli*	

Female genitourinary system

NF/pathogens*	IMD
Staphylococcus	Urinary catheters
Streptococcus	Intrauterine devices
Neisseria	
E. coli	
Candida	
Staphylococcus*	
Streptococcus*	
Enterococcus*	
E. coli*	
Candida*	

Male genitourinary system

NF/pathogens*	IMD
Staphylococcus	Urinary catheters
E. coli	Penile implants
Streptococcus	
Neisseria	
Candida	
Staphylococcus*	
Enterococcus*	
E. Coli*	
Candida*	

FIGURE 14.1
The human body is colonized by a mix of normal and pathogenic microbes. Above is an anatomical diagram of the microbial residents throughout the body (IMD-associated pathogenic microbes are denoted with an asterisk, while the normal flora (NF) microbes are without asterisk).

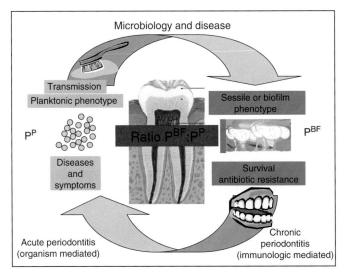

FIGURE 14.2
Ratio of biofilm phenotype to planktonic phenotype.

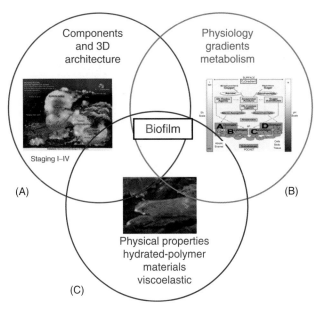

FIGURE 14.3
Three components of biofilms.

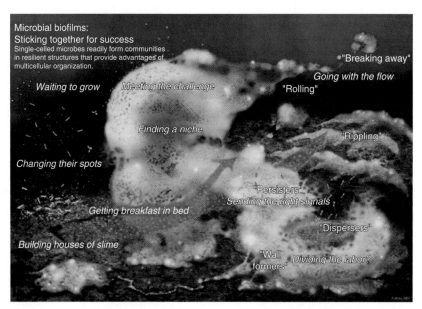

FIGURE 14.4
Mushroom structure of a biofilm. (From Hall-Stoodley, L., Costerton, J.W., and Stoodley, P., *Nat. Rev. Microbiol.*, 2, 95, 2004.)

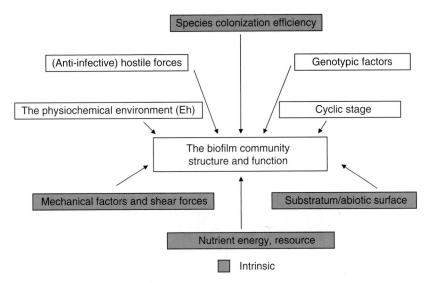

FIGURE 14.5
Eight factors involved in biofilm architecture.

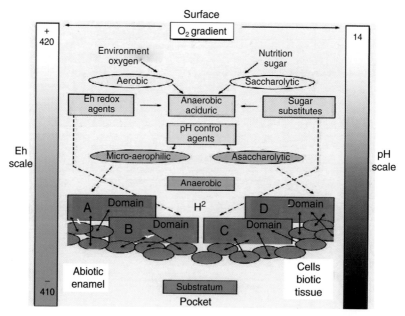

FIGURE 14.6
Eh and pH gradients of biofilms.

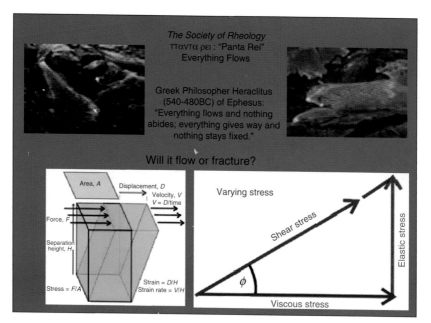

FIGURE 14.7
Rheology/viscosity of biofilms.

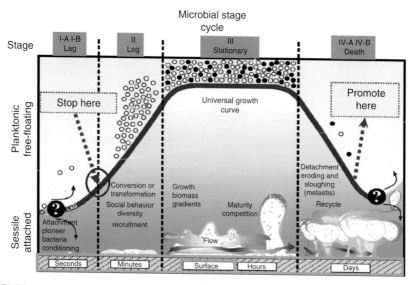

FIGURE 14.8
Four stages of biofilm growth: I Attachment (lag), II growth (log), III maturity (stationary), and IV dispersal (death).
Note: Intra oral sessile (P^{BF}) (▬) and planktonic (P^P) (▬) life forms are not mutually exclusive, but biofilms are the preferred growth vehicle. (From John Thomas, West Virginia University, unpublished.)

Ratio of planktonic/sessile during antimicrobial therapy

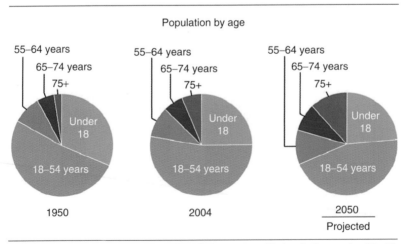

FIGURE 14.9
Effect of antimicrobial therapy on planktonic phenotype.

TABLE 14.2

Distribution of U.S. Population by Age and Year

Population by age

Source: Centers for Disease Control and Prevention, National Center for Health Statistics, Health, United States, 2005, Figure 2.

TABLE 14.3

Number of Coronary Artery Stent Procedures Based on Age and Year

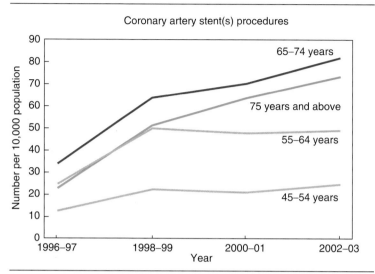

Source: Centers for Disease Control and Prevention, National Center for Health Statistics, Health, United States, 2005, Figure 25.

TABLE 14.4

Age (18–64 years) Distribution of Chronic Conditions Causing Limitation of Activity

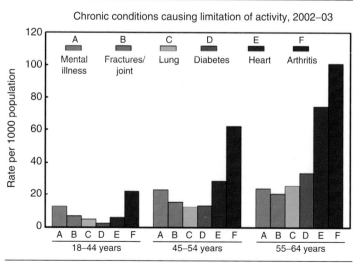

Source: Centers for Disease Control and Prevention, National Center for Health Statistics, Health, United States, 2005, Figure 19.

TABLE 14.5

Age (64–85+ years) Distribution of Chronic Conditions Causing
Limitation of Activity

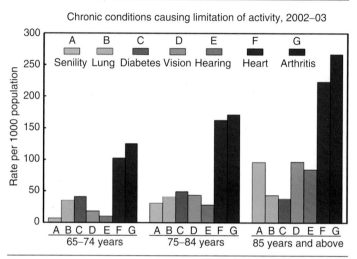

Source: Centers for Disease Control and Prevention, National Center for
Health Statistics, Health, United States, 2005, Figure 20.

TABLE 14.10

Loss Per Incidence for Medicare Charges for Infectious Disease Related
DRGs Over 4 Years

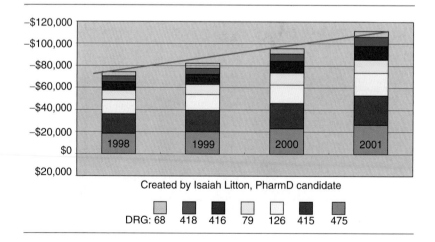

TABLE 14.12

Comparison of Planktonic and Biofilm Phenotypes

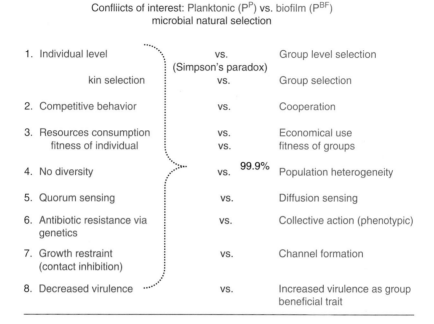

Confliicts of interest: Planktonic (P^P) vs. biofilm (P^{BF}) microbial natural selection

1. Individual level	vs.	Group level selection
	(Simpson's paradox)	
kin selection	vs.	Group selection
2. Competitive behavior	vs.	Cooperation
3. Resources consumption	vs.	Economical use
fitness of individual	vs.	fitness of groups
4. No diversity	99.9% vs.	Population heterogeneity
5. Quorum sensing	vs.	Diffusion sensing
6. Antibiotic resistance via genetics	vs.	Collective action (phenotypic)
7. Growth restraint (contact inhibition)	vs.	Channel formation
8. Decreased virulence	vs.	Increased virulence as group beneficial trait

TABLE 14.13

Percent of Culturable Prokaryotes in Different Habitats

Habitat	Cultured (%)
Seawater	0.001–0.1
Freshwater	0.25
Mesotrophic lakes	0.1–1
Estuarine waters	0.1–3
Activated sludge	1–15
Sediments	0.25
Soil	0.3

Source: Daims, H., University of Vienna (Powerpoint Presentation, unpublished).

Note: The majority of prokaryotes is unculturable and the numbers are based on direct cell counts.

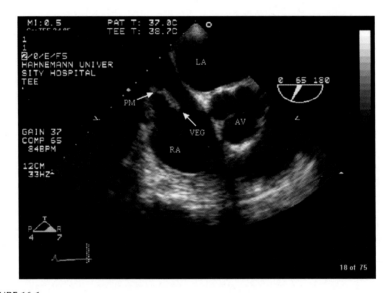

FIGURE 16.1

Transesophageal echocardiographic view of a vegetation on a pacemaker lead. LA = left atrium, PM = pacemaker lead, AV = aortic valve, VEG = vegetation, RA = right atrium.

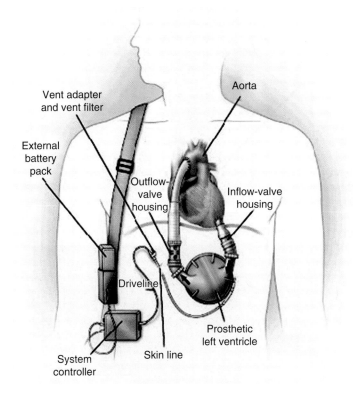

FIGURE 16.2
Schematic of components of one type of left ventricular assist device.

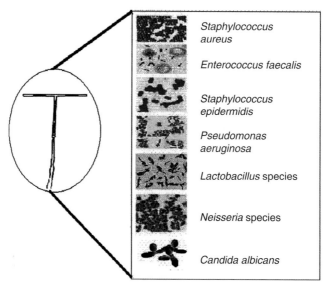

FIGURE 17.3
Microflora associated with infected IUDs in Indian women.

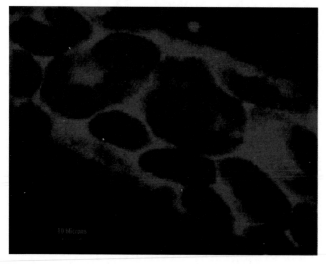

FIGURE 17.6
Confocal laser scanning microscope (CLSM) picture of biofilm formed on IUD string as visualized by fluorescent staining with propidium iodide showing fluid filled channels.

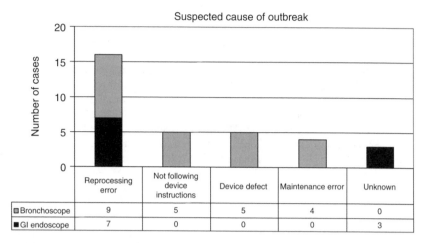

FIGURE 10.12
Summary of suspected outbreak cause.

Acknowledgment

Graphics by Gregory A. Dick and Charles P. Pachono. Graphic designers from the Research and Development, Department of STERIS Corporation, Mentor, Ohio.

References

1. Hirschowitz, B.I., Endoscopic examination of the stomach and duodenal cap with the fiberscope, *Lancet*, I, 1074, 1961.
2. Conner, R. and Reno, D., *Standards, Recommended Practices, and Guidelines*, 2006 edition, AORN, Denver, CO, 2006.
3. Gorse, G.J. and Messner, R.L., Infection control practices in gastrointestinal endoscopy in the United States: A national survey, *Gastroenterol. Nursing*, 14, 72, 1991.
4. Nelson, D.B., Jarnis, W.R., Rutala, W.A., Foxx-Orenstein, A.E., Isenberg, G., Dash, G.P., Alvarado, C.J., Ball, M., Griffin-Sobel, J., Peterson, C., Ball, K.A., Henderson, J., and Stricof, R.L., Multi-society guideline for reprocessing flexible gastrointestinal endoscopes, *Dis. Colon Rectum*, 47, 413, 2004.
5. http://www.cancer.org/docroot/PED/content/PED_2_3XEndoscopy.asp?sitearea=PED
6. http://www.bellevueurology.com/procedures/proureteroscopy.htm
7. http://www.healthatoz.com/healthatoz/Atoz/ency/hysteroscopy.jsp
8. http://digestive.niddk.nih.gov/ddiseases/pubs/ercp/
9. http://www.tddctx.com/enter.html
10. From various Olympus, Pentax, and Fujinon operator manuals.

11. Spach, D.H., Silverstein, F.E., and Stamm, W.E., Transmission of infection by gastrointestinal endoscopy and bronchoscopy, *Ann. Intern. Med.*, 118, 117, 1993.
12. Weber, D.J. and Rutala, W.A., Lessons from outbreaks associated with bronchoscopy, *Infect. Control Hosp. Epidemiol.*, 22, 403, 2001.
13. Nelson, D.B., Infectious disease complications of GI endoscopy: Part II, Exogenous infections, *Gastrointest. Endosc.*, 57, 695, 2003.
14. Miyaji, H., Kohli, Y., Azuma, T., Ito, S., Hirai, M., Ito, Y., Kato, T., and Kuriyama, M., Endoscopic cross-infection with *Helicobacter pylori*, *Lancet*, 345, 464, 1995.
15. Wu, M.S., Wang, J.T., Yang, J.C., Wang, H.H., Sheu, J.C., Chen, D.S., and Wang, T.H., Effective reduction of *Helicobacter pylori* infection after upper gastrointestinal endoscopy by mechanical washing of the endoscope, *Hepatogastroenterology*, 43, 1660, 1996.
16. Lo Passo, C., Pernice, I., Celeste, A., Perdichizzi, G., and Todaro-Luck, F., Transmission of *Trichosporon asahii* oesophagitis by a contaminated endoscope, *Mycoses*, 44, 13, 2001.
17. Davis, A.R., Pink, J.M., Kowalik, A.M., Wylie, B.R., and McCaughan, G.W., Multiple endoscopies in a Sydney blood donor found positive for hepatitis B and C antibodies, *Med. J. Aust.*, 164, 571, 1996.
18. Andrieu, J., Barny, S., Colardelle, P., Maisonneuve, P., Grand, V., Robin, E., Breart, G., and Coste, T., Prevalence et facteurs de risqué de l'infecttion par le virus de l'hepatite C dans une population hospitalisee en gastroentrerlogie. Role des biopsies per endoscopques, *Gastroenterol. Clin. Biol.*, 19, 340, 1995.
19. Bronowicki, J.P., Venard, V., Botte, C., Monhoven, N., Gastin, I., Chone, L., Hudziak, H., Rihn, B., Delanoe, C., LeFaou, A., Bigard, M.A., and Gaucher, P., Patient-to-patient transmission of hepatitis C virus during colonoscopy, *N. Engl. J. Med.*, 337, 237, 1997.
20. Le Pogam, S., Gondeau, A., and Bacq, Y., Nosocomial transmission of hepatitis C virus, *Ann. Intern. Med.*, 131, 794, 1999.
21. Federman, D.G. and Kirsner, R.S., Leukocytoclastic vasculitis, hepatitis B, and the risk of endoscopy, *Cutis*, 63, 86, 1999.
22. Esteban, J., Gadea, I., Fernandez-Roblas, R., Molleja, A., Calvo, R., Acebron, V., and Soriano, F., Pseudo-outbreak of *Aeromonas hydrophila* isolates related to endoscopy, *J. Hosp. Infect.*, 41, 313, 1999.
23. Fraser, T.G., Reiner, S., Malczynski, M., Yarnold, P.R., Warren, J., and Noskin, G.A., Multidrug-resistant *Pseudomonas aeruginosa* cholangitis after endoscopic retrograde cholangiopancreatography: Failure of routine endoscope cultures to prevent an outbreak, *Infect. Control Hosp. Epidemiol.*, 25, 856, 2004.
24. Ranjan, P., Das, K., Ayyagiri, A., Saraswat, V.A., and Choudhuri, G., A report of post-ERCP *Pseudomonas aeruginosa* infection outbreak, *Indian J. Gastroenterol.*, 24, 131, 2005.
25. Agerton, T., Valway, S., Gore, B., Pozsik, C., Plikaytis, B., Woodley, C., and Onorato, I., Transmission of a highly drug-resistant strain (strain W1) of *Mycobacterium tuberculosis*. Community outbreak and nosocomial transmission via a contaminated bronchoscope, *JAMA*, 278, 1073, 1997.
26. Michele, T.M., Cronin, W.A., Graham, N.M., Dwyer, D.M., Pope, D.S., Harrington, S., Chaisson, R.E., and Bishai, W.R., Transmission of *Mycobacterium tuberculosis* by a fiberoptic bronchoscope. Identification by DNA fingerprinting, *JAMA*, 278, 1093, 1997.
27. Schelenz, S. and French, G., An outbreak of multidrug-resistant *Pseudomonas aeruginosa* infection associated with contamination of bronchoscopes and an endoscope washer–disinfector, *J. Hosp. Infect.*, 46, 23, 2000.

28. Kramer, M.H., Krizek, L., Gebel, J., Kirsh, A., Wegan, E., Marklein, G., and Exner, M., Bronchoscopic transmission of *Pseudomonas aeruginosa* due to a contaminated disinfectant solution from an automated dispenser unit. Final Program of the 11th Annual Scientific Meeting of the Society of Healthcare Epidemiology of America, Toronto, Ontario, Canada April 1–3, 2001, Abstract 118.

29. Sorin, M., Segal-Maurer, S., Mariano, N., Urban, C., Combest, A., and Rahal, J.J., Nosocomial transmission of imipenem-resistant *Pseudomonas aeruginosa* following bronchoscopy associated with improper connection to the Steris system 1 processor, *Infect. Control Hosp. Epidemiol.*, 22(7), 409–413, 2001.

30. Ramsey, A.H., Oemig, T.V., Davis, J.P., Massey, J.P., and Torok, T.J., An outbreak of bronchoscopy-related *Mycobacterium tuberculosis* infections due to lack of bronchoscope leak testing, *Chest*, 121, 976, 2002.

31. Srinivasan, A., Wolfenden, L.L., Song, X., Mackie, K., Hartsell, T.L., Jones, H.D., Diette, G.B., Orens, J.B., Yung, R.C., Ross, T.L., Merz, W., Scheel, P.J., Haponik, E.F., and Perl, T.M., An outbreak of *Pseudomonas aeruginosa* infections associated with flexible bronchoscopes, *N. Engl. J. Med.*, 348, 221, 2003.

32. Kirschke, D.L., Jones, T.F., Craig, A.S., Chu, P.S., Mayernick, G.G., Patel, J.A., and Schaffner, W., *Pseudomonas aeruginosa* and *Serratia marcescens* contamination associated with a manufacturing defect in bronchoscopes, *N. Engl. J. Med.*, 348, 214, 2003.

33. Cetre, J.C., Nicolle, M.C., Salord, H., Perolm, M., Tigaud, S., David, G., Bourjault, M., and Vanhems, P., Outbreaks of contaminated broncho-alveolar lavage related to intrinsically defective bronchoscopes, *J. Hosp. Infect.*, 61, 39, 2005.

34. Corne, P., Godreuil, S., Jean-Pierre, H., Jonquet, O., Campos, J., Jumas-Bilak, E., Parer, S., and Marchandin, H., Unusual implication of biopsy forceps in outbreaks of *Pseudomonas aeruginosa* infections and pseudo-infections related to bronchoscopy, *J. Hosp. Infect.*, 61, 20, 2005.

35. Hagan, M.E., Klotz, S.A., Bartholomew, W., Potter, L., and Nelson, M., A pseudoepidemic of *Rhodotorula rubra*: A marker for microbial contamination of the bronchoscope, *Infect. Control Hosp. Epidemiol.*, 16, 727, 1995.

36. Wang, H.C., Liaw, Y.S., Yang, P.C., Kuo, S.H., and Luh, K.T., A pseudoepidemic of *Mycobacterium chelonae* infection caused by contamination of a fibreoptic bronchoscope suction channel, *Eur. Respir. J.*, 8, 1259, 1995.

37. Mitchell, D.H., Hicks, L.J., Chiew, R., Montanaro, J.C., and Chen, S.C., Pseudoepidemic of *Legionella pneumophila* serogroup 6 associated with contaminated bronchoscopes, *J. Hosp. Infect.*, 37, 19, 1997.

38. Center for Disease Control and Prevention (CDC), Bronchoscopy-related infections and pseudoinfections—New York, 1996 and 1998, *MMWR Morb. Mortal Wkly. Rep.*, 48, 557, 1999.

39. Strelczyk, K., Pseudo-outbreak of acid-fast bacilli, *Am. J. Infect. Control*, 27, 198 (Abstract), 1999.

40. Wilson, S.J., Everts, R.J., Kirkland, K.B., and Sexton, D.J., A pseudo-outbreak of *Aureobasidium* species lower respiratory tract infections caused by reuse of single-use stopcocks during bronchoscopy, *Infect. Control Hosp. Epidemiol.*, 21, 470, 2000.

41. Chaturvedi, V.N. and Narang, P., *Mycobacterium* in bronchoscopic aspirate and bronchoscope: An iatrogenic infection, *Natl. Med. J. India*, 13(2), 109–110, 2000.

42. Larson, J.L., Lambert, L., Stricof, R.L., Driscoll, J., McGarry, M.A., and Ridzon, R., Potential nosocomial exposure to *Mycobacterium tuberculosis* from a bronchoscope, *Infect. Control Hosp. Epidemiol.*, 24, 825, 2003.

43. Kressel, A.B. and Kidd, F., Pseudo-outbreak of *Mycobacterium chelonae* and *Methylobacterium mesophilicum* caused by contamination of an automated endoscopy washer, *Infect. Control. Hosp. Epidemiol.*, 22, 414, 2001.

44. Rossetti, R., Lencioni, P., Innocenti, F., and Tortoli, E., Pseudoepidemic from *Mycobacterium gordonae* due to a contaminated automatic bronchoscope washing machine, *Am. J. Infect. Control*, 30, 196, 2002.

45. Singh, N., Belen, O., Leger, M.M., and Campos, J.M., Cluster of *Trichosporon mucoides* in children associated with a faulty bronchoscope, *Pediatr. Infect. Dis. J.*, 22, 609, 2003.

46. Silva, C.V., Magalhaes, V.D., Pereira, C.R., Kawagoe, J.Y., Ikura, C., and Ganc, A.J., Pseudo-outbreak of *Pseudomonas aeruginosa* and *Serratia marcescens* related to bronchoscopes, *Infect. Control Hosp. Epidemiol.*, 24, 195, 2003.

11

Microbiological Testing of Disinfectants and Decontaminants for Critical Surfaces

Daniel A. Klein

CONTENTS

11.1 Introduction

This chapter is not intended to review every microbiological method in detail for testing disinfectants and decontaminants. There are entire books and scientific journals dedicated to the subject. Instead, the intent of this chapter is to outline some of the better choices for characterizing these materials, both for product development and for end users, provide references, and outline some of the pitfalls to avoid.

Perhaps, even more than other scientific disciplines, microbiological testing is highly dependent on the selection of test methodologies. Because the challenge test article can easily comprise millions of individual microorganisms in a heterogeneous state, results can often be surprisingly unpredictable. Additionally, results can be highly variable from test to test even when using the same test method. This is because, no two individual populations containing millions of microorganisms will ever be entirely identical, just as no two human populations ever will. For this reason, it is important to take an iterative approach to microbiology testing that employs multiple methods and multiple tiers of testing over several test dates using several different microorganism populations.

This is especially true when evaluating the microbiological efficacy of disinfectants on critical surfaces. For the purposes of this chapter, a critical surface will be classified as either a hard surface or a skin surface. There are millions of different surface types, but when focusing on those critical surfaces requiring disinfection and decontamination, hard surfaces can be broadly defined as those porous and nonporous surfaces found in hospitals, pharmaceutical manufacturing environments, and anyplace where disinfection is important. Skin surfaces will be defined as human skin in the healthcare environment or wherever the transfer of microorganisms by transient colonization of the skin results in adverse events. In this chapter, disinfection refers to the reduction of microorganisms from hard surfaces, and decontamination or antisepsis refers to the reduction of microorganisms on skin. Highly porous surfaces, such as fabric and carpet, are beyond the scope of this chapter and will not be reviewed as they often present unique challenges for decontamination.

While microbiological testing options remain endless, as in most scientific disciplines, the proper evaluation of disinfectants on surfaces can primarily be segmented into three categories. The first category is the screening and background studies, the second category is regulatory and modified regulatory studies, and the third category of microbiological test for testing disinfectants is the in situ trials designed to evaluate real-world performance.

11.2 Hard Surface Testing

11.2.1 Screening Studies

Minimum inhibitory concentration (MIC) testing is a frequently used method for testing microbial susceptibility for many microbiological applications. However, for the testing of disinfectants and antiseptics, this screening method is of very little value. An MIC test uses dilutions of an active ingredient to determine at what concentration this ingredient is able to inhibit microbial growth. While useful for evaluations of actives, especially antibiotics, formulated disinfectants and antiseptics are typically much more complex than a single diluted active ingredient. Therefore, efficacy depends greatly on formulation and the proper combination of ingredients other than the active. Because of this, even initial screening tests should focus on using formulated products.

One of the quickest ways to evaluate a formulated product is through a time-kill suspension test. Alternatively called a *D*-value test or a kill-curve test, this method uses a suspension of the test article inoculated with the challenge microorganism in order to measure the reduction in viable bacteria over time. While inherently straightforward, many variations of time

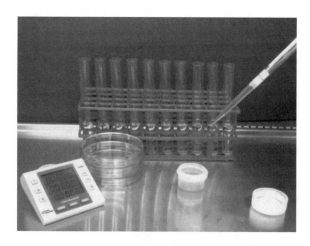

FIGURE 11.1 (See color insert following page 210)
Typical time-kill setup.

kills exist. A guide published by the American Society for Testing and Materials (ASTM E2315, 2003) [1] outlines several of these options providing insight into proper experimental design. The main benefit of time-kill testing is the relative ease with which it can be conducted. The ability to test many samples and variables allows an experimenter to obtain a great deal of information with a minimal input of resources. Although easier to conduct and less technique-sensitive than many of the other methods described in this chapter, errors are possible and do occur. To minimize these errors, one should focus on controlling key variables including inoculum preparation, contact conditions, and neutralization (Figure 11.1).

One of the most common errors in time-kill testing is the inconsistent preparation of the inoculum for use in the study. Depending on the class of microorganism to be tested, this can result in highly variable and misleading results. Broth cultures tend to be easier to work with than resuspended agar cultures, but may also be less challenging to the product in terms of disinfectant susceptibility [2]. Microorganisms that have a tendency to clump such as fungi or mycobacteria must be manipulated before use in the study. This requires the inclusion of a maceration step with a well-paired macerator. In general, it is always worth the extra time to evaluate a culture microscopically before deciding on a homogenization technique for culture preparation. Ensuring a uniform and well-dispersed culture is always important because of the inherent problems in standard plate count methods and the tendency of a clump of cells to present as a single colony. Another source of misleading results occurs during testing of hard-to-neutralize antimicrobial agents such as triclosan [3]. In a time-kill study, it is essential to stop the activity of the disinfectant at a precise moment in order to adequately calculate log reduction or D-values. One should follow a comprehensive method to characterize neutralizers such as ASTM 1054 and perform the neutralizer assay using a highly susceptible microorganism under worst-case conditions [4]. Ideally, neutralizer effectiveness tests should be performed any time one tests a new product. During product development or sample screening, it may be difficult to evaluate every formulation for neutralizer effectiveness. Under these circumstances, it is important to select representative formulations and carefully watch for problems. A major indication of an improperly neutralized product is observed when the dilution series is not serial. This is often evident when a lower dilution plate contains fewer colonies than a higher dilution plate and occurs when the test system requires the additional neutralization inherent in the serial plating in order to achieve neutralization. Occasionally, neutralization cannot be achieved with a standard 1:10 dilution in neutralizing broth and subsequent use of neutralizing agar. In these cases, one can still maintain their limits of detection by plating a greater volume of the initial neutralization tube in several petri dishes. Although neutralizer selection and characterization can be a lot of work, failure to control neutralization can easily mislead investigators into overstating the product's performance.

Unfortunately, neutralization problems or other experimental errors are not always as easy to recognize as nondilutional serial dilutions. This emphasizes the need for utilizing proper internal controls within a study to gain quick insight into a flawed study. These controls should include well-characterized test articles in addition to any experimental formulations. In addition to including the proper internal controls, the experimenter should also control the various other components of the laboratory test conditions. Although small changes in temperature, humidity, or incubation time tend to impact the results more subtly than other possible mistakes, they still should be standardized to minimize variability.

While other methods exist for preliminary characterization of antimicrobial formulations, properly controlled time-kill testing is still the most applicable in vitro test. New and emerging methods continue to challenge this paradigm and should be considered as potential screening tools depending on the needs of the investigator.

11.2.2 Regulatory Studies

Although time-kill suspension studies are very useful for acquiring a snapshot of a product's performance, especially when comparing multiple products or formulations, their application for official government-regulated studies is limited. A time-kill study, by definition, tests an article in solution. While very useful, this method does not give all the information one might need on the actual activity of disinfectants on hard surfaces. To characterize this activity, an investigator should use standardized methods adapted by the presiding regulatory agencies in their regions and, when possible, modify these methods to meet their actual needs.

When considering anything on a global basis, there is undoubtedly great variation from region to region and country to country. Not surprisingly, the regulatory requirements for the testing of disinfectants vary greatly as well. While much of the world defers to either the United States or the European Union for methods to support the testing of disinfectants, one should always be aware of exceptions. Global harmonization efforts, while a noble endeavor, have not been successful to date and must overcome many challenges including the different legal considerations of each country, variation in use patterns, and preferences and unique needs throughout the world.

In Europe, the approach to disinfectant testing follows a logical sequence of iterative studies. Using a combination of phases and steps within each phase, the EU authorities outline multiple tiers of testing with suspension studies designed to initially characterize the activity of disinfectants and quantitative hard surface studies to further delineate activity. The most commonly utilized methods for characterizing a disinfectant for the EU member nations are the EN 1276 for bacteria, EN 1650 for fungi, EN 14348 for mycobacteria, EN 14476 for viruses, and EN 13697 for hard surfaces [5–9]. Each of these methods presents some unique challenges; however, because the 1276, 1650, and 14348 are suspension tests, the challenges are similar to those encountered during time-kill testing. Conversely, the 13697 presents challenges similar to hard surface methods used in the United States, and similar strategies and controls should be employed.

In the United States, hard surface disinfectants are regulated by the Environmental Protection Agency (EPA) unless they are developed for use on medical devices, where they fall under the jurisdiction of the Food and Drug Administration (FDA). EPA and FDA share many of the methods that are required for registration. Although these methods are drawn from consensus sources or from the agencies themselves, the most useful source for hard surface disinfectant methods remains the Association of Official Analytical Chemists (AOAC) International. Chapter 6 of the 15th edition of the *Official Methods of Analysis*, published in 1990, remains the main repository for the key methods used for disinfectant testing in the U.S. [10]. For hard surface testing against bacteria, the most commonly utilized method is the AOAC use-dilution test (UDT) [11]. This method utilizes stainless steel penicylinders and can be adapted for most species of bacteria. To perform this test, the penicylinders are coated with bacteria and then placed into the disinfectant for the prescribed contact time. Following the contact time, the carriers are aseptically transferred to individual tubes of neutralizing growth media and incubated for growth. The flaws in this method are numerous, including the limited information one can achieve from a qualitative study and the technique-sensitive nature of aseptically transferring penicylinders at the end of a wire hook from tube to tube. Because of the numerous flaws in this 40 plus-year old method, anything an experimenter can do to improve the test outcome should be employed. One such task is the careful monitoring of the carriers used in testing. It is critically important to inspect each carrier before use in the study. Although the method contains a biological screen, a visual screen using a dissecting microscope can be very useful for detecting flaws in the carrier that can give a false-positive result. Carrier variability is a problem in all of the AOAC qualitative procedures and should be carefully considered when using any of these methods [12,13]. Because of these problems with the current method, activities continue to replace this method with more reproducible quantitative methods of analysis. Much progress has been made; however, this is a lengthy process involving legal precedent, regulatory policy, and scientific analysis.

In the interim, a basic way to add more reproducibility to this method is to enumerate the carriers after removal from the disinfectant. This can be accomplished through the combination of sonication and use of a vortex mixer to remove the bacteria from the penicylinder after disinfectant contact. Removal and quantification of the bacteria can be challenging and the approaches are varied among laboratories. Although exact counts are highly elusive, 5 min sonication followed by 2 min vortex seems to be sufficient to give an indication of the number of bacteria on a carrier. These conditions are not strictly defined and many labs omit the sonication or shorten the vortex time [14]. These changes may not be appropriate for regulatory submissions, but essential to increase the overall understanding of the product.

A similar method, the AOAC sporicidal activity test is utilized for testing spores against disinfectants [15]. This method is conceptually very similar to the AOAC UDT except in the number of replicates and the use of porcelain penicylinders and fabric suture loops with the spore populations. Subsequently, this test includes many of the same potential sources of variability as the UDT. A quantitative step can be added, but care should be taken at overinterpreting the results from the inoculated suture loops used in the study, as the spores may be difficult to remove from this material.

For tuberculocidal testing, the most efficient evaluation is the quantitative tuberculocidal method (QTB), a quantitative method that uses *Mycobacterium bovis* BCG [16]. Although this method works well, it may not be as challenging as a hard surface method and its applicability for submission to regulatory agencies for some active classes is questionable. The accepted hard surface test for TB testing of all disinfectants in the United States remains the AOAC tuberculocidal method [17]. This method has limited application as a screening tool because of a 90 day incubation period and the numerous inherent problems. While the incubation period can be shortened, this method still provides little usable information. Tests using other mycobacteria species with faster growth cycles, such as *M. smegmatis*, also serve little purpose as these microorganisms have been shown to be more sensitive to disinfectants than *M. bovis* [18].

For virucidal testing, the EPA allows for any suitable virology method to be utilized. The most commonly utilized method follows the procedure outlined in the ASTM method E1053-97 [19]. This method consists of a viral test suspension dried on an inanimate, nonporous surface, usually the bottom of a petri dish, to evaluate disinfectants when compared to a cytotoxicity control.

For fungicidal products, several possible tests exist. The simplest and most effective study to conduct is the AOAC fungicidal test [20]. This method uses *Trichophyton mentagrophytes* tested in suspension. Because

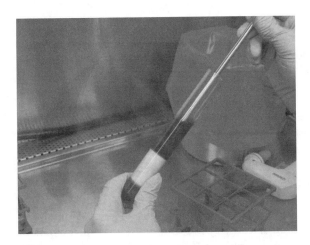

FIGURE 11.2 (See color insert following page 210) Homogenization of fungal culture.

suspension tests tend to be less challenging than hard surface studies, this method is not always a good indicator of overall fungicidal activity. Additionally, it is important to note that this species of fungus does not represent the most difficult to disinfect fungal species. Conidiospore-forming fungi, such as *Aspergillus niger*, have been shown to be significantly harder to kill [21]. Although the use of standard methods is key for using a performance standard to help ameliorate some of the variability in microbiology testing, not all standard methods were designed to utilize the most resistant organism (MRO) during testing. Even for those methods that did have this objective in mind, the MRO will be highly dependent on the active tested [22].

For each of these methods, the inoculum preparation remains critical. Not only does media selection influence results and warrant careful consideration but also does the resuspension method after incubation. This is especially true for mycobacteria, fungi, or other microorganisms that may clump together during growth. If desired, one can add polysorbate 80 at a low concentration after resuspension of a mycobacterial culture to prevent clumping after maceration. Again, the two components of the macerator must be well paired and fit together perfectly to get optimal results. If detergents are used to prevent clumping, care must be taken to ensure that these detergents do not negatively impact the test article (Figure 11.2).

11.2.3 In Situ Studies

Even though the EPA, FDA, and other regulatory bodies recommend standardized methods, these methods are not intended to serve as the sole source of characterization of a disinfectant product. The ultimate test of a product's performance should be final qualification testing on actual facility surfaces. For hospitals and healthcare facilities, the label claims on the disinfectant should be strictly followed as little testing is feasible in an individual hospital. Because of the myriad of factors contributing to hospital-acquired infections, few studies exist that definitively determine the impact of disinfectant choice on infection rates. Although multiple studies certainly suggest a cause and effect relationship of some types of nosocomial infections and proper disinfection, the ultimate study design remains elusive [23]. Furthermore, the impact of disinfection will vary by pathogens and the vectors they use for transmission. Ultimately, hospital users must depend on the claims and methods granted by the regulatory agencies.

The situation is different for pharmaceutical and life sciences consumers. Because of the use patterns of disinfectants in these facilities, validation studies on disinfectants are standard procedures for these consumers. While testing is done by most pharmaceutical companies, the means by which one conducts the validation testing can vary significantly. The best means to validate a disinfectant for pharmaceutical use is to follow the principles of the United States Pharmacopeia (USP) <1072> [24]. While this document includes screening studies similar to those outlined in this chapter, it also includes an important recommendation for companies to perform surface challenge testing. This recommendation calls for the inoculation of a 2 in. by 2 in. coupon made of materials encountered in the use area, with microorganisms that may be encountered in the facility. To be effective by these criteria, a disinfectant must show a 2 $\log_{10}$ reduction for bacterial spores and a 3 $\log_{10}$ reduction for vegetative bacteria. When performing these methods, it is important to work with the manufacturer of the disinfectant to design the experiments. Many potential problems can be avoided through this simple step.

Perhaps, even more important than the validation testing is the real-world monitoring of environmental surfaces for contamination during the use of a disinfectant. Although no perfect sampling method exists, a consistent and well-designed evaluation of the bioburden in an area and the types of microorganism encountered can provide key insight into the effectiveness of a chosen disinfectant [25].

Hard surface disinfectants vary as widely in scope and formulation as do those areas in which they are used. Because of this, there is no universal plan for ensuring the formulation or selection of the right one. Through a testing program that includes multiple tiers and levels of testing, one maximizes the possibility of ending up with the right product.

11.3 Skin Testing

The other category of critical surface is the skin. Few could argue with its inclusion as a critical surface or the importance of designing a technical test plan with at least the equivalent rigor as that incorporated for critical hard surfaces. To that end, a similar approach of screening studies, regulatory-type tests, and in situ evaluations are recommended.

11.3.1 Screening Studies

Suspension tests can be useful for screening and selecting skin decontaminants. As in hard surface product testing, time-kill studies can offer a great deal of information about products designed for the skin. This is especially useful when comparing the activity of multiple products or formulations or when comparing the relative activity of products against multiple microbial isolates. However, since many of the products formulated for use on the skin do not possess the same characteristics in a suspension as they do when interacting with human skin, it is essential to use more than just a time-kill study to screen these formulations or products.

Live human skin is clearly the best means to test topical products, but clinical studies are expensive and time consuming and must be limited to unequivocally ensure the safety of the test population. Without the ability to test products exclusively with human subjects, many investigators turn to the next best thing, in vitro human skin. Excised skin models using cadaver skin or skin removed during plastic surgery procedures can provide excellent results and closely mimic human trials [26]. Unfortunately, there are problems with this approach. Human skin is expensive and its availability is limited. Furthermore, there are concerns regarding the proper disposal of human remains and consent issues with skin from surgery. These challenges necessitate the use of models other than human skin.

One test model that has been used repeatedly as a surrogate for human skin is excised pigskin. Although pigs lack eccrine sweat glands, many characteristics of pigskin are remarkably similar to human skin. The thickness of the epidermis, sparseness of hair, and other factors make pigskin an excellent indicator for human skin [27]. Because of this, pigskin studies have been used for years as a surrogate test model to evaluate the antimicrobial efficacy of skin decontaminants [28,29]. Pigskin tests work well for basic efficacy evaluations and can also be employed for other types of studies including residual evaluations, compatibility studies, and transfer experiments.

It is important to pay attention to several key variables when performing pigskin testing. After processing and storing in the refrigerator for a certain period, the skin will need to be equilibrated before testing. Care should be taken to monitor the temperature and dryness of the skin during a study. Failure to do so can result in variable results, especially for products impacted by dilution. Another important factor when testing using a pigskin surrogate is to properly control the source and treatment of the pigskin before testing. An optimal situation is a relationship with a local farm in which the testing laboratory can acquire the skin soon after slaughter of the animal for meat processing. As this skin is rarely used for any other purpose, this skin can be relatively inexpensive. The skin does need processing before testing as fat on the underside and dirt and debris on the skin surface must be removed before testing. A good relationship with the processing facility allows for this material to be removed using only pressurized water and prevents the use of chemical cleaning methods. Also, this relationship allows one to get the genetics and history of the animal before testing. Although antibiotic feed treatment does not seem to have an impact on the testing attributes of the skin, treatment with harsh soaps and detergents during the cleaning process may do so [18]. Other important considerations include the careful inspection of the skin before testing and standardization of the region of the pigskin used. Typically, the back skin of the pig provides the best samples and ease of use since pigskin can vary significantly from different parts of the animal and results can vary from site to site.

When performing tests using the pigskin model, any number of modifications are possible. Most use shaved pigskin and includes inoculation of both pieces of a pair of skin with a low volume of pathogenic agent. These skins, glued by epoxy to a plastic cap, are then treated with an antimicrobial soap or alcohol sanitizer and rubbed together to simulate product use. After the treatment, the skins are sampled using a sterile sampling cup, Teflon policeman, and a neutralizing debriding solution.

The pigskin model is predominantly used for bacteria, especially antibiotic-resistant pathogens, but works for other types of microorganisms as well. If desired, it can be adapted easily for use with fungi, spores, or viruses. Although the pigskin model works well at predicting activity under real-world conditions, especially when one wants to use pathogenic microorganisms, the best method of evaluation still employs the hands of human volunteers (Figures 11.3 and 11.4).

11.3.2 Regulatory Studies

Topical antimicrobial products for use in the United States are classified as drugs and regulated products by the FDA. The FDA is currently refining a

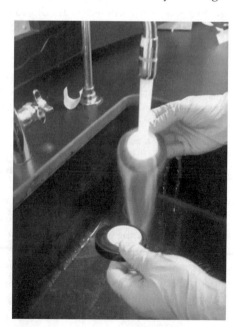

FIGURE 11.3 (See color insert following page 210)
Washing of pigskins.

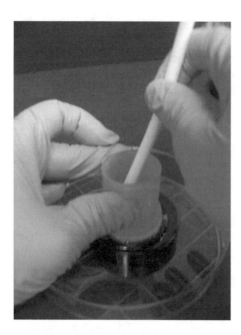

FIGURE 11.4 (See color insert following page 210)
Debriding with Teflon policeman.

tentative final monograph (TFM) that outlines indications for healthcare personnel handwash (HCPHW) products, patient preoperative skin (POP) preparation products, and surgical hand scrub (SS) products [30]. The conduct of these clinical studies requires a great deal of time and expense, which makes it essential that any potential risks for mistakes are mitigated.

The HCPHW method, which typically utilizes a red pigmented bacterium, *Serratia marcescens*, as an indicator of product activity, measures the ability of a decontaminant to remove the transient flora found on the hands of healthcare professionals. Because this method requires multiple applications of microorganism and product to simulate multiple patient contacts over the course of a healthcare worker's day, the design of the study is integral. Important considerations include adequate population size and diversity, uniform preparation of culture, and careful monitoring of clinical subjects. A thorough, recent review of some of these challenges along with others facing topical antimicrobials can be found in Paulson's *Handbook of Topical Antimicrobials* [31]. Although trade organizations have published versions of this method, the official method is still exclusively available directly from the FDA.

The surgical scrub study also uses the hands of volunteer subjects, but evaluates a product's ability to reduce the resident flora of the skin instead of a transient inoculation. This eliminates some of the concerns with inoculum preparation that are present in the HCPHW, but the challenges of working with clinical subjects remain.

The aim of preoperative skin preparation method is to reduce the amount of bioburden on the patient, in order to lessen the risk of pathogenic microorganisms being introduced into the incision site. While this study also uses resident flora instead of an artificial inoculum, it presents unique challenges because of the sites sampled. Instead of evaluating the hands as the other two main methods, this technique uses the groin and abdomen to represent surgical sites. Because baseline bacteria numbers must be of a specified titer on each site to enable the investigator to show a significant reduction from the test product, finding qualified subjects can be difficult. In order to find sufficient qualified subjects, it is important to maintain a large pool of possible volunteers and remain diligent for subject deviations from required inclusion criteria.

In the United States, there are no accepted standardized virucidal tests for demonstrating antiviral activity on the skin. Other countries, however, have had the prescience to understand the importance of antiviral activity on the skin and will accept data from appropriate methods. The best method for antiviral testing on human skin utilizes the fingerpads of human adult volunteers and follows ASTM method E1838 [32].

Virucidal evaluations are not the only way that testing of skin decontaminants in other countries differs from that of the United States. The European Union relies heavily on the use of the EN 1500 method, which compares products with a reference alcohol product. This method utilizes *E. coli* and standardizes the handrub procedure to increase consistency. The EN 1500 successfully simplifies the test method relative to the U.S. methods [33–35]. Nonetheless, equivalent care should be taken in conducting and interpreting clinical studies using human subjects for any regulatory application (Figures 11.5 and 11.6).

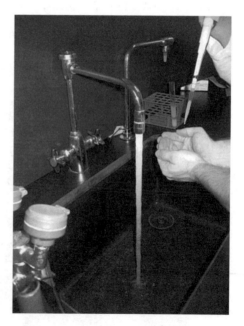

FIGURE 11.5 (See color insert following page 210)
Inoculum application in a HCPHW.

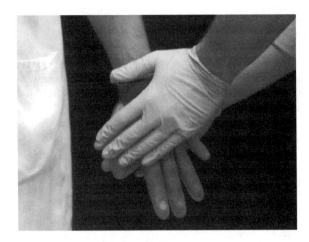

FIGURE 11.6 (See color insert following page 210) Sampling using a glove juice procedure.

11.3.3   In Situ Studies

Clinical tests on human subjects are an excellent indicator of real-world activity, but are limited by the selection of microorganisms that do not cause disease in healthy adult volunteers. To determine the impact of a particular antimicrobial in actual use against real pathogenic agents, a study designed to evaluate infection rates or lost time must be designed. These studies are amazingly complicated to design, and the number of influential factors is endless. If one does undertake such a study, the quintessential element is the selection of the right product. Not all antimicrobial products are equal. For example, a product formulated for home use may not have the same efficacy profile as one created for hospital use.

Like testing in any scientific discipline in any industry, testing on disinfectants and decontaminants for products used on critical surfaces is complicated, expansive, and often includes a steep learning curve to begin to get it right. Hopefully, by avoiding some common mistakes, along with a well-designed, iterative approach to testing, this learning process can be made easier.

References

1. American Society for Testing and Materials (ASTM), Standard guide for assessment of antimicrobial activity using a time-kill procedure ASTM, West Conshohocken, PA, 2003.
2. Russell, A., Factors influencing the efficacy of antimicrobial agents, in *Principles and Practices of Disinfection, Preservation, and Sterilization*, Russell, A.D., Hugo, W.B., and Ayliffe, G.A.J. (Eds.), Blackwell Scientific Publications, Oxford, 1999, pp. 95–123.
3. McDonnell, G., Importance of neutralization in the evaluation of triclosan-containing products, *J. Ind. Microbiol. Biotechnol.* 21, 184–186, 1998.

4. American Society for Testing and Materials (ASTM), Standard test methods for evaluation of inactivators of antimicrobial agents ASTM, West Conshohocken, PA, 2002.
5. European Committee for Standardization, Chemical disinfectants and antiseptics— Quantitative suspension test for the evaluation of bactericidal activity of chemical disinfectants and antiseptics used in food, industrial, domestic, and institutional areas—Test method and requirements (phase 2, step 1), 1997.
6. European Committee for Standardization, Chemical disinfectants and antiseptics— Quantitative suspension test for the evaluation of fungicidal activity of chemical disinfectants and antiseptics used in food, industrial, domestic and institutional areas—Test method and requirements (phase 2, step 1), 1998.
7. European Committee for Standardization, Chemical disinfectants and antiseptics— Quantitative non-porous surface test for the evaluation of bactericidal and/or fungicidal activity of chemical disinfectants used in food, industrial, domestic and institutional areas—Test methods and requirements without mechanical action (phase 2, step 2), 2001.
8. European Committee for Standardization, Chemical disinfectants and antiseptics— Quantitative suspension test for the evaluation of mycobactericidal activity of chemical disinfectants in the medical area including instrument disinfectants— Test methods and requirements (phase 2, step 1), 2005.
9. European Committee for Standardization, Chemical disinfectants and antiseptics— Virucidal quantitative suspension test for chemical disinfectants and antiseptics used in human medicine—Test methods and requirements (phase 2, step 1), 2005.
10. AOAC International, Disinfectants, Chapter 6, 15th ed., *Official Methods of Analysis of the Association of Official Analytical Chemists*, AOAC International, Arlington, VA, 1990.
11. AOAC International, Use-dilution methods, Chapter 6, in *AOAC Official Methods of Analysis*, 15th ed., Arlington, VA, 1990.
12. Ascenzi, J.M., Ezzell, R.J., and Wendt, T.M., Evaluation of carriers used in the test methods of the Association of Official Analytical Chemists, *Appl. Environ. Microbiol.* 51 (1), 91–94, 1986.
13. Cole, E.C., Rutala, W.A., and Alfano, E.M., Comparison of stainless steel penicy-linders used in disinfectant testing, *J. Assoc. Off. Anal. Chem.* 71 (2), 288–289, 1988.
14. Tomassino, S., Enumeration procedure for monitoring test microbe populations on inoculated carriers in AOAC use-dilution methods, *JAOAC Int.* 89 (6), 1629–1634, 2006.
15. AOAC International, Sporicidal activity of disinfectants, Chapter 6, in *AOAC Official Methods of Analysis*, 15th ed., Arlington, VA, 1990.
16. Ascenzi, J.M., Ezzell, R.J., and Wendt, T.M., A more accurate method for measurement of tuberculocidal activity of disinfectants, *Appl. Environ. Microbiol.* 53 (9), 2189–2192, 1987.
17. AOAC International, Tuberculocidal activity of disinfectants, Chapter 6, in *AOAC Official Methods of Analysis*, 15th ed., Arlington, VA, 1990.
18. Klein, D., Unpublished data, 1998.
19. American Society for Testing and Materials (ASTM), Standard test method for efficacy of virucidal agents intended for inanimate environmental surfaces ASTM, West Conshohocken, PA, 2002.
20. AOAC International, Fungicidal activity of disinfectants, Chapter 6, in *AOAC Official Methods of Analysis*, 15th ed., Arlington, VA, 1990.

21. Klein, D., Dell'Aringa, B., Karanja, P., Bartnett, C., Barteau, T., Macauley, J., and Bucher, J., Disinfectant testing against *Aspergillus niger*, General meeting of the American Society for Microbiology, May 2006.
22. Block, S.S., *Disinfection, Sterilization, and Preservation*, 5th ed., Lippincott Williams and Wilkins, Philadelphia, 2001.
23. Hota, B., Contamination, disinfection, and cross-colonization: are hospital surfaces reservoirs for nosocomial infection? *Clin. Infect. Dis.* 39 (8), 1182–1189, 2004.
24. United States Pharmacopeia, Disinfectants and Antiseptics.
25. Cundell, A.M., Bean, R., Massimore, L., and Maier, C., Statistical analysis of environmental monitoring data: Does a worst case time for monitoring clean rooms exist? *PDA J. Pharm. Sci. Technol.* 52 (6), 326–330, 1998.
26. Messager, S., Goddard, P.A., Dettmar, P.W., and Maillard, J.Y., Comparison of two in vivo and two ex vivo tests to assess the antibacterial activity of several antiseptics, *J. Hosp. Infect.* 58 (2), 115–121, 2004.
27. Vardaxis, N.J., Brans, T.A., Boon, M.E., Kreis, R.W., and Marres, L.M., Confocal laser scanning microscopy of porcine skin: implications for human wound healing studies, *J. Anat.* 190 (Pt 4), 601–611, 1997.
28. Bush, L.W., Benson, L.M., and White, J.H., Pig skin as test substrate for evaluating topical antimicrobial activity, *J. Clin. Microbiol.* 24 (3), 343–348, 1986.
29. McDonnell, G., Haines, K., Klein, D., Rippon, M., Walmsley, R., and Pretzer, D., Clinical correlation of a skin antisepsis model, *J. Microbiol. Methods* 35 (1), 31–35, 1999.
30. Food and Drug Administration, Tentative final monograph for health care antiseptic products: proposed rule, *Fed. Regist.*, 1994, pp. 31401–31452.
31. Paulson, D., *Handbook of Topical Antimicrobials Industrial Applications in Consumer Products and Pharmaceuticals*, Marcel Dekker, Inc., New York, 2003.
32. American Society for Testing and Materials (ASTM), Standard test method for determining the virus-eliminating effectiveness of liquid hygienic handwash and handrub agents using the fingerpads of adult volunteers ASTM, West Conshohocken, PA, 2002.
33. European Committee for Standardization, Chemical disinfectants and antiseptics—Hygienic handrub—Test method and requirements (phase 2, step 2), 1997.
34. Pietsch, H., Hand antiseptics: Rubs versus scrubs, alcoholic solutions versus alcoholic gels, *J. Hosp. Infect.* 48 Suppl. A, S33–S36, 2001.
35. Rotter, M.L., European norms in hand hygiene, *J. Hosp. Infect.* 56 Suppl. 2, S6–S9, 2004.

12

Regulatory Constraints on Disinfectants and Decontamination

Michael G. Sarli

CONTENTS

12.1 Introduction

The broad spectrum of cutting edge, technical advancements made in disinfection and decontamination are a testament to the human desire to continuously strive to learn about and control the negative impact of our environment upon us. These efforts are highly laudable and provide a worthy challenge to our best and brightest. However, our best and brightest must also be fully cognizant of the impact of our technological advancements upon humans, animals, and our environment; inventors do not have the luxury of "caveat emptor." Even in this age of instant and ubiquitous communication, consumers cannot make their own value judgment in determining the safety and effectiveness of decontamination and disinfection technologies. This is where the regulators step in as advocates and protectors of the public, armed with an intricate, far-reaching, and oftentimes confusing arsenal of regulations. The regulators and the regulated community philosophically share the same vocation of protecting public health, but disagreements over the implementation of that philosophy do arise.

To the creative mind, regulations are at best a hindrance and at worst a death knell to creativity itself. The title of this chapter, which was preordained, endorses that sentiment. Nevertheless, the biotech, pharmaceutical, and medical device industries are and will continue to be highly regulated. Awareness of impending regulatory trends and developing products and services that meet those regulatory constraints while simultaneously meeting the customer's needs is the key to success. Avoiding or fighting the regulatory realities may provide short-term benefits, but the risk is unmanageable for the long-term. Industry and even the public may not agree with the policies developed by the agencies but their responsibility endures.

Most if not all presentations on regulatory related topics begin with reference to the challenge of navigating the maze of an alphabet soup of acronyms. The regulated world is full of cryptic references such as, but certainly not limited to, TSCA, FIFRA, FDCA, PDUFA, 510(K), 483, CERCLA, and BPD. Any industry regulatory professional worth their salt can speak for hours without uttering an actual word. The alphanumeric soup analogy is usually followed by a long-winded historical background to the particular regulation or regulatory organization to rival Homer. The reader will be spared any inadequate attempt to place Odysseus in the realm of the Environmental Protection Agency (EPA) and the Food and Drug Administration (FDA).

However, trite are the hackney descriptions of regulatory mazes, they are accurate. The first and perhaps greatest challenge is determining which regulatory agency's or agencies' maze is applicable to your product or service.

There is a concept most often associated with hazardous waste regulations described as "cradle to grave," wherein the generator of hazardous waste is responsible from the formation of the waste until its ultimate disposition or termination. Whoever else participates in handling your waste shares in responsibility, but you can never alleviate yourself of responsibility of the waste material. This same concept can be applied to the regulations impacting disinfection and decontamination products and services. From the selection of ingredients used in formulating a product to its ultimate disposition, there is some form of regulation (often overlapping and sometimes conflicting) impacting the product. One needs to be cognizant of the regulations for the full life of the product or service and how those regulations may impact the developer and the user.

In the United States, for the developer/marketer and the user of an antimicrobial, regulatory oversight may come from the U.S. Environmental Protection Agency (U.S. EPA), the FDA, or both, depending on how and where the antimicrobial product is applied. EPA regulates antimicrobial products as pesticides under authority of the Federal, Insecticide, Fungicide, and Rodenticide Act[1] (FIFRA) and FDA regulates antimicrobials as drugs or medical devices. Products intended for the control of fungi, bacteria, viruses, or other microorganisms in or on living humans or animals are considered drugs. Products intended to reprocess medical devices, i.e., high-level disinfection and terminal sterilization are considered medical devices. There are as always some exceptions to the rule, which will be discussed later in the chapter. In the recent past, EPA and FDA have coordinated and clarified their seemingly confusing overlapping responsibilities specifically through the enactment of the Food Quality Protection Act[2] (FQPA) of 1996 but some overlap continue. The EPA and the FDA do not necessarily speak a common language regarding disinfection and decontamination. There are differences in terminology that confuse everyone. Sometimes the distinctions are negligible but sometimes the distinctions are significant.

A significant point of difference is how the various regulatory bodies classify and interpret hazards associated with chemicals and chemical products. A material may be corrosive by EPA definition but not corrosive by the Department of Transportation regulations. The definition of flammability is different between the EPA, the Occupational Safety and Health Administration (OSHA), and the transportation authorities.

A summary of regulatory classifications, requirements, and approaches for the regulation of antimicrobials will be presented by each agency. The focus will be upon the U.S. regulatory framework followed by some general comments on the international arena, specifically Canada and Europe. Emphasis is placed upon the U.S. EPA regulation of chemicals since the biopharmaceutical and cosmetic industries are seemingly much more familiar with the FDA regulations. This chapter is intended as a survey of applicable regulations. Further details should be sought out through the references provided and through the regulatory agencies identified.

In the United States, Congress enacts legislations that are called acts, such as the Federal Food Drug and Cosmetic Act,[3] which dictate what is required by law. Various governmental departments such as the FDA publish regulations in the Code of Federal Regulations (CFR), which define the requirements to meet compliance with the act. The applicable sections of the CFR will be identified with the associated agency and further detailed by subject area when appropriate such as the FDA sections for drugs vs. devices. Additionally, URLs for the various agencies will be listed as appropriate (Figure 12.1).

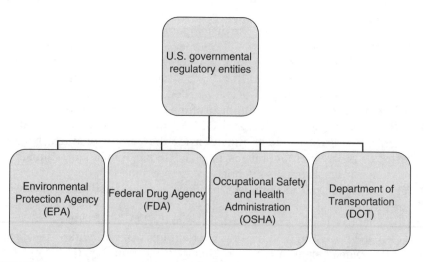

FIGURE 12.1
U.S. governmental regulatory entities.

12.2 Environmental Protection Agency

EPA's Web site can be found at www.epa.gov. Regulations of the EPA can be found in 40 CFR.

EPA regulations impact the full spectrum of the use of chemicals.

The Federal Insecticide, Fungicide, and Rodenticide Act[1] (FIFRA) regulates the formulation and use of antimicrobials on hard surfaces.

The Toxic Substance Control Act[4] (TSCA) regulates the existence of chemicals.

The Clean Water Act[5] regulates chemicals introduced into the water.

The Clean Air Act[6] regulates chemicals introduced into the air.

The Resource Conservation and Recovery Act[7] (RCRA) regulates the disposal of waste.

The Comprehensive Environmental Response, Compensation, and Liability Act[8] (CERCLA) and the Superfund Amendments and Reauthorization Act[9] (SARA) regulate hazardous waste contaminated sites and release of hazardous materials into the environment (Figure 12.2).

12.2.1 Federal Insecticide, Fungicide, and Rodenticide Act

EPA's regulations, information, and guidance covering pesticides can be found at www.epa.gov/pesticides/. The regulations governing pesticides can be found at 40 CFR § 150–180.

EPA most directly impacts antimicrobial products under FIFRA. FIFRA requires that before any person can distribute any pesticide in the United States, they must obtain registration from the U.S. EPA.

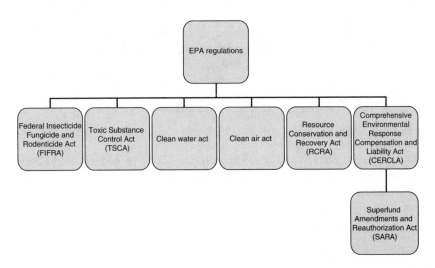

FIGURE 12.2
Environmental Protection Agency (EPA) regulations.

Pesticide is defined under FIFRA[1] as "any substance or mixture of substances intended for preventing, destroying, repelling or mitigating any pest,...Pest is defined as any insect, rodent, nematode, fungus, weed or any other form of terrestrial or aquatic plant or animal life or virus, bacterial or other microorganism (except viruses, bacteria or other microorganisms on or in living man or other living animals) which the administrator declares to be a pest." Antimicrobial agents are substances or mixtures of substances used to destroy or suppress the growth of harmful microorganisms whether bacteria, viruses, or fungi on inanimate objects and surfaces. Antimicrobial products contain about 300 different active ingredients and are marketed in several formulations: sprays, liquids, concentrated powders, and gases.

To obtain registration of a pesticide, manufacturers of antimicrobial products must demonstrate that the product will not cause unreasonable adverse effects to human health or the environment, and that product labeling and composition comply with the requirements of FIFRA. Registrants are required to submit or cite in support of registration detailed information on the chemical composition of their product, the chemical characteristics of the formulation, effectiveness data to support their claims against specific microorganisms, and safety or toxicity data. Much of the labeling is prescriptive and reflective of the chemistry, safety, and efficacy data. (See EPA Label Review Manual www.epa.gov/oppfead1/labeling/lrm/labelreview manual.pdf)

Registration decisions are made by the EPA on a "risk vs. benefit" approach. It is understood that antimicrobials are biocides and often present risk to nontarget organisms including humans. EPA evaluates the safety of the product and application process by assessing worker safety, food safety (when applicable), nontarget organisms, and surfaces by reviewing acute/chronic safety studies, and exposure risk assessments of both active ingredients and the finished product.

Efficacy claims divide antimicrobial products into two categories: nonpublic health claims and public health claims. Nonpublic health claims are for the control and growth of algae; odor-causing bacteria; bacteria that cause spoilage, deterioration, or fouling of materials; and microorganisms infectious only to animals. This general category includes products used in cooling towers, jet fuel, paints, and treatments for textile and paper products. Public health claims are for the control of microorganisms infectious to humans in any inanimate environment.

In addition to the regulations found in the Code of Federal Regulations, EPA communicates requirements and guidance through PR notices, label review manuals, harmonized test guidelines, and other vehicles that can be found on the EPA Pesticide Web site.

12.2.1.1 FIFRA—Registrations

EPA requires companies seeking to hold product registrations or to package registered products (but not necessarily hold any product registrations) to file for establishment licenses. The establishment number is assigned sequentially by EPA and is further uniquely identified by an extension to that number with a two digit abbreviation of the state where the establishment is located and then by a facility identifier. The packager's full EPA establishment number is required to be placed on the product label.

Product registration numbers include the establishment number followed by a sequentially assigned number that uniquely identifies each product registration. Types of product registrations that are granted by the EPA include Section 3, Section 5 (Experimental Use Permit), Section 18 (Emergency Use Exemption), and Section 24C registration (Special Local Need). The sections refer to FIFRA. There is also a reregistration process for Section 3 registrations.

Section 3 registrations can be simply referred to as standard registrations. Supplemental registrations, "Me-Too" registrations, and restricted use products are variations of Section 3 registrations.

A Section 5 registration (Experimental Use Permit) is rarely if ever required for an antimicrobial product. A Section 5 registration requires much of the same supportive data as a Section 3 registration and is typically utilized to obtain EPA permission to conduct field experiments on agricultural pesticides where there are environmental and ecological implications far beyond what is typical for an antimicrobial.

Section 18 (Emergency Use Exemption) registrations are used in emergency situations where technologies are either not available or unapproved for the intended application and there is a specific public health need for the application. Application is submitted by a government agency which will be identified in this discussion as the petitioner. The registration is limited in scope and duration. Special reporting in the way of interim and final reports is required to be submitted by the petitioning agency.

EPA is typically more flexible with data requirements under either emergency situations or limited use areas managed under state control. The basic registrant must be informed of the intent of the petitioner. A Section 18 registration may be extended, but a commitment from the registrant to ultimately pursue an amendment to the registration consistent with the Section 18 labeling is required.

Section 24C (Special Local Need) is filed with state authorities and approved by the governor.

States notify EPA of their intent to approve the use of a pesticide in an unapproved application and EPA has authority to veto. Otherwise, the requirements for a Section 24C registration are substantially similar to the requirements for a Section 18 registration.

Supplemental registrations are essentially private-label versions of an existing Section 3 registration. A Section 3 registrant may grant access to another party to sell the registrants' product under the other party's brand name. The supplemental registration must be identical in formulation and precautionary label language. The supplemental registrant may omit certain label directions for use but cannot change or add to the directions for use. This type of registration does not require an extensive review by EPA. It is essentially an administrative function wherein the two parties to the supplemental registration are dually responsible for compliance with FIFRA.

Me-Too registration defines a process of obtaining a Section 3 registration citing data submitted under an existing registration that is substantially similar in formulation and labeling to the Me-Too registration. This process requires less in-depth data review than that which occurs with an original product registration application.

Some Section 3 registrations are classified as restricted use products (RUP). A product, or its uses, may be classified as "restricted use" typically due to the toxicity of the product, but also at times due to the complexity involved in properly applying the material such as a fumigant used for a whole-house termiticide treatment. Restricted use products may only be applied by a certified pesticide applicator or under the direct supervision of a certified applicator. State regulatory agencies oversee the certification process for pesticide applicators. Information on restriction of use of a pesticide is found in the Code of Federal Regulations (Chapter 40, Part 152.160–175)

EPA is reviewing older pesticides (those initially registered before November 1984) under FIFRA to ensure that they meet current scientific and regulatory standards and that they can be used without causing unreasonable adverse effects to human health or the environment. This process, called reregistration, considers the human health and ecological effects of pesticides and results in actions to reduce risks that are of concern. EPA also is reassessing tolerances (pesticide residue limits in food) to ensure that they meet the safety standard established by the Food Quality Protection Act (FQPA) of 1996. EPA is responsible for the development and issuance of reregistration eligibility decision documents (REDs) for all chemicals with mainly antimicrobial uses. The RED document formally presents the agency's evaluation of the database supporting the reregistration of a pesticide (conclusions regarding which uses are eligible for reregistration under specific condition and requirements, etc.). The antimicrobial reregistration status page contains the list of the antimicrobial REDs, their schedules, and completed documents and available fact sheets.

After the completion of the reregistration process, EPA is developing a system of periodic reviews as assurance of providing EPA with the "state-of-the-art" supportive data to preserve the integrity of the risk vs. benefit registration decision.

12.2.1.2 FIFRA—Product Types

Product types include technical grade active ingredients (TGAI), manufacturing-use products (MUP), and end-use products (EUP). There are different data requirements for each type of product. End-use products may rely upon or bridge certain data submitted for registered TGAI and registered

MUP that are used to formulate the EUP. If a registered source is not utilized in the manufacture of an EUP, then the registrant of the EUP is responsible for a larger spectrum of data at the time of registration and in the future such as at the time of reregistration.

12.2.1.3   FIFRA—Supportive Data

Data required in support of registration includes product chemistry, acute toxicity, chronic toxicity, efficacy, environmental fate, and ecological toxicity. Additional data is required to establish residue tolerances or exemption from tolerances. All must be conducted under good laboratory practices (GLP) 40 CFR 160. EPA has authority to audit data filed in a registration package and any new data that is generated or otherwise comes to light that might effect the risk vs. benefit decision made by the EPA, including failed efficacy data related to product label efficacy claim required to be filed under provision 6(a)2 of FIFRA.

There are various approaches or methods of data support:

- Owner generated, wherein the registrant generates the data or contracts with a laboratory to conduct the data.
- Cite-all, wherein data that has been submitted previously to EPA in support of other registrations is cited by the applicant. The applicant must make an offer to pay for compensation to the original data submitter.
- Selective method, wherein the applicant can submit their own data, cite other data, or have obtained approval from another data submitter for rights to utilize that data in support of registration.

The Office of Prevention, Pesticides, and Toxic Substances' (OPPTS) harmonized testing guidelines have been developed for use in the testing of pesticides and toxic substances and the development of test data that must be submitted to the agency for review under federal regulations. The purpose for harmonizing these guidelines into a single set of OPPTS guidelines is to minimize variations among the testing procedures that must be performed to meet the data requirements under TSCA and FIFRA. The guidelines for safety (Series 870) and product chemistry (Series 820) studies are essentially complete. Efficacy data that is not required under TSCA is in the process of being formalized as guidelines, but in the immediate future the following still holds.

12.2.1.4   FIFRA—Antimicrobial Efficacy

Performance claims require the generation of efficacy data under controlled testing conditions. Data generated in support of claims for microorganisms that are pathogenic to humans must be submitted to EPA for their review and approval. Data generated for microorganisms not considered pathogenic to

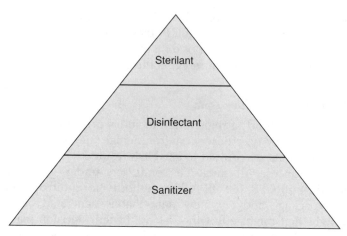

FIGURE 12.3
EPA hierarchy of antimicrobial efficacy (compare Figures 12.4 and 12.6).

humans does not need to be submitted to the agency but must be on file and available to EPA upon request (Figure 12.3).

Generation of data to support label claims and product registration may be conducted under the following approaches: recognized consensus methods, EPA approved protocols, or published peer reviewed data (rarely used).

Test data is generated against specific microorganisms or surrogate organisms identified by EPA as acceptable marker organisms. EPA has recognized or established standard methodology and continues to work cooperatively with interested parties to develop improved consensus methods.

12.2.1.4.1 Indications

The term indication is not typically a part of EPA parlance but is relevant to the device and drug industry and best describes the label efficacy claims that EPA allows to be made on antimicrobial product labeling. The general efficacy indications recognized by EPA are as follows:

- Sterilizers (sporicides) are used on hard inanimate surfaces and objects to eliminate all forms of microbial life including fungi, viruses, and all forms of bacteria and spores. Spores are considered the most difficult form of microorganism to destroy. Therefore, EPA considers the term sporicide to be synonymous with sterilizer. All ethylene oxide products for antimicrobial uses (including medical device reprocessing) are regulated by the EPA as a pesticide.

- Disinfectants are used on hard inanimate surfaces and objects to eliminate or irreversibly inactivate infectious bacteria but not necessarily their spores. Disinfectant products are divided into two major categories: hospital and general use. Hospital-type disinfectants are the most critical to infection control and are used in health-care settings. General disinfectants are the major source of products used in households, swimming pools, and water purifiers. EPA treats the terms germicide and bactericide as synonymous with disinfectant.

- Sanitizers are used to reduce, but not necessarily eliminate, microorganisms from the inanimate environment to levels considered safe as determined by public health codes or regulations. Sanitizers include food contact and non-food contact products.
- Fungicide is an agent that destroys fungi (including yeasts) and fungal spores pathogenic to man or other animals in the inanimate environment.
- Tuberculocide is an agent that destroys or irreversibly inactivates tubercle bacilli in the inanimate environment.
- Virucide is an agent that destroys or irreversibly inactivates viruses in the inanimate environment.

12.2.1.4.2   Efficacy Testing

Currently the efficacy guidelines are available under the Product Performance/Efficacy Test Guidelines; 91A: Public Health Use Antimicrobial Agent Product Performance and 91B: Non-public Health Use Antimicrobial Agent Product Performance last published in 1982, and the Disinfectant Technical Science Section (DIS/TSS) documents (www.epa.gov/oppad001/sciencepolicy.htm). Alternative methods necessary for special application methods or unique organism testing where standard methods are not appropriate require EPA review and approval of protocols prior to generation and submission of the data.

The test guidelines are summarized by the label claim or indication:

- Sterilizer claim requirements: AOAC sporicidal test (60 carriers each on two surfaces) (porcelain penicylinders and silk suture loops) against spores of *Bacillus subtilis* (ATCC 19659) and *Clostridium sporogenes* (ATCC 3584) (three samples representing three lots, one lot 60 days old, killing all 720 carriers).
- Disinfectants (limited efficacy) requirements: AOAC use-dilution method or AOAC germicidal spray products test against *Salmonella choleraesuis* ATCC (10708) or *Staphylococcus aureus* (ATCC 6538) (60 carriers testing three samples representing three different lots, one lot 60 days old killing 59 out of each set of 60 carriers).
- Disinfectants (general) requirements: AOAC use-dilution method or AOAC germicidal spray products test against *Salmonella choleraesuis* (ATCC 10708) and *Staphylococcus aureus* (ATCC 6538) (60 carriers testing three samples representing three different lots, one lot 60 days old killing 59 out of each set of 60 carriers).

- Disinfectants (hospital or medical environment) requirements: AOAC use-dilution method or AOAC germicidal spray products test against *S. choleraesuis* (ATCC 10708), *S. aureus* (ATCC 6538) and *Pseudomonas aeruginosa* (ATCC 15442) (60 carriers testing three samples representing three different lots, one lot 60 days old killing 59 out of each set of 60 carriers).

- Fungicide requirements: AOAC fungicidal test or (versions of the AOAC use-dilution method or germicidal spray products test modified with appropriate elements in the AOAC fungicidal test) against *Trichophyton mentagrophytes* (ATCC 9533) (10 carriers testing two samples representing two different lots killing all fungal spores).

- Virucide requirements: Carrier methods as modifications of either the AOAC use-dilution method or the AOAC germicidal spray products test against the particular virus with a recoverable virus titer of at least 10^4 from the test surface (two different lots of four determinations per each dilution showing inactivation of virus at all dilutions when no cytotoxicity is observed and at least a 3-log reduction in viral titer for both samples when cytotoxicity is present).

- Tuberculocide requirements: Tuberculocidal activity method or the AOAC germicidal spray products test modified to meet the requirements of the tuberculocidal activity method against *Mycobacterium tuberculosis* var. *bovis* (BCG) (two samples representing two different lots killing all of the test microorganism on all carriers and no growth in any of the inoculated tubes of two additional media). Alternative method—quantitative tuberculocidal activity test method (Four log kill required). Products with tuberculocidal claims that are formulated with quaternary ammonium compounds may be evaluated for tuberculocidal efficacy using any one of the test methods listed above. However, validation data is required for any test method chosen. Validation data must be developed by testing one additional sample of the product by a laboratory of the registrant's choice (other than the laboratory which developed the original efficacy data) using the same optional test procedure and test conditions as the original laboratory.

- Non-food contact sanitizer requirements: Guideline 91–30 Method No. 8—against *Staphylococcus aureus* (ATCC 6538) and either *Klebsiella pneumoniae* (ATCC 4352) or *Enterobacter aerogenes* (ATCC 13048) on representative surfaces depending on the proposed uses including but not limited to glass, metal, and unglazed or glazed ceramic tile or vitreous china showing a bacterial reduction of at least 99.9% over the parallel control count within five minutes.

- Food-contact sanitizer requirements: For halide chemical products—AOAC available chlorine germicidal equivalent concentration method against *Salmonella typhi* (ATCC 6539) (One test on each of three samples representing three lots, one which is at least 60 days old showing product concentrations equivalent in activity to 50, 100, and 200 ppm of available chlorine). For other chemical products, i.e., quaternary ammonium compounds, chlorinated trisodium phosphate, and anionic detergent-acid formulations—AOAC germicidal and detergent sanitizers method against *Escherichia coli* (ATCC 11229) and *Staphylococcus aureus* (ATCC 6538) (one sample from each of three different lots, one of which is at least 60 days old demonstrating 99.999% reduction in the number of each test organism within 30 s).

- Additional organism requirements (applies to specific microorganisms other than those named by the AOAC use-dilution method, AOAC germicidal spray products test, AOAC fungicidal test, and AOAC tuberculocidal activity method and not including viruses): AOAC use-dilution method or AOAC germicidal spray products test against the specific organism (10 carriers testing two samples representing two different lots killing all carriers).
- Other more specific claims include residual self-sanitizing activity of dried chemical residue, towelettes, air sanitizers, laundry additives, carpet sanitizers, drinking water, swimming pool water, and preservatives.
- Other circumstances and variables to consider are confirmatory efficacy testing, organic soil load (one-step application) claim, and hard water (400 ppm) claim.

12.2.1.5 *FIFRA—Ingredients*

Pesticide products contain both active and inert ingredients.

An active ingredient is one that prevents, destroys, repels, or mitigates a pest, or is a plant regulator, defoliant, desiccant, or nitrogen stabilizer. By law, the active ingredient must be identified by name on the label together with its percentage by weight. Most active ingredients are registered as technical grade active ingredients (TGAI) or in some instances as the active ingredients are formulated into a manufacturing-use product (MUP). In these instances where the active ingredient is a registered source, a formulator or registrant of an end-use product (EUP) can claim the "formulator's exemption" which relieves the EUP formulator or registrant from the data requirements such as chronic toxicity that exist for the active ingredient. The EUP registrant would still be responsible for product chemistry, acute toxicity, and efficacy data on the EUP itself. If the active ingredient used in an EUP is not from a registered source, then the EUP registrant must address the data requirements for the active ingredient.

An inert ingredient means any substance (or group of structurally similar substances if designated by the agency), other than an active ingredient, which is intentionally included in a pesticide product. Inert ingredients play a key role in the effectiveness of a pesticidal product. For example, inert ingredients may serve as a solvent, allowing the pesticide's active ingredient to penetrate an organism's outer surface. In some instances, inert ingredients are added to extend the pesticide product's shelf-life or to protect the pesticide from degradation due to exposure to sunlight. Pesticide products can contain more than one inert ingredient, but federal law does not require that these ingredients be identified by name or percentage on the label. Only the total percentage of inert ingredients is required to be on the pesticide product label.

Inert ingredients are not held to the same scrutiny by EPA as active ingredients, but if the inert ingredient has not been used in an approved pesticide formulation previously, EPA minimally requires full chemical disclosure of the ingredient. If the ingredient's full chemical identity is confidential, then the supplier can submit the disclosure directly to EPA as trade-secret information. If EPA has any concerns regarding the safety of the ingredient, they may require additional data prior to allowing use of the ingredient in the pesticide formulation and may limit where and how the ingredient may be used. Status of chemical ingredients may be found at the EPA's substance registry system (www.epa.gov/srs/).

12.2.1.6   FIFRA—Tolerances

EPA sets limits on the amount of pesticides that may remain in or on foods. These limits are called tolerances. The tolerances are set based on a risk assessment and are enforced by the Food and Drug Administration.

EPA is reassessing tolerances (chemical residue limits in food) and exemptions from tolerances for inert ingredients in pesticide products to ensure that they meet the safety standard established by the FQPA in 1996. FQPA requires the reassessment of inert ingredient tolerances and tolerance exemptions that were in place prior to August 3, 1996. Since the passage of FQPA, EPA has been reassessing inert ingredients. Completed inert ingredient tolerance reassessment decision documents are available on the tolerance reassessment status page (http://www.epa.gov/opprd001/inerts/decisiondoc_a2k.html).

Refer to the U.S. Government Printing Office issued Code of Federal Regulations (CFR) for complete and current information on inert (other) ingredient tolerances and tolerance exemptions. The CFR describes how each inert ingredient may be used.

12.2.1.7   FIFRA—Pesticide Registration Improvement Act

The recent legislation, Pesticide Registration Improvement Act[10] (PRIA), establishes pesticide registration service fees for registration actions in three pesticide program divisions: antimicrobials, biopesticides and pollution prevention, and the registration divisions. The category of action, the amount of the pesticide registration service fee, and the corresponding decision review periods by year are prescribed in the statute.

12.2.1.8   FIFRA—Postmarket Surveillance

EPA holds the authority to sample registered product samples that are released for sale for the purpose of assaying the active ingredient concentrations against label claim and notified certified limits, and to reassess efficacy data through the same methodology utilized in support of the product registration. EPA works cooperatively with state regulatory agencies that are responsible for pesticides on the state level. The state inspectors under federal credentials will collect the samples and associated documentation. EPA conducts some assays themselves as well as utilizes a number of university or state laboratories.

12.2.1.9   State Pesticide Registrations

Every state, Puerto Rico, and Washington DC (for this discussion, Puerto Rico and Washington DC will be considered "states") require registration of FIFRA pesticides accompanied by a registration fee to allow the product to be sold or used in their state. The state agency entrusted with the regulatory purview for pesticides is usually the state department of agriculture, but some states have separate environmental or pesticide agencies. For most states, registration is an administrative function and a fee generator and they work cooperatively with the U.S. EPA. California, under the California Department of Pesticide Regulation (www.cdpr.gov) and the California Environmental Protection Agency Cal-EPA (www.calepa.ca.gov), requires submission and their review and approval of supporting data along the same lines as EPA. There are occasions where Cal-EPA and U.S. EPA do not arrive at the same conclusions. The two agencies have worked more cooperatively with each other recently lessening the disparity in regulatory decisions impacting industry but differences do continue to exist.

12.2.2   Toxic Substance Control Act

EPA administers the Toxic Substance Control Act[4] (TSCA), which requires testing, as well as regulation of production, use, and disposal of new and existing chemicals other than chemicals that are ingredients in food, food additives, tobacco, drugs, cosmetics, medical devices, and pesticides which are regulated by other legislation.

TSCA requires that manufacturers and importers submit information on all new chemical substances before manufacture for commercial purposes; requires that manufacturers and processors collect, maintain, and possibly submit information on chemical substances; and regulates chemical substances (both new and existing) that are expected to present or are presenting unreasonable risks to human health and the environment.

Manufacturers are responsible for determining whether a substance is a new chemical substance under TSCA. From 1977 through 1979, EPA established a chemical substance inventory defining which chemical substances are "existing" in U.S. commerce. Since that time, new substances have been added following premanufacture review and through corrections of the initial inventory reports and premanufacturing notices. A chemical included in the chemical substance inventory is often referred to as "TSCA listed."

12.2.3 Resource Conservation and Recovery Act

The U.S. EPA regulates the disposal of solid waste including solid waste classified as hazardous waste under the Resource Conservation and Recovery Act[7] (RCRA). Solid waste that is not classified as hazardous waste can be disposed off in regular trash collection.

Certain chemicals are "listed" wasted and should be disposed off under strict guidelines. Substances or products can also be classified as hazardous waste due to the following characteristics: ignitability, corrosivity, reactivity, or toxicity. Most antimicrobial products do not contain a specific listed waste as defined by RCRA. Containers themselves are not hazardous waste. Containers that originally contained a hazardous waste (either listed or by characteristic) are considered hazardous waste unless they meet the criteria for "RCRA empty" defined by regulation under 40 CFR § 261.7.

12.2.4 Clean Water Act

EPA regulates under the Clean Water Act[5] drinking water quality, effluent discharge, and storm water. The local Publicly Own Treatment Works (POTW) agency, oftentimes the local sewer district, will monitor the effluent from industrial and commercial facilities. There are established limitations set for priority pollutants as well as chemical characteristics such as pH. Local regulations can exceed the federal standards, for example, the lower effluent levels established for mercury by the greater Boston regulatory authority for their regional area of authority.

Certain ingredients in antimicrobial products per se may be primary pollutants and users must contend with effluent limits on the discharge of spent end-use products disposed to the sanitary sewer.

12.2.5 Clean Air Act[6]

EPA regulates the release of chemicals defined as hazardous air pollutants to the environment. End-users of antimicrobials are not typically impacted by this act, but manufacturers of large volumes of product may be impacted. Monitoring of emissions may be required for certain chemicals.

12.2.6 CERCLA and SARA

The Comprehensive Environmental Response, Compensation, and Liability Act[8] (CERCLA), commonly known as Superfund, created a tax on the chemical and petroleum industries and provided broad federal authority to respond directly to releases or threatened releases of hazardous substances that may endanger public health or the environment. The act established prohibitions and requirements concerning closed and abandoned hazardous waste sites; provided for liability of persons responsible for releases of hazardous waste at these sites; and established a trust fund to provide for cleanup when no responsible party could be identified.

CERCLA also enabled the revision of the National Contingency Plan (NCP). The NCP provided the guidelines and procedures needed to respond to releases and threatened releases of hazardous substances, pollutants, or contaminants. CERCLA was amended by the Superfund Amendments and Reauthorization Act[9] (SARA). SARA required EPA to revise the Hazard Ranking System (HRS) to ensure that it accurately assessed the relative degree of risk to human health and the environment posed by uncontrolled hazardous waste sites that may be placed on the National Priorities List (NPL).

SARA Title III also known as the Emergency Planning and Community Right to Know Act[11] (EPCRA) was designated to help local communities protect public health and safety and the environment from chemical hazards.

To implement EPCRA, Congress required each state to appoint a state emergency response commission (SERC). Each SERC is required to divide their state into emergency planning districts and to name a local emergency planning commission (LEPA) for each district.

Broad representation by fire fighters, health officials, government and media representatives, community groups, industrial facilities, and emergency managers ensure that all necessary elements of the planning process are represented.

Facilities that are required to prepare or have available material safety data sheets (MSDS) for a hazardous chemical or mixture under OSHA are considered "covered facilities" by EPCRA. Covered facilities are required to submit an MSDS for each hazardous chemical or mixture to the LEPA, SERC, and the local fire department with jurisdiction over the covered facility. The covered facilities are also responsible for providing chemical inventory information for the OSHA-defined hazardous chemicals to the same three bodies.

For certain covered facilities, EPCRA requires that they report toxic chemical release information for quantities of certain chemicals that have exceeded a threshold quantity.

12.3 Food and Drug Administration

The FDA's Web site can be found at www.fda.gov. Regulations of the FDA can be found in 21 CFR.

The FDA regulates medical devices and drugs under authority of the Federal, Food, Drug, and Cosmetic Act[3] (FFDCA). FDA is separated into eight centers or divisions. The Center for Devices and Radiological Health (CDRH) www.fda.gov/cdrh/index.html regulates medical devices and the Center for Drug Evaluation and Research (CDER) www.fda.gov/cder/index.html regulates drugs.

12.3.1 Medical Devices

12.3.1.1 Classification of Medical Devices

The FDA regulations on medical devices can be found in 21 CFR § 800–1271.

The FFDCA defines a "medical device" as an instrument, apparatus, implement, machine, contrivance, implant, in vitro reagent, or other similar or related article, including any component, part, or accessory, which is

- Recognized in the official National Formulary, or the United States Pharmacopoeia, or any supplement to them
- Intended for use in the diagnosis of disease or other conditions, or in the cure, mitigation, treatment, or prevention of disease, in man or other animals
- Intended to affect the structure or any function of the body of humans or other animals, and which does not achieve its primary intended purposes through chemical action within or on the body of humans or other animals and which is not dependent upon being metabolized for the achievement of its primary intended purposes

The FDA classified medical devices that were in commercial distribution prior to the 1976 amendments to the FDCA for medical devices, or the so-called pre-amendments devices, into one of three regulatory classes: Class I, II, or III. The classification depends on the intended use of the device. The class establishes the regulatory controls that are necessary to provide reasonable assurance of the device safety and effectiveness. Class I devices are subject to general controls. Class II devices are subject to general controls and any FDA established special controls (as amended by the Safe Medical Devices Act of 1990). Class III devices are subject to premarket approval procedures. Class I devices, for example, include elastic bandages, examination gloves, and hand-held surgical instruments. Class II devices include, for example, powered wheelchairs, infusion pumps, surgical drapes, autoclaves, washer or disinfectors, and biological indicators. Class III devices include, for example, implantable pacemaker pulse generators and endosseous implants.

- Class I—General controls apply to all devices, irrespective of class. Requirements include the registration of manufacturers, record keeping, labeling, compliance with GMP Regulations (21 CFR § 820), Medical device reporting. A 510 (K) notification may or may not be required. A 510 (K) is an application to market a medical device that is based on proving that the device is "substantially equivalent" to a previously marketed (predicate) device. The device and predicate must have the same intended use, must have the same technological characteristics, and must not raise any new questions of safety or effectiveness.

- Class II—Special controls apply to devices for which general controls are not sufficient to ensure safety and effectiveness. All the general controls listed above apply with the addition of performance standards, postmarket surveillance, patient registries, guidelines, recommendations. A 510 (K) is required. The sponsor must demonstrate that the device is safe and effective.

- Class III—Premarket approval requires FDA approval before the device can be marketed. Premarket approval (PMA) is based on FDA determination that PMA contains sufficient valid scientific evidence that provide reasonable assurance that the device is safe and effective for intended uses.

12.3.1.2 Quality System Regulation

All sites involved in the production and distribution of medical devices intended for marketing or leasing in the United States are required to register with the FDA. Manufacturing sites must follow the quality system regulation (QSR) requirements that cover methods and controls used for the design, manufacture, packaging, storage, and installation of devices. The QSR includes requirements for organization and personnel; design practices and procedures; design of labels and packaging; control for components, processes, packaging; distribution and holding; evaluation, device, and manufacturing records; complaint handling; and audits (both internal and of suppliers).

12.3.1.3 MedWatch

Medical device firms that receive complaints of device malfunctions, serious injuries, or death must notify the FDA of the incident through the through the MedWatch program (www.fda.gov/medwatch/).

12.3.1.4 Guidance—Liquid Chemical Sterilants/High-Level Disinfectants

Content and format of premarket notification [510 (K)] submissions for liquid chemical sterilants/high-level disinfectants, January 3, 2000 can be found at http://www.fda.gov/cdrh/ode/397.pdf.

When the FDA classified the general hospital and personal use devices (45 FR 69678–69737, October 21, 1980), liquid chemical germicides were not included. At that time, the FDA regulated only those liquid chemical germicides with labeled indications for use on specific devices (e.g., hemodialyzers). Because the FDA considered liquid chemical germicides to be accessories to the devices they were used to process, the FDA regulated them in the same class as the primary device. Thus, the same liquid chemical germicide could be regulated as a Class I, Class II, and Class III device.

In the early 1990s, the FDA began actively regulating all liquid chemical germicides with health-care indications. In order to avoid the potential problem of regulating the same product under multiple classes, the FDA decided to regulate liquid chemical germicides as a separate type of medical device and determined that these were unclassified devices. Additionally, the FDA adapted the terminology and classification scheme described by Spaulding (1970) for devices (i.e., critical, semicritical, and noncritical) and the four levels of processing as proposed by the Centers for Disease Control and Prevention (CDC), namely, sterilization, high-level disinfection, intermediate-level disinfection, and low-level disinfection (Favero and Bond, 1993), to categorize medical devices. Furthermore, the FDA developed criteria to support efficacy claims for the processing levels (Figure 12.4).

The FDA defines a high-level disinfectant as a germicide that demonstrates efficacy as a sterilant per results of the Association of Official Analytical Chemists (AOAC) Official Methods 966.04 Sporicidal Activity of Disinfectants (AOAC sporicidal test) at a longer contact time than that needed for high-level disinfection.

The FDA defined three types of liquid chemical germicides for processing medical devices: sterilant or high-level disinfectant, intermediate-level disinfectant, and low-level disinfectant.

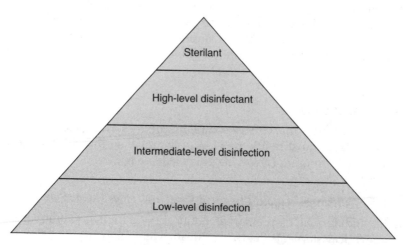

FIGURE 12.4
FDA hierarchy of antimicrobial efficacy (compare Figures 12.3 and 12.6).

From a regulatory perspective, the FDA divided these products into two categories:

1. Liquid chemical sterilants or high-level disinfectants for processing critical and semicritical devices
2. General purpose disinfectants that include intermediate-level disinfectants and low-level disinfectants for processing noncritical devices and medical equipment surfaces

The EPA regulates liquid chemical germicides as pesticides under FIFRA.

In an effort to ease the burden of this dual regulation, the FDA and the EPA signed a memorandum of understanding (MOU) (FDA, 1993, 1994), which gave the FDA primary responsibility for premarket efficacy and safety data review of liquid chemical sterilants/high-level disinfectants and the EPA primary responsibility for premarket efficacy and safety data review of general purpose disinfectants. The MOU also provided interim procedures to eliminate dual efficacy and safety data reviews until completion of the EPA rule-making process to exempt liquid chemical sterilants/high-level disinfectants from regulation under FIFRA and the FDA classification process to exempt general purpose disinfectants from 510 (K) requirements.

In 1996, the Food Quality Protection Act[2] (FQPA) exempted liquid chemical sterilants/high-level disinfectants used to process critical and semicritical medical devices from the definition of a pesticide under FIFRA and no longer regulates them. The FDA now has sole regulatory jurisdiction over liquid chemical sterilants/high-level disinfectants used to process reusable critical and semicritical medical devices. The FQPA did not affect the regulatory authority over general purpose disinfectants; therefore, the MOU remains in effect for these products and the dual regulatory requirements.

The FDA currently classifies liquid chemical sterilants/high-level disinfectants as Class II devices (general and special controls) and that general purpose disinfectants as Class I devices (general controls) and be exempted from 510 (K) requirements.

12.3.2 Drugs

FDA regulations on drugs can be found in 21 CFR § 200–369.
The FFDCA defines a drug as

- An article recognized in the U.S. Pharmacopoeia, the U.S. Homeopathic Pharmacopoeia, or the National Formulary
- An article intended for use in the diagnosis, cure, mitigation, treatment, or prevention of disease in humans or other animals

- An article (other than food) intended to affect the structure or any function of the body of humans or other animals
- An article intended for use as a component of any article specified in the three previous points, and obtains its desired effects by chemical means

A drug is further classified as prescription or over-the-counter (OTC).

FDA is charged with assuring that drug products are safe and effective for its intended uses.

12.3.2.1 Premarketing Requirements

Manufacturers are required to support claims made with respect to the intended uses of the drug. Clinical trials are used to generate data to support claims. Clinical trials are performed under good clinical practice (GCP) regulations.

Manufacturers must comply with current good manufacturing practice (cGMP), which can be found in 21 CFR § 210/211). The cGMPs are designed to ensure the quality and consistency of drug products for human use. Key components include requirements for personnel, buildings or facilities, control of components and drug product, containers or closures, production or process controls, packaging and labeling controls, holding and distribution, laboratory controls, documentation, and management of returns or salvaged drug products.

Labels include written, printed, or graphic material on the immediate container of a drug product. Labels must be clear, accurate, and easy to understand. "Drug Facts" is a new standardized label system that includes requirements for identification of active ingredients, followed by uses, warnings, directions, and inactive ingredients. The rule also sets minimum type sizes and other graphic features.

12.3.2.2 Postmarketing Requirements

Manufacturers are required to comply with the dictates of adverse drug reaction reporting through the MedWatch program (www.fda.gov/medwatch/), manage complaints, effectively manage labeling changes, and manage recalls when necessary.

12.3.2.3 Over-the-Counter Drug Products

OTC drug products are those drugs that are available to consumers without a prescription. As with prescription drugs, CDER oversees OTC drugs to ensure that they are properly labeled and that their benefits outweigh their risks.

OTC drugs provide easy access to certain drugs that can be used safely without the direct oversight of a health-care practitioner.

Most OTC drug products have been marketed for many years, prior to the laws that require proof of safety and effectiveness before marketing. For this reason, FDA has been evaluating the ingredients and labeling of these products as part of "The OTC Drug Review Program." The goal of this program is to establish OTC drug monographs for each class of products.

OTC drug monographs establish acceptable ingredients, doses, formulations, labeling, and establish appropriate testing to demonstrate safety and effectiveness. OTC drug monographs are continually updated to add additional ingredients and labeling as needed. Products conforming to a monograph may be marketed without FDA preapproval, whereas those that do not must undergo a separate review and approval through the new drug application (NDA) process.

The NDA process, and not the monograph process, is also used for new ingredients entering the OTC marketplace for the first time. The newer OTC drug products (previously available only by prescription) are first approved through the NDA process and their "switch" to OTC status is approved via the NDA process.

12.3.2.4 *Topical Antimicrobial Drug Products for Over-the-Counter Human Use; Health-Care Antiseptic Drug Products*

The topical antimicrobial drug products for over-the-counter human use; Health-care antiseptic drug products which can be found at www.fda.gov/cder/otcmonographs/Antimicrobial/new_antimicrobial.htm is a monograph that has had a long and inconclusive history. Initiated in 1972, it has not yet been issued as a final rule. On June 17, 1994, a tentative final rule (typically referred to as tentative final monograph or TFM) was issued, which still reflects the official position of FDA. Extensions have been granted and hearings have been held before expert advisory panels, but the finalization date is as yet unknown.

The monograph defines an antiseptic drug as "The representation of a drug, in its labeling, as an antiseptic shall be considered to be a representation that it is a germicide, except in the case of a drug purporting to be, or represented as, an antiseptic for inhibitory use as a wet dressing, ointment, dusting powder, or such other use as involves prolonged contact with the body." A "health-care antiseptic" is defined as an antiseptic containing drug product applied topically to the skin to help prevent infection or to help prevent cross contamination.

Three indications are covered by the monograph:

- *Antiseptic hand wash or health-care personnel hand wash*: It is a drug product which is an antiseptic containing preparation designed for frequent use; it reduces the number of transient microorganisms on intact skin to an initial baseline level after adequate washing, rinsing, and drying; it is broad spectrum, fast acting, and if possible, persistent.

- *Patient preoperative skin preparation*: It is a drug product which is a fast acting, broad spectrum, and persistent antiseptic containing preparation that significantly reduces the number of microorganisms on intact skin.

- *Surgical hand scrub drug*: It is a product that is an antiseptic containing preparation that significantly reduces the number of microorganisms on intact skin; it is broad spectrum, fast acting, and persistent.

The monograph will establish active ingredients for each indication at appropriate concentrations as safe and effective. The monograph further establishes the appropriate label language for directions and precautions associated with each indication and each active ingredient as appropriate.

12.3.2.4.1 Topical Antimicrobial Drug Products for Over-the-Counter Human Use; Health-Care Antiseptic Drug Products—Efficacy Requirements

The monograph requires broad spectrum activity as evidenced by in vitro activity against a variety of gram negative bacteria, gram positive bacteria, and yeast listed in the monograph, as demonstrated by in vitro minimum inhibitory concentration (MIC) determinations and time-kill studies. The MIC method recommended is the National Committee for Clinical Laboratory Standards' "Methods for dilution antimicrobial susceptibility test for bacteria that grow aerobically." The active ingredient, the vehicle, and the finished product are all to be tested.

12.3.2.4.2 Topical Antimicrobial Drug Products for Over-the-Counter Human Use; Health-Care Antiseptic Drug Products—In Vivo Efficacy Requirements

In addition to the in vitro efficacy data, each indication is required to pass the criteria of effectiveness in clinical studies following methodology detailed in the monograph. The following test methods are cited in the TFM by indication:

- American Society for Testing and Materials' "Standard method for evaluation of surgical hand scrub formulation, designation E 1115"
- American Society for Testing and Materials' "Standard method for evaluation of health-care hand wash formulation, designation E 1174"
- American Society for Testing and Materials' "Standard method for evaluation of a patient preoperative skin preparation, designation E 1173"

The surgical hand scrub method requires reduction of baseline resident flora over a number of uses of the product over five days on hands. The baseline counts for subjects must equal or exceed 1.5×10^5. The criteria for effectiveness is 1-log 10 reduction of resident flora on each hand within 1 min of product use with the count of resident flora on each hand not subsequently exceeding baseline within 6 h on the first day, and 2-log 10 reduction of the microbial resident flora on each hand within 1 min of product use by the end of the second day, and a 3-log 10 reduction of the resident microbial flora on each hand within 1 min of product use by the end of the fifth day. The test formulation is to be used according to label directions including the use of nail cleaner and brush if so indicated. If labeling directions do not exist, the monograph provides standard directions for use. A positive control of any FDA-approved surgical scrub formulation is required.

The health-care personnel hand wash method requires reduction of indicator organism, *Serratia marcescens* (ATCC No. 14756), over 10 uses of the product on the hands. The criteria for effectiveness is a 2-log 10 reduction of indicator organism, on each hand within 5 min after the first wash and a 3-log 10 reduction of indicator organism on each hand within 5 min after the tenth wash. The test formulation is to be used according to label directions. If labeling directions do not exist, the monograph provides standard directions for use. A positive control of any FDA approved health-care personnel hand wash formulation is required.

The patient preoperative skin preparation method includes criteria based upon two separate sets of directions for use. The baseline counts for each site must equal or exceed 3-log 10. A positive control of any FDA-approved surgical scrub formulation is required.

The first set of directions describes the preparation of the skin prior to surgery. For this use of the product, the method requires reduction of baseline resident flora on a representative dry surgical site, the abdomen, and a representative wet surgical site, the groin, after one application of the product. The criteria for effectiveness is a 2-log 10 per square centimeter reduction of resident flora on the dry site within 10 min of product use with the count of resident flora on the site not subsequently exceeding baseline within 6 h after product use, and a 3-log 10 per square centimeter reduction of resident flora on the wet site within 10 min of product use with the count of resident flora on the site not subsequently exceeding baseline within 6 h after product use.

The second set of directions, only for products that contain alcohol, describes the preparation of the skin prior to an injection. For this use of the product, the method requires reduction of baseline resident flora on a representative dry surgical site, such as the arm from the shoulder to the elbow or the posterior surface of the hand and below the wrist after one application of the product. The criterion for effectiveness is a 1-log 10 per square centimeter reduction of resident flora on the dry site within 30 s of product use.

12.3.2.5 *Alcohol Topical Antiseptic Products*

The publication "Centers for disease control and prevention guideline for hand hygiene in health-care settings: Recommendations of the health-care infection control practices advisory committee and the HICPAC/ SHEA/APIC/IDSA Hand Hygiene Task Force. MMWR 2002; 51(No. RR-16)" makes recommendations encouraging the use of alcohol-based topical antiseptic products (or also referred to as alcohol-based hand rubs) in health-care settings. This recommendation has increased the placement of alcohol products that are typically flammable in hallways and patient rooms of hospitals and other health-care settings.

The flammable nature of the products has presented a challenge to fire protection authorities. Models have been created to assess the hazard associated with these products and revision have been made to the National Fire Protection Agency (NFPA) Life Safety Code (NFPA 101). On March 7, 2006, the Joint Commission (JCAHO) announced their official stance on the use of alcohol-based hand rubs (ABHR) dispensers in egress corridors in their article "Using alcohol-based hand rubs to meet national patient safety goal 7" published in the March 2006 edition of *Joint Commission Perspectives*. Additional work is underway to fully understand the implications of pressurized versions of alcohol-based topical antiseptic products.

The Centers for Medicare & Medicaid Services (CMS) issued an "Amendment to fire safety requirements for certain health-care facilities"—File code CMS-3145-IFC, in the federal register. This final rule, effective May 24, 2005 (following a public comment period), adopts the substance of the April 15, 2004 temporary interim amendment (TIA) 00-1 (101) "Alcohol-based hand rub solutions," an amendment to the 2000 edition of the Life Safety Code, published by the NFPA. This amendment will allow certain health-care facilities to place alcohol-based hand rub dispensers in egress corridors under specified conditions.

The American Society for Healthcare Engineering maintains a Web site that provides the latest developments at www.ashe.org/ashe/codes/ handrub/index.html.

12.4 Occupational Safety and Health Administration

OSHA administers the Hazard Communication and Right-to-Know (29 CFR § 1910.1200) regulations in the United States regarding the evaluation of potential chemical hazards in the workplace, communicating the hazards, and identifying the proper protective measure for the safe handling of hazardous chemicals.

Employers are required to develop and maintain a written hazard communication program for the workplace, including lists of hazardous chemicals present, proper labeling, and the distribution of material safety data sheets (MSDS) for addressing hazards of chemicals and protective measures. Manufacturers are required to develop and distribute MSDS for any of their products that meet the definition of hazardous material. Labeling requirements for medical devices, drugs, and pesticides are regulated under separate legislations (i.e., FFDCA and FIFRA, respectively) and supersede OSHA regulations.

12.5 Transportation

The transportation of hazardous materials is regulated worldwide. In the United States, the Department of Transportation (DOT) regulates the transport of materials. Transportation by air overseas is regulated by the International Civil Aviation Organization/International Air Transport Agency (ICAO/IATA). Transportation by water overseas is regulated by the International Maritime Dangerous Goods Code (IMDG). The regulations include instructions on classification of hazards, proper packaging, labeling, placarding, and transportation documents, (i.e., bills of lading, airbills, etc.).

In the recent past, great strides toward harmonization of the transportation regulations have been accomplished. However, differences still exist between the three sets of regulations. For example, the DOT classification of "consumer commodity" is recognized with some minor additions to the package marking by ICAO/IATA but is not recognized by the IMDG. Packages that are acceptable for ground transportation in the United States may not be suitable for air transport.

12.6 International

12.6.1 Europe

The EU has been recognized as the single authority for regulations in Europe but not every nation in Europe is a member of the EU. Nonmember nations' regulations may and often do conflict with EU regulations. Many national regulations of member states are still in effect until a harmonized EU-wide regulation comes into effect (http://ec.europa.eu/).

12.6.1.1 *Dangerous Substances Directive[12] and Dangerous Preparations Directive[13]*

Regulations have been established that provide a system for identifying and classifying hazards of chemical substances and chemical mixtures referred to as preparations. The hazard classification translates into applicable precautionary icons, safety phrases, and risk phrases that are to appear on product labeling and safety data sheets (http://ec.europa.eu/environment/dansub/home_en.htm).

12.6.1.2 Medical Device Products Directive

The Medical Device Directive (93/42/EEC) (MDD) is applicable to instruments, apparatus, appliances, materials, or other articles to be used for human beings for the purpose of elimination or reduction of risks, diagnosis, monitoring, treatment or alleviation of or compensation for an injury or handicap; investigation, replacement, or modification of the anatomy or of a physiological process; contraception; and which do not achieve their principal intended action in or on the human body by pharmacological, immunological, or metabolic means. However, they may be assisted in their function by such means. Included would be biological indicators, and autoclaves. Certain medical devices are regulated under the Active Implantable Medical Device Directive (90/383/EEC) or the In-Vitro-Diagnostic-Device Directive (98/79/EC).

Each member state of the EU establishes a competent authority to oversee compliance with the national legislation version of the MDD. All medical devices are analyzed and classified according to risk. To gain access to the European market they must meet the requirements of the MDD and manufacturers have to follow the specified conformity assessment procedure typically involving a third party "notified body." The notified body assures compliance with the MDD; however, certain information is required to be submitted to the competent authority in the country in which the business is located or where a representative to a foreign-based company is located.

There are four classifications determined by risk of the intended use of the product. The classes are in order of increased risk and accompanying increased requirements are Class I, IIA, IIB, and III.

The essential requirements of conformity include general requirements addressing safety and reduction of risk and requirements regarding design and construction such as chemical, physical, and biological properties and supportive clinical data (http://ec.europa.eu/enterprise/medical_devices/index_en.htm).

12.6.1.3 Detergent Regulations

The EU regulation 648/2004 on detergents requires that surfactants and detergents containing surfactants must meet the criteria for ultimate aerobic biodegradation to stay on market. The regulation stipulates requires that a dossier of test results and an ingredients list are available upon request by regulators and medical professionals. It also requires that MSDSs be updated to address compliance with the regulation. Additional regulations may be promulgated on detergents and their ingredients (http://ec.europa.eu/enterprise/chemicals/legislation/detergents/index_en.htm).

Additional issues facing detergent products include potential regulations on phosphates and ethylenediaminetetraacetic acid (EDTA).

The EU is to determine if legislation should be proposed on the use of phosphates with a possible gradual phase-out or restriction on specific applications. Currently, member states are permitted to maintain or create rules concerning the use of phosphates in detergents (http://ec. europa.eu/environment/water/phosphates.html).

EDTA is banned in Switzerland (not a member of the EU). The Detergents Product Regulation requires that labeling provisions for detergents sold to the general public must list the presence of EDTA content if the content is greater than 0.2% by weight. There are no specific requirements placed upon EDTA when used in detergents for industrial use. Free movement clause (Article 14) would not allow Member States specific restrictions relating to EDTA.

12.6.1.4 Biocidal Products Directive

The Biocidal Products Directive (98/8/EC) (BPD) is an approval process for a broad spectrum of biocides that includes antimicrobials used for inanimate surfaces as well as hand sanitizing products. The regulation encompasses both active substances and end-use formulations.

The first stage of the directive was to identify biocide active ingredients. Secondly, industry was required to notify the EU of the intent to generate the required data to obtain approval of the specific substance.

Data (technical dossier) is to be submitted on the active substance to the appropriate rapporteur Member State has been identified based upon the classification of biocide. The rapporteur will review the data, communicate with the submitters, and ultimately make the approval decision.

Once the active substance is approved, then the end-use formulations must submit data (product dossier) in support of registration of their product. National legislation that is in place is in effect until the full review of the active substance and product is complete (http://ecb.jrc.it/biocides/).

12.6.1.5 General Chemicals—REACH

REACH[14] stands for the registration, evaluation, and authorization of chemicals which is legislation intended to be a single registration system for both existing and new chemicals with requirements for the generation of safety data and effective management of the hazards associated with the chemical.

The regulation came into effect recently and is being met with resistance from both industry and member state governments. Of specific concern are chemicals that demonstrate carcinogenicity, mutagenicity, reproductive toxicity, and persistence or bioaccumulation in the environment. The requirement for registration will probably also be determined by the volume of chemical in use (http://ecb.jrc.it/REACH/ and http://ecb.jrc.it/REACH-IT-INFORMATICS/).

12.6.2 Canada

12.6.2.1 Health Canada

Health Canada regulates essentially all antimicrobial products including topical antiseptics, hard-surface disinfectants, and chemical sterilants used

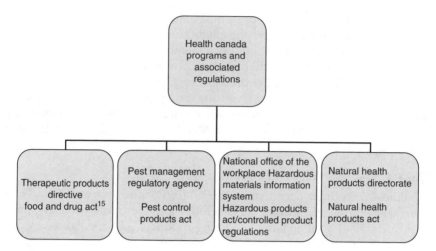

FIGURE 12.5
Health Canada programs and associated regulations.

to reprocess medical devices. This single entity is ostensibly the counterpart to U.S. EPA, U.S. FDA, and U.S. OSHA. One exception is requirements of the Canadian Environmental Protection Act[16] (CEPA). Ingredients must be listed on the approved list, the DSL, or else information may have to be submitted to Environment Canada (http://www.hc-sc.gc.ca/) (Figure 12.5).

12.6.2.1.1 *Drugs*

12.6.2.1.1.1 Drug—Category IV Monographs Many topical antiseptic products fall under the Category IV Monograph for medicated skin cleansers. The monograph identifies ingredients and concentrations that are acceptable and defines appropriate labeling and efficacy standards. A submission is required that essentially includes certification statements of compliance with the dictates of the monograph. Data is not required to be submitted but must be on file and available for review by the regulators if requested. These products are subject to all GMP requirements for drug products (http://www.hc-sc.gc.ca/dhp-mps/prodpharma/applic-demande/guide-ld/cat-iv-mono/antisept_cat4_e.html).

Many hard-surface antimicrobials fall under the Category IV Monograph for disinfectants (http://www.hc-sc.gc.ca/dhp-mps/prodpharma/applic-demande/guide-ld/cat-iv-mono/surfac_cat4_e.html). The monograph identifies ingredients and concentrations that are acceptable, defines appropriate labeling and efficacy standards. A submission is required that essentially includes certification statements of compliance with the dictates of the monograph. Data is not required to be submitted but must be on file and available for review by the regulators if requested. Hard-surface disinfectants are exempt from Canadian GMP requirements but follow a voluntary quality assurance standard.

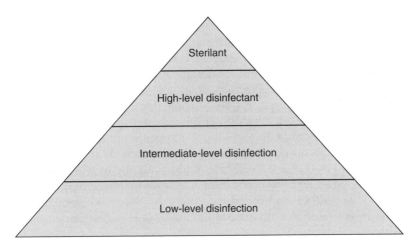

FIGURE 12.6
Health Canada hierarchy of antimicrobial efficacy (compare Figures 12.3 and 12.4).

12.6.2.1.1.2 Drug—Hard-Surface Antimicrobials Hard-surface disinfectants are regulated as drug products, requiring product approval. Hard-surface disinfectants are exempt from Canadian GMP requirements but follow a voluntary quality assurance standard.

The classification system is substantially similar to FDA's for medical device reprocessing (see Section 3.1.4). Health Canada essentially recognizes the same AOAC-based efficacy methodology that EPA recognizes for evaluation criteria (see Section 2.1.4.2) (http://www.hc-sc.gc.ca/dhp-mps/ prodpharma/applic-demande/guide-ld/disinfect-desinfect/disinf_desinf_e. html) (Figure 12.6).

12.6.2.1.1.3 Good Manufacturing Practices The importer or distributor for the products such as the drug products would be required to obtain a Drug Establishment License, involving implementing Canadian GMPs and undergoing a Health Canada Inspection. The GMPs are similar to the FDA cGMP regulations with a critical addition of the requirement of in-country testing and release (http://www.hc-sc.gc.ca/dhp-mps/compli-conform/ gmp-bpf/index_e.html).

12.6.2.1.2 Pest Control Products Act[17]—Antimicrobials

Certain applications of antimicrobials are regulated as pesticides. This includes antimicrobial pest control products labeled with only sanitizer claims, as well as disinfectant products, such as greenhouse disinfectants, material preservatives, algaecides, swimming pool biocides, and wood preservatives, to be used for purposes other than prevention of a human or animal disease.

12.6.2.1.3 Hazardous Product Act—WHMIS

Requirements for classification, labeling, and MSDS for hazardous chemicals are established under the federal Hazardous Products Act[18] which includes Workplace Hazardous Materials Information System (WHMIS) and associated Controlled Products Regulations.[19] All of the provincial, territorial, and federal agencies responsible for occupational safety and health have established WHMIS employer requirements within their respective jurisdictions. The regulations require employers to ensure that controlled products used, stored, handled, or disposed off in the workplace are properly labeled, that MSDSs are made available to workers, and that workers receive education and training to ensure the safe storage, handling, use, and disposal of controlled products in the workplace. MSDS is to be made available in both official Canadian languages, i.e., English and French.

12.6.2.1.4 Natural Health Products

The Natural Health Products[20] (NHP) Act is a recently promulgated regulation that regulates vitamins and minerals, herbal remedies, homeopathic medicines, traditional medicines such as traditional Chinese medicines, probiotics, and other products like amino acids and essential fatty acids.

The alcohol hand washes are regulated as natural health products rather than drug products. The requirements for NHP are twofold. The act requires product registration and site licensing. The product registration requirements include mandatory labeling and efficacy requirements. Efficacy requirements are waived for alcohol-based hand washes if the products meet certain formulation criteria (http://www.hc-sc.gc.ca/dhp-mps/prodnatur/index_e.html).

12.6.3 Harmonization

Strides have been made toward internationally harmonized regulations. Both FDA and EPA have made strides in this area. FDA has recognized ICH standards and their reciprocity among U.S., EU, and Japanese regulations. EPA has recognized harmonized toxicity testing protocols. However, the process is slow and we are far from true harmonization.

There are a number of countries around the world that have developed standards requiring transmittal of information to users or handlers of chemicals called the Globally Harmonized System for Hazard Communication. Although similar to requirements in the United States, the variations result in different labels and material safety data sheets for the same chemicals.

The official text of the completed system can be found on the UN Web site at http://www.unece.org/trans/danger/publi/ghs/ghs_rev00/00files_e.html.

12.7 Glossary of Terms

EPA pesticides: www.epa.gov/pesticides/glossary/index.html
FDA drugs: www.fda.gov/cder/drugsatfda/glossary.htm

12.8 Conclusion

Space limitation does not allow presentation of every detail required to fully comprehend the impact of all applicable regulations to industry. A broad overview of existing regulations has been provided instead as evidence of the scope and complexity of regulations that impact industry. These regulations are here to stay and will continue to evolve and expand. In the time after publication of this manuscript, it is a certainty that changes have been made in the regulations per se or the implementation of these regulations. As technology evolves and expands so will regulation of those technologies.

Industry must be ever vigilant in monitoring the development and implementation of regulations. As we move closer and closer to a true global economy, it is becoming a necessity to understand the regulatory framework in every region and every country. Great strides have been made in harmonization of regulations but great differences in approach and substance still exist from region to region and country to country. Improvements in communication technology has made the job of monitoring easier, but changes to existing regulations and development of new regulations match the speed at which technology expands. Information is easier to find but there is more and more information to monitor.

Technologies that provide disinfection and decontamination often provide risk to the public health per se. A balance must be struck by regulators between the benefit provided by the technology and the risk provided by that same technology. Regulators require information from all stakeholders including industry to develop regulations that meet their responsibility of protecting public health. Full comprehension and effective communication by industry of both the benefits and the risks associated with the developed technology is an absolute necessity.

References

1. Federal, Insecticide, Fungicide and Rodenticide Act, United States Code Title 7, chapter 6.
2. Food Quality Protection Act, Public Law 104–170.
3. Federal Food, Drug and Cosmetic Act, United States Code Title 21, chapter 9.
4. Toxic Substance Control Act, United States Code Title 15, chapter 53.
5. Clean Water Act, United States Code Title 33, chapter 26.

6. Clean Air Act, United States Code Title 42, chapter 85.
7. Resource Conservation and Recovery Act Code Title 42, chapter 82.
8. Comprehensive Environmental Response, Compensation, and Liability Act, United States Code Title 42, chapter 103.
9. Superfund Amendments and Reauthorization Act, United States Code Title 42, chapter 103.
10. Pesticide Registration Improvement Act, Public Law 108–199.
11. Emergency Planning and Community Right to Know Act, United States Code Title 42, chapter 116.
12. Dangerous Substances Directive, 67/548/EEC.
13. Dangerous Preparations Directive, 99/45/EEC.
14. Registration, Evaluation, and Authorization of Chemicals Directive, 2006/121/EC.
15. Food and Drug Act (R.S., 1985, C.F-27).
16. Canadian Environmental Protection Act (1999, c.33).
17. Pest Control Products Act (2002, c.28).
18. Hazardous Products Act (SOR/88-66).
19. Controlled Products Regulations (SOR/88-66).
20. Natural Health Products Act, (SOR/2003-196).

13

Biofilms: A Summarized Account and Control Strategies

Kanavillil Nandakumar and Kurissery R. Sreekumari

CONTENTS

13.1 Introduction

13.1.1 History

The preference of marine bacteria to grow on surfaces was first mentioned by ZoBell [1]. Subsequently, this aspect was extended to freshwater ecosystems and a variety of microbial ecosystems such as the surfaces of eukaryotic tissues [2]. Early era of biofilm studies, i.e., until late 1980s, perceived biofilms as simple sheath- or slab-like structures on surfaces [3]. Modern era of biofilm studies began with the use of environmental scanning electron microscope and confocal laser scanning microscope, which significantly enhanced our understanding of the biofilm architecture. Biofilms are now described as a complex boundary layer community with bunches of mushroom-like structures and water channels running in between [4]. This revelation resolved the perturbing question of how nutrients and oxygen are supplied in biofilms. The supply of nutrients and removal of waste products help biofilms grow in thickness and complexity [5]. The vertical profiling of the nutrient and oxygen concentrations in biofilms showed that they decrease from the surface to the interior. The species distribution showed colonization of anaerobic forms in the interior portion of biofilms, whereas aerobic forms occupied the outer surface.

Thus, biofilm now is described as a dense assemblage of cells and their by-products on hard substrata. They are usually heterotrophic in nature [6]. In a multispecies biofilm, organisms are juxtaposed and are thus in a position to benefit from the interspecies substrate exchange or mutual end-product removal [7]. The individual cells in a mature multispecies biofilm, therefore, typically live in a unique microniche, where the nutrients are provided and waste products are removed by the neighboring cells by diffusion [3].

13.1.2 Biofilm Extracellular Polymeric Substances

One of the distinctive features of biofilms is that they are embedded in a structurally complex matrix called extracellular polymeric substances (EPS). EPS gives structural support to biofilms. The nature, composition, and role of EPS vary with the species that secretes it as well as with the physicochemical properties of the water in which it is formed. The EPS can be of two types according to the nature and its cell association: the cell-associated EPS or capsular EPS, and unbound EPS or the slime. The capsular EPS is connected to bacterial cells with diacyl glycerol moiety or other types of lipid compounds. EPS is composed of several compounds such as polysaccharides, proteins, and nucleic acids [5]. Among these, polysaccharides represent the most important and dominant component [8]. However, there are reports showing the predominance of other compounds in EPS such as proteins and humic acids [9]. Depending on the environmental conditions in which microorganisms live, the structure of EPS produced by the same species varies [10]. Thus, the same species of bacteria can produce more than one type of EPS. The structure and chemistry of EPS show variation. There are polysaccharides with 80–100 carbohydrate monomers per molecule and there are others such as cellulose with more than 3000 monomers per molecule. When the EPS is composed of only one type of monomer, it is called a homopolysaccharide; however, when it is composed of more than one type of monomer, it is called a heteropolysaccharide [11]. While a homopolysaccharide such as cellulose with repeating glucose units imparts anionic or neutral charge to the EPS, the heteropolysaccharides generally result in only anionic charges.

13.1.3 Growth of Biofilms

The growth of biofilms occurs by one or all of the following three processes: (1) through binary fission of attached cells on the surface, thus, gradually spreading over the surface two dimensionally or three dimensionally [12], (2) by the redistribution of already attached cells with the help of flagella (surface motility) or by the twitching motility aided with Type IV pili [13], and (3) by the continuous recruitment of cells from bulk media, aiding in the development of biofilms [14]. Because of the above-mentioned reasons, biofilm maturity takes varying periods extending to several days, depending on the type of species present. The overall process of development of biofilm on immersed surfaces follows a series of sequential processes. This starts with the formation of a conditioning film composed of nutrients and organic matter. The subsequent accumulation of nutrients attracts chemosensitive bacteria toward the surfaces and negotiates a reversible phase of attachment. The entire process can occur within minutes to hours. Once the bacteria secure attachment, the phase of irreversible attachment occurs with the help of EPS secreted by them. Secretion of EPS thus consolidates the attachment of bacteria with the surfaces. Biofilms grow to maturity by following the growth processes mentioned above. This leads to the formation of mushroom-shaped structures with water channels running in between. When the growth process continues, biofilm bacteria also shed cells into the bulk media, and these cells will be carried away by water currents aiding in dispersal. Microorganisms secrete excess EPS in the presence of toxic substances such as antifouling compounds [15]. The secretion of EPS helps biofilms to grow in thickness and restricts the entry of toxic substances into them. This results in the protection of bacterial cells from toxic chemicals [3,16].

It should be noted that EPS need not always facilitate attachment of bacteria. As reported by Hanna and others [17], colanic acid of *Escherichia coli* impairs initial attachment even though colanic acid production was found to be necessary for the architecture of a mature *E. coli* biofilm [18]. Similarly, EPS secreted by the mucoid isolate of *Pseudomonas aeruginosa* has also been shown to impair attachment of planktonic bacteria.

13.1.4 Nature and Formation of Biofilms

The literature on biofilms shows that during the early stages work has been primarily done on its architecture and developmental kinetics [3,19–22]. Even though biofilms are known to be composed of organisms belonging to various phyla and other taxa, the exact nature of their interactions is not well understood. There are a few reports that explain the interaction between algae and bacteria in a biofilm [23–25]. According to Mitchell and Krichman [26], the chemosensitivity of algae to the compounds produced by the biofilm bacteria plays a significant role in the algal settlement. This observation supported the earlier observation of Tosteson and Corpe [27] who found that significant attachment of *Chlorella vulgaris* occurs on surfaces with a dense bacterial film.

Factors influencing the development and composition of biofilm community are a matter of concern. Some say that the most important factors are the physical parameters of the water column such as flow rates and velocities [28], while, others describe substratum roughness as the most important factor deciding the biofilm formation [29]. It is certain that in addition to biological parameters, physical properties of the water also play an important role in biofilm formation [30]. For example, biofilms grown under higher shear and stress will be smoother than the ones grown under less shear and stress. Besides, biofilms grown under high shear and stress conditions will be more rigid and stronger than those developed under less shear and stress conditions. Genetic studies showed the importance and need of certain genes for biofilm formation. This was made possible by gene knockout studies [31,32]. However, a closer examination showed that in many of these cases biofilm formation is not totally prevented by knocking out a specific gene; instead, the growth is either retarded or reduced [33]. Various laboratory studies showed that biofilms display viscoelastic properties, that is, they display elasticity and viscosity according to the physical parameters of the water column. The factor responsible for the viscoelastic properties of biofilms is the EPS with the influence of multivalent cations. Cations are responsible for the cross-linking of polymers in EPS, and depending on the type of EPS, the viscoelastic properties of biofilms change. Thus, EPS necessarily determines the cohesive strength as well as the structure of biofilms [34]. The physical strength of biofilms is contributed by a number of weak interactions such as hydrogen bonding, electrostatic interactions, and van der Waals forces. As EPS contains many functional groups capable of having these types of interactions with the surfaces, their force is multiplied several orders of magnitude per macromolecule to increase the overall binding force of EPS with the surface [35].

In a single-species biofilm, bacteria alter gene expression as compared to the planktonic forms. This is to maximize their survival in the microenvironment [36,37]. In a multispecies biofilm, bacteria are distributed in such a way that the coexistence can enhance the survival by having a symbiotic existence [38]. This shows that bacteria in a multispecies biofilm are not randomly distributed, instead, with an organized mandate to meet the needs and challenges of each species. Molecular biological studies of biofilm bacteria showed that bacteria in a multispecies biofilm communicate with each other through a mechanism now widely known as quorum sensing or cell signaling. This is largely done by the secretion of substances by one type of cell that enters other cells. This communication between cells in biofilm provides evidence that microbes function as a coordinated group [39,40], though controversy still exists [41]. An example of such quorum sensing chemical is acyl-homoserine lactones in natural and cultured biofilms [42,43]. Some other compounds that can act as quorum sensing molecules are bacterial metabolites, genetic materials such as DNA or RNA, proteins, etc. Such intercellular substances can be either beneficial or detrimental to other species. Thus, depending on the nature of compounds produced by bacteria in a multispecies biofilm, they take up the roles of diversity and distribution factors.

Biofilm compositions show that 5%–25% is made up of living organisms and the rest is matrix. However, 95%–99% of the matrix volume is composed of water [15]. The dry weight is mainly due to the acidic exopolysaccharides secreted by the organisms.

13.1.5   Beneficial and Detrimental Roles of Biofilms

Biofilms assume significant roles in a variety of environmental processes such as nutrient recycling, aquatic productivity, and detoxification. Depending on the nature and processes, they can be either beneficial or detrimental. The notable beneficial roles of biofilms are in bioremediation of polluted sites and wastewater treatment plants. Some of the detrimental effects are biofouling of engineering surfaces, microbiologically influenced corrosion of metals and alloys, fouling of heat transfer equipment of power generation plants, contamination of drinking water systems, colonization on surfaces of food-processing units, bacterial infection on dental surfaces, colonization on medical implants, and blockage of catheters. The equilibrium thickness of the biofilms in the cooling water system components varies anything from 500 to 1000 μm. A 250 μm thickness biofilm on the heat exchangers is reported to reduce the heat exchanging capacity by 50% [44]. A great deal of money is being spent worldwide to counter these menaces.

Irrespective of whether biofilm formation results in harmful or useful scenarios, it is considered as a useful mechanism for bacteria to tide over the stressful environmental conditions. In addition, it also provides bacteria with additional nutrition as well as new genetic materials that are acquired by genetic transfer from neighboring cells. A proper treatment regime is needed to prevent biofilm formation on industrial surfaces or to dislodge the same if formed on surfaces. This will help prevent spreading of infection as well as problems in industries. The following section briefly describes the important strategies that are followed to get rid of biofilms from important surfaces, industrial as well as medical and dental.

13.2   Control Techniques

Biofilm control techniques depend on the type of substratum to be protected or the environment in which they are in. For example, control of biofilms on moving objects such as hulls of ships is achieved mostly by applying anti-fouling coatings. However, structures such as condenser tubes or heat exchangers of power-generating industries are protected by adding biocides in the cooling water. In case of human implants and organs, the general method used is the administration of antibiotics, whereas manual cleaning is often practiced for dental plaques. Control techniques of biofilms can be broadly grouped into biological/molecular biological, i.e., by genetic inter-vention, chemical, physical, electrochemical, and mechanical. In certain circumstances, a combination of some of the above-mentioned methods is also used. In summary, the strategies that could be developed as effective control techniques can target the following: stopping or preventing micro-bial attachment, restricting or preventing the growth of attached bacteria, disrupting the cell-to-cell communication, or disintegrating the biofilm mat-rix [45]. A brief account of the state-of-the art techniques is given below.

13.2.1   Surfaces That Resist Biofilm Formation

Since the first step in biofilm formation is the attachment of bacteria on to substratum surfaces, the earliest opportunity to deter biofilm formation is by intervention of the process at this stage. One of the ways to affect this is by modifying the substratum surface so that the attachment can be pre-vented or altered. However, as the diversity of bacteria increases, the like-lihood of this approach to succeed in preventing biofilm formation diminishes. Nonmetallic antibacterial materials are available in plenty in the market. There were also a few attempts to make metals antibacterial. Some important studies that deserve mention here are of Nakamura and others [46] about the Cu-containing stainless steels and of Yokota and others [47] and Sreekumari and others [48] about the Ag-containing stainless steels exhibiting antibacterial properties. The trend of using silver as an antibac-terial agent has a very old history as that of the human race, since it was used to combat microbial infections. This was mainly due to the nontoxic nature of silver to the human beings while exhibiting excellent antibacterial properties. The antibacterial activity of Ag ions is several hundred times stronger than that of Cu or Ni [47]. Silver was also reported to be used as a water purifier in swimming pools and space shuttles. Studies on the anti-bacterial properties of silver-containing AISI type 304 stainless produced excellent results in preventing bacterial attachment. This material was also observed to be resistant against microbially influenced corrosion, another detrimental effect of biofilm formation on metallic surfaces. It has also been demonstrated that Ag ions infiltrate through the cell membranes more rapidly than Cu or Ni ions [49]. This makes the amount of Ag required for antibacterial effect minimal in comparison to other elements. During these studies, the area of bacterial adhesion on control coupons of AISI type 304 stainless steel was much higher than that observed on silver-containing stainless steels [48].

Altering the composition of conditioning film is another important step to intervene biofilm formation. The composition of conditioning film can change the subsequent process of biofilm formation. Thus, modification of substratum surface to inhibit the adsorption of specific classes of molecules is another way to control attachment of microbes. In this context, the polymer surface coating that attracted much attention is polyethylene oxide (PEO)–polyethylene glycol (PEG). The PEO–PEG surfaces continue to show promise by minimizing the adhesion of biomolecules [50]. There are studies that show PEG with short polymer chains capable of providing full strength resistance to bioadhesion. This polymer coating appeared to be more effective in preventing attachment of proteins than other types of coatings based on phospholipids, polyacrylamide, and polysaccharide. Thus, the prevention of attachment of biomolecules would result in an altered scenario on the substratum surface that can deter the formation of biofilm.

Another important property of substratum surface in relation to fouling or biofilm formation is the free surface energy. The surface energy affects biofilm formation as described below: the higher the surface energy, the better the chance of biofilm formation. This is true with the roughness of the surfaces as well. Therefore, degree of biofouling on surfaces varies based on the surface tension of the coating/surface, the chemistry of surface, and the biological agent (microbe). Teflon with a low surface energy, thus, decreases biofouling but is not capable of preventing it. Protein, the universal glue with which the organisms largely establish attachment, can recognize the surfaces and over a period of time can make the attachment happen. However, a surface with a low surface free energy, even though prone to attachment of biological agents, can be easily cleaned as the strength of attachment with the surface is weak.

13.2.2 Possibilities of Biological/Molecular Biological Intervention of Biofilm Formation

Although full fledged genetic intervention of biofilm formation is far from reality, studies are conducted in this direction. The basic concept for genetic intervention of biofilm formation is derived from the following aspects. Bacterial attachment and formation of biofilm are often mediated by biological means, i.e., with the help of proteinaceous appendages such as pili, flagella, and fimbriae. Therefore, bacterial attachment can be interfered by disrupting the functions of these appendages. For example, *Pseudomonas aeruginosa* performs twitching motility over the surface with the help of Type IV pili. When an iron restriction was imposed, *P. aeruginosa* did not stop twitching motility, there by, resulting in no attachment or biofilm formation. Thus, a surface capable of iron chelation could effectively result in slowing down biofilm formation by *P. aeruginosa* [51]. Another important concept is the existence of cell-to-cell communication among biofilm-forming bacteria. Davies and others [43] reported that cell-to-cell communication or quorum sensing between bacterial cells is a prerequisite for a normal biofilm formation. The role of quorum sensing molecules (diffusible chemical molecules secreted by bacterial cells) in microcolonies of biofilm raises enough prospects to intervene the biofilm development by manipulating the metabolic pathways leading to the formation of such chemicals in bacteria. The documentation of chemical communication between bacteria in the lungs of people with cystic fibrosis described the importance of bacteria in the form of biofilms causing life-threatening pneumonia [52]. Subsequently, in another study [53], the effectiveness of furanones isolated from a marine algae *Delisea pulchra* in interfering the cell-to-cell communications in bacteria and thus to prevent the biofilm formation is described. This group has also shown that mice injected with furanones undermine the infectious biofilm bacteria as compared to mice that are not injected with furanones. These observations open up avenues for genetic intervention of biofilm formation. Another important aspect is the synthesis of matrix in biofilms. Matrix containing EPS gives cohesiveness as well as gluing ability to bacteria to establish a firm attachment with the surfaces. Thus, genetic interventions of matrix formation or designing chemicals that can disintegrate biofilm matrix are other possible avenues to prevent biofilm formation. Some encouraging results in these directions are available, i.e., bacteria that are unable to modify the polymer backbone by acetylation are less capable of biofilm formation [54,55].

Bacteria in biofilms are reported to be physiologically in a similar state with the stationary phase bacteria of a laboratory developed batch culture. In stationary phase, gram-negative bacteria are known to develop stress-response resistance regulated through the induction of sigma-factor gene known as *RpoS* [56]. Starvation is one of the common types of stresses that bacteria are exposed to and is known to induce several stress-response networks such as *rpo*H-, *rpo*S-, and *oxy*R-dependent regulons [56–59]. *RpoS* is now seen as the master regulator of general stress responses. It is triggered by the carbon starvation signals, and in many circumstances may provide cells with the ability to tolerate the stresses [60]. *RpoS* is also reported to control the quorum sensing genes that are considered responsible for the cell-to-cell communication in bacteria. It is true that when planktonic bacteria are exposed to stressful environments, such as the presence of antimicrobials or pollutants, they tend to form biofilm, thus, enhancing the possibility of survival. Therefore, formation of biofilms can also be interpreted as a stress-response mechanism in bacteria. This leads to the notion that *RpoS* gene assumes a significant role in the biofilm formation and deletion of this gene thus could reduce the biofilm formation in bacteria. Several studies were conducted to test the role played by *RpoS* gene in biofilm formation in several bacterial species such as *P. aeruginosa, E. coli,* and *Moraxella* sp. The results were inconclusive as the outcome differed with species [61–65]. The *rpoS* mutants of *P. aeruginosa* and *E. coli* showed increased biofilm biomass and a higher resistance to the antibiotic tobramycin [66,67]. The *RpoS* mutant of *Moraxella* strain also formed denser and thicker biofilm on the stainless steel surfaces as compared to its wild type [65]. However, a decreased biofilm formation in *rpoS* mutant of *E. coli* was reported by Adams and McLean [68]. This study thus does not agree with the earlier-mentioned observations. Therefore, the role of *RpoS* gene in biofilm formation is still uncertain and hence the possibility of genetic intervention via *RpoS* gene to prevent biofilm formation is still under investigation.

13.2.3 Antifouling Coatings

13.2.3.1 Toxic Coatings

Antifouling paints are generally applied to the hulls of ships and boats. The concentration of toxic species and its nature and the leaching rate determine the efficiency of an antifouling paint [69,70]. These are also the factors that determine the life span of antifouling coatings. Some commonly used toxic species are copper thiocyanate, copper powder, cuprous oxide, and more recently organotin compounds such as tributyl-tin oxide (TBTO), tributyl-tin fluoride (TBTF), triphenyl-tin fluoride (TPTF), and triphenyl-lead acetate (TPLA) [71]. The paints are composed of either soluble matrix, also called conventional or ablative type paint, or insoluble matrix. However, a more recent development is the physically tough vinyl-chlorinated rubber and epoxy resin paints. Paints with matrix of self-polishing polymers are also available, which keeps the paint surface smooth. This type of paint is called a self-polishing copolymer paint. The disadvantage of antifouling paints is that they are subjected to wearing, and hence a repeated application is required for continued protection [71]. The efficacy of antifouling paints to deter biofilm formation did not attract much attention because of its limitation at the users' end. This technique is widely used to prevent macrofouling growth on moving objects.

13.2.3.2 Nontoxic Foul-Release Coatings

The nontoxic foul-release coatings rely on reduced surface energy [71]. This is based on the principle that the strength of fouling attachment depends on the surface energy of the solid surface, because the adhesive strength is the sum of the surface free energy of the solid surface and the surface tension of the water column minus the interfacial surface tension between the solid and the liquid [72]. Therefore, when the surface energy of the substratum is less than the surface tension of the liquid (in this case the adhesive of fouling organisms), the strength of attachment will be weaker. Thus, the lower the surface energy, weaker is the strength of attachment. Fluorinated polyurethane and silicone coatings are working on this principle.

Silicon elastomers are found to be highly efficient in deterring biofouling. In case of low-energy surface coatings, attachment occurs, but the strength of attachment will be so weak that with the water flow in good speed, the attached organisms get detached. This technique is effective mainly to prevent the growth of macrofouling organisms and not biofilms; however, the same principle could be applied to prevent biofilm growth as well.

13.2.4 Treatment with Biocides

Studies on the efficacy of various sanitizers, biocides, and detergents on biofilm prevention/removal showed that biofilms once formed on surfaces are difficult to remove using chemical treatments. Complete removal becomes tricky especially when EPS is formed on the surface [73]. Biocides are reported to be much less effective against biofilm as compared to their planktonic counterparts [74,75]. It has been speculated that lower sensitivity of biofilm bacteria to biocides is due to the physiological status of the cells in biofilm and also due to the lesser permeability in biofilm matrix [76]. Thus, for an effective chemical treatment to control biofilm, it is important that the EPS is broken down so that the chemical agents can reach up to bacterial cells living in the interior part of biofilm. To remove biofilms from surfaces, various chemical agents in combination with other techniques were tried with varying success. The majority of chemical cleaning agents that are used in food industries are of alkaline origin. They are often used in combination with sequestrants such as sodium phosphate derivatives or chelators such as EDTA. Nonionic wetting agents are used because they are good emulsifiers and control foaming [73]. Jackson [77] recommended the use of detergents to remove biofilms in pipelines carrying milk at temperature above 70°C and above 77°C to clean up heat exchangers and pasteurizers. It is noted that such cleaning processes can remove more than 90% of bacteria from biofilms but may not kill them. Wirtanen and others [78] found that removal of *Bacillus* biofilm was influenced by flow rate, contact time, and temperature of the cleaning agent, as well as the presence of chelators. In another study with *Pseudomonas* and *Staphylococcus* biofilms, Gibson and others [79] found that alkali and acid cleaners are ineffective as they found only a 1 log reduction of microorganisms in biofilms after treatment. These studies thus suggest that a prolonged treatment with biocides is necessary to have an appreciable impact on biofilm.

Chlorine is the most commonly used oxidizing biocide. This is due to its high sanitization power, disinfection capacity, low cost, and easy handling [80,81]. The hypochlorous acid (HOCl) responsible for biocidal action is generated from hypochlorite when pH is between 4 and 7. As chlorine diffuses through the cell membrane, it oxidizes the cellular components and interrupts the metabolic activities resulting in cell mortality. However, the penetration of chlorine through EPS is found to be not good. In a mixed biofilm of *Pseudomonas aeruginosa* and *Klebsiella pneumoniae* of thickness 150–200 μm, the chlorination for a period of 105 h was found to be largely inefficient in penetrating biofilm [81]. They observed a reduction of only 1–2 orders of magnitude in biofilm bacterial counts after chlorination. There were localized areas where the penetration was lesser and these locations acted as local sources of bacteria that can regrow and reestablish biofilms over a short period of time. Studies showed that in water distribution systems and in osmotic membranes, a residual chlorine concentration of 15–20 ppm was needed to remove biofilms. Frank and Chmielewski [82] noted that the removal of *Listeria* biofilms from stainless steel surfaces was difficult using chlorine. However, chlorine and other anionic sanitizers were found to remove *Listeria* and *Salmonella* biofilms from stainless steel surfaces more efficiently than by the nonoxidizing biocides such as quaternary ammonium compounds (QACs). As the contact time and residual concentration of chlorine increased, the effectiveness to remove biofilm increased. One of the chloroorganic derivatives of chlorine, the monochloramines, when dissolved in water, was found to have better penetration in biofilms than the residual chlorine. However, it required a prolonged contact time to achieve this level [83].

The guidelines issued by EPA restrict chlorine level during water treatment so as to minimize the environmental impacts. Chloroorganics are known carcinogens. The EPA guideline stipulates the maximum residual chlorine level not to exceed 0.2 ppm for any given 2 h. Industries follow various types of chlorination regimes to meet this guideline. One of these is by the targeted chlorination [84], where a part of the condenser tube is treated with a high dose of chlorine at a time to remove the biofouling organisms. In another instance, shock dose chlorination is practiced by administering a high concentration of residual chlorine such as 1–2 ppm residual for a limited period of time to get rid of the biofouling organisms. However, a high dose of chlorination could initiate corrosion of Cu and mild steel components of the industries [85].

Bromine is used as an alternative to chlorine because of various reasons. It is less expensive, its residuals decay quickly, hypobromous acid (HOBr) is a weak acid, it is stable in ammoniacal conditions, and it has superior disinfection properties [86]. Bromine is used in the form of bromine chloride, activated bromide, etc. Hypobromous acid is responsible for the biocidal action.

Another alternative to chlorine and bromine is ozone. Ozone is a strong oxidant and hence is considered as a more efficient biocide. But, the ozonizer is expensive and requires a large space at the point of application. Besides, its lower effectiveness in the presence of halides in the water is considered as a disadvantage. However, the strong ability of ozone gas to penetrate polysaccharide matrix of biofilms is considered as a huge advantage in controlling biofilms. The affinity of ozone to hydrophilic environment, as well as to the highly polar polysaccharides, makes biofilm an easy target of ozone. The breaking down of EPS in biofilm is considered the main mechanism by which ozone acts on biofilms [87].

One of the commercially available nonoxidizing biocides is Calmtrol (Betz Dearborn) that is being used in several places in Europe and in some Asian countries. Some other chemicals such as QACs, tannic acid and acrylamides (repellents), acrolin, etc. are also tested and used with varying success to control biofouling buildup [88]. The QACs are cationic surfactant sanitizers and are often applied as foam that provides a longer contact time with biofilms on surfaces. The QACs are reported to be effective against gram-positive and gram-negative bacteria, yeasts, and molds [89]. The results of the studies with QACs on biofilms showed a high concentration of above 400 ppm for a contact time of more than 5 min that was needed to inactivate *Listeria monocytogenes* biofilm [90]. However, in another study, Frank and Koffi [91] noted that more than 20 min contact time was not sufficient to produce a desired level of inactivation in *Listeria* biofilm.

Hydrogen peroxide is another broad-spectrum sanitizer that was tested for its efficacy to remove biofilms. It is toxic against bacteria and its endospores [92]. This was found to be more effective in controlling *L. monocytogenes* and *Salmonella* sp. biofilm than the hypochlorite solutions [93]. Rossoni and Gaylarde [94] reported that hypochlorite solution was much more effective to mollify biofilm than peracetic acid, another sanitizer studied for controlling biofilms. Acid-anionic sanitizers such as phosphoric and sulfamic acid were found to be acting quickly on yeast and viruses but their action was quite slow on bacteria [95]. The foregoing description thus explains the efficacies and inefficacies of certain types of sanitizers in controlling biofilm. In summary, the selection of sanitizer or biocide to deactivate a biofilm depends on the types of major organisms present in the biofilm, as well as the organic load of the biofilm, as efficacy of biocides varies with organisms and also the amount of organic compounds present in the water column.

13.2.5 Treatment with Antibiotics

The primary function of EPS matrix is protection. Thus, most bacteria tend to form biofilm and secrete EPS to increase its chances of survival. This is the reason why majority of bacteria live in biofilms. Biofilms serve as a source of nutrition and protection for bacteria. The protection could be true biologically, as well as physicochemically. Biologically, the microniche provides protection from immediate predators such as single-celled eukaryotes and immunological agents such as leucocytes, whereas physicochemically, the EPS acts as a barrier for chemicals and toxins. However, there are reports stating the ability of leucocytes to penetrate EPS of *Staphylococcus aureus* biofilms [96]. A host of information is available on the protective nature of EPS matrix against various agents such as desiccation, toxins, UV irradiation, chemical disinfectants, temperature, and osmotic shock [97]. Studies show EPS can act as a charged barrier for certain cationic antibiotics as EPS is largely anionic in nature. For example, there are studies showing interaction of alginate content of EPS with aminoglycoside tobramycin [98,99]. However, many other antibiotics are demonstrated to be affected by the EPS of the laboratory biofilm [3,100–102]. The studies of Nield-Gehrig [103] showed that to kill bacteria in biofilms, antibiotics concentration had to be increased between 50 and 1500 times that of the amount required to kill normal planktonic bacteria.

In a recent study conducted with bacterial isolates from clinical cases of infections with cattle, sheep, pigs, chickens, and turkeys, Olson and others [104] showed that most of the bacterial isolates were highly resistant to a variety of antibiotics when they form biofilms. Invariably, in almost all the above cases, the same bacteria when in planktonic form were highly susceptible to antibiotics. However, biofilms of *Streptococcus dysgalactiae* and *S. suis* were sensitive to penicillin, ceftiofur, cloxacillin, ampicillin, and oxytetracycline, biofilm of *E. coli* was sensitive to entrofloxacin and gentamicin and biofilms of *Pseudomonas aeruginosa* and *Salmonella* spp. were sensitive to entrofloxacin. In another study on antibiotics susceptibility of *P. aeruginosa* biofilms isolated from cystic fibrosis, it was showed that these bacteria are highly resistant to two or three combinations of the following antipseudomonal antibiotics tobramycin, amikacin, meropenem, piperacillin, ceftazidime, and ciprofloxacin [105]. Amorena and others [106] noted that the efficacy of antibiotics decreased with the age of the biofilm; thus, biofilms of younger age were more susceptible to a certain dose of antibiotics compared to biofilms of older age.

Similar to the studies that examined the efficacy of combinations of antibiotics in controlling biofilm formation, there are other combinations of methods such as bioelectric effect on the enhancement of the efficacy of antibiotics in deterring biofilm formation. This is basically done by applying a weak electric current on the surface with biofilm while exposing them to antibiotics. The mild electric current of <100 $\mu A/cm^2$ enhanced the efficacy of antibiotics in killing biofilm bacteria to a great level that the concentration of antibiotics required to eradicate biofilm bacteria was considerably reduced to a level that is required for killing planktonic forms [22]. They also observed that the application of electric current does not result in localized generation of antimicrobial molecules or ions. In another study, Blenkinsopp and others [107] showed that the application of a mild electric current (electric current density 15 $\mu A/cm^2$ and field strength of 1.5–20 V/cm) could completely eliminate the inherent resistance by biofilm bacteria to biocides and antibiotics [108].

However, it is reasonably clear that EPS matrix plays a significant protective role against antimicrobials such as reactive oxygen species. In this case, matrix acts as a sink to the reactive species such as superoxide radicals or hydroxyl ions. In an interesting study, Schembri and others [109] demonstrated the production of biofilm and aggregation of cells in the presence of reactive oxygen species. *E. coli* regulates the surface adhesion and aggregation by antigen 43, which is produced by the regulatory protein OxyR. When *E. coli* is exposed to reactive oxygen species, OxyR induces the production of antigen 43, thereby initiating the biofilm production and aggregation. The aggregated cells and biofilm were found to be highly resistant to reactive oxygen, thus enabling these bacteria to tide over the stressful condition.

13.2.6 Other Recent Developments in Nontoxic Methods

13.2.6.1 Electrochemical Methods

Rosenberg [110] found that electrolysis products from a platinum electrode can inhibit cell division in *E. coli*. Similarly, many other electrochemical methods have been used to deter biofouling growth [111–113]. In the electrochemical methods, the surface to be protected is made anode. This is done by the deposition of a thin film of platinum. An electrical current of 5 mA/cm^2 was found to restrict the fouling growth considerably for a long duration [114]. Alternatively, fouling control was achieved by the use of conductive paint electrode on fishing nets and other surfaces [113]. In a laboratory study, when a potential of 1.2 V vs. a saturated calomel electrode was applied to the conductive paint electrode, *Vibrio alginolyticus* cells attached on the electrode were completely eliminated. Thus, the application of a negative potential of the surface resulted in the removal of attached bacterial cells from the surfaces. However, in the filed study, when an alternating potential of 1.2 V vs. Ag/AgCl for 60 min and −0.6 V vs. Ag/AgCl for 10 min was applied to the electrode there was significant inhibition of bacterial attachment. This was found to be effective for a period of 5 months. In addition, those bacteria attached on the electrode were killed by this alternating potential [111,112,115]. Even though most of these studies were concentrated on deterring biofilm formation, attachment of organisms such as the larvae of oysters and barnacles was also found to be significantly inhibited by this method. Therefore, it is believed that, this technique could be extended to prevent macrofouling growth as well.

13.2.6.2 Irradiation Techniques

The use of laser irradiation as a tool to remove biofilm formed on hard surfaces was studied by Dobson and Wilson [116]. Although there are a few reports available on the laser irradiation impacts on bacteria lawned on agar plates, studies on biofilms developed on hard surfaces are sparse. In this context, the study on the impact of CO_2 laser on calculus-laden root surfaces is important [117]. The ability of laser to cause bacterial mortality has been exploited to devise a laser-based sterilization technique of surgical tools [118–121] used in surgery and in medical and dental fields [122,123]. Laser irradiation is known to cause molecular changes in bacterial cells [124–126] and cell membranes [127] by producing reactive radicals that are harmful to bacteria [128]. Wilson [128] described in his review that the laser impacts depended on laser fluence and accordingly the impacts were grouped into four categories: photochemical, photothermal, photoablative, and photomechanical. Various types of lasers, based on the lasing medium used to generate the output, have been tested for their efficacy to kill bacteria [128].

It is reported that pulsed laser irradiation of fluence 0.1 J/cm^2 for 10–15 min is capable of significantly removing biofilm from both metallic and nonmetallic surfaces [121]. The removal of biofilm, however, depended on the duration of laser irradiation and the fluence. The count of biofilm bacteria also showed a significant reduction. The reduction in biofilm bacteria in comparison with the reduction in the area cover of biofilm after irradiation showed that even though some parts of biofilm remained on the coupon surface they were mainly composed of dead bacteria. Another interesting observation was the dislodging of bacteria from biofilm to the bulk medium (sterile water) by the pulsed laser irradiation. In an experimental study that measured the number of bacteria in the bulk media containing coupons with biofilm before and after laser irradiation showed a significant increase immediately after irradiation. This shows, in addition to bacterial mortality, laser irradiation also resulted in the dislodging of biofilms from the surface [129].

Studies on UV impacts of sunlight reaching the Earth's surface have been on the focus in recent times due to the depletion of ozone layer in the stratosphere. Several studies on the impact of UV radiation on the primary productivity in water systems are also available [130–132]. However, reports on the UV impacts on bacterial biofilms are not many. One of the studies showed that laser irradiation caused a higher bacterial mortality in biofilms than by UV, certainly not in terms of the dose delivered [65]. This was true for the biofilm removal capacity as well. Visible laser light removed biofilms from the surfaces more efficiently than UV [65]. The reason for this is described as the efficacy of visible laser light of 532 nm to penetrate through the water column and the upper layers of biofilm, whereas the higher attenuation of UV radiation by the surface layers of bacterial cells and the matrix of biofilm resulting in lesser impacts. In any case, laser radiation in the visible range was found to be a promising way to remove biofilms from surfaces and might be useful in many industries such as dental cleaning or in the treatment of ballast water tanks of ships.

13.2.7 Physical and Mechanical Methods

The most common physical method employed to remove biofouling is the manual cleaning of the surfaces using scrubbers. High water turbulence and flow rates are also used to remove biofilms. Among the mechanical methods, the Amertrap systems (American MAN system) attracted attention [133,134]. This system employs sponge balls to clean condensers of power generation plant. The Amertrap system of cleaning is widely used in Europe; however, this system is of little use for plate or fin-type condenser systems. Raising temperature of the water column can control biofilm formation; however, temperature at which this occurs differs with the type of microorganisms present in biofilms. Studies show water with very high temperature such as 125°C for 30 min as the most effective method to remove biofilms, however, biofilms of 3 days old with high EPS content was found to be difficult to remove completely even at this temperature [78]. The circulation of temperature-elevated cooling water to control fouling growth is followed in some power plants such as California Edison Company in California, La Spezia in Italy, and by some power plants in Russia and the Netherlands.

Here, the hot effluent from the condenser is diverted to the precondenser cooling water system, and, as it once again circulates through the condenser tubes, picks up more heat. It is reported that retention of water with temperature of 40°C for 1 h can eliminate large fouling organisms such as mussels and barnacles. However, in order to prevent the slime formation in these parts, the required temperature increase is very high.

The speed of water flow is used as a method to deter biofouling on industrial surfaces. The principle that works here is that the shear stress of water flow exceeds the shear strength of the sessile organisms. In Vado Ligure Power Station in Italy, the precondenser water channels are kept free of biofouling by maintaining a water flow speed of 11 m/s [135]. In a laboratory study using *E. coli* on the hydrodynamic influences on biofilm formation and growth, Kraigsley and others [136] found that when the velocity gradient of water near the surface is below 20/s, *E. coli* whose average swimming speed is 20 μm/s [137] can orient itself at a distance equal to its body length of 1 μm from the surface without getting dispersed. This means *E. coli* can establish attachment when the gradient of water flow is below 20/s. However, if the velocity gradient increases, the attachment of cells and the nature of the biofilm get disturbed. This shows the possibility of deterring biofilm formation by increasing the water flow velocity. The general conclusion of the effect of flow regime on the architecture of biofilm is that when biofilms grow under turbulent flow, the higher the velocity, the thinner and denser the biofilm will be [138]. In the studies of Brading and others, when *Pseudomonas fluorescens* biofilms were subjected to different flow velocities under a laminar flow regime, bacteria produced more exopolysaccharides when the fluid velocity was higher, thus portraying that the biological growth within the biofilm is flow dependent. A similar result was obtained for *P. fluorescens* biofilm when studied by Pereira and others [139].

Various other techniques that are being tested in laboratory with varying degrees of success are the osmotic control method, sonic methods, magnetic field method, bioelectric method, and surface modifications. In the osmotic control method, marine biofoulers are killed with a diluted seawater circulation. Even though this was found to be successful, it was uneconomical [140]. Similarly, deoxygenation of the water was tested by shutting down the circulation system for several hours to create an anaerobic condition. The drawback of this method is the initiation of pitting corrosion [134]. In summary, even though all these physical and mechanical methods were reported to have a certain degree of success, they were found not useful in the long run.

13.3 Summary and Conclusions

Majority of bacteria on Earth are believed to be living in biofilms. Biofilms are formed on wet surfaces. These are mushroom-like structures formed by bacteria and their EPS. Water channels running between these structures aid in nutrient supply as well as waste removal. The shedding of cells from biofilms results in their dispersal. Bacteria living in biofilms are found to have several orders higher tolerance toward antibiotics, biocides, or toxins than those living as planktonic forms. The tolerance displayed by biofilm bacteria is mainly due to the EPS produced by them. EPS restricts the entry of chemicals and reactive radicals into biofilms. Biofilm formation is considered as a stress-response mechanism of planktonic bacteria. Biofilms once established are difficult to remove. This is due to the adhesive nature of EPS. Biofilms are known to display various interesting properties such as viscoelastic properties, polyionic properties, EPS production, etc. These properties along with the types of bacterial species involved in biofilm formation, and the physicochemical properties of the water column determine the effectiveness of a biofilm abating technique. This means, a particular biofilm abating technique effective in a certain circumstance may not be successful in deterring biofilm formation in another area. Biofilm formation involves a series of steps starting from the formation of conditioning film, reversible and irreversible attachment of bacteria, and their subsequent growth. This takes varying periods. Molecular biological studies on biofilms show bacteria in biofilms communicate with each other by cell-to-cell signaling now known as quorum sensing. Thus, these bacteria are believed to be acting as a functional group. Biofilms depending on the place of its formation have detrimental and beneficial impacts on human activities. Various biofilm prevention and removal strategies are available. They are surface modifications (chemical and physical modification of the surfaces to be protected), application of toxic coatings, application of biocides and antibiotics, genetic intervention of bacterial attachment, and cell-to-cell communication, electrochemical techniques, bioelectric methods, using various types of radiation and mechanical or physical methods. Each of them has its own merit and demerit, and their application is chiefly site specific. In addition, combinations of one or more of the above-mentioned techniques are also tried with varying success. All these reveal that the fight against biofilm is an ongoing phenomenon and will continue as long as bacteria with their innate nature of mutation, as well as acquisition of genetic materials from other species coexisting in biofilms that enable them to tide over the abating techniques, continue.

Acknowledgments

Kanavillil Nandakumar acknowledges the support and advice from Dr. Heidi Schraft and Dr. Kam T. Leung, Department of Biology, Lakehead University, Thunder Bay, Ontario, Canada.

References

1. ZoBell, C.E., The effect of solid surfaces upon bacterial activity, *J. Bacteriol.*, 46, 39, 1943.
2. Costerton, J.W., Stewart, P.S., and Greenberg, E.P., Bacterial biofilms: A common cause of persistent infections, *Science*, 284, 1318, 1999.
3. Costerton, J.W. et al., Bacterial biofilms in nature and disease, *Annu. Rev. Microbiol.*, 41, 435, 1987.
4. Lawrence, J.R. et al., Optical sectioning of microbial biofilms, *J. Bacteriol.*, 173, 6558, 1991.
5. Stoodley, P. et al., Biofilms as complex differentiated communities, *Annu. Rev. Microbiol.*, 56, 187, 2002.
6. Geesey, G.G. et al., Microscopic examination natural sterile bacterial population from an alpine stream, *Can. J. Microbiol.*, 23, 1733, 1977.
7. MacLeod, F.A., Guiot, S.R., and Costerton, J.W., Layered structure of bacterial aggregates produced in an upflow anaerobic sludge bed and filter reactor, *Appl. Environ. Microbiol.*, 56, 1598, 1990.
8. Christensen, B.E., The role of extracellular polysaccharides in biofilms, *J. Biotechnol.*, 10, 181, 1989.
9. Jahn, A. and Nielsen, P.H., Cell biomass and exopolymer composition in sewer biofilms, *Water Sci. Technol.*, 37, 17, 1998.
10. Sutherland, I., Biofilm exopolysaccharides: A strong and sticky framework, *Microbiology*, 147, 3, 2001.
11. Starkey, M. et al., A sticky business: The extracellular polymeric substance matrix of bacterial biofilms. In: *Microbial Biofilms*, Ghannoum, M. and O'Toole, G.A., Eds., ASM Press, Washington, DC, 2004, 174.
12. Heydorn, A. et al., Quantification of biofilm structures by the novel computer program COMSTAT, *Microbiology*, 146, 2395, 2000.
13. Dalton, H.M., Goodman, A.E., and Marshall, K.C., Diversity in surface colonization behaviour in marine bacteria, *J. Ind. Microbiol.*, 17, 228, 1996.
14. Tolker-Nielson, T. et al., Development and dynamics of *Pseudomonas* sp. biofilms, *J. Bacteriol.*, 182, 6482, 2000.
15. de Laubenfels, E., Control of biofilm in aquatic systems. Selected paper in *Biofilm and Biodiversity*. Virtue special reports: http://www.mdsg.umd.edu/wit/paper/htm, 2001.
16. Watnick, P. and Kolter, R., Biofilm, city of microbes, *J. Bacteriol.*, 182, 2675, 2000.
17. Hanna, A. et al., Role of capsular colonic acid in adhesion of uropathogenic *Escherichia coli*, *Appl. Environ. Microbiol.*, 69, 4474, 2003.
18. Danese, P.N., Pratt, L.A., and Kolter, R., Exopolysaccharide production is required for development of *Escherichia coli* K-12 biofilm architecture, *J. Bacteriol.*, 182, 3593, 2000.
19. Characklis, W.G., Fouling biofilm development: A process analysis, *Biotechnol. Bioeng.*, 23, 1923, 1981.
20. Characklis, W.G. and Cooksey, K.E., Biofilms and microbial fouling, *Adv. Appl. Microbiol.*, 29, 93, 1983.
21. Marshall, K.C., *Microbial Adhesion and Aggregation*, Springer Verlag, Berlin, New York, 1984, 424.
22. Costerton, J.W. et al., Biofilms, the customized microniche, *J. Bacteriol.*, 176, 2137, 1994.

23. Escher, A. and Characklis, W.G., Algal–bacterial interactions in aggregates, *Biotechnol. Bioeng.*, 24, 2283, 1982.
24. Haack, T.K. and McFeters, G.A., Nutritional relationships among microorganisms in an epilithic biofilm community, *Microb. Ecol.*, 8, 115, 1982.
25. Murray, R.E., Cooksey, K.E., and Priscu, J.C., Stimulation of bacterial DNA synthesis by algal exudates in attached algal–bacterial consortia, *Appl. Environ. Microbiol.*, 52, 1177, 1986.
26. Mitchell, R. and Krichman, D., The microbial ecology of marine surfaces. In: *Marine Biodeterioration: An Interdisciplinary Study*, Costolow, J.D. and Tipper, R., Eds., U.S. Naval Institute, Annapolis, MD, 1984, 49.
27. Tosteson, T.R. and Corpe, W.A., Enhancement of adhesion of marine *Chlorella vulgaris* to glass, *Can. J. Microbiol.*, 21, 1025, 1975.
28. Lamb, L.A. and Lowe, R.L., Effects of current velocity on the physical structuring of diatom (Bacillariophyceae) communities, *Ohio J. Sci.*, 87, 72, 1987.
29. Characklis, W.G., McFeters, G.A., and Marshall, K.C., Physiological ecology in biofilm systems. In: *Biofilms*, Characklis, W.G. and Marshall, K.C., Eds., John Wiley & Sons Inc., New York, 1990, 341.
30. Liu, Y. and Tay, J.H., Metabolic response of biofilm to shear stress in fixed-film culture, *J. Appl. Microbiol.*, 90, 337, 2001.
31. Caiazza, N.C. and O'Toole, G.A., Alphatoxin is required for biofilm formation by *Staphylococcus aureus*, *J. Bacteriol.*, 185, 3214, 2003.
32. Whitchurch, C.B. et al., Extracellular DNA required for bacterial biofilm formation, *Science*, 295, 1487, 2002.
33. Purevdorj-Gage, L.B. and Stoodley, P., Biofilm structure, behavior, and hydrodynamics. In: *Microbial Biofilms*, Ghannoum, M. and O'Toole, G.A., Eds., ASM Press, Washington, DC, 2004, 160.
34. Flemming, H.C. et al., Cohesiveness in biofilm matrix polymers. In: *Community Structure and Cooperation in Biofilms*, Allison, D. et al., Eds., SGM Symposium Series 59, Cambridge University Press, Cambridge, U.K., 2000, 87.
35. Flemming, H.C., The forces that keep biofilms together, *DECHEMA Monogr.*, 133, 311, 1996.
36. Huang, C.T. et al., Spatial patterns of alkaline phosphatase expression within bacterial colonies and biofilms in response to phosphate starvation, *Appl. Environ. Microbiol.*, 64, 1526, 1998.
37. Xu, K.D. et al., Spatial physiological heterogeneity in *Pseudomonas aeruginosa* biofilm is determined by oxygen availability, *Appl. Environ. Microbiol.*, 64, 4035, 1998.
38. Moller, S. et al., In situ gene expression in mixed-culture biofilms: Evidence of metabolic interactions between community members, *Appl. Environ. Microbiol.*, 64, 721, 1998.
39. Parsek, M.R. and Greenberg, E.P., Acylhomoserine lactone quorum sensing in gram negative bacteria: A signaling mechanism involved in associations with higher organisms, *Proc. Natl. Acad. Sci. USA*, 97, 8789, 2000.
40. Kolenbrander, P.E. et al., Communication among oral bacteria, *Microbiol. Mol. Biol. Rev.*, 66, 486, 2002.
41. Kjelleberg, S. and Molin, S., Is there a role for quorum sensing signals in bacterial biofilms? *Curr. Opin. Microbiol.*, 5, 254, 2002.
42. Allison, D.G. et al., Extracellular products as mediators of the formation and detachment of *Pseudomonas fluorescnes* biofilms, *FEMS Microbiol. Lett.*, 167, 179, 1998.

43. Davies, D.G. et al., The involvement of cell-to-cell signals in the development of a bacterial biofilm, *Science*, 280, 295, 1998.
44. Goodman, P.D., Effect of chlorination on material for seawater cooling system: A review of chemical reactions, *Br. Corros. J.*, 22, 56, 1987.
45. Stewart, P.S. and Costerton, J.W., Antibiotic resistance of bacteria in biofilms, *Lancet*, 358, 135, 2001.
46. Nakamura, S. et al., Antimicrobial activity and basic properties of "NSSAM-1" antimicrobial ferritic stainless steel, *Nisshin Steel Tech. Rep.*, 76, 48, 1997.
47. Yokota, T. et al., Development of silver superdispersed type 304 stainless steel with antibacterial activity, Proceedings of 1st International Symposium on Environmental Materials and Recyling, Osaka University, Osaka, Japan, 69, 2001.
48. Sreekumari, K.R. et al., Silver containing stainless steel as a new outlook to abate bacterial adhesion and microbiologically influenced corrosion, *ISIJ (Iron and Steel Institute of Japan) Int.*, 43, 1799, 2003.
49. Kul'skill, L.A., "Water with silver ions," Nitsuo News Agency, Japan, 7, 1987.
50. Kingshott, P. and Griesser, H.J., Surfaces that resist bioadhesion, *Curr. Opin. Solid State Mat. Sci.*, 4, 403, 1999.
51. Singh, P.K. et al., A component of innate immunity prevents bacterial biofilm development, *Nature*, 407, 762, 2002.
52. Singh, P.K. et al., Quorum sensing signals indicate that cystic fibrosis lungs are infected with bacterial biofilms, *Nature*, 407, 762, 2000.
53. Giviskov, M., The presence of AHL signals and the effects of signal inhibitors in *Pseudomonas aeruginosa* involved in cystic fibrosis infections infectious. American Society of Microbiology, 101st General Meeting, Orlando, FL, May 2001.
54. Nivens, D.E. et al., Role of alginate and its O acetylation in formation of *Pseudomonas aeruginosa* microcolonies and biofilms, *J. Bacteriol.*, 183, 1047, 2001.
55. Recht, J. and Kolter, R., Glycopeptidolipid acetylation affects sliding motility and biofilm formation by *Mycobacterium smegmatis*, *J. Bacteriol.*, 183, 5718, 2001.
56. Hengge-Aronis, R., Survival of hunger and stress: The role of *rpoS* in stationary phase gene regulation in *Escherichia coli*, *Cell*, 72, 165, 1993.
57. Jenkins, D.E., Schultz, J.E., and Matin, A., Starvation-induced cross protection against heat or H_2O_2 challenge in *Escherichia coli*, *J. Bacteriol.*, 170, 3910, 1988.
58. Kolter, R., Siegele, D.A., and Tormo, A., The stationary phase of the bacterial life cycle, *Annu. Rev. Microbiol.*, 47, 855, 1993.
59. Nystrom, T., Starvation, cessation of growth and bacterial aging, *Curr. Opin. Microbiol.*, 2, 214, 1999.
60. Hengge-Aronis, R., Signal transduction and regulatory mechanisms involved in control of the σ^s (*RpoS*) subunit of RNA polymerase, *Microbiol. Mol. Biol. Rev.*, 66, 373, 2002.
61. Loewen, P.C. and Hengge-Aronis, R., The role of the sigma factor σ^s (KatF) in bacterial global regulation, *Annu. Rev. Microbiol.*, 48, 53, 1994.
62. Loewen, P.C. et al., Regulation in the *rpoS* regulon of *Escherichia coli*, *Can. J. Microbiol.*, 44, 707, 1998.
63. Ventrui, V., Control of rpoS transcription in *Escherichia coli* and *Pseudomonas*: Why so different? *Mol. Microbiol.*, 49, 1, 2003.
64. Schuster, M. et al., The *Pseudomonas aeruginosa rpoS* regulon and its relationship to quorum sensing, *Mol. Microbiol.*, 51, 973, 2004.
65. Nandakumar, K. et al., Visible laser and UV-A radiation impact on a PNP degrading *Moraxella* strain and its *rpoS* mutant, *Biotechnol. Bioeng.*, 94, 793, 2006.

66. Whiteley, M. et al., Gene expression in *Pseudomonas aeruginosa* biofilms, *Nature*, 413, 860, 2001.

67. Corona-Izquierdo, F.P. and Membrillo-Hernandez, J., A mutation in *rop*S enhances biofilm formation in *Escherichia coli* during exponential phase of growth, *FEMS Microbiol. Lett.*, 211, 105, 2002.

68. Adams, J.L. and McLean, R.J.C., Impact of *rpo*S deletion on *Escherichia coli* biofilms, *Appl. Environ. Microbiol.*, 65, 4285, 1999.

69. Preiser, H.S., Ticker, A., and Bohlander, G.S., Coating selection for optimum ship performance. In: *Marine Biodeterioration: An Interdisciplinary Study*, Costlow, J.D. and Tipper, R.C., Eds., Naval Institute Press, Annapolis, MD, 223, 1984.

70. Fisher, E.C. et al., Technology for control of marine biofouling—a review. In: *Marine Biodeterioration: An Interdisciplinary Study*, Costlow, J.D. and Tipper, R.C., Eds., Naval Institute Press, Annapolis, MD, 261, 1984.

71. Nair, K.V.K., Marine biofouling and its control with particular reference to condenser-cooling circuits of power plants—an overview, *J. Indian Inst. Sci.*, 79, 497, 1999.

72. Lindmer, E., *Recent Developments in Biofouling Control*, Oxford and IBH Publishing Co. Pvt. Ltd., New Delhi, 1994, 305.

73. Chmielewski, R.A.N. and Frank, J.F., Biofilm formation and control in food processing facilities, *Comp. Rev. Food Sci. Food Safety*, 2, 22, 2003.

74. Chen, C.I., Griebe, T., and Characklis, W.G., Biocide action of monochloramine on biofilm systems of *Pseudomonas aeruginosa*, *Biofouling*, 7, 1, 1993.

75. Nichols, W.W., Susceptibility of biofilms to toxic compounds. In: *Structure and Function of Biofilms*, Characklis, W.G. and Wilderer, P.A., Eds., John Wiley and Sons, Inc., New York, 321, 1989.

76. Brown, M.R.W. and Gilbert, P., Sensitivity of biofilms to antimicrobial agents, *J. Appl. Bacteriol. Symp.* 74, suppl 87, 1993.

77. Jackson, A.D., Cleaning of food processing plant. In: *Developments in Food Preservation-3*, Thorne, S., Ed., Elsevier Science Publishers, New York, 95, 1985.

78. Wirtanen, G., Husmark, U., and Matilla-Sandholm, T., Microbial evaluation of the biotransfer potential from surfaces with *Bacillus* biofilms after rinsing and cleaning procedures in closed food-processing systems, *J. Food Prot.*, 59, 727, 1996.

79. Gibson, H. et al., Effectiveness of cleaning techniques used in the food industry in terms of the removal of bacterial biofilms, *J. Appl. Microbiol.*, 87, 41, 1999.

80. Satpathy, K.K., Biofouling control measures in power plant cooling systems: A brief overview. In: *Marine Biofouling and Power Plants*, Nair, K.V.K. and Venugopalan, V.P., Eds., IGCAR Press, Kalpakkam, India, 1990, 153.

81. de Beer, D., Srinivasan, R., and Stewart, P.S., Direct measurement of chlorine penetration into biofilms during disinfection, *Appl. Environ. Microbiol.*, 60, 4339, 1994.

82. Frank, J.F. and Chmielewski, R.A., Effectiveness of sanitation with quaternary ammonium compound or chlorine on stainless steel and other domestic food-preparation surfaces, *J. Food. Prot.*, 60, 1, 1997.

83. LeChevallier, M.W., Cawthon, C.D., and Ramon, R.G., Inactivation of biofilm bacteria, *Appl. Environ. Microbiol.*, 54, 2492, 1988.

84. Moss, B.L., Tovey, D., and Court, P., Kelps as fouling organisms on North Sea platforms, *Bot. Mar.*, 24, 207, 1981.

85. Pisigan, R.A. and Singley, J.E., Influence of buffer capacity, chlorine residual and flow rate on corrosion of mild steel and copper, *J. Am. Water Works Assoc.*, 2, 62, 1987.

86. Jones, B. and Slifer, G., Cooling water treatment using a liquid gas feeder and bromine chloride. 46th Annual Meeting, International Water Conference, Pennsylvania, IWC-85-42, 1, 1985.

87. Barnes, R.L. and Caskey, D.K., Using ozone in the prevention of bacterial biofilm formation and scaling, *Water Conditioning and Purification*, October 2002.

88. Little, B.J. and Depalma, J.R., Marine biofouling. In: *Materials for Marine Systems and Structures*, Hasson, D.F., Ed., Academic press, London, 1988, 90.

89. Carsberg, H., Selecting your sanitizers, *Food Qual.*, 2, 35, 1996.

90. McCarthy, C.A., Attachment of *Listeria monocytogenes* to chitin and resistance to biocides, *Food Technol.*, 46, 84, 1992.

91. Frank, J.F. and Koffi, R., Surface adherence growth of *Listeria monocytogenes* is associated with increased resistance to surfactant sanitizers and heat, *J. Food. Prot.*, 53, 550, 1990.

92. McDonnell, G. and Russell, A.D., Antiseptics and disinfectants: Activity, action, and resistance, *Clin. Microbiol. Rev.*, 12, 147, 1999.

93. Harkonen, P. et al., Development of a sample in-vitro test system for the disinfection of bacterial biofilm, *Water Sci. Technol.*, 39, 219, 1999.

94. Rossoni, E.M. and Gaylarde, C.C., Comparison of sodium hypochlorite and per acetic acid as sanitizing agents for stainless steel food processing surfaces using epifluorescence microscopy, *J. Food Microbiol.*, 61, 81, 2000.

95. Giese, I., Sanitization: The key to food safety and public health, *Food Technol.*, 45, 74, 1991.

96. Leid, J.G. et al., Human leukocytes adhere to, penetrate, and respond to *Staphylococcus aureus* biofilms, *Infect. Immun.*, 70, 6339, 2002.

97. Costerton, J.W. et al., Microbial biofilms, *Annu. Rev. Microbiol.*, Ornston, L.N., Balows, A., and Greenberg, E.P., Eds., Annual Reviews Inc., Palo Alto, CA, 49, 711, 1995.

98. Nichols, W.W. et al., Inhibition of tobramycin diffusion by binding to alginate, *Antimicrob. Agents Chemother.*, 32, 518, 1988.

99. Nichols, W.W. et al., The penetration of antibiotics into aggregates of mucoid and non mucoid *Pseudomonas aeruginosa*, *J. Gen. Microbiol.*, 135, 1291, 1989.

100. Andrel, J.N., Franklin, M.J., and Stewart, P.S., Role of antibiotic penetration limitation in *Klebsiella pneumoniae* biofilm resistance to ampicillin and ciprofloxacin, *Antimicrob. Agents Chemother.*, 44, 1818, 2000.

101. Stone, G. et al., Tetracycline rapidly reaches all the constituent cells of uropathogenic *Escherichia coli* biofilms, *Antimicrob. Agents Chemother.*, 46, 2458, 2002.

102. Zeng, Z. and Stewart, P.S., Penetration of rifampin through *Staphylococcus epidermidis* biofilms, *Antimicrob. Agents Chemother.*, 46, 900, 2002.

103. Nield-Gehrig, J.S., *Periodontal Concepts for the Dental Hygienist*, Lippincott Williams & Eilkins, Baltimore, MT, 480, 2002.

104. Olson, M.E. et al., Biofilm bacteria: Formation and comparative susceptibility to antibiotics, *Can. J. Vet. Res.*, 66, 86, 2002.

105. Aaron, S.D. et al., Single and combination antibiotic susceptibilities of planktonic, adherent, and biofilm-grown *Pseudomonas aeruginosa* isolates cultured from sputa of adults with cystic fibrosis, *J. Clin. Microbiol.*, 40, 4172, 2002.

106. Amorena, B. et al., Antibiotic susceptibility assay for *Staphylococcus aureus* in biofilms developed in vitro, *J. Antimicrob. Chemother.*, 44, 43, 1999.

107. Blenkinsopp, S.A., Khoury, A.E., and Costerton, J.W., Electrical enhancement of biocide efficacy against *Pseudomonas aeruginosa* biofilms, *Appl. Environ. Microbiol.*, 58, 3770, 1992.

108. Khoury, A.E. et al., Prevention and control of bacterial infections associated with medical devices, *Am. Soc. Artif. Int. Organ. J.*, 36, M174, 1992.
109. Schembri, M.A. et al., Differential expression of the *Escherichia coli* autoaggregation factor antigen 43, *J. Bacteriol.*, 185, 2236, 2003.
110. Rosenberg, B., Inhibition of cell division in *Escherichia coli* by electrolysis products from a platinum electrode, *Nature*, 205, 698, 1965.
111. Nakasono, S. and Matsunaga, T., Electrochemical sterilization of marine bacteria adsorbed onto a graphite-silicone electrode by application of an alternating potential, *Denki Kagaku*, 61, 899, 1993.
112. Matsunaga, T. et al., Prevention of marine biofouling using a conductive paint electrode. *Biotechnol. Bioeng.*, 59, 374, 1998.
113. Nakayama, T. et al., Use of a titanium nitride for electrochemical inactivation of marine bacteria, *Environ. Sci. Technol.*, 32, 798, 1998.
114. Smith, A.P. and Kretschmer, T.R., Electrochemical control of fouling. In: *Marine Biodeterioration: An Interdisciplinary Study*, Costlow, J.D. and Tipper, R.C., Eds., Naval Institute Press, Annapolis, MD, 1984, 250.
115. Nakasono, S. et al., Electrochemical prevention of marine biofouling with a carbon chloroprene sheet, *Appl. Environ. Microbiol.*, 59, 3757, 1993.
116. Dobson, J. and Wilson, M., Sensitization of oral bacteria in biofilms to killing by light from a low-power laser, *Arch. Oral. Biol.*, 37, 883, 1992.
117. Tucker, D. et al., Morphologic changes following in vitro CO_2 laser treatment of calculus-ladened root surfaces, *Lasers Surg. Med.*, 18, 150, 1996.
118. Andrian, J.C. and Gross, A., A new method of sterilization: The carbon dioxide laser, *J. Oral. Pathol.*, 8, 60, 1979.
119. Powell, G.L. and Whisenant, B., Comparison of three lasers for dental instrument sterilization, *Lasers Surg. Med.*, 11, 69, 1991.
120. Coffelt, D.W. et al., Determination of energy density threshold for laser ablation of bacteria, *J. Clin. Periodontol.*, 24, 1, 1997.
121. Nandakumar, K. et al., Inhibition of bacterial attachment by Nd:YAG laser irradiations: An in-vitro study using marine biofilm forming bacteria *Pseudoalteromonas carrageenovora*, *Biotechnol. Bioeng.*, 80, 552, 2002.
122. Deckelbaum, L.I., Cardiovascular applications of laser technology. *Lasers Surg. Med.*, 15, 315, 1994.
123. Lee, J.S. et al., CO_2 laser sterilization in the surgical treatment of infected median sternotomy wounds, *South. Med. J.*, 92, 380, 1999.
124. Ito, T., Cellular and sub-cellular mechanisms of photodynamic action: The $^1O_2$ hypothesis as a driving force in recent research, *Photochem. Photobiol.*, 28, 493, 1978.
125. Ito, A. and Ito, T., Possible involvement of membrane damage in the inactivation by broad-band near-UV radiation in *Saccharomyces cerevisiae* cells, *Photochem. Photobiol.*, 37, 395, 1983.
126. Kelland, L.R., Moss, S.H., and Davis, D.J.G., An action spectrum for ultraviolet radiation induced membrane damage in *Escherichia coli* K-12. *Photochem. Photobiol.*, 37, 301, 1983.
127. Davi, S.K., 1986. Pulsed laser damage thresholds and laser treatment energy parameters, in vivo, of human aphakic intraocular membranes. *Lasers Surg. Med.*, 6, 449, 1986.
128. Wilson, M., Photolysis of oral bacteria and its potential use in the treatment of caries and periodontal disease, *J. Appl. Bacteriol.*, 75, 299, 1993.

129. Nandakumar, K., Laser ablation of natural biofilms, *Appl. Environ. Microbiol.*, 70, 6905, 2004.

130. Bracher, A.U. and Wiencke, C., Simulation of the effects of naturally enhanced UV radiation on photosynthesis of Antarctic phytoplankton, *Mar. Ecol. Prog. Ser.*, 196, 127, 2000.

131. Hader, D.-P. et al., Aquatic ecosystems: Effects of solar ultraviolet radiation and interactions with other climatic change factors, *Photochem. Photobiol. Sci.*, 2, 39, 2003.

132. Prezelin, B.B., Moline, M.A., and Matlick, H.A., Antarctic Sea ice: Biological processes, interactions and variability, *Antarc. Res. Ser.*, Lizotte, M.P. and Arrigo, K.R., Eds., American Geophysical Union, Washington, DC, 73, 45, 1998.

133. Kawabe, A. and Treplin, F.W., Control of macrofouling in Japan, existing and experimental methods, *Res. Rep.*, 23, 1, 1986.

134. Little, B.J. and Depalma, J.R., Marine biofouling. In: *Materials for Marine Systems and Structures*, Hasson, D.F., Ed., Academic Press, London, 90, 1988.

135. Whitehouse, J.W. et al., Marine fouling and power stations. A Collaborative Research Report by CEGB, EDF, ENEL and KEMA, 1, 1985.

136. Kraigsley, A., Ronney, P.D., and Finkel, S.E., Hydrodynamic influences on biofilm formation and growth, http://carambola.usc.edu/research/biophysics/biofilms4Web.html

137. Berg, H.C., Motile behaviour of bacteria, *Phys. Today*, 53, 24, 2000.

138. Brading, M.J., Boyle, J., and Lappin-Scott, H.M., Biofilm formation in laminar flow using *Pseudomonas fluorescens* EX101, *J. Ind. Microbiol.*, 15, 297, 1995.

139. Pereira, M.O. et al., Effect of flow regime on the architecture of a *Pseudomonas fluorescens* biofilm, *Biotechnol. Bioeng.*, 78, 164, 2002.

140. Fisher, E.C., Technology for control of marine biofouling—a review. In: *Marine Biodeterioration: An Interdisciplinary Study*, Costlow, J.D. and Tipper, R.C., Eds., Naval Institute Press, Annapolis, MD, 1984, 261.

14

Consequences of Biofilms on Indwelling Medical Devices: Costs and Prevention

John G. Thomas, Lindsay Nakaishi, and Linda Corum

CONTENTS

14.1 Introduction

The use of indwelling medical devices and biomaterials has increased dramatically as a consequence of several factors, two of which are foremost: an aging population and the number of applications in new devices. However, this advance is complicated by the previously unrecognized form of microbial attachment: The biofilm. Today, biofilms are responsible for 50% of nosocomial infections.

This chapter is organized into four parts. Its uniqueness hinges on the fact that each of the sections emphasizes anatomic location with the potential exposure of medical devices to the resident or normal flora within that area. Simultaneously, the diseases are organized via the anatomic location, again emphasizing the risk to colonization and infection from resident flora in a proximal setting. This also allows for the categorization of cost, a major feature relative to anatomic location of a disease and potential outcomes with the infecting organisms. There is a very significant link between the anatomy and the positioning of the indwelling medical device. To standardize the discussion of diseases in comparison costs, diagnostic related groups (DRGs) are used. The DRG system was developed by medicare to classify hospital cases into groups ($\sim$500 groups) in order to assign cost or payment.

The four sections discuss (1) the use of medical devices, (2) the consequences focusing on outcomes, (3) the kinds of coatings and device coverings used to reduce colonization, and (4) the evidence reviewed as to the clinical impact of these devices and neutralizing the potential consequence of biofilm.

14.2 Uses of Indwelling Medical Devices

Approximately four million medical devices are implanted annually in the United States. In spite of many advances in biomaterials, a significant proportion of devices become colonized by bacteria and become the focus of a chronic device-related infection. Of the two million nosocomial infections diagnosed annually, as many as 50% are associated with implanted devices, contributing substantially to rising health-care costs. In most cases, the cost increase due to implant-associated infections surpasses the cost of the original device. Medical devices will be discussed in general, first by their age-related use and their relationship to population demographics, second by their use in inpatient and outpatient settings, including the home environment, and third by an anatomical grouping of the devices, highlighting their association with the normal host flora bioburden in each of the four anatomic reservoirs: the skin, the oral cavity/respiratory tract, the gastrointestinal (GI) tract, and the genitourinary (GU) tract.

14.2.1 Population Demographics

Table 14.1 lists the various types of implanted medical devices (IMDs) and their frequency of use. As one reads through the list of various catheters, implants, and devices listed in Table 14.1, it becomes apparent that the majority of patients needing joint replacements, cochlear implants, dental implants, and cardiac devices are the elderly.

The population that is 55–64 years of age will be the fastest growing segment of the adult population over the next 10 years. This projected growth of the elderly population is illustrated in Table 14.2, which shows the proportional increase in age groups for 1950, 2004, and 2050 (projected). In 2004, this group numbered about 29 million people. Over the next 10 year period, the 55–64 year age group will increase by 11 million to an estimated 40 million. It is projected that by 2050, 12% or about 1 in 8 Americans will be 75 years of age or older. The continued aging of the population will result in an increase in the number of indwelling medical devices and a parallel rise in health-care costs [1].

The retention of teeth into advanced age has increased the incidence of caries and periodontitis. Approximately 1%–3% of people attaining the age of 65 will eventually require hip replacements. A total of 300,000 hip replacements are performed annually in the United States, and this number is projected to rise to 600,000 by 2015. The number of coronary artery stents placed annually in the United States for the treatment of coronary artery disease is 600,000–800,000. Table 14.3 shows the number of coronary artery stent procedures performed from 1996–2003 on patients in age groups 45–54, 55–64, 65–74, and above 75. Of the 35 million Americans over 65 years of age, 400,000 suffer from profound hearing loss. Many of them may become candidates for cochlear implants. The need for peritoneal

TABLE 14.1

Annual Use of Common IMDs

Indwelling Medical Device	Annual Use in the United States
Surgical site devices	27,000,000[a]
Endotracheal tubes	
Central venous catheters	5,000,000[b]
Joint prostheses and fracture-fixation devices	6,000,000[c]
Dialysis catheters (hemodialysis or peritoneal)	243,000[b]
Cardiovascular devices (total)	8,35,700[e]
Mechanical heart valve	85,000[e]
Vascular graft	450,000[e]
Pacemaker-defibrillator	300,000[e]
Ventricular assist device	700[e]
Central nervous system devices	40,000[f]
Urinary catheters	30,000,000[b]
Penile implants	15,000–20,000[b]
Intrauterine devices	1,000,000 (worldwide)
Breast implants	130,000[e]

Source: From [a]Centers for Disease Control and Prevention, National Center for Health Statistics, Health, United States, 2005. http://www.cdc.gov/nchs/hus.htm; [b]Kojic, E.M. and Darouiche, R.O., *Clin. Micro. Rev.*, 17, 255, 2004; [c]Hedrick, T.L., Adams, J.D., and Sawyer, R.G., *J. Long-Term Eff. Med. Implants*, 16, 83, 2006; [d]Thomas, J.G., Ramage, G., and Lopez-Ribot, J.L., in *Microbial Biofilms*, M. Ghannoum and G. O'Toole, Eds., ASM Press, Washington, 2005, 269–293; [e]Darouiche, R.O., *N. Engl. J. Med.*, 350, 1422, 2004; [f]Darouiche, R.O., *Clin. Infect. Dis.*, 33, 1567, 2001.

TABLE 14.2 (See color insert following page 210)

Distribution of U.S. Population by Age and Year

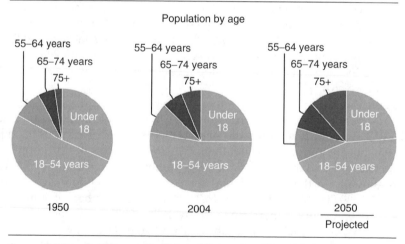

Source: Centers for Disease Control and Prevention, National Center for Health Statistics, Health, United States, 2005, Figure 2.

TABLE 14.3 (See color insert following page 210)

Number of Coronary Artery Stent Procedures Based on Age and Year

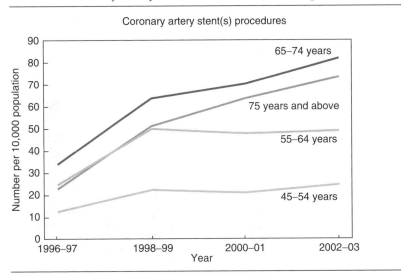

Coronary artery stent(s) procedures

Source: Centers for Disease Control and Prevention, National Center for Health Statistics, Health, United States, 2005, Figure 25.

dialysis and hemodialysis for the treatment of end-stage renal disease increases with age, as does the need for parenteral nutrition and urinary catheterization. Tables 14.4 and 14.5 indicate the number of people, grouped by age, who have chronic conditions that cause the limitation of activity. These tables illustrate the increasing demand for the medical devices related to each of these conditions. Along with the increased demand and use is the increased expense and cost [1].

14.2.1.1 Inpatient vs. Outpatient

Infections associated with indwelling medical devices are chronic infections that are not easily controlled by conventional antibiotic therapy. It is apparent that these devices are inserted or implanted for varying lengths of time. Many are considered to be permanent implants such as orthopedic, cardiac, and central nervous system devices and dental, penile, breast, and cochlear implants. Other devices are more temporary: endotracheal tubes and dialysis, central venous, and urinary catheters. Most patients receive their implants in an inpatient facility and then return to the community as an outpatient, shifting the care and maintenance required for each device from the hospital/inpatient setting to the community, nursing home, physician's office, or home environment, i.e., the outpatient setting. Table 14.6 lists the types of indwelling medical devices associated with inpatient and outpatient status. Table 14.6

TABLE 14.4 (See color insert following page 210)

Age (18–64 years) Distribution of Chronic Conditions Causing Limitation of Activity

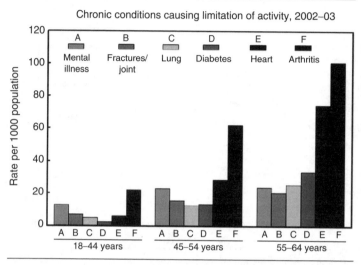

Source: Centers for Disease Control and Prevention, National Center for Health Statistics, Health, United States, 2005, Figure 19.

TABLE 14.5 (See color insert following page 210)

Age (64–85+ years) Distribution of Chronic Conditions Causing Limitation of Activity

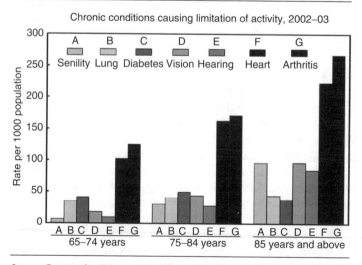

Source: Centers for Disease Control and Prevention, National Center for Health Statistics, Health, United States, 2005, Figure 20.

TABLE 14.6

Inpatient and Outpatient IMDs

IMDs Associated with Inpatient Use	IMDs Associated with Outpatient Use
Sutures	Sutures
Orthopedic devices (fixators, nails, screws)	Orthopedic devices (fixators, nails, screws)
Intravenous catheters[a]	Contact lenses
Endotracheal tubes[a]	Mechanical heart valve
Urinary catheters[a]	Voice prosthesis
Central venous catheters[a]	Penile implants
	Pacemakers/automated intracardiac devices
	Prosthetic joints
	Breast implants
	Dental implants
	Peritoneal dialysis catheters

[a] Indicates that use of the device has shifted from primarily inpatient use to increased use in an outpatient environment.

illustrates the increasing number of patients with tracheotomy tubes, indwelling central venous catheters (CVCs), and urinary catheters, and patients receiving intravenous therapy for total parenteral nutrition (TPN), who are being cared for at home or in an outpatient environment. In the United States, more patients are in long-term than in acute-care facilities [2].

The Centers for Disease Control and Prevention (CDC) surveillance definitions for infection include two categories: nosocomial and community-acquired. An increasing number of patients are cared for at home, or receive treatments in nursing homes, rehabilitation centers, dialysis centers, and physicians' offices. These facilities are really health-care associated rather than community-associated facilities and they are associated with "health-care environments" that are very different microbially from community-based or home-based environments [3].

About 5%–10% of residents of long-term care facilities have urinary drainage managed with chronic indwelling catheters. These residents have increased morbidity from urinary infections compared to bacteriuric residents without chronic catheters. The most effective means to prevent infection is to limit chronic catheter use [2].

In the observational study described by Friedman et al. (methicillin-resistant *Staphylococcus aureus*) MRSA was the cause of 20% of nosocomial infections, 10% of health care–associated infections and only 2% of community-acquired infections. Ampicillin-resistance and vancomycin-resistance strains were more common in nosocomial bloodstream infections (BSIs) than in health care–associated BSIs. Neither was associated with community-acquired BSIs. Community-acquired BSI infections were more frequently caused by *Escherichia coli* and *Streptococcus pneumoniae* [3].

An evaluation of BSIs in patients receiving home infusion therapy done by Tokars et al. found a higher occurrence of polymicrobial BSIs (20%)

compared to hospital-acquired BSIs (14%) and outpatient-related BSIs (11%). The term nosohusial has been suggested to describe infections acquired via home care. Gram-positive cocci caused 58% of the nosohusial BSIs, 32% were attributed to Gram-negative rods, 10% were due to *Candida*, and 6% were corynebacterial infections [4].

Additional problems encountered in nursing homes and long-term care facilities involve functional impairments such as incontinence, immobility, confusion, or dementia. Incontinence and impaired mental status were associated with higher rates of asymptomatic urinary tract infections (UTIs). Infections involving MRSA were more frequent in bed-bound patients with pressure ulcers, fecal incontinence, and the presence of a feeding tube or urinary catheter [2].

14.2.1.2 Home-Based Care

Efforts to decrease the length of hospital stay and shift care to ambulatory settings, as well as patient and family preference to receive care at home, have contributed to the substantial growth of home care in the past decade. As life expectancy in the United States population continues to increase and patients with chronic illnesses live longer, home care will continue to expand. Most patients involved in home care are elderly and have chronic conditions requiring skilled nurses and aides. Home care may involve infusion therapy, tracheotomy care and ventilator support, dialysis, and other highly invasive procedures. Post-acute care patients are also released to receive outpatient or home care. This category includes postoperative patients, postpartum mothers and their newborns, and patients with recent strokes. As the health-care system shifts the delivery of care from hospitals to other settings, infection control and treatment will be greatly affected [5].

14.2.2 IMDs Related to Anatomic Location

The following is a review of implanted medical devices organized relative to the four anatomic reservoirs of normal host flora and their bioburden: the skin, the oral cavity/respiratory tract, the GI tract, and the GU tract as illustrated in Figure 14.1.

14.2.2.1 Oral Cavity/Respiratory Tract

14.2.2.1.1 Dentures and Implants

Oral flora is an incredibly diverse ecosystem, containing at least 700 different species. The widespread use of dentures has resulted in the study of denture-induced stomatitis and periodontal disease. Periodontal disease was one of the earliest biofilm associated infections to be characterized and studied, and has become a model for the study of pathophysiology in a multispecies sessile community [6]. The titanium dental implant was first introduced in 1969 and has become the treatment of choice for the replacement of missing teeth. Two consequences of implant infections are peri-implant mucositis

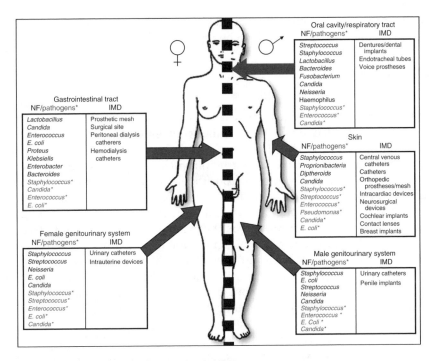

FIGURE 14.1 (See color insert following page 210)
The human body is colonized by a mix of normal and pathogenic microbes. Above is an anatomical diagram of the microbial residents throughout the body (IMD-associated pathogenic microbes are denoted with an asterisk, while the normal flora (NF) microbes are without asterisk).

and peri-implantitis. Peri-implant mucositis can be treated with oral hygiene instruction and local debridement, whereas peri-implantitis requires oral hygiene instruction, local debridement, and antibiotic therapy [6,7].

14.2.2.1.2 Endotracheal Tubes

Health-care associated pneumonia (HAP) alone accounts for 15% of all nosocomial infections and 13%–48% of infections in nursing homes. This translates to 150,000–300,000 cases annually. The primary risk factor for development of HAP is mechanical ventilation using an endotracheal tube. Mortality rates for ventilator-associated pneumonia (VAP) can be as high as 24% in ICUs and 87% for patients with VAP due to high-risk pathogens such as *Acinetobacter* spp. or *Pseudomonas aeruginosa*. The literature review done by Garcia provided strong evidence for a link between bacterial colonization of the oropharynx and dental plaque with the occurrence of nosocomial pneumonia in community, long-term care, and hospital settings [6,8].

14.2.2.1.3 Voice Prostheses

Voice prostheses are implanted in patients who have undergone laryngectomies. These prostheses often fail within months of placement because the biofilm causes malfunction (increased airflow resistance and leakage) of the valve and the prostheses needs to be replaced [6].

14.2.2.2 Skin

14.2.2.2.1   Central Venous Catheters

CVC infections are the leading cause of nosocomial BSIs. They pose a greater risk of infection than any other indwelling medical device, with infection rates of 3%–5% [9]. Virtually all indwelling CVCs are colonized by microorganisms organized in a biofilm matrix. They gain access to the catheter by migrating externally from the skin along the exterior catheter surface or internally from the catheter hub or port. Colonization and biofilm formation may occur within 3 days. Catheters in place for less than 10 days have biofilm formation primarily on the external surface. Those in place for longer than 30 days have more biofilm on the inner surface. Gram-positive organisms are associated with greater than 75% of catheter-related bloodstream infections (CRBSIs), primarily due to coagulase-negative staphylococci and *S. aureus*. *Candida* spp. are the third leading cause and are associated with the highest overall crude mortality rate [6,9].

14.2.2.2.2   Peritoneal Dialysis Catheters

Peritoneal dialysis catheters are used when the kidneys are incapable of removing excess fluid and wastes like urea and potassium from the blood (renal failure). This type of catheter requires access to the peritoneum, and thus breaking the normal skin barrier. Infection of peritoneal dialysis catheters leading to peritonitis is a common complication in patients with end-stage renal disease. *S. aureus* is the leading cause of infection [6].

14.2.2.2.3   Prosthetic Mesh

Mesh repair has become the standard of care in abdominal wall hernia surgery. Approximately 750,000 patients undergo herniorrhaphy in the United States annually, with infection rates as high as 14% [7].

14.2.2.2.4   Orthopedic Prostheses

Approximately six million Americans have either an internal fixation device or an artificial joint, making orthopedic prostheses the most common prostheses implanted annually. Arthroplasty is now the standard of care for patients over 55 with intractable pain and disability from knee arthritis. Although the risk of infection is minimal, only 1%–2%, the volume of patients with orthopedic hardware makes the portion with infection substantial [7,10].

14.2.2.2.5   Neurosurgical Devices

Three types of neurosurgical devices are implanted: ventriculoperitoneal (VP) shunts, intraspinal catheters, and neurostimulator devices. VP shunting is the most common procedure performed for the treatment of hydrocephalus.

Approximately 18,000–33,000 VP shunts are used annually in the United States. Infection rates vary from 1% to 30% depending on the institution, the surgeon's experience, and the technique used to implant the device. Once the shunt becomes infected, bacteria seed the peritoneal cavity. The most effective method of treatment involves implant removal in combination with antimicrobial therapy [7].

Implanted intraspinal catheters are being used to treat patients with refractory pain, especially those with advanced malignancies. Two types of infection occur: those involving the central nervous system (CNS) and those not involving the CNS [10].

Neurostimulator devices are used primarily for management of intractable neuropathic pain (via spinal cord or peripheral nerve stimulation) and treatment of movement disorders such as Parkinson's disease (via deep brain stimulation). Neurostimulator devices consist of leads, with electrodes at their tips that are surgically placed in the targeted area of the nervous system, and a neurostimulator that is implanted under the skin near the collarbone for deep brain stimulators or in the wall of the abdomen for spinal cord stimulators [10].

14.2.2.2.6   Cochlear Implants

Cochlear implants are used in patients with profound bilateral sensorineural deafness. Approximately 60,000 devices have been implanted worldwide. The majority of complications associated with cochlear implants are infections, wound dehiscence, or skin flap necrosis [7,10].

14.2.2.2.7   Contact Lenses

Twenty-five to twenty-seven million Americans wear contact lenses—thin, plastic disks that cover the cornea in order to correct vision problems. Contact lenses are used to correct the same problems that eyeglasses correct: myopia (nearsightedness), hyperopia (farsightedness), astigmatism (distorted vision), and presbyopia (need for bifocals) [11]. Biofilms have been shown to develop on contact lenses as well as storage cases. In fact, the lens case has been implicated as the primary source of organisms for contaminated lens disinfectant solutions and lenses. It has been estimated that 80% of contact lens users had contaminated storage cases. *Acanthamoeba* has also been identified as a component of contact lens-associated biofilms [9].

14.2.2.2.8   Intracardiac Devices

The major surgically implanted intracardiac devices are prosthetic cardiac valves, pacemakers, defibrillators, and ventricular assist devices (VADs). Cardiac pacemakers and defibrillators are used for the management of arrhythmias, decreasing mortality in patients with known ventricular arrhythmias, as well as for those at risk for ventricular arrhythmias due to ischemic cardiomyopathy. The use of these devices is expected to continue to increase. Pacemakers and defibrillators consist of a generator inserted sub-cutaneously into the chest or abdominal wall, as well as lead wires that pass through the right side of the heart and terminate with electrodes implanted in either or both the right atrial and right ventricular endocardium [9,12].

Left ventricular assist devices (LVADs) are temporary mechanical circu-latory support devices that act as a bridge for patients awaiting cardiac transplantation. The LVAD replaces the pump function of the damaged left ventricle. The device consists of a pump, an inflow cannula from the left ventricle, and an outflow cannula inserted into the ascending aorta [10].

14.2.2.2.9 Breast Implants

It is estimated that over two million women have undergone breast aug-mentation or implantation in the United States. In 2001, The American Society of Plastic Surgeons reported 206,354 breast augmentation surgeries, making it the most popular plastic surgery operation performed that year [13]. Breast augmentation is usually requested to increase the size of small breasts, correct a size difference between breasts, and part of reconstruction after a mastectomy for breast cancer. The incidence of infection for patients undergoing augmentation is 1%–2%. Traditionally, infections required implant removal, antibiotic treatment, and delayed reimplantation. However, salvaging procedures are being developed and implemented [7,10].

14.2.2.3 Genitourinary System

14.2.2.3.1 Urinary Catheters

Urinary catheters are inserted in more than five million patients in acute-care hospitals and extended-care institutions annually [48]. The percentage of patients undergoing indwelling urinary catheterization was 13.2% for hospital patients, 4.9% for nursing homes, and 3.9% for patients receiving home care [6,9,14].

14.2.2.3.2 Intrauterine Devices

Two types of intrauterine devices (IUDs) are commonly used: IUDs made of a nonabsorbable material, such as polyethylene, impregnated with barium sulfate, and IUDs that release a chemically active substance, such as copper or a progestational agent. IUDs usually have a tail that facilitates locating the device for removal. The tails are composed of a plastic monofilament surrounded by a nylon sheath. This tail portion may be a primary source of contamination. IUDs can become rapidly colonized (within 2 weeks) after insertion through the vagina and contact with resident microflora. IUD infections can lead to pelvic inflammatory disease (PID) [6,9].

14.2.2.3.3 Penile Implants

For the last 20 years, penile implants have been the solution to impotency and erectile dysfunction with approximately 15,000 penile implants inserted annually in the United States. Infection of penile implants necessitates implant removal and antibiotic treatment. Reimplantation may occur within 4–6 months but is often complicated by significant scar formation [7,10].

14.3   Consequences of Biofilms and IMDs

In order to meet today's health-care demands and growing aged population, medical technology surges forward with advances, particularly in the development of IMDs. Biofilm formation on IMDs is a clinically relevant process that must be acknowledged and addressed by clinicians to avoid infection caused by biofilms that colonize on medical devices [14]. The medical consequences of device-related infections include local site infection, life-threatening systemic infections that develop from the spread of infection beyond the device-site (e.g., CRBSI, septic thrombophlebitis, endocarditis, lung abscess, brain abscess, osteomyelitis, and endophthalmitis) [1], device malfunction that necessitates removal and thus increased risk of infection, and tissue destruction [15].

Approximately two million patients (5%–10% of patients admitted to acute-care hospitals) acquire nosocomial infections, 90,000 of which result in death. Nosocomial, or hospital-acquired, infections cost the United States an additional $4.5 to $5.7 billion per year in health-care costs. With such high rates of incidence and cost, infection control is critical for improved patient care and reduced health-care costs. Of all nosocomial infections, more than 80% are due to device-related infections that involve catheter-associated urinary tract infections (CAUTIs), surgical site infections, intra-vascular device associated BSIs, and VAP. Thus, increased awareness, understanding, and prevention of IMD infections could drastically change the landscape of patient health care [16].

Table 14.7 illustrates the annual use of the most common IMDs, their rate of infection, and the cost consequences of treating infections associated with IMDs. The magnitude of monetary consequences is apparent since the average cost of combined medical and surgical treatment for IMD-associated infections ranges from $35,000 to $50,000.

14.3.1   Consequences

14.3.1.1   Mortality and Morbidity

Although infection can result in mortality, mortality rates are independent of infection rates. Urinary catheters cause infection in 10%–30% of patients, yet the attributable mortality rate of urinary catheter-related infections is less than 5%. In contrast, for patients with vascular CRBSIs there is a low infection rate (3%–8%), but a high attributable mortality rate (5%–25%). Device location plays a critical role in a patient's attributable mortality

TABLE 14.7

Indwelling Medical Devices and Annual Use

Indwelling Medical Device	Annual Use in the United States	Average Rate of Infection (%)	Estimated Average Cost of Combined Medical and Surgical Treatment ($)
Surgical site devices	27,000,000	N/A	N/A
Endotracheal tubes	N/A	N/A	N/A
Central venous catheters	5,000,000	3–8	N/A
Joint prostheses and fracture-fixation devices	6,000,000	1–2	45,000
Dialysis catheters	N/A	N/A	N/A
Cardiovascular devices (total)	8,35,700		
Mechanical heart valve	85,000	4	50,000
Vascular graft	450,000	4	40,000
Pacemaker-defibrillator	300,000	4	35,000
Ventricular assist devices	700	40	50,000
Central nervous system devices	40,000	6	50,000
Urinary catheters	30,000,000	10–30	N/A
Penile implants	15,000–20,000	3	35,000
Dental implants	1,000,000	5–10	N/A
Cochlear devices	60,000	1.7–3.3	N/A
Intrauterine devices	N/A	N/A	N/A
Artificial voice prosthesis	N/A	N/A	N/A
Breast implants	130,000	1–2	N/A

Source: Adapted from Darouiche, R.O., *N. Engl. J. Med.*, 350, 1422, 2004; Darouiche, R.O., *Clin. Infect. Dis.*, 33, 1567, 2001; Kojec, E.M. and Darouiche, R.O., *Clin. Micro. Rev.*, 17, 255, 2004; Hedrick, T.L., Adams, J.D., and Sawyer, R.G., *J. Long-Term Eff. Med. Implants*, 16, 83, 2006.

rate—there is only a 1%–5% chance of infection with intravascularly placed medical devices (prosthetic heart valves and aortic grafts), yet mortality rates are extremely high and all infections associated with intravascular devices are considered life threatening [17,18].

Bacterial colonization on IMDs not only causes infection, but also malfunction of the medical device such as blockage of biliary stents, contracture of mammary implants, and failure of dental implants [19]. When a device becomes inefficient and ineffective, it must either be completely or partially removed or replaced, putting the patient at a greater risk since the rate of infection increases with reimplanted devices and partially explanted infected implants [18].

Some infections may have low mortality rates, but high risk of morbidity. Infections caused by medical devices implanted in sexual organs (i.e., penile and mammary implants) are rarely life threatening; however, there is great risk for deformity and psychological trauma [18].

14.3.1.2 *Outcomes and Cost*

Patients with IMD-related infections are managed with hospitalizations, prolonged antibiotic treatment, and surgical intervention, all reducing patient quality of life and increasing health-care costs. Outcomes and cost assessment are critical components of infectious disease care. Thomas et al. describe the three tiers of outcomes in the ECHO model:

Economic: costs (laboratory, pharmacy, total) and length of stay (LOS)

Clinical: ± cultures, body temperature, and other values

Humanistic: patient, physician, nursing, satisfaction, and health related

Outcomes: quality of life (QOL)

The ECHO model has been refined further to involve three quantifiable values: cost of disease, mortality or life years lost, and reduced QOL or quality adjusted life year (QALY). Clinical cost is determined via DRGs—classifications of diagnoses from patients with similar lengths of stay and resource use that are grouped together for hospital reimbursement purposes. QOL/QALY is a combination of mortality or life years lost and reduced quality of life [6].

IMDs are particularly relevant to medical outcomes because health-care costs increase considerably when a device-related infection occurs due to the increased LOS and when additional medical procedures, consultations, or treatments are necessary to address the infection [19]. Often the morbidity, mortality, and cost associated with IMD infections are greater than the insertion of the device itself [7].

Table 14.8 lists infectious disease DRGs and the indwelling medical devices, such as endotracheal tubes or catheters, which may be associated with the diseases. Table 14.9 illustrates the magnitude of the monetary outcomes that infectious diseases (and their related IMDs) have on health-care costs. For example, DRG 475 is respiratory diagnosis with ventilator support and is directly related to VAP and endotracheal tubes. Treatment for DRG 475 is more than $4 billion and the hospital loss is approximately $19,000 per case. The urgency to reduce IMD-related costs is evident in Table 14.10, which demonstrates the steady increase in cost over 4 years.

The cost of device-related infections can best be illustrated with intravascular catheter bloodstream-related infections due to central vascular catheters and peripheral intravascular devices [49]. In the United States, there are more than 15 million CVC days per year among patients. For every thousand catheter days, there can be up to five catheter-related infections—resulting in 75,000 CVC infections per year [19]. The total costs of a CRBSI include the cost of one catheter tip culture, three blood cultures and antibiotic susceptibility tests, 7 days of intravenous vancomycin therapy, five additional days in the ICU, two additional days on the general hospital ward, and professional fees [20]. Depending on the severity of the infection,

TABLE 14.8

IMDs Related to Infectious Disease DRG

DRG Number	Disease Description	Related IMDs
20	Nervous system infection except viral meningitis	Central nervous system devices
79	Respiratory infections and inflammations (age > 17 w cc)[a]	Endotracheal tubes
80	Respiratory infections and inflammations (age > 17 w/o cc)[b]	Endotracheal tubes
81	Respiratory infections and inflammations (age 0–17)	Endotracheal tubes
89	Simple pneumonia/pleur (age > 17 w cc)	Endotracheal tubes
90	Simple pneumonia/pleur (age > 17 w/o cc)	Endotracheal tubes
91	Simple pneumonia/Pleur (age 0–17)	Endotracheal tubes
126	Acute and subacute endocarditis	Cardiovascular devices
238	Osteomyelitis	Joint prostheses, fracture-fixation device
242	Septic arthritis	Joint prostheses, fracture-fixation device
277	Cellulitis (age > 17 w cc)	Surgical site devices, dialysis catheters, urinary catheters
278	Cellulitis (age > 17 w/o cc)	Surgical site devices, dialysis catheters, urinary catheters
279	Cellulitis (age 0–17)	Surgical site devices, dialysis catheters, urinary catheters
320	Kidney and urinary tract infections (age > 17 w cc)	Urinary catheters
321	Kidney and urinary tract infections (age > 17 w/o cc)	Urinary catheters
322	Kidney and urinary tract infections (age > 17 w cc)	Urinary catheters
415	OR procedure for infectious and parasitic diseases	Surgical site devices
416	Septicemia (age > 17)	N/A
417	Septicemia (age > 17)	N/A
418	Postoperative and posttraumatic infections	Surgical site devices
419	Fever of unknown origin age > 17 w cc	N/A
420	Fever of unknown origin age > 17 w/o cc	N/A
423	Other infections and parasitic diseases	N/A
475	Respiratory system diagnosis with ventilator support	Endotracheal tubes

[a] With co-morbidity conditions.
[b] Without co-morbidity conditions.

a CVC infection can cost from $3,700 to $29,000 per patient. Thus, the United States sees an increase in health-care costs from $300 million to $2.3 billion each year from CVC infections alone [19].

TABLE 14.9
The Cost Consequences of Infectious Disease Related DRGs

DRG	Frequency	LOS (Days)	Total Charges (TC) ($)	Medicare Reimbursement (MR) ($)	Loss (TC-MR) ($)	Loss Per Day ($)	Loss Per Incidence ($)
475	109,976	11.2	4,309,597,761	2,123,367,727	2,186,230,034	195,199,110.18	19,879.16
415	40,384	14.2	1,5575,736,15	788,214,118	787,522,033	55,459,298.10	19,500.84
79	184,393	8.5	3,117,071,295	1,393,200,239	1,723,871,056	202,808,359.53	9,348.90
416	197,736	7.4	3,146,282,803	1,396,098,205	1,750,184,598	236,511,432.16	8,851.12
417	40	5.9	556,129	264,568	291,561	49,417.12	7,289.03
89	528,045	6	5,664,833,052	2,405,365,156	3,259,467,896	543,244,649.33	6,172.71
418	22,545	6.1	244,893,747	111,745,816	133,147,931	21,827,529.67	5,905.87
320	183,451	5.4	1,647,218,607	686,061,458	961,157,149	177,992,064.63	5,239.31
277	84,755	5.7	749,730,593	306,160,447	443,570,146	77,819,323.86	5,233.56
91	53	3.3	441,889	194,669	247,220	74,915.15	4,664.53
90	51,793	4.2	338,049,234	125,225,236	212,823,998	50,672,380.48	4,109.13
278	28,935	4.3	165,577,731	59,918,784	105,658,947	24,571,848.14	3,651.60
321	28,549	3.9	166,258,354	62,024,040	104,234,314	26,726,747.18	3,651.07
322	83	4	578,828	295,537	283,291	70,822.75	3,413.14
TOTAL	11,659,885	6.1	178,059,683,139	79,362,721,125	98,696,962,014	16,179,829,838.36	8,464.66

TABLE 14.10 (See color insert following page 210)

Loss Per Incidence for Medicare Charges for Infectious Disease Related DRGs Over 4 Years

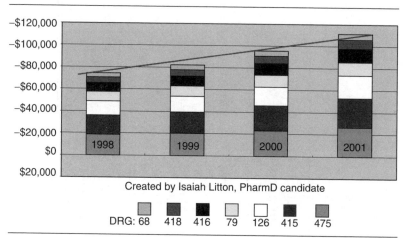

Created by Isaiah Litton, PharmD candidate

DRG: 68   418   416   79   126   415   475

Another large economical consequence is the cost of the most common of all nosocomial infections—VAP. If a ventilated patient develops VAP, LOS is increased by 4–9 days. In the United States, 7%–8% of the total health-care costs are attributable to VAP, resulting in about $1.5 billion. VAP is especially costly in the ICU, where, on average, 20%–34% of hospital costs are devoted to ventilated patients [21].

14.3.2 Pathogenic Organisms and Anatomical Location

Device location is a key feature of infection control that can be correlated with the microorganisms that colonize the IMDs. Although there is normal colonization of microorganisms throughout the body, infection occurs when organisms that are not normal inhabitants are in a new region of the body or in sterile regions of the body that are foreign to pathogens. Table 14.11 is organized to highlight the greater prevalence of Gram-positive organisms such as *Staphylococcus epidermidis* and *Candida* spp., compared to the smaller frequency of Gram-negative organisms. For example, the organisms isolated from CVC biofilms originate from the patient's skin microflora, external microflora from the health-care professional, or contaminated equipment. The more common IMD microorganisms are *S. epidermidis*, *S. aureus*, and *Candida albicans*, which have unique survival mechanisms that favor their ability to cause IMD-associated infections. *S. epidermidis* produces an extracellular slime substance that aids survival on foreign materials, *S. aureus* thrives inside blood vessels because of the protein-coagulation system interaction, and *C. albicans* demonstrates a particular capacity for adherence to extracellular matrix proteins like fibrogen and fibronectin [22].

TABLE 14.11

Organisms Associated with Indwelling Medical Devices

Indwelling Medical Devices	*Staphylococcus epidermidis*	*Staphylococcus aureus*	Coagulase-negative staphylococci	*Streptococcus* spp.	*Candida* spp.	*Enterococcus* spp.	Enteric gram-species	*Escherichia coli*	*Pseudomonas aeruginosa*
	Gram-Positive Organisms						Gram-Negative Organisms		
Central venous catheters	X	X	X̲		X				
Joint prostheses		X	X	X		X		X	X
Cardiovascular devices (i.e., prosthetic hear valve)	X	X	X̲	X	X	X			
Urinary catheters	X		X		X	X		X̲	
Intrauterine devices	X	X		X	X	X			
Artificial voice prosthesis	X			X	X				
Endotracheal tubes		X		X	X	X̲	X		
Dialysis catheters		X			X				X
Central nervous system devices	X	X				X			
Penile implants						X		X	
Cochlear devices					X				X
Surgical site devices	X̲	X							
Contact lenses	X̲	X		X̲					

Source: Thomas, J.G., Ramage, G., and Lopez-Ribot, J.L., in M. Ghannoum and G. O'Toole, Eds., *Microbial Biofilms*, ASM Press, Washington, 2005. With permission; Also adapted from Donlan, R.M., *Emerg. Infect. Dis.*, Special Issue, 7, 277, 2001.

Note: Underline (X̲) denotes principal organisms.

The nature or composition of the IMD also plays a part in microorganism colonization since flow rate and composition of the medium in the device influence biofilm formation. Gram-negative aquatic organisms are able to maintain growth better than Gram-positive organisms if the IMD is a vesicle for nutrient-poor fluids (i.e., catheters) [12].

14.3.2.1 Oral Cavity/Respiratory Tract

14.3.2.1.1 Denture/Oral Implant Infections

Peri-implant infections are classified into two groups based on severity. Peri-implant mucositis is a reversible inflammatory reaction of the soft tissues surrounding the implant. Peri-implantitis is an inflammatory reaction with loss of supporting bone in the tissues surrounding the implant. It is the leading factor in the development of late implant failure and occurs at a rate of 0.6% per year. The development of peri-implant infections is a result of the accumulation of Gram-negative anaerobes and spirochetes. *Candida* sp. has become increasingly involved in periodontal and implant-related infections [6,7].

14.3.2.1.2   Ventilator-Associated Pneumonia (Endotracheal Tubes)

Although there is normally a continuous inoculation of bacteria, the structure of the respiratory system allows the normal lung to be sterile distal to the central carina and also protected from airborne particles. These protective mechanisms include the nose, pharynx, respiratory bronchioles, and alveoli acting as natural sieves for inspired particles; the reflex closure of the glottis to prevent aspiration of liquids; and the mucociliary system's removal of foreign materials in the airways. However, for intubated patients, these defense mechanisms are eliminated by the presence of the endotracheal tube [23]. VAP is a common outcome for intubated patients and is due to bacterial colonization of the aerodigestive tract and aspiration of contaminated secretions. The sources of aspirated secretions include the stomach, upper airway, teeth, artificial airway, and nasal sinuses [21].

Thomas et al. observed that the primary colonization of endotracheal tubes is associated with organisms originating in the oral pharynx that gain access to the device. The organisms implicated in VAP for a ventilation period of less than 5 days include methicillin-susceptible, coagulase-negative staphylococci and methicillin-susceptible *S. aureus* and *Haemophilus influenzae*. After 5–7 days, this core group is joined by *Acinetobacter* spp., *C. albicans*, methicillin-resistant *S. aureus*, and nonresistant Gram-negative enterics including *E. coli*, *Klebsiella*, and *Proteus* spp. Patients on ventilation for longer periods of time may become colonized by multidrug-resistant Gram-negative enterics (*E. coli*, *Klebsiella*, and *Proteus* spp.), *P. aeruginosa*, *C. albicans*, methicillin-resistant *S. aureus*, and *Stenotrophomonas maltophilia* [6].

14.3.2.1.3   Voice Prosthesis–Associated Infections

Voice prosthesis devices are subject to rapid microbial colonization and subsequent biofilm formation, most notably by Gram-positive organisms and *Candida* spp. [6].

14.3.2.2   Skin

14.3.2.2.1   Catheter-Associated Infections

Organisms colonizing CVCs include coagulase-negative staphylococci, 50%; *S. aureus* (both methicillin-susceptible and -resistant strains), 25%; *Candida*, 10%; enterococci, 5%; *E. coli*, *Klebsiella*, and *Proteus* spp., 5%; *Pseudomonas cepacia*, 2%; *Acinetobacter*, 1%; and *Bacillus* spp., 1%. Organisms originate from the patients' own resident microflora, exogenous microflora from health-care personnel, or contaminated infusates [6,9].

Urinary catheter organisms run parallel to CVC organisms; however, *E. coli* and *Proteus mirabilis* (urinary pathogens) are two exceptions that typically frequent urinary catheters more than CVCs. The urogenital microflora, *Staphylococcus epidermidis, Enterococcus faecalis, E. coli, Proteus mirabilis, P. aeruginosa, K. pneumoniae,* and other Gram-negative organisms, colonize these devices. About 10%–15% of nosocomial UTIs are caused by *Candida*. The organisms that attach to the catheter and develop the microfilm originate from several sources. They can be introduced into the urethra or bladder as the catheter is inserted. Organisms can enter through the sheath or exudate that surrounds the catheter, or they can travel intraluminally from the inside of the tubing or collection bag. All leg bag designs support biofilms and serve as the primary reservoir for catheter contamination. The ascent up the catheter to the bladder occurs within 1–3 days. Initially, catheters are colonized by single species of urogenital microflora, such as *S. epidermidis, Enterococcus faecalis, E. coli,* or *P. mirabilis.* As the catheter remains in place, the number and diversity of organisms increase and mixed communities develop consisting of *Providencia stuartii, Pseudomonas aeruginosa, Proteus mirabilis,* and *Klebsiella pneumoniae.* About 10% of nosocomial UTIs are caused by *Candida* spp. For short-term urinary catheterization (up to 7 days), 50% of the patients may develop an associated UTI. Almost all patients undergoing long-term catheterization (greater than 28 days) develop a UTI. The risk of catheter-associated infection increases by approximately 10% for each day the catheter is in place [6,9,14].

The medium in which the device resides is also important to consider: an increase in urinary pH and ionic strength enhances bacterial adhesion [9].

14.3.2.2.2 Prosthetic Mesh–Associated Infections

Mesh-related infections fall into two categories. Superficial wound infections occur early in the postoperative period and are independent of mesh placement. Deep infections of the mesh occur late in the postoperative period (weeks to months) in approximately 0.6%–6% of patients. Superficial wound infections respond well to antibiotics and local drainage, whereas deep-seated mesh infections almost invariably require mesh removal [7].

14.3.2.2.3 Orthopedic Prosthesis–Associated Infections

Prosthetic joint infections are classified into three categories based on the onset of symptoms. Early infections (29%) develop within 3 months of surgery and are caused by *S. aureus* or Gram-negative bacilli. Delayed infections (41%) occur within 3–24 months. Delayed infections are usually associated with less virulent pathogens like coagulase-negative staphylococci and *Proprionibacterium acnes.* Both early and delayed infections are the result of intraoperative contamination. Late infections (30%) occur more than 24 months post surgery and are the result of hematogenous seeding. Treatment involves a combination of medical and surgical procedures [7,10].

14.3.2.2.4 Neurosurgical Device–Associated Infections

Infections associated with CNS shunts are primarily due to colonization of coagulase-negative staphylococci and other Gram-positive organisms. *S. epidermidis* has developed a particularly advantageous survival mechanism for attachment to foreign bodies (shunts): the production of a mucoid or slime substance. The slime generated by coagulase-negative pathogens is the glycocalyx of a biofilm that promotes attachment and provides protection from host defenses by inhibiting T-cell function and phagocytosis. Once a biofilm has formed on the CNS shunt and infection has occurred, the only effective treatment is device removal [24].

VP shunt infections usually occur within the first 6 months following insertion and involve *S. epidermidis* and *S. aureus*. Infection rates for intraspinal catheters are similar to those for VP shunts. Most of the infections occur within the first 2 weeks after insertion and are caused by skin flora (streptococcus spp., coagulase-negative staphylococci, and corynebacterium) [7].

14.3.2.2.5 Cochlear Implant–Associated Infections

The rate of infection has been reported as 1.7%–3.3%. In 2002, cochlear implants were associated with 52 cases of bacterial meningitis, resulting in 12 deaths. The infections were attributed to *Streptococcus pneumoniae* and were associated with a certain type of electrode, which has been discontinued by the manufacturer. Since July 2003, there have been 90 cases of meningitis associated with cochlear implants [7,10].

14.3.2.2.6 Contact Lens–Associated Infections

Bacteria readily adhere to both soft and hard contact lens surfaces causing microbial keratitis. Organisms involved include *P. aeruginosa*, *S. aureus*, *S. epidermidis*, *Serratia* spp., *E. coli* and *Proteus* spp., and *Candida* spp. [9].

14.3.2.2.7 Intracardiac Device–Associated Infections

Among valve recipients, 3%–5.7% develop prosthetic valve endocarditis (PVE). The mortality rate is highest for PVE that develops within the first few months after surgery, ranging from 33% to 45%. PVE is divided into three stages: early onset (within the first 2 months), delayed (2–12 months after surgery), and late (more than 12 months after surgery). The early PVEs are primarily due to coagulase-negative staphylococci, followed by methicillin-resistant *Staphylococcus epidermidis* and *S. aureus*. Late PVE reflects community acquisition and is caused by viridans group streptococci, *S. aureus*, and enterococci. Delayed PVEs are caused by the same pathogens involved in early and late PVE [9,14].

Infection rates for pacemakers and defibrillator devices range from 1% to 7%. Early infection occurs because of intraoperative contamination of the device or the pocket at the time of implantation and is caused by *S. aureus*.

Late infections usually result from exposure of the device after skin erosion and involve coagulase-negative staphylococci [12,14].

Whether temporary or permanent, infection of LVADs remains a major complication occurring in 25%–50% of patients. The high-risk time for infections is between 2 and 6 weeks post implantation. One-third of LVADs become infected within 3 months. The infections involve biofilm formation and tend to be caused by microorganisms from the health-care environment, especially methicillin-resistant staphylococcal spp. [10].

14.3.2.2.8   Breast Implant–Associated Infections

Although the infection rate for augmentation is 1%–2%, the infection rate is higher (24%) for patients undergoing breast reconstruction following mastectomy. The majority of early-onset infections are caused by staphylococcal spp. The most common complication of breast implantation is subclinical infection due to biofilm formation resulting in chronic pain and capsule contracture. 47% of these infections are due to *P. acnes*, 41% are due to *S. epidermidis*, and 1.5% are due to *E. coli* [7,10].

14.3.2.2.9   Surgical Site Infections

The pathogens and infections associated with surgery depend on the type of surgical procedure. Most surgical site infections are caused by aerobic Gram-positive cocci while aerobic Gram-negative bacilli (*E. coli*, *Proteus mirabilis*, and *Pseudomonas aeruginosa*) are less common. In clean surgical procedures (no penetration of the GI, gynecologic, and respiratory tracts) the causes of infection are generally *S. aureus* from the environment or microorganisms from the patient's skin. However, in clean-contaminated, contaminated, and dirty surgical procedures, the infection causing pathogens resembles the microflora (aerobic and anaerobic) of the excised organ [25].

14.3.2.3   Genitourinary System

14.3.2.3.1   Intrauterine Device–Associated Infections

The following organisms have been isolated from IUDs removed from PID-associated infections: *S. epidermidis*, *S. aureus*, *E. coli*, group B streptococci, beta-hemolytic streptococcus, *Corynebacterium* spp., *Micrococcus* spp., *Candida albicans*, *Enterococcus* spp., and anaerobic lactobacilli [6,9].

14.3.2.3.2   Penile Implant–Associated Infections

The incidence of penile prosthesis-related infection is 1%–3% for first time insertions and 7%–18% for revisions. Skin flora microbes, most commonly staphylococci, account for the majority of infections whereas Gram-negative bacteria account for 20% [6,10].

14.4 Biofilms and IMDs

14.4.1 Introduction and Definitions

The term "biofilm" was originally coined by engineers who were evaluating the impact of adhered microcolonies to various surfaces, interfering with heat exchange, filtration, and other engineering devices. Essentially, biofilms had limited or no apparent application in health care outcomes. Most publications and findings were evident in environmental, food, and industrial journals. Microbial nature and composition of the biofilm was less important than was the contribution to dysfunction of the engineering or instrumentation impact.

Today, biofilms are loosely defined as a community of microbes attached to a surface, living and supporting the members in a protected matrix enclosed environment. The term "microbial communities" have become more prominent in recent medical literature, addressing the pathogenic features associated with biofilms: (1) a survival characteristic associated with increased population diversity in chronic infections and simultaneously (2) significant increase in resistance to anti-infectives. The organism make-up of the "community" is reflective of the environment or anatomic location of the colonized surface.

As subtleties and nuances of the biofilm have been elaborated, the definitions have had to mirror these changes. Basically, however, the best definition was that proposed by Donlan and Costerton [9], who stated that biofilms were microbially derived sessile communities characterized by

- Cells irreversibly attached to a substratum, or interface, or to each other
- Embedment in a matrix of extracellular polymeric substance they produced
- Exhibited an altered phenotype with respect to the growth rate and gene transcription

Functionally, biofilms could be characterized as

- A primitive type of developmental biology in which spatial organization of the cells within the matrix optimizes the utilization of the nutritional resources available.
- An immobilized enzyme system in which the milieu and the enzyme activities are constantly changing and evolving to appropriate steady state.
- The steady state can be radically altered by applying physical factors such as high sheer.

The components of a medical biofilm reflect their location anatomically, but basically all biofilms have the following relative compositions:

Component	% of Matrix
Water	Up to 97%
Microbial cells	2%–5% (many species)
Polysaccharide	1%–2% (neutral and polyanionic)
Proteins	1%–2%
DNA and RNA	1%–2%
Ions	?
Host	Fibrin, RBCs, WBCs

It is important to recognize that all organisms have the capacity to exist in one of two phenotypes; a planktonic or free-floating phenotype or an attached sessile (biofilm) phenotype. It is equally important to recognize that organisms have a predilection to grow in the biofilm phenotype, and when given the opportunity, 99.9% preferentially select attachment for reasons listed in Table 14.12. Equally important is the recognition that either phenotype is not mutually exclusive, and that the same organisms in the environment or human disease state have a population ratio of biofilm to

TABLE 14.12 (See color insert following page 210)

Comparison of Planktonic and Biofilm Phenotypes

Confliicts of interest: planktonic (P^P) vs. biofilm (P^{BF}) microbial natural selection

1. Individual level	vs.	Group level selection
	(Simpson's paradox)	
kin selection	vs.	Group selection
2. Competitive behavior	vs.	Cooperation
3. Resources consumption	vs.	Economical use
fitness of individual	vs.	fitness of groups
4. No diversity	vs. 99.9%	Population heterogeneity
5. Quorum sensing	vs.	Diffusion sensing
6. Antibiotic resistance via genetics	vs.	Collective action (phenotypic)
7. Growth restraint (contact inhibition)	vs.	Channel formation
8. Decreased virulence	vs.	Increased virulence as group beneficial trait

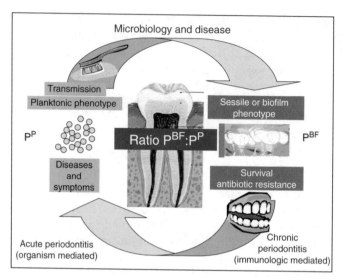

FIGURE 14.2 (See color insert following page 210)
Ratio of biofilm phenotype to planktonic phenotype.

planktonic (Figure 14.2). Lastly, it is important to recognize that the biofilm phenotype contributes to greater than 50% of microbial communities in different habitats which are "viable, but non-cultivable," by standard laboratory techniques (Table 14.13).

TABLE 14.13 (See color insert following page 210)

Percent of Culturable Prokaryotes in Different Habitats

Habitat	Cultured (%)
Seawater	0.001–0.1
Freshwater	0.25
Mesotrophic lakes	0.1–1
Estuarine waters	0.1–3
Activated sludge	1–15
Sediments	0.25
Soil	0.3

Source: Daims, H., University of Vienna (Powerpoint Presentation, unpublished).

Note: The majority of prokaryotes is unculturable and the numbers are based on direct cell counts.

14.4.2 Architecture and Structure

In describing a biofilm, it is easiest to unmask its features as an interface of three components. These are shown in Figure 14.3 which highlight (A) architecture and structure, (B) the physiology and metabolism linked via gradients, and (C) physical properties described by engineering terms recognizing that the biofilm acts as a hydrated polymer.

Architecture and 3D structure of mature biofilms can be visualized by a number of qualitative and quantitative imaging techniques, the most recent of which has been confocal laser scanning microscopy (CLSM) and environmental scanning electron microscope (SEM). The general consensus is that a mature biofilm looks like a "kelp bed" with multiple mushroom or club-shaped structures protruding upward from the abiotic surface. These are quite flexible and are able to bend according to the physical forces of the environment. This flexibility is also recognized as a dispersal characteristic, allowing the biofilm to "ripple" or "move-down" a surface as in endotrachs, or in fact, roll as a tumbleweed, still attached but moving with direction. A 3D rendition of the mature biofilm is shown in Figure 14.4.

The intricacies and complexity of the 3D architecture is defined by an interaction of eight parameters. These are shown in Figure 14.5. Although the importance of all of them is significant, four of these are particularly critical in defining its structure and function. These four are (1) the species of a colonizing organism, (2) the substratum or abiotic surface, (3) nutritional

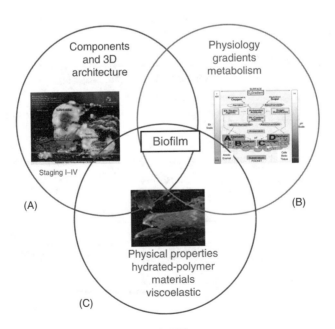

FIGURE 14.3 (See color insert following page 210)
Three components of biofilms.

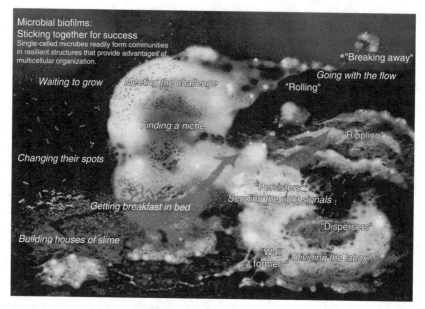

FIGURE 14.4 (See color insert following page 210)
Mushroom structure of a biofilm. (From Hall-Stoodley, L., Costerton, J.W., and Stoodley, P., *Nat. Rev. Microbiol.*, 2, 95, 2004.)

energy or resources, and (4) the mechanical or shear forces in which the biofilm is forming.

It is important to recognize that not every organism of the same general species has the same adherent capability and will form the same robust

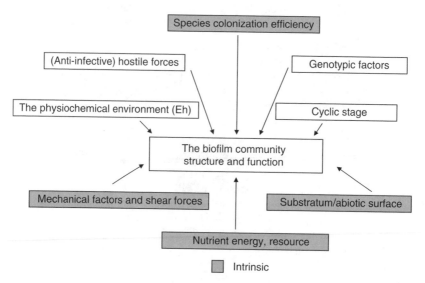

FIGURE 14.5 (See color insert following page 210)
Eight factors involved in biofilm architecture.

biofilm structure. The "efficiency of attachment" is defined by a number of interacting loci and that different strains of the same species will clearly adhere differently. The abiotic surface is critical, given that regular (smooth) surfaces are less likely to encourage biofilm structure than irregular surfaces and surfaces that are hydrophilic will be more likely to allow colonization than those that are hydrophobic. The energy source is often overlooked or misunderstood as a key feature of biofilm growth. A starvation nutrient environment or one with less carbohydrate sources may, in fact, induce a greater upregulation from the planktonic organism to the biofilm phenotype. Hence, it is important to recognize that a 1:10 dilution of nutrient broth may be a better medium to favor biofilm formation than a full strength medium used routinely in clinical laboratories.

The most important factor in defining the 3D architecture of a biofilm is the mechanical force or shear provided by that environment. Shear force is measured in Reynold's units and describes parallel (<5000) vs. vortexing (>5000) forces; it is almost universal that parallel lines of force vs. rippling or turbulent by vortexing will create less Reynold's or less depth or height to the biofilm than increased shear associated with a higher Reynold's force.

14.4.3 Physiology and Metabolism

Figure 14.6 defines, graphically, the interaction of pH and Eh and the concept of "mixed-gradients" within the biofilm, which defines the organism

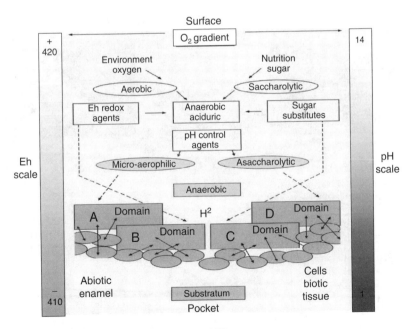

FIGURE 14.6 (See color insert following page 210)
Eh and pH gradients of biofilms.

location. Eh (green) is associated with oxygen tension and correlates with, in a positive Eh, aerobic metabolism and in a negative Eh, anaerobic metabolism. The pH (red scale) is associated with by-products of metabolism and the two acting synergistically define gradients that are favorable for certain combinations of microbes; these microbial communities linked by common metabolic pathways are called "domains." In domains, recent investigations discuss cross-talk, co-adherence, and nutrient gradients. Basically, microbes interact synergistically to metabolize complex host molecules and food webs are established, enabling the efficient cycling of nutrients; bacteria communicate via diffusible small molecules and "cross-talk" with host cells as part of a domain. The environmental heterogenicity of the biofilm encourages genotypic and phenotypic diversity (marsh), which enhances their ability to persist.

14.4.4 Physical Properties

Figure 14.7 highlights the most unrecognized feature of a biofilm or the microbial community; it is best defined by the physical parameters associated with engineering metrics, since the community behaves as a hydrated polymer. A hydrated polymer is best defined by viscoelasticity and rheology. The benefits of viscoelasticity are its dual nature to be (1) viscous (liquid-like) or (2) elastic (amorphous, solid-like) and dispense harmful energy in a

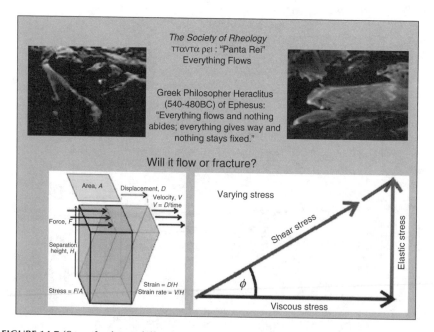

FIGURE 14.7 (See color insert following page 210)
Rheology/viscosity of biofilms.

manner which will allow for greater community survival. Interestingly, the Society of Rheology was formed in 540 BC with the concept that "everything moves and nothing stays fixed." Today, in the biofilm lexicon, rheology describes that the "flow" is often associated with a lava-like movement and the detachment or fracture (Stage IV) would allow for transmission of microbial communities as potential resistant pathogenic aggregate or sessile fragments.

Figure 14.8 highlights two key features of biofilms. First, that biofilms have a cyclical nature and that there is a staging (stages I–IV) that may have significant impact into potential treatment (stage I or stage IV). Second, staging is part of a growth curve that mirrors the universal growth curve of planktonic bacteria (red line). This universal growth curve states that microbes, independent of phenotype (planktonic or biofilm), have a lag (stage I), log (stage II), stationary (stage III), and death phase (stage IV). Normally, domains within the biofilm will have a growth cycle of approximately 72–96 h, but their ultimate nature and time of maturity (stage III) is defined by (1) the shear forces that promote its 3D structure, (2) its microbial diversity, and (3) hostile forces in the environment such as antibiotics, white cells, and biocides. Evident, also, in Figure 14.8 is a reaffirmation that planktonic (yellow) and biofilm (blue) phenotypes are not mutually exclusive but rather the ratio may be an important feature of chronicity and associated

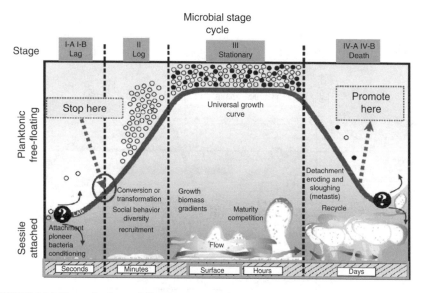

FIGURE 14.8 (See color insert following page 210)
Four stages of biofilm growth: I Attachment (lag), II growth (log), III maturity (stationary), and IV dispersal (death).
Note: Intra oral sessile (P^{BF}) (■■) and planktonic (P^P) (▬▬) life forms are not mutually exclusive, but biofilms are the preferred growth vehicle. (From John Thomas, West Virginia University, unpublished.)

pathogenicity. Additionally, if biofilms are associated with higher antibiotic resistance, interference with the up-regulation at stage I to II or promotion of stage IV to the planktonic phenotype, may be features in biofilm disease management. As the biofilm matures there is an increase in biodiversity and population density coordinated through "quorum sensing" via small communicator molecules.

14.4.5 Pathogenicity

It is important to recognize that biofilms play a central role in survivability and providing means for organism diversity, growth, and disease process. Biofilms are well implicated in industrial and environmental problems (sick building syndrome, sick instrument syndrome), the dental community (periodontal disease), and most recently, the medical community (otitis media, endocarditis, and indwelling medical devices). CDC states that as many as 80% of chronic infections and hospitalized infections are associated with biofilms. The mechanism of pathogenicity, however, is not so universal and a number of schemes have been devised based on unique characteristics of biofilms, i.e., "structure equals function." These include increased organisms and density per unit of measurement, linked genes to pathogenic products including enzymes, motility and new virulence markers, increased genetic transfer within a protected environment, and loss of susceptibility, i.e., increased antibiotic resistance up to 1000-fold and increased resistance to immunologic capabilities such as white cell phagocytosis.

Perhaps the most unrecognized is the emerging concept that the greater the ratio of biofilm phenotype to planktonic phenotype, the greater the pathogenicity of the community. This may be illustrated in Figure 14.9, which

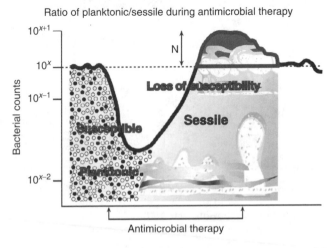

FIGURE 14.9 (See color insert following page 210)
Effect of antimicrobial therapy on planktonic phenotype.

underscores early (planktonic) vs. late (sessile) in the presence of antibiotic therapy. Therapy reduces the susceptible planktonic isolates, enhances up-regulation of a higher organism concentration and density of a resistant biofilm phenotype—only reversed by removal of selective pressure via therapy.

14.5 Coatings and Anti-Infective Devices

Strategies used to counter device-related infections are directed against the three stages in the pathogenesis of the infection process: bacterial adhesion, bacterial accumulation or colonization, and biofilm formation (Figure 14.8). IMDs provide an abiotic surface or substratum, which is one of the eight key ingredients in establishing a biofilm of firmly bound sessile, single, or multispecies communities (Figure 14.5) [6].

In the mid-1980s it became obvious that biofilms, chronic infections, and indwelling medical devices were beginning to have a significant impact on individual cases and on public health. One of the most pursued approaches for the prevention of infections associated with IMDs is inhibiting adherence. Accordingly, a variety of materials have been attempted to be impregnated, coated, or attached in some manner that will neutralize the colonization or adherence of organisms to abiotic surfaces.

Strategies aimed at preventing bacterial adhesion include modifications to the materials used for the medical device and modification of the material surface. The chemical composition of the device material is an important consideration because the characteristics of the IMD materials (i.e., PVC, Teflon, polystyrene) play a crucial, but often unrecognized, role in adhesion (Table 14.14). The goal is to produce a surface texture that would reduce bacterial adherence as well as protein and platelet adherence. Polystyrene surfaces modified with copolymers such as poly(ethylene oxide) and poly(propylene oxide) have reduced bacterial adherence from 70% to 95%. Various hydrophilic surfaces, including the addition of surfactants to increase water uptake, have also been shown to reduce bacterial adherence. However, it has become increasingly clear that it is probably impossible to obtain a surface with zero adhesion [26].

There is a long history of incorporating anti-infective agents, such as antibiotics, disinfectants, antimicrobial biomaterials, and metals, into medical devices. Table 14.15 shows the most common of anti-infectives and is segregated by the type of anti-infective used on the IMD. The anti-infective products are designed to interfere at different steps involved with biofilm architecture and growth. Others prevent the process from occurring all together. It is also important to recognize that the anti-infective activity needs to impact both bacterial phenotypes—planktonic and biofilm. It may not be important to sterilize the environment, but rather reduce the bacterial challenge to not be pathologic since there may be a threshold, below which organisms are not associated with disease.

TABLE 14.14

Chemical Products Used in Medical Devices

Common Name	Chemical Name	Specific Gravity	Stiffness (Flex Mod; $\times 10^3$ psi)	Cost/lb	Use(s)
PVC	Polyvinyl chloride (flexible)	1.25	<10	$0.70	IV fluid lines, bag material, stamped sheet protectors
PE	Polyethylene	0.92–0.96	22	$0.60	Tubing, injection molded parts
Teflon	Polytetrafluoroethylene	2.15		$7.50	Joints, tubing, sleeving, wire insulation
PVDF (Kynar)	Polyvinylidene difluoride	1.77	260	$7.25	Wire insulation
PP	Polypropylene	0.90	190	$0.75	
PS	Polystyrene	1.05	45	$0.70	
PA66	Polyamide (nylon) 66	1.14	200		Encasements, fittings, brackets
PC	Polycarbonate	1.20			Luers, fittings
ABS	Acrylonitrile butadiene styrene	1.05	300	$0.75	Encasements
PETG	Polyethylene terephthalate glycol	1.27	300	$0.72	Packaging
TPU	Polyurethane	1.02			Tubing, catheters

14.5.1 Antimicrobial Agents

The development of antimicrobial polymers has focused on prevention of microbial colonization rather than microbial adhesion [26]. The antimicrobial agent can be bound superficially to the device or directly incorporated into the structure of the polymer. The antimicrobial is then released into the vicinity of the device and the activation of the antimicrobial is triggered when it comes into contact with an aqueous environment. This eliminates adherent microorganisms on the device and in the surrounding area as long as the antimicrobial is present at sufficient concentrations. This tactic is usually effective for early-onset infections.

Recently, the medical industry has aimed its efforts toward antimicrobial-coated catheters. The antimicrobial substance is coated on the exterior of the device, coated in the lumen of the device, or incorporated into the device polymer. Investigations (in vitro and animal) of Hydrocath (a central venous, hydrophilic-coated polyurethane catheter) showed the prevention of colonization of *S. epidermidis* and *S. aureus*. Raad et al. determined that when compared with a control, untreated catheter, the Cook Spectrum catheter coated on the inner and outer surfaces with minocycline and rifampicin decreased colonization (26%–8%) of Gram-negative, Gram-positive, and yeast organisms due to the synergistic effect of the combined antimicrobials.

TABLE 14.15

Anti-Infectives Commonly Used on IMDs

Anti-Infective Coating/Material	Indwelling Medical Device
Metals	
Silver	Endotracheal tubes, urinary catheters, intravascular catheters
Antibiotics	
Minocycline–rifampin	Impregnated catheters
Vancomycin (with or without heparin)	Catheter locks
Dicloxacillin	Intravascular catheter
Clindamycin	Intravascular catheter
Cefazolin with benzalkonium chloride	Catheter
Triclosan	Sutures
Anti-infective biomaterials/polymers	
Polypeptides	
Chlorohexidine	Sponge dressings
Chlorohexidine–silver sulfadiazine	Impregnated catheters
Iodinated alcohol	Catheter hubs
Fusidic acid	Intravascular catheter
Hydrophilic-coated polyurethane with glycopeptides teicoplanin	Central venous catheter
Irgasan (disinfectant)	Catheter
Hydrophilic hydromer-coating	CNS shunts
Other	
Polypeptides	Various experimental biomaterials

Mermel found that infection was reduced at least 5% by using minocycline–rifampin-impregnated catheters, then for every 850 catheters there would be a savings of $500,000 [27].

Although most of antimicrobial coated catheters have decreased early-onset infection, there is still the risk for developing resistance—particularly if the antibiotic used on the catheter is also a first-line drug for infection therapy [26].

14.5.2 Anti-Infective Biomaterials/Polymers

von Eiff et al. found that with a polyvinylpyrrolidone-complex version of the Hydrocath catheter, in vitro adherence of *Staphylococcus* spp., *E. coli*, and *Candida* spp. were inhibited when the iodine was released. Several investigations are reporting the most efficient catheter to be the CHSS catheter, which has an external and internal coating of the antiseptic chlorhexidine with silver sulfadiazine. Although this catheter is still being tested in clinical trials, early reports indicated that there is significant reduction in colonization and infection [26]. Mermel found that chlorhexidine–silver sulfadiazine impregnated catheters reduced infections from 5.2% to 3%, which (for every 300 used) resulted in a savings of $60,000, the prevention of seven CRBSIs and the prevention of one death [27].

Europe is using 3% iodinated alcohol in catheter hubs that release a small amount of iodine (0.024 mg) into the bloodstream when the hub is punctured. Mermel reported on a prospective, randomized trail that resulted in a fourfold reduction in infection when the hub was in place for 15–16 days [27].

14.5.3   Metals

Silver has long been known to have potent antimicrobial activity without causing bacterial resistance. It is effective against fungi, yeast, and algae, as well as bacteria. A company known as AgION Technologies has encased silver ions in a powdery ceramic called zeolite. The silver ions are released from the zeolite by exchange with other positive ions in the environment. The silver-containing zeolite can be incorporated into fibers or coatings for medical devices [28].

14.5.4   Other

14.5.4.1   *Polymicrobial Peptides*

Antimicrobial peptides are naturally occurring host defense peptides that have demonstrated to have activity against a wide spectrum of infectious bacterial, viral, fungal, and parasitic pathogens. They also function as immune modulators. They occur in all forms of life ranging from bacteria to plants, invertebrates to vertebrates, and mammals. Antibacterial peptides are called bacteriocins and are extremely potent. They tend to have some characteristic properties in common. They are usually cationic and have an overall positive charge; they are amphipathic, hydrophobic, and flexible, enabling them to attack the bacterial cell wall [29].

Several antimicrobial peptides have been developed and have entered into clinical trails. As yet, none have been granted FDA approval for clinical use. An indolicidin-based antimicrobial peptide variant, MBI-226 (Omiganin) had been developed by Migenix for the treatment of catheter-related infections. Catheter colonization was reduced by 40% and tunnel infections were reduced by 50%. The peptide is also being tested in a gel-based format. A variant of histatin, a naturally occurring peptide in saliva, is being tested by Pacgen and an inhalation treatment for *Pseudomonas aeruginosa* infections in cystic fibrosis patients. Other antibacterial peptides are being tested in animal models for osteomyelitis, wound healing, and polymicrobial oral mucositis [29].

In an attempt to target multidrug-resistant organisms, such as *Pseudomonas*, MRSAs, and vancomycin-resistant *S. aureus*, companies such as Ceragenix are constructing new peptides like the squalamine-based cationic steroid antibiotics that share the biophysical properties of antimicrobial peptides, such as net positive charge, amphipathicity, and selectivity for prokaryotic membranes [29].

A new and creative approach to battle orthopedic, device-related infections is the development of a novel, nanocoating-based antimicrobial therapy. Multilayer nanocoatings of oppositely charged polypeptides, poly(L-lysine) and poly(L-glutamic acid), can be applied to a stainless steel orthopedic device. The polypeptide nanocoatings self assemble, layer-by-layer, on the device. The primary target for this strategy is *S. aureus*. Therefore the antibiotic, cefazolin, was chosen. However, any combination of drugs or antibiotics can be incorporated into the nanocoatings. The cefazolin ions bind to the polyelectrolytes in the coatings and are released by modulating the pH. The rationale is that as the cefazolin is released from the polyelectrolyte layers, planktonic bacteria in the vicinity of the device will be killed before they can initiate biofilm formation. This technology has the advantage of controlled antibiotic release into the vicinity of the device. This would enable the use of effective antibiotics that are not suitable for systemic use due to toxicity or to the adverse effects of the high concentrations that would be required to treat antibiotic-resistant biofilms. In addition, the polypeptides themselves produce no adverse reactions and are biodegradable [30].

14.5.4.2 Antibiotic Lock Therapy

CVCs have become an integral part of medical management for a variety of patients, including those requiring long-term parenteral nutrition, chemotherapy, or hemodialysis. Infection of the catheter or catheter hub and catheter-related septicemia are major complications for patients with permanent or semi-permanent indwelling catheters. The replacement of an infected catheter is not a feasible option for many patients. Systemic and local antibiotic therapy is often used to prevent catheter removal in patients relying on indwelling devices for long-term medical management. One of every 20 CVCs inserted results in at least one systemic infection. Catheter-related infections are reported to occur in 2%–40% of patients with indwelling devices. These infections are associated with a mortality rate of up to 25%, an estimated cost of $28,690 for each episode of catheter-related sepsis, and an average increase of 6.5 days of hospitalization for critically ill patients. The number of catheter-related bacteremia cases is estimated to be more than 35,000 per year [31].

The antibiotic lock technique is a controversial method used to prevent and treat catheter-related infections. This technique involves instilling an antibiotic solution into the catheter hub and allowing the solution to dwell for a certain length of time in order to sterilize the catheter lumen and eradicate organisms from an infected line, or to prevent a line from becoming colonized and developing into an infected device. The goal of this therapy or treatment is to prolong the life of the catheter while reducing morbidity and the costs of managing catheter-related infections. In addition, this technique may avoid the toxic effects of systemic antibiotic administration, while also reducing the risk of bacterial resistance. The CDC includes the use of the antibiotic lock technique as a therapeutic option for infected CVCs where the device is not removed and as prophylaxis in selected patient populations [31].

Bestul and VandenBussche [31] have conducted an extensive literature review on antibiotic lock technique covering the period from January 1972 to January 2004. The studies are divided into two groups: in vitro and clinical. The clinical studies are further separated into trials designed to evaluate prevention of CVC infections and trials designed to evaluate treatment of infected CVCs. The studies employed a wide range of infectious models and assessment methods. Study populations also varied significantly. The in vitro studies evaluated various concentrations, dosages, and combinations of selected antibiotics used for varying dwell times, and their activity against causative organisms. Most of the antibiotics evaluated maintained stability or efficacy in polyurethane CVCs for at least 7 days. Antibiotics such as aminoglycosides and fluoroquinolones, known as concentration-dependent killers, are optimal at concentrations 8–10 times the minimum inhibitory concentration (MIC). It was not necessary to maintain high concentrations of these antibiotics throughout dosing intervals or dwell times within a catheter. In comparison, beta lactams and vancomycin require maintenance of a concentration above the targeted organism's MIC for most of the dosing interval for optimal killing and are known as time-dependent killers. Antibiotics used in antibiotic lock technique are most effective if high drug concentrations are maintained for extended periods of time in order to penetrate the microbial biofilm. Stability and compatibility become important factors when selecting appropriate antibiotics. Another highly controversial issue is the use of heparin in the antibiotic lock solution. Based on study results, heparin may increase the efficacy of the antibiotics used [31].

Most of the clinical studies conducted did not include heparin in the antibiotic lock solutions and most of the prevention trials used antibiotic flush solutions rather than true antibiotic lock therapy. The flush solutions are infused systemically as opposed to being contained within a catheter lumen and withdrawn before utilizing the catheter for drug administration. In spite of these inconsistencies, preventive therapy was successful in inhibiting catheter colonization when administered with daily flush solutions or 1 h dwell times every 1–2 days. Evidence supported the use of vancomycin 25 μg/ml in combination with heparin 9.75 U/ml to prevent Gram-positive infections with the addition of ciprofloxacin 2 μg/ml to prevent Gram-negative infections [31].

The antibiotic lock technique is well tolerated and generally effective in treating CVC-related infections that do not involve soft tissue or that are not fungal in origin, without removing the indwelling device. Catheter clearance may be achieved after 1–2 weeks of antibiotic lock therapy alone or in combination with systemic antibiotics [31].

14.6 Evidence-Based Benefits

14.6.1 Strategies Designed to Prevent Biofilm Formation in IMDs

The challenge presented by IMD-related infection has required the combined efforts of clinicians, surgeons, dentists, microbiologists, immunologists, and pharmacists working in all sectors of medical research: academic, clinical, public, private, and commercial. This combined effort has resulted in a multitude of products, treatments, drugs, therapy regimens, etc. We will approach this information by first discussing IMD-related protocols that have undergone clinical trials and are currently in use. We will present a table of selected clinical studies followed by a discussion of IMD-related infections for the following devices: endotracheal tubes, surgical site wounds, CVCs, and urinary catheters.

In the second part of this section, we will discuss strategies that have been published in the literature and are currently being developed and evaluated in various research facilities, but have not yet proceeded to clinical trails.

14.6.1.1 *Oral and VAP-Associated Studies*

The major route for acquiring endemic VAP is oropharyngeal colonization by the endogenous flora or by pathogens acquired exogenously from the intensive care unit environment. The stomach represents a potential site of secondary colonization and reservoir of nosocomial Gram-negative bacilli. Aspiration of microbe-laden oropharyngeal, gastric, or tracheal secretions around the cuffed endotracheal tube into the normally sterile lower respiratory tract results in most cases of endemic VAP. Strategies to eradicate oropharyngeal and intestinal microbial colonization, such as chlorhexidine oral care, prophylactic aerosolization of antimicrobials, or selective aerodigestive mucosal antimicrobial decontamination, have all been shown to reduce the risk of VAP. In spite of the continued systemic use of broad-spectrum antibiotics for prophylaxis (short-term) and treatment (long-term), VAP remains the most formidable of all nosocomial infections and continues to occur at a high rate [32] (Table 14.16).

Selective digestive tract decontamination is a method where topical non-absorbable antibiotics are applied to the oropharynx and stomach. This method is primarily aimed at the prevention of VAP. The rationale for SDD is that VAP usually originates from the patients' own oropharyngeal microflora. The combination of topical and systemic agents more effectively reduces the incidence of VAP. Even though the incidence of VAP is reduced, mortality is often not reduced in individual trials. The most important drawback of SDD is the development of resistance and an increased selection pressure toward Gram-positive pathogens following SDD regimens [33].

The administration of nebulized antibiotics has demonstrated moderate success. The use of nebulized antibiotics in VAP allows attainment of higher

TABLE 14.16

Clinical Trials for Oral and VAP-Associated Medical Devices

Reference	Device	Treatment	Result
ORAL			
[56]	Ankylos implant	Antibiotic prophylaxis at time of implantation	No advantage to antibiotic prophylaxis
VAP			
[57]	Mechanical ventilation and Endotracheal intubation	Gingival and dental plaque decontamination	Decontamination decreased oropharyngeal colonization, but not infection rates
[58]		Initial antibiotic monotherapy vs. combination therapy	Initial therapy with antipseudomonal penicillins plus beta-lactamase inhibitors lowered mortality rates
[59]		Intravenous levofloxacin treatment for early-onset VAP	Levofloxacin may represent a valid alternative to non-Pseudomonal beta-lactams or aminoglycosides
[60]		Monotherapy of cefepime vs. combination therapy with cefepime with and levofloxacin	Cephalosporin monotherapy was as effective as combination therapy with cephalosporin and aminoglycoside or fluoroquinolone
[34]		Nebulized gentamicin vs. parenteral cefotaxime or defuroxime	Nebulized gentamicin was more effective in preventing biofilm formation
[61]		Selective decontamination of digestive tract (SDD) and aerosolized antibiotics	SDD appears to be effective, but remains controversial Aerosolized antibiotics appeared to be effective
[62]		Selective decontamination of digestive tract (SDD)	SDD spears to reduce the incidence of VAP, but should be restricted to specific circumstances and pop

pulmonary drug concentrations than parenteral administration of antimicrobials. When compared to SDD, it may be less expensive and easier to administer. Other potential benefits may include reduced morbidity and exposure to systemic antimicrobials and improved survival rates [34].

These studies have been somewhat disappointing. The magnitude of the impact has been relatively minor and subdued. This may reflect the multi-species biofilm which may be evident. The presence of mucus in the airway that binds to the endotracheal tube provides bacteria with a surface that promotes adherence and the formation of a biofilm that compromises the effectiveness of antibiotic therapy. Furthermore, culture results have been based on planktonic recovery and may not really reflect the growth of the organism on the device.

Since no strategy has demonstrated proven success, it may be that a multifaceted approach is necessary and that combinations of preventive measures are required.

14.6.1.2 Wound Studies

Staphylococcus aureus is the most common nosocomial pathogen and is responsible for approximately one-third of hospital-acquired bacteremias. The emergence of strains with multidrug resistance, including resistance to vancomycin, presents the medical community with a major public health problem. Recently, the emergence of a more virulent strain associated with community-acquired infections threatens to become a global pandemic. Over the past 50 years, *S. aureus* has been a dynamic human pathogen and has earned the deepest respect from both microbiologists and clinicians. It possesses "weapons of mass destruction" and uses them effectively as part of an "anti-immunization insurgency" [35]. The strategies listed above represent several unique approaches to this problem (Table 14.17).

TABLE 14.17

Clinical Trials of Wound-Associated Medical Devices

Reference	Device	Treatment	Results
[63]	Staph VAX vaccine for *Staphylococcus aureus*	Phase III clinical trail to determine efficacy of vaccine	Initially, vaccine was effective for up to 10 months. Final result: not effective.
[64]	MRSA surgical site infection	Linezolid vs. vancomycin therapy	Intravenous or oral linezolid was superior to vancomycin for MRSA infections.
[65]	Wound healing/infection	Vacuum-assisted closure wound therapy	This therapy had a positive effect on wound healing, but did not reduce bacterial load.
[37]	Wound matrix device	Matrix device made from intestinal submucosa	Effective in treatment of dermal wounds and possibly serving as a barrier to infection.

The conjugate vaccine, StaphVAX, included the two most prevalent capsular polysaccharides, types 5 and 8, coupled to a carrier-protein efficient in promoting a Th2 response. Although early reports, such as the reference above, were favorable, StaphVAX ultimately failed the Phase III clinical trial. We have included it here because of the success of other protein–polysaccharide conjugate vaccines that have been highly effective against *Haemophilus influenzae* type b, *Streptococcus pneumoniae*, and *Neisseria meningitides* [36]. We have also included this study to emphasize the importance of controlled clinical trials as opposed to pilot studies, and to emphasize the desperate need for development of an effective vaccine.

Vacuum-assisted closure has become a new technique in the challenging management of contaminated, acute, and chronic wounds. Although the technique has a positive effect on wound healing, its effect on wound infection rates needs to be established. The study cited above found that wounds treated with the vacuum-assisted closure technique showed a significant decrease in nonfermentative Gram-negative bacilli, but a significant increase in *S. aureus*. This is a serious issue that warrants further study.

The development of a biologically derived matrix for the regeneration of tissue structures is known as "smart modeling." This phenomenon is remarkable because the new tissue generated by the host is specific to the site of implantation. The submucosal layer consists primarily of a collagen-based extracellular matrix that provides structural support, stability, and biochemical signals to the regenerating mucosal cell layer. All of the resident cells are lysed and the biomaterial is sterilized by treatment with ethylene oxide and lyophilized. This allows long-term storage without destroying its ability to support wound healing and tissue repair. It is biocompatible and forms a partial barrier to dehydration and infection. Further studies are needed to evaluate the anti-infective properties of this biomaterial [37].

14.6.1.3 *Central Venous Catheter Studies*

Duration of catheter placement was the strongest risk factor for both colonization and CRBSI. Catheters used for TPN had a significantly lower level of colonization (1.6%) compared to catheters used for monitoring, administration of fluids or medication, or dialysis (12%–21%). Catheters placed in the subclavian vein, rather than the internal jugular vein, had lower rates of infection [38]. This study is in contradiction to previous studies which have repeatedly shown that catheters used for TPN are at higher risk for infection (Table 14.18).

The randomized controlled trial conducted by Bong et al. was designed to evaluate silver iontophoretic catheters in a long-term study (greater than 10 days). Silver ions, in addition to carbon and platinum particles, are impregnated into the structure of the OLIGON polyurethane catheters. The silver

TABLE 14.18

Clinical Trials for Central Venous Catheter-Related Bloodstream Infections

Reference	Device	Treatment	Results
[50]	Central venous catheter (CVC)	Chlorhexidine–silver sulfadiazine-coated catheter vs. standard catheter	Antiseptic-coated catheter significantly reduced colonization and infection
[39]	CVC	Efficacy of silver iontophoretic CVC	No significant difference between silver-coated and control catheters
[40]	Triple lumen CVC	Heparin-coated vs. chlorhexidine and silver sulfadiazine-coated CVCs	Chlorhexidine and silver sulfadiazine-coated CVCs reduced colonization of gram-positive cocci and *Candida* spp.
[51]	CVC	Comparison of CVC-related infection rates for total parenteral nutrition (TPN) vs. other uses	Colonization rates were lowest in CVCs used for TPN only
[52]	CVC	Effect of chlorhexidine and silver sulfadiazine-coated CVC on bloodstream infection	Antiseptic-coated catheters were less likely to be colonized at time of removal than controls

particles interact with the body fluid, resulting in the movement of electrons from silver to platinum, producing a sustained release of silver ions [39].

In the comparison between heparin-coated and chlorhexidine/silver sulfadiazine-coated catheters, Carrasco et al. found that the heparin-coated catheters had an incidence of CRBSIs per 1000 catheter days of 3.24. The chlorhexidine and silver sulfadiazine-coated catheters had an incidence of 2.6 [40].

The antimicrobial venous catheters have demonstrated some success in reducing bloodstream infections primarily due to Gram-positive cocci. The results are encouraging, but bacterial adherence and colonization still occur at an unacceptable level.

14.6.1.4 *Urinary Catheter Studies*

UTIs account for 30%–40% of nosocomial infections resulting in morbidity, mortality, and increased length of hospital stay [41]. The number of patients with urinary catheters is so large that catheter-associated UTI is the most common cause of infections in hospitals and other health-care facilities [42]. The results of the clinical trials listed above indicate that the antimicrobial-coated catheters currently in use appear to reduce the incidence of UTI and were shown to be cost-effective in one study. There has been no direct comparison study between nitrofurazone-coated and silver alloy-coated catheters. The nitrofurazone-coated catheters reduced infection rates slightly more than the silver-alloy coated catheters in the separate trials [43] (Table 14.19).

Stickler [42] has raised concerns about the reintroduction of chlorhexidine use since extensive use of chlorhexidine in the past resulted in chlorhexidine-resistant and multidrug-resistant strains *Proteus mirabilis*. In addition, chlorhexidine was not effective against the Gram-negative organisms that colonize the perineal skin. *P. mirabilis* is an aggressive pathogen with a predilection for the upper urinary tract. They are able to invade the kidneys and induce pyelonephritis. Chronic infection can result in the formation of bladder and kidney stones. All types of catheters are susceptible to rapid blockage and encrustation by crystalline *P. mirabilis* biofilms. Stickler also points out that *Pseudomonas aeruginosa* and *P. mirabilis* are nitrofurazone-resistant [42].

Vascular catheters coated with minocycline and rifampin or silver sulfadiazine and chlorhexidine have been somewhat successful in preventing bloodstream infection that is primarily caused by staphylococci. Their success in the urinary tract, where there is a mixed population of resistant Gram-negative organisms, is an entirely different scenario. The urinary tract presents a much greater challenge to biocide-containing devices.

TABLE 14.19

Clinical Trials for Catheter-Associated Urinary Tract Infections (CAUTIs)

Reference	Device	Treatment	Result
[53]	Urinary catheter	Efficacy of silicon-based, silver ion-impregnated Foley catheter	This catheter was not effective in preventing CAUTIs
[66]	Urinary catheter	Comparison of nitrofurazone-coated catheter compared to silicone control	No difference in rates of UTI, but lowered infection in elderly and catheterization for 5–7 days
[54]	Urinary catheter	Efficacy of nitrofuroxone-coated catheter for prevention of CAUTIs	Antimicrobial catheter did reduce CAUTIs in orthopedic and trauma patients
[41]	Urinary catheter	Efficacy of silver-coated catheters	Silver-coated catheters reduced CAUTIs by 32% and were cost-effective
[55]	Urinary catheter	Comparison between catheters coated with hydrogel/silver salts and control	No significant difference in the rate of CAUTIs between the two groups
[43]	Urinary catheter	Meta analysis of antimicrobial urinary catheters	Antimicrobial catheters can reduce CAUTIs during short-term use

14.6.2 Published Strategies Currently Being Developed and Evaluated

In the future, treatments that inhibit the transcription of biofilm-controlling genes might be able to inhibit these infections [44]. It may also be possible to disrupt transcription of the genes responsible for multiple drug-resistant pumps. A possible strategy to prevent biofilm formation is disruption of transcription of the genes involved in quorum sensing or cell-to-cell communication. Signaling molecules, such as acyl-homoserine lactones, are involved in biofilm architecture and detachment. It may be possible to disrupt these quorum-sensing systems by using antagonists such as furanones [9].

Research is also being conducted to try to understand how some organisms control other organisms, such as the interaction between *P. aeruginosa* and *C. albicans* in cystic fibrosis infections [45].

The idea of employing colonization of a catheter or medical device by a benign microorganism has been exploited by Trautner et al. Urinary catheters were coated with a colicin-producing strain of *E. coli* K12. This bacteriocin production prevented colonization of the catheter by an uropathogenic strain of *E. coli* obtained from clinical isolates that were colicin-susceptible [46].

Attempts have been made to physically dislodge organisms from the biofilm. Ultrasound treatment was shown to remove 95% of *Pseudomonas diminuta* attached to ultrafiltration membranes. Ultrasound was also used to disrupt *P. aeruginosa* biofilms on steel and then followed by treatment with gentamicin. Disruption of the biofilm greatly increased the efficacy of the antibiotic. The application of low-strength electric fields combined with a low current density has also increased antibiotic efficacy [9].

Mixtures of enzymes, such as proteinase K and dispersin, have been used to eradicate biofilms consisting of multiple organisms in vitro. If it is possible to identify the specific polysaccharides produced by the organisms in the biofilm, then the corresponding enzymes could be used to digest the biofilm. In this manner, alginate lyase was used to permit more effective diffusion of gentamicin and tobramycin through alginate, the biofilm polysaccharide produced by *P. aeruginosa* [9].

14.7 Conclusions and the Future

Current treatment strategies for IMD infections still primarily depend on a combination of antimicrobial therapy and surgical intervention, which is both inadequate and costly. The ultimate solution is to prevent the infection from occurring. This review of current research and future avenues of investigation reveals how much progress has been made in the study and understanding of biofilm formation and development at the molecular or genetic level. Hopefully, this new information will be translated into useful therapeutic strategies that will prevent bacterial adherence and colonization of IMDs via the biofilm phenotype.

The use of indwelling medical devices will continue to grow, energized by an aging population who can benefit from these devices and the simultaneous development of new devices. Conversely these devices offer abiotic surfaces that are natural habitats for communities of microbes and biofilms. The biofilms may either alter or eliminate the function of the device or act as a reservoir for organisms and seeding of surrounding tissue. The resultant infection can be catastrophic for the patient in both morbidity and mortality and bring a significant cost factor to the health-care system.

It is important to separate monospecies biofilms from multispecies biofilms and the length that the device will be used. Most nosocomial associated infections are monospecies biofilms, but this may be a reflection of the proximity of the device to normal flora with the potential acquisition of multiple species producing multiple co-aggregates and co-biofilms. The longer the insertion, the potential higher risk of co-biofilms; and hence the establishment of the 30 day benchmark as a tool to evaluate usage of impregnated devices. Multispecies biofilms offer a much different challenge to management than single species biofilms.

The use of DRGs was an attempt to standardize comparisons and relative costs to various devices that become colonized and produce a subsequent infection. It is important to remember that there are at least two costs associated with the outcome: the cost of the device being placed into service for a particular disease category, and the cost of removing it or managing it due to an infection. The cost can also be described as a "disease burden" and involves the total patient characterization, the cost of the insertion, cost of the infection, and cost of the quality of life (QOL) to the individual associated. Calculating the burden of disease is, perhaps, the best way to describe the impact and of total health care of an individual following the biofilm associated infection.

There is a problem associated with biofilms and the device colonization. Present-day methodology in microbiology does not emphasize biofilm recovery or establish the guidelines for biofilm detection either through viable methods or molecular methods that will take into account "viable, but not cultivable." With the recognition of significant cost and multispecies biofilms, it is imperative that better techniques and methods be established for the recovery of this community of microbes in diagnostic microbiology.

References

1. Center for Disease Control and Prevention, National Center for Health Statistics, Health, United States, 2005, http://www.cdc.gov/nchs/hus.htm
2. Nicolle, L.E., Preventing infections in non-hospital settings: Long-term care. *Emerg. Infect. Dis.*, 7, 205, 2001.

3. Friedman, N.D. et al., Health care-associated bloodstream infections in adults: A reason to change the accepted definition of community-acquired infections, *Ann. Intern. Med.*, 137, 791, 2002.

4. Tokars, J.I. et al., Prospective evaluation of risk factors for bloodstream infection in patients receiving home infusion therapy, *Ann. Intern. Med.*, 131, 340, 1999.

5. Rhinehart, E., Infection control in home care, *Emerg. Infect. Dis.*, 7, 208, 2001.

6. Thomas, J.G., Ramage, G., and Lopez-Ribot, J.L., Biofilms and implant infections, in *Microbial Biofilms*, Ghannoum, M. and O'Toole, G., Eds., ASM Press, Washington, 2005, Chapter 15, pp. 269–293.

7. Hedrick, T.L., Adams, J.D., and Sawyer, R.G., Implant-associated infections: an overview, *J. Long-Term Eff. Med. Implants*, 16, 83, 2006.

8. Garcia, R., A review of the possible role of oral and dental colonization on the occurrence of health care-associated pneumonia: underappreciated risk and a call for interventions, *Am. J. Infect. Control*, 33, 527, 2005.

9. Donlan, R.M. and Costerton, J.W., Biofilms: survival mechanisms of clinically relevant microorganisms, *Clin. Microbiol. Rev.*, 15, 167, 2002.

10. Vinh, D.C. and Embil, J.M., Device-related infections: a review, *J. Long-Term Eff. Med. Implants*, 15, 467, 2005.

11. The University of Michigan Kellogg Eye Center, online source: http://www.kellogg. umich.edu/patientcare/conditions/contact.lenses.html 2006.

12. Donlan, R.M., Biofilm formation: a clinically relevant microbiological process. *Clin. Infect. Dis.*, 33, 1387, 2001.

13. Didie, M.A. and Sarwer, D.B., Facts that influence the decision to undergo cosmetic breast augmentation surgery, *J. Women's Health*, 12, 241, 2003.

14. Donlan, R.M., Biofilms and device-associated infections, *Emerg. Infect. Dis.*, Special issue, 7, 277, 2001.

15. Kojic, E.M. and Darouiche, R.O., Candida infections of medical devices, *Clin. Microbiol. Rev.*, 17, 255, 2004.

16. Burk, J.P., Infection control—a problem for patient safety, *N. Eng. J. Med.*, 7, 651, 2003.

17. Darouiche, R.O., Device-associated infections: a macroproblem that starts with microadherence, *Clin. Infect. Dis.*, 33, 1567, 2001.

18. Darouiche, R.O., Treatment of infections associated with surgical implants, *N. Engl. J. Med.*, 350, 1422, 2004.

19. Weinstein, R.A., Adding insult to injury: device-related infections, in *38th Annual Meeting of the Infectious Diseases Society of America*.

20. Chaiyakunapruk, N. et al., Vascular catheter site care: The clinical and economic benefits of chlorhexidine gluconate compared with povidone iodine, *CID*, 37, 764, 2003.

21. Kollef, M.H., The prevention of ventilator-associated pneumonia, *N. Eng. J. Med.*, 340, 627, 2006.

22. Sheretz, R.J., Pathogenesis of vascular catheter infection, in *Infections Associated with Indwelling Medical Devices*, 3rd ed., Waldvogel, F.A. and Bisno, A.L., Eds., ASM Press, Washington, 111, 2000.

23. Denver, L.L. and Johanson, W.G., Infections associated with endotracheal intubation and tracheostomy, in *Infections Associated with Indwelling Medical Devices*, 3rd ed., Waldvogel, F.A. and Bisno, A.L., Eds., ASM Press, Washington, 307, 2000.

24. Yogev, R. and Bisno, A.L., Infections of the central nervous system shunts, in *Infections Associated with Indwelling Medical Devices*, 3rd ed., Waldvogel, F.A. and Bisno, A.L., Eds., ASM Press, Washington, 231, 2000.

25. Nichols, R.L., Preventing surgical site infections: a surgeon's perspective, *Emerg. Infect. Dis.*, 7, 220, 2001.
26. von Eiff, C. et al., Infections associated with medical devices, *Drugs*, 65, 179, 2005.
27. Mermel, L.A., New technologies to prevent intravascular catheter-related bloodstream infections, *Emerg. Infect. Dis.*, 7, 197, 2001.
28. Kher, U., Hygiene's silver bullet. *Time*. Bonus Section, A2, January, 2006.
29. Jenssen, H., Hamill, P., and Hancock, R.W., Peptide antimicrobial agents, *Clin. Microbiol. Rev.*, 19, 491, 2006.
30. Li, B. et al., unpublished data, 2006.
31. Bestul, M.B. and VandenBussche, H.L., Antibiotic lock technique: review of the literature, *Pharmacotherapy*, 25, 211, 2005.
32. Safdar, N., Crnich, C.J., and Maki, D.G., The pathogenesis of ventilator-associated pneumonia: its relevance to developing effective strategies for prevention, *Resp. Care*, 50, 725, 2005.
33. McCrory, R. et al., Pharmaceutical strategies to prevent ventilator-associated pneumonia, *J. Pharm. Pharmacol.*, 55, 411, 2003.
34. Adair, C.G. et al., Eradication of endotracheal tube biofilm by nebulized gentamicin. *Intensive Care Med.*, 28, 426, 2002.
35. Projan, S.J., Staphylococcal vaccines and immunotherapy: To a dream the impossible dream, *Curr. Opin. Pharmacol.*, July 2006.
36. Craven, D.E. and Shapiro, D.S., *Staphylococcus aureus*: times, they are a-changing, *CID*, 42, 179, 2006.
37. Brown-Etris, M.B., Cutshall, W.D., and Hiles, M.C., A new biomaterial derived from small intestine submucosa and developed into a wound matrix device, *Wounds*, 14, 150, 2002.
38. Dimick, J.B. et al., Risk of colonization of central venous catheters: catheters for total parenteral nutrition vs. other catheters, *Am. J. Crit. Care*, 12, 328, 2003.
39. Bong, J.J. et al., Prevention of catheter-related bloodstream infection by silver iontophoretic central venous catheters: a randomized controlled trial, *J. Clin. Pathol.*, 56, 731, 2003.
40. Carrasco, M.N. et al., Evaluation of a triple-lumen central venous heparin-coated catheter versus a catheter coated with chlorhexidine and silver sulfadiazine in critically ill patients, *Intensive Care Med.*, 30, 633, 2004.
41. Karchmer, T.B. et al., A randomized crossover study of silver-coated urinary catheters in hospitalized patients, *Arch. Intern. Med.*, 160, 3294, 2000.
42. Stickler, D.J., Susceptibility of antibiotic-resistant gram-negative bacteria to biocides: a perspective from the study of catheter biofilms, *J. Appl. Micro. Symposium* Suppl., 92, 163S, 2002.
43. Johnson, J.R. et al., Systematic review: antimicrobial urinary catheters to prevent catheter-associated urinary tract infection in hospitalized patients. *Ann. Intern. Med.*, 144, 116, 2006.
44. Costerton, J.W., Montanaro, L., and Arciola, C.R., Biofilm in implant infections: its production and regulation, *Int. J. Artif. Organs*, 28, 1062, 2005.
45. Pierce, G.E., *Pseudomonas aeruginosa, Candida albicans*, and device-related nosocomial infections: Implications, trends, and potential approaches for control, *J. Ind. Microbiol. Biotechnol.*, 32, 309, 2005.
46. Trautner, B.W., Hull, R.A., and Darouiche, R.O., Colicins prevent colonization of urinary catheters, *J. Antimicrob. Chemother.*, 56, 413, 2005.
47. Hall-Stoodley, L., Costerton, J.W., and Stoodley, P., Bacterial biofilms from the natural environment to infections diseases, *Nat. Rev. Microbiol.*, 2, 95, 2004.

48. Maki, D.G. and Tambyah, P.A., Engineering out the risk for infection with urinary catheters, *Emerg. Infect. Dis.*, 7, 342, 2001.
49. Chaiyakunapruk, N. et al., Vascular catheter side care: the clinical and economic benefits of chlorhexidine gluconate compared with povidone iodine, *CID*, 37, 764, 2003.
50. Brun-Buisson, C. et al., Prevention of intravascular catheter-related infection with newer chlorhexidine-silver sulfadiazine-coated catheters: a randomized controlled trial, *Intensive Care Med.*, 30, 837, 2004.
51. Dimick, J.B. et al., Risk of colonization of central venous catheters: catheters for total parenteral nutrition vs other catheters, *Am. J. Crit. Care*, 12, 326, 2003.
52. Rupp, M.E., Effect of a second-generation venous catheter impregnated with chlorhexidine and silver sulfadiazine on central catheter-related infections: a randomized, controlled trial, *Ann. Intern. Med.*, 143, 136, 2005.
53. Srinivasan, A. and Ray, A.K., Novel silicon-carbon fullerence-like nanostructures: an Ab initio study on the stability of Si54C6 and Si60C6 clusters, *J. Nanosci. Nanotechnol.*, 6, 43, 2006.
54. Al-Habdan, I. et al., Assessment of nosocomial urinary tract infections in orthopaedic patients: a prospective and comparative study using two different catheters, *Int. Surg.*, 88, 152, 2003.
55. Thibon, P. et al., Randomized multi-centre trial of the effects of a catheter coated with hydrogel and silver salts on the incidence of hospital acquired urinary tract infections, *J. Hosp. Infect.*, 45, 117, 2000.
56. Morris, H.F., Ochi, S., Plezia, R., Gilbert, H., Dent, C.D., Pilulski, J., and Lambert, P.M., AICRG, Part III: The influence of antibiotic use on the survival of a new implant design, *J. Oral. Implantol.*, 30, 144, 2004.
57. Fourrier, F., Dubois, D., Pronnier, P., Herbecq, P., Leroy, O., Desmettre, T., Pottier-Cau, E., Boutigny, H., DiPompeo, C., Durocher, A., and Roussel-Delvallez, M., PIRAD Study Group, Effect of gingival and dental plaque antiseptic decontamination on nosocomial infections acquired in the intensive care unit: a double-blind placebo-controlled multicenter study, *Crit. Care Med.*, 33, 1728, 2005.
58. Fowler, R.A., Flavin, K.E., Barr, H.J., Weinacker, A.B., Parsonnet, J., and Gould, M.K., Variability in antibiotic prescribing patterns and outcomes in patients with clinically suspected ventilator-associated pneumonia, *Chest*, 123, 835, 2003.
59. Pea, F., Pavan, F., DiQual, E., Brollo, L., Nascimben, R., Baldassarre, M., and Furlanet, M., Urinary pharmacokinetics and theoretical pharmacodynamics of intravenous levofloxacin in intensive care unit patients treated with 500 mg b.i.d. for ventilator-associated pneumonia, *J. Chemother.*, 15, 563, 2003.
60. Damas, P., Garweg, C., Monchi, M., Nys, M., Canivet, J.L., Ledoux, D., and Preiser, J.C., Combination therapy versus monotherapy: a randomized pilot study on the evolution of inflammatory parameters after ventilator associated pneumonia [ISRCTN31976779], *Crit. Care*, 10, R52, 2006.
61. McCrory, R., Jones, D.S., Adair, C.G., and Gorman, S.P., Pharmaceutical strategies to prevent ventilator-associated pneumonia, *J. Pharm. Pharmacol.*, 55, 411, 2003.
62. Bonten, M.J. and Krueger, W.A., Selective decontamination of the digestive tract: cumulating evidence, at last? *Semin. Resp. Crit. Care Med.*, 27, 18, 2006.
63. Fattom, A.I., Horwith, G., Fuller, S., Propst, M., and Naso, R., Development of StaphVAX, a polysaccharide conjugate vaccine against *S. aureus* infection: from the lab bench to phase III clinical trials, *Vaccine*, 22, 880, 2004.

64. Weigelt, J., Kaafarani, H.M., Itani, K.M., and Swanson, R.N., Linezolid eradicates MRSA better than vancomycin from surgical-site infections, *Am. J. Surg.*, 188, 760, 2004.
65. Moues, C.M., Vos, M.C., van den Bemd, G.J., Stijnen, T., and Hovius, S.E., Bacterial load in relation to vacuum-assisted closure wound therapy: a prospective randomized trial, *Wound Repair Regen.*, 12, 11, 2004.
66. Lee, S.J., Kim, S.W., Chu, Y.H., Shin, W.S., Lee, S.E., Kim, C.S., Hong, S.J., Chung, B.H., Kim, J.J., and Yoon, M.S., A comparative multicentre study on the incidence of catheter-associated urinary tract infection between nitrofurazone-coated and silicone catheters, *Int. J. Antimicrob. Agents*, 24 (Suppl. 1), S65, 2004.
67. Daims, H., University of Vienna, Powerpoint Presentation, unpublished.

15

Infection-Resistant Implantable Devices: Biofilm Problems and Design Strategies

Anthony W. Smith and Semali Perera

CONTENTS

15.1 Introduction

Advancements in medical and surgical technologies mean that indwelling devices are being used increasingly, either as permanent implants such as prosthetic joints and coronary arterial stents or for maintaining access to sites in the short term such as venous and urinary catheters. Increasing attention is being focussed on implantable matrices as devices to control the delivery of drugs to a particular target site. Inevitably, implanted devices of whatever composition provide an unwanted focus for infection through development of microbial biofilms. Biofilms represent a significant clinical challenge and frequently result in device failure. Here, we review the clinical challenge posed by microbial biofilms and present some of the strategies that have been proposed to reduce or prevent biofilm formation on implantable devices.

15.2 Medical Devices and Biofilms

There is now widespread recognition of the contribution of biofilms to device-related infections such as those associated with artificial joints, prosthetic heart valves, and catheters. Despite perioperative antimicrobial prophylaxis and a laminar airflow surgical environment, around 1% of hip and shoulder replacements and 2% of knee replacements become infected after surgery resulting in significant health care costs [1]. Indwelling device-related infections now account for approximately 50% of nosocomial infection in the United States [2].

A microbial biofilm is broadly defined as adherent microorganisms within a polymeric matrix, typically comprising exopolysaccharide, that develops into a complex community [3]. Once thought of as an unusual mode of growth by a small number of microbial species, it is now clear that this multicellular existence is much more the norm than was thought by microbiologists even just a few years ago. Growth within this multicellular community offers its members a number of crucial survival advantages including protection from environmental insults. In the clinical context, antimicrobial agents must be included on this list of environmental insults and it is increasingly apparent that microbial biofilms make a major contribution to the challenge of resistance to antimicrobial agents [2]. Whilst much attention is now focussed on the biofilm mode of growth and identification of the signals and conditions for colonization of a surface, development of the biofilm matrix, and detachment from the biofilm, a complex picture is emerging of significant heterogeneity [3]. The type of biofilm formed will depend on the surface and the surrounding environment—and indeed there are significant differences among the laboratory systems used to model biofilm development [4–6]. So in flow-cells, submerged biofilms comprise glycocalyx-enclosed microorganisms in stalk- or mushroom-like structures with water channels occurring between them, whereas more complex community organization is seen in floating biofilms or pellicles that lack a solid support and are not subject to the shear stresses imposed under flow conditions. Conventional colony growth on agar-solidified medium is also useful to examine multicellular organization and development of the extracellular matrix [7]. The extracellular matrix or glycocalyx is a key element in determining the architecture of the biofilm and typically comprises extracellular polysaccharides and proteins as well as DNA [7]. Even dead cells may contribute to the properties of the matrix.

15.2.1 Biofilm Resistance: Overview

Growth as a biofilm almost always leads to a large increase in resistance to antimicrobial agents compared with cultures grown in suspension (planktonic) in conventional liquid media, with up to 1000-fold decreases in susceptibility reported [8–10]. Currently, there is no generally agreed mechanism to account for the broad resistance of microbial biofilms [6,11–14]. However, the general resistance of microbial biofilms is largely phenotypic, that is when the same cells are grown planktonically then they become susceptible to antimicrobials. Sterilization of an infected surface by antimicrobial treatment requires elimination of all the bacteria, which in vivo will be assisted by host defense mechanisms. Hence, biofilm resistance can be determined by the susceptibility of the most resistant cell. There is significant heterogeneity even within single species biofilms and it is not the case that all cells within a biofilm are always highly resistant [11,15]. There is likely a spectrum of physiologies within the biofilm as a consequence of gradients of nutrient availability resulting in some cells growing rapidly and others not at all [16]. But the most resistant members of a biofilm population are typically orders of magnitude more resistant than similar members of a planktonic population and subculture often fails to show the existence of resistant mutants. Some studies have proposed a role for persister cells within the biofilm contributing to antimicrobial resistance [17,18]. These are perhaps most easily seen as phenotypic variants that have differentiated into an inactive but highly protected state. Mathematical modeling, based on the assumptions of persisters being incapable of growth and generated at a fixed rate, independent of the presence of substrate or antimicrobial agent, indicated persister numbers increased in numbers in the biofilm in regions of substrate limitation due to the slow conversion of normal cells to persister cells. When antimicrobial treatment was simulated, bacteria near the biofilm surface were killed, whereas persister cells were not. After treatment ceased, persister cells reverted allowing the biofilm to regrow [19].

15.2.2 Biofilm Resistance: Possible Mechanisms

A complete review of biofilm resistance is beyond the scope of this chapter, not least because there is no compelling evidence for a unique contribution by any of the well-characterized resistance mechanisms [13,20]. In offering a brief overview of some of these mechanisms, it is worth remembering that the simplest strategy will be to design devices which do not support or favor biofilm formation. Whilst simple in concept, the reality is a very significant challenge.

Many papers have investigated the possible lack of antimicrobial penetration as an explanation of biofilm resistance [11,12,21–23]. However, given that in many circumstances biofilm communities have aqueous channels flowing through them, impenetrability seems highly unlikely [24]. Reaction–diffusion limited penetration may result in only low levels of antimicrobial exposure in the deeper regions of the biofilms; it has been proposed that such sheltered cells could enter an adapted-resistant state if the time for adaptation is faster than that for disinfection. In this model, effective disinfection would require the biocide concentration to be increased quadratically or exponentially with biofilm thickness [25]. When the antimicrobial agent reacts rapidly with an exopolymer or is adsorbed to it, then the net effect is that of having the appearance of a penetration barrier. This has only been demonstrated as problematic with the highly reactive oxidizing biocides such as chlorine and hydrogen peroxide. Molecular genetic studies are now being used to identify biofilm-modulated genes and one study has identified a *Pseudomonas aeruginosa* mutant which forms apparently normal biofilms, but exhibited enhanced susceptibility to various antimicrobial agents, including tobramycin [26]. The mutation mapped to the *ndvB* gene is required for synthesis of periplasmic glucans. The authors proposed that transcription within the biofilm resulted in trapping of antimicrobials such as tobramycin within the periplasm.

Antimicrobial agents can induce the production of inactivating enzymes in bacteria [27]. The relatively large amounts of antibiotic-inactivating enzymes such as β-lactamases, which accumulate within the biofilm matrix can result in concentration gradients across it and protection of underlying cells [28,29], whilst in other systems, enzyme inactivation appears to play little or no part [30,31].

Efflux has also been reviewed [32], and the contribution of efflux to antimicrobial resistance of biofilms is uncertain. Recent work has highlighted the contributions of oxygen deprivation and anaerobic growth to antibiotic resistance. There is evidence of *P. aeruginosa* growing anaerobically within airway mucus of cystic fibrosis patients [33], and microelectrode studies on colony biofilms indicated that oxygen penetrated only to approximately 25% of the depth of the biofilm [34]. Moreover, large anoxic regions developed as the biofilm aged, and regions of active protein synthesis occurred only in a narrow band at the air interface [35]. When challenged with antimicrobials, 4-h-old colony biofilms growing in the presence of air were susceptible; however, similar aged biofilms grown anaerobically were much less susceptible [34]. The authors calculated that oxygen limitation could explain 70% or more of the protection of old biofilm cells [27].

Much of our understanding of the properties of microbial biofilms, as with most other aspects of microbiology, is based on studies with single species in pure culture. This notwithstanding, there is widespread recognition, not least in theory that most biofilm communities will comprise a mixture of species and the properties of the community may not simply be the sum of the properties of the component organisms. A recent study has found that when four species that had been identified growing together within a natural ecosystem were cultured in the laboratory, then the amount of biofilm biomass formed was much greater and more resistant to the antibacterial agents, hydrogen peroxide or tetracycline, compared with single species biofilms of the component isolates [36]. The term "sociomicrobiology" has been coined to describe the connections between quorum sensing and biofilms [37]. Sociomicrobiology is complicated, particularly when it is recognized that the notion of a "quorum" may differ in a biofilm community from that in the more frequently studied planktonic culture, that other members of the community may respond to each others' signals and that some members may inactivate or destroy other members' signals. Interest in quorum sensing, or bacterial intercellular signaling and biofilm eradication increased with the observation that biofilms formed by a quorum sensing-deficient mutant of *P. aeruginosa* were more easily displaced by treatment with the detergent sodium dodecyl (lauryl) sulfate, than wild-type biofilms [38].

15.3   Antibiofilm Approaches: Design Strategies

A number of the key elements in the infectious process and formation of biofilm have been proposed for application of novel technologies and drug delivery implants—prevention of colonization and biofilm formation, accumulation at the biofilm surface, and drug penetration into the biofilm (Figure 15.1). Recent developments will be reviewed here. Given the increasing use of relatively invasive medical and surgical procedures, the material properties of medical devices have received much attention as have strategies to target antimicrobials to device-related infections.

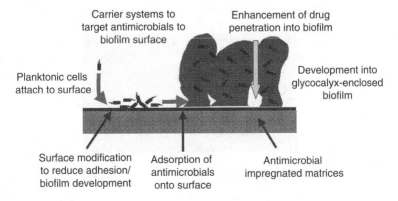

FIGURE 15.1
Antibiofilm strategies.

15.3.1 Material Modification to Prevent Biofilm Formation

Given the difficulties of eradicating a biofilm once it has become established, preventing biofilm formation is an obvious strategy. However, at present there seems to be no such thing as an unadulterated surface. Any surface, synthetic or otherwise, becomes coated with constituents of the local environment: first water and electrolytes, then organic material. Microorganisms may or may not be present at this early stage and in the context of using indwelling drug delivery devices will depend on the site of use and insertion techniques employed. Initial electrostatic interactions between microbes and the conditioned surface will be weak and reversible resulting from Brownian motion, gravitation, diffusion, or microbial motility [39]. Evidence is now gathering for the inverse regulation of genes involved in motility and extracellular matrix production whereby bacteria appear to give up the ability to move by means of flagella propulsion in order to settle and establish on a surface [40]. The observation that surfaces immediately become coated with a conditioning film can thwart coating-based strategies to reduce infection resulting from hematogenous seeding of medical devices such as intravascular catheters. Put simply, the coating becomes coated. However, given that many catheter-related infections are due to skin organisms acquired at the time of catheter insertion, then anticolonization strategies are still worth exploring. For example, surfaces containing immobilized long-chain N-alklyated polyvinylpyridines and structurally-unrelated N-alklyated polyethylenimines were lethal to *Staphylococcus aureus, Staphylococcus epidermidis, P. aeruginosa,* and *Escherichia coli* [41]. There are many instances where broad-spectrum antimicrobials have been incorporated into the device [42–46]. These are then eluted with the aim of preventing biofilm formation by killing early colonizing bacteria. As noted by others, such a strategy is not without its problems [47]. Sufficient antibiotic must be incorporated for the "user-lifetime" of the device and such incorporation must not damage the properties of the material, for example, lubrication, lifetime, and host compatibility. There is also the nagging concern that subinhibitory levels of antimicrobials could favor acquisition of antibiotic-resistant organisms [48] and enhance biofilm development [49].

Alternative approaches to antibiotic incorporation have also been used. For example, rifampin and amoxicillin have been adsorbed on polyurethanes exhibiting acidic or basic properties, the binding affinity increasing with the introduction of polymer side chain functional groups and with matrix hydrophilicity [50]. Biodegradable polyurethanes have been synthesized incorporating the fluoroquinolone antibiotic ciprofloxacin which is released upon the action of the macrophage-derived enzyme cholesterol esterase. Since this enzyme recognizes hydrophobic moieties, drug release could be achieved by formulating polyurethanes synthesized with 1,12 diisocyanatododecane with long hydrophobic monomers immediately adjacent to the ciprofloxacin [51]. An alternative approach, effective in inhibiting bacterial adhesion for up to 7 days has been to coat silicone catheter material with a poly(ethylene glycol [PEG])-gelatin hydrogel incorporating liposomal-encapsulated ciprofloxacin [52]. Antibiotic encapsulation in the matrix could be used to prevent infections around such commonly used devices as catheters as well as more permanent implants, such as pacemakers [53].

Numerous agents such as surfactants, heparin, nonsteroidal anti-inflammatory agents have also been applied to the external surface of catheters in an attempt to prevent the adherence of bacteria [53–59]. Silver coatings have been shown to be quite effective in preventing catheter infection when incorporated in a catheter cuff [53]. Until recently, these have been very difficult to apply to the plastic catheter surface itself. Recent advances in biomaterial technology have made it possible to coat catheters effectively with silver preparations [60]. Catheter-related infection has also been minimized by coating catheters with a silicon or gold layer loaded with silver nanoparticles [61,62]. In vitro tests of the polyurethanes matrix impregnated with silver nanoparticles demonstrated its hydrophilic character and its antibacterial activity. The adherence of bacteria to the surface of such catheters was minimal, proliferation of bacteria adhered to it was prevented, as was formation of a biofilm [62]. The alterations of chemical, biological, immunological, and physical properties of polymeric materials used for central venous catheters have been extensively investigated to prevent their colonization by bacteria. The current state-of-the-art strategies to minimize the risk of catheter-related blood stream infections are analyzed. Progress made in developing infection- and thrombosis-resistant dialysis catheters has been described [63,64]. A novel biodegradable fluoroquinolone polymer has been developed. It releases nalidixic acid slowly and continuously and prevents catheter-induced infection during drainage of cerebrospinal fluid. It was prepared from diisopropyl-carbodiimide, poly(caprolactone) (PCL), and nalidixic acid. The released antibiotic and its derivatives are active for 3 months against *E. coli*, *S. aureus*, and *Salmonella typhi*. Application of this polymer will enable manufacture of drainage catheters that will resist catheter-induced infection by delivering antibiotics for more than 1 month [65]. Emergence of multidrug-resistant pathogens is always a risk when antibiotics are used both in prophylaxis and long-term therapy as well as in conventional acute therapy. Hence, there is interest in inhibiting biofilm formation on polymer surfaces using novel compounds that are not otherwise used. One such compound is usnic acid, a secondary lichen metabolite that possesses antimicrobial activity against a number of gram-positive bacteria, including *S. aureus*, *Enterococcus faecalis*, and *E. faecium*. When loaded into modified polyurethane and placed in infection-simulating flow-cells, confocal laser scanning microscopy indicated that whilst adhesion of *S. aureus* occurred, usnic acid inhibited development of biofilm [66]. *P. aeruginosa* biofilm did form on usnic acid-coated polymer, although its morphology differed from the biofilm formed on untreated polymer, leading the authors to suggest that signaling pathways within the biofilm may have been perturbed.

15.3.2 Polymer-Based Antimicrobial Delivery Systems

Many reviews have highlighted the use of biodegradable polyesters as effective drug carriers including nano- or microparticles, hydrogels, micelles, and fibrous scaffolds [67–74]. Inevitably, there are advantages and disadvantages associated with each delivery system; however, these experimental approaches have been investigated in a number of infections, including periodontitis and osteomyelitis, as well as intracellular infections such as tuberculosis and brucellosis. The advantages of localized biodegradable drug therapy include high, local antibiotic concentration at the site of infection, as well as obviation of the need for removal of the implant after treatment. Biodegradable polymers were first introduced to the medical industry in the 1960s, in the form of bioabsorbable sutures [75,76]. Because the biocompatibility of these polymers is well established, they have been extensively used in biomaterial applications such as tracheal replacement, ligament reconstruction, surgical dressings and dental repairs, and as carriers in drug delivery systems [77,78]. Today, these materials are increasingly finding use in other medical applications [79], including orthopaedic fixation devices such as pins, rods and screws, and tissue engineering scaffolds [80–82].

Biodegradable polymers have become increasingly popular as carriers in the design of drug delivery systems since the mid-1970s. As these polymers degrade in the body to small molecular weight compounds that are either metabolized or excreted, they obviate the need for removal of the carrier after the device is exhausted. Although there are many classes of biodegradable polymers available, the most widely studied belong to the polyester family. These include the poly(α-hydroxy acids) i.e., poly(lactic acid) (PLA), poly(glycolic acid) (PGA), and their copolymers, poly(D,L-lactide-co-glycolide) (PLGA), which have a long history of safe clinical use as resorbable sutures and are approved by the U.S. Food and Drug Administration (FDA). Other FDA-approved biodegradable polymers include homopolymers and copolymers of caprolactone, poly-p-dioxanone, and trimethylene carbonate [80]. One of the advantages is that the PLGA polymers are esters of α-hydroxy acetic acid and break down by bulk hydrolysis of ester bonds to their constituent monomers, lactic, and glycolic acids, which are eliminated from the body through Krebs cycle.

An important attribute of biodegradable polymers is the possibility of fabricating them into various shapes of microcapsules, implants or films, and pore morphologies for different systems. More importantly, it is possible to tailor their mechanical properties and degradation kinetics to suit the application (e.g., by changing the chemical composition, molecular weight, or glass transition temperature) to achieve precise delivery of drugs [83].

15.3.2.1 *Poly(glycolic acid)*

Poly(glycolic acid) is a linear aliphatic polyester synthesized by ring-opening polymerization of glycolide (glycolic acid dimer). It is highly crystalline (46%–50%) and has a glass transition temperature (T_g) of 35°C–40°C [70].

This means that at temperatures below the T_g, the polymer will act like glass but at higher temperatures it will have a rubbery texture. Fibers from PGA exhibit high strength but they degrade quite rapidly (6–12 months). The final degradation product is glycolic acid, a natural metabolite which is eliminated from the body through the Krebs cycle.

15.3.2.2 Poly(lactic acid)

Lactic acid exists as two enantiomeric forms, L(+)-lactic acid and D(−)-lactic acid [76–78]. The polymerized form, L-PLA, is commonly used as it is preferentially metabolised in the body. It is semicrystalline (37%) with a glass transition temperature of 60°C–65°C. However, the degradation time of L-PLA is quite long, requiring more than 2 years to be completely absorbed. This is due to its hydrophobic nature caused by the presence of methyl groups; hence, there is higher resistance to hydrolytic attack of its bonds [79].

For a more rapid degradation time, DL-PLA can be used. It has a lower initial crystallinity and consists of a random distribution of both isomeric forms of lactic acid. This results in an amorphous structure which enhances the rate of polymer degradation (12–16 months for full degradation) [79]. PLA degrades to lactic acid which is eventually converted to carbon dioxide and water by the Krebs cycle and released through respiration [76–80].

15.3.2.3 Poly(lactide-co-glycolide)

Copolymers of PLA and PGA can be produced to yield faster degradation rates, as they are less crystalline than their constituent homopolymers. The rate at which the PLGA degrades is dependent on the ratio of lactic acid to glycolic acid. Since PLA is more hydrophobic than PGA, and hence breaks down less rapidly, a higher ratio of lactic acid in the PLGA copolymer will result in slower degradation times [84]. The most unstable copolymer is 50/50 PLGA with a degradation period of less than 2 months [85]. Increasing the ratio of lactic acid to glycolic acid increases the time for degradation, with 85/15 PLGA taking up to 6 months to be completely absorbed. However, the tensile strength and elongation of the copolymers remain similar regardless of its acid ratio. Biodegradable microspheres prepared from poly(lactide) (PLA) and its copolymers with PLGA and PCL can release encapsulated drugs in a controlled manner, depending on the method of microencapsulation and the physicochemical properties of the polymer and drug. Among the polymers discussed so far, 85:15 PLGA perhaps has the best potential for a controlled drug release implant, as the rate of degradation is not too high and the increase in porosity seems gradual. This could allow a complete release of drug from its pores before completely degrading thus reducing the possibility of dose dumping. Small biodegradable microspheres are useful alternatives to liposomes for targeting drugs to the monocyte–macrophage system. They tend not to suffer from the same difficulties of low encapsulation efficiency and stability on storage typically exhibited by liposomal formulations.

15.3.2.4 Poly(caprolactone)

PCL is another aliphatic polyester that has been used in biomedical applications. It is a semicrystalline polymer with a low-glass-transition temperature ($-62°C$) which gives it a rubbery texture [80]. The degradation time of homopolymer PCL is considerably longer than other polyesters, often in the order of 2 years, due to its hydrophobic character and crystallinity [86]. PCL is able to form blends with other polymers such as PLA and PGA leading to amorphous structures with enhanced degradability [87]. Copolymerizing PCL with PGA also produces a fiber with reduced stiffness compared with pure PGA fibers. This polymer is ideal for producing drug-encapsulated patches. The typical degradation path of the PCL matrix is shown in Figure 15.2.

The hydrolytic degradation of PCL results in the release of 6-hydroxycaproic acid which is also metabolized by the Krebs cycle. Compared with PLA and PGA, PCL does not generate as acidic an environment during degradation and has a high permeability to small drug molecules. All of these factors make it suitable as a long-term implantable drug delivery system, such as Capronor, an FDA-approved contraceptive device [88].

Another very promising candidate for implants is chitosan. It is a nontoxic, biodegradable, biocompatible natural polymer, fully soluble at pH 5 in a very slightly acidic aqueous medium, making it ideal as a drug carrier. It can be formed into fibers in various coagulating solutions. Chitosan fibers, used for scaffold preparation seemed to combine adequate porous structure

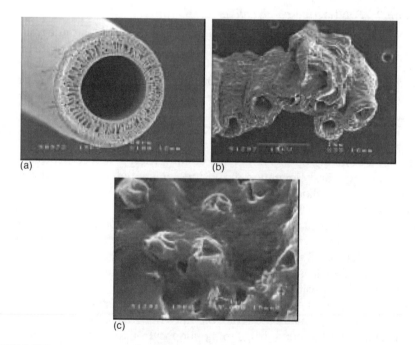

(a)   (b)

(c)

FIGURE 15.2

(a) PCL microfibers, (b) fibers after 6 weeks degradation in phosphate buffer solution at 37°C, and (c) blister formation on the surface.

with sufficient degradability and mechanical properties. They were produced by a wet spinning technique and treated with methanol to improve their mechanical strength. They also showed bioactive behavior and can achieve controlled delivery of antibiotics during important bone repair period [89]. Although PLA, PCL, or PLGA implant dosage forms have been widely investigated, in most cases, controlled release periods have been long and drug dosage levels have been too low, and these devices have been prone to infections [90–94]. In some cases such a slow release may limit the applicability of PLA, PCL, and PLGA as drug carriers. Namely, for drugs for which the dose amount required per administration is large, such a slow release leads to a very bulky dosage form; such a bulky dosage form is a burden on patients and cannot be used practically. Hence still there is an enormous need for better design and development of antibacterial medical implants.

15.3.3 Importance of Early Treatment

It is essential to eradicate bacteria as early as possible since any delay in implementing antibiotic therapy, even 1 or 2 days after the biofilm is established, may result in the failure of the antibiotic treatment [90,91]. Commonly used devices such as catheters account for about 50,000 hospital deaths in the United States each year, many of them because of infection [92]. The current strategy to prevent biomaterial-related infections is to treat the patients systemically with antibiotics at a high concentration, although studies show that this practice is of limited efficiency [91]. A better strategy would be to provide antibiotics immediately after implantation at the site where they are most needed, namely at the interface where the biomaterial interacts with the body and the colonizing bacteria. This following section provides information on the development of biopolymeric vehicles that would release in a controlled manner a low molecular weight antibiotic in the near-wall zone to prevent bacterial colonization [91–95].

15.3.4 Implantable Matrices, Microparticles, Fibrous Scaffolds, and Thermoreversible Gels

Prevention and treatment of osteomyelitis, particularly associated with orthopaedic implant surgery, have been the focus of many studies. Systems implanted at the same time as the prosthesis may be either nonbiodegradable or biodegradable. Polymethylmethacrylate beads impregnated with gentamicin have been available for about 20 years; however, since they are nonbiodegradable they have to be removed, usually about 4 weeks after insertion [96]. A number of osteoconductive/biodegradable alternatives have been studied, including calcium phosphates such as hydroxyapatite whose chemical composition is similar to the bone mineral phase. Studies with ciprofloxacin incorporated into hydroxyapatite and poly(D,L-lactide) formulations implanted in the femur of rabbits, indicated that therapeutic bone levels were achieved over 6 weeks with release enhanced by erosion—disintegration and bone in-growth into the implant [97]. Other studies using the glycopeptide antibiotic teicoplanin, effective against *S. aureus*, indicate that it too can be effective over several weeks when incorporated into microspheres prepared from PLGA (75:25) (Mw 136,000) polymer [98,99]. Other materials studied against implant-related osteomyelitis include poly(3-hydroxybutyrate-co-3-hydroxyvalerate) (PHBV). The release behavior of sulbactam:cefoperazone from rods comprising 7%, 14%, or 22% (mol) 3-hydroxyvalerate were representative of typical monolithic devices where a rapid early release phase is followed by a slower and prolonged phase. With PHBV 22 rods, this extended phase lasted for up to 2 months, making it a promising controlled release vehicle since treatment of device-related infections is typically up to 6 weeks [100]. Release studies of antibiotics adsorbed onto the surface of hydroxyapatite cylinders having a bimodal pore size distribution also have a prolonged duration of release, attributed to the small pores, combined with favorable osteoconduction properties into the large pores [101]. A number of matrices have been formed using PLGA, including discs [102], and electrospun nanofibrous scaffolds [103]. Some of the typical drug-loaded (20% vancomycin) biodegradable polymer configurations are shown in Figure 15.3 [83]. The drug-encapsulated microbeads

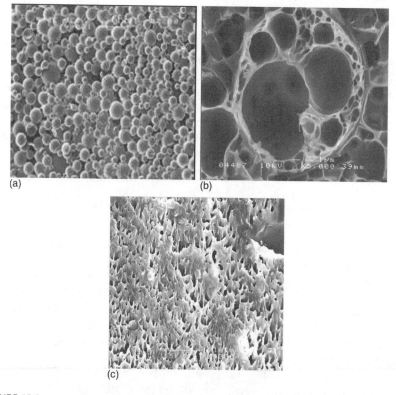

(a)

(b)

(c)

FIGURE 15.3

Biodegradable implants. (a) Drug-loaded PLGA microspheres, (b) cross section of microparticles made from single emulsion, and (c) drug-loaded PLA sheet made by casting.

are generally formed by single (Figure 15.3a and 15.3b) or double emulsion technique and solvent evaporation [64–67]. The sheets are formed by casting followed by phase inversion of the polymer (Figure 15.3). In an effort to achieve the ideal drug release pattern of no lag time and zero-order release, lactide monomer or glycolide monomer have been incorporated into PLGA discs loaded with gentamicin. The idea here is that channels will form in the disc following dissolution of the monomer to aid the release of gentamicin. Discs containing 10% monomers showed nearly zero-order release kinetics for more than 1 month [102]. Evidence for the channel-forming properties of the monomers came from the water uptake by the discs. After 7 days, the amount of water absorbed by the control discs was 20%, compared with 60% in monomer-containing discs. In another study biodegradable chitosan has been cross-linked by reaction of its amino groups with PEG600 terminated at its two ends with carboxyls. Thus, covalently cross-linked networks are obtained forming pH sensitive hydrophilic hydrogels which can deliver antibiotics. The in vivo studies revealed that such antibiotic doped hydrogels facilitated bone regeneration and prevent infection. When the degradation of the implant was completed, bone defects filled with such hydrogels were completely regenerated [104]. Tests conducted in vivo on sheep, with micro-fracture defects in the knees, showed that implants of chitosan–glycerol phosphate plus full autologous blood can improve the structural and compositional properties of repaired cartilage [105]. Other applications of PLGA have included incorporation of antibiotic-loaded microparticles into an injectable collagen sponge, resulting in local drug delivery combined with the tissue regeneration properties of collagen [105]. Several other matrices have been used, including the aluminosilicate material halloysite. Whilst chemically similar to kaolin, halloysite has a tubular structure that can be loaded with drug. Moreover, surface charge neutralization using cationic polymers can place an additional level of control on drug release. In one study of tetracycline incorporation into halloysite for the treatment of periodontitis, an initial burst of tetracycline release was followed by a dramatic reduction in release. Coating with the cationic polymer chitosan reduced the "burst" release from 45% to 30% and the total release over 9 days from 88% to 78% [106]. A key concern was delivery of the halloysite to the gingival pocket and its subsequent retention. The polyoxyethylene-polyoxypropylene copolymer poloxamer 407 was used for its thermoreversible properties of being liquid below 20°C and forming a hydrogel at higher temperatures, if sufficiently concentrated, that is also biodegradable and relatively nontoxic. The ideal here is that the delivery system should be a liquid at room temperature, so avoiding the need for refrigeration, and a device-retaining gel at the temperature within the gingival crevice. In this particular study, the transition temperature range was narrowed by the addition of polyethylene glycols; however, this caused the storage modulus (solidity) of the gel to drop as well [106]. This was overcome by the addition of 1% octyl cyanoacrylate (OCA), a powerful tissue adhesive that polymerizes at neutral pH. For storage, the pH of the system was held at 4 by addition of glacial acetic acid, and upon application to the gingival crevice the natural buffer capacity of the tissue slowly brought the pH of the formulation up to neutral, allowing the OCA to polymerize. Thermoreversible polymers have also been studied as delivery matrices in their own right [107–109]. For example, pluronic F-127 has been used to deliver vancomycin to treat methicillin-resistant *S. aureus* otitis media [110]. Based on the sol-gel phase transition measurements, a 25% w/v pluronic solution was loaded with vancomycin and injected through a 26 gauge needle. It changed to a gel at 36°C–37°C that consisted of large populations of the micelles and aqueous channels from which the vancomycin was released. Bioabsorption studies followed over 50 days indicated that the gel gradually disappeared leaving open spaces that initially showed some inflammatory exudate, but by day 50, normal tissues were observed without any inflammation, fibrosis, or open spaces from the gel.

15.4 State-of-the-Art Prosthetic Implants

As we discussed earlier biomaterials-related infections are often observed with prosthetic implants and in many cases result in the failure of the devices. Acrylic bone cement is generally used as a delivery system for depot administration of antibiotics to the infected bones [111]. However, it is not biodegradable and requires secondary procedure for its removal. Though many biodegradable polymeric and natural materials have been investigated, none has yet been approved for clinical use in the United States [111]. In a recent study, an implant made of PLA with the carbonated calcium phosphate/$CaCO_3$, was prepared by combination of hot pressing and gas forming. It contained the macroporous, fast degradable poly(D,L-lactic acid)/$CaCO_3$, on the inside, which promoted the in-growth of bone cells. To ensure mechanical stability and protection of the implant, it contained the antibiotic-loaded slowly degradable compact poly(L-lactic acid)(PLLA) and the carbonated calcium phosphate on the outside. Calcium carbonate neutralized lactic acid formed as a result of degradation of PLA. Good proliferation of the human osteoblastic cells was observed in vitro, testifying to the biocompatibility of this implant for cranial reconstruction [112]. Satisfactory results were obtained, when PLLA implants were used for healing 23 human patients, with scaphoid failure of broken bones [113]. The application of implants, prepared from 80:20 poly([L/DL]-lactide) instead of the commercially available 30:70 poly([L/DL]-lactide), was investigated for fixation of fractures of bones in regions of low load. They were tested in vivo to promote bone regeneration of circular cranial defects in white rabbits. Short-term studies demonstrated that there was almost complete regeneration of cranial defects after an 8 week period. However, long-term studies must be conducted to confirm the expected advantages of this formulation [113]. Most of the studies indicate that to design a biomedically useful polymer vehicle that is intrinsically infection-resistant, it is important to develop an antibiotic (e.g., ciprofloxacin, vancomycin)-loaded polymer matrix that releases antibiotic locally at the implant surface, thereby minimizing bacterial accumulation. The recent report by Ratner and others [60] showed that they have developed a method of fabrication and formulation for making an antibiotic-loaded device, as well as evidence of their bactericidal properties [60,114]. The antibiotic ciprofloxacin has been combined with a polymer polyethylene glycol—an approved food additive—and mixed with the polyurethane used to make medical implants. The plasma process has been used to coat the material with an ultra thin layer of butyl methacrylate to control the rate of drug release. When the device is implanted, fluids pass through the thin, permeable outer coating and dissolve the polyethylene glycol, which makes the polyurethane porous. The antibiotic then leaches out of the polyurethane (BIOSPAN PU). The coating acts as a barrier to the antibiotic, controlling the rate at which it is released to the surface of the device. An optimum formulation consisting of Biospan PU, PEG, and ciprofloxacin offered the longest effective period of sustained release (5 days) [95]. The bactericidal efficacy of the released ciprofloxacin against *P. aeruginosa* was four times that of the control PU without antibiotics. This bactericidal efficiency was due to an increase in the *PA* detachment from the surface. These observations suggested that the released ciprofloxacin was biologically active in preventing the bacteria from permanently adhering to the substrate, and thus decreasing the possibility of biofilm-related infection.

Ramchandani and Robinson [115] reported the results of in vitro and in vivo release studies of ciprofloxacin from PLGA 50:50 implants. The group investigated a biodegradable, implantable delivery system containing ciprofloxacin hydrochloride (HCl) for the localized treatment of osteomyelitis and to study the extent of drug penetration from the site of implantation into the bone. Osteomyelitis is an inflammatory bone disease caused by pyogenic bacteria and involves the medullary cavity, cortex, and periosteum. PLGA 50:50 implants were compressed from microcapsules prepared by nonsolvent-induced phase-separation using two solvent–nonsolvent systems, viz., methylene chloride–hexane (nonpolar) and acetone–phosphate buffer (polar). In vitro dissolution studies showed the effect of manufacturing procedure, drug loading, and pH on the release of ciprofloxacin HCl. The extent of penetration of the drug from the site of implantation has also been studied using a rabbit model [115]. The results of in vitro studies illustrated that drug release from implants made by the nonpolar method was more rapid compared with implants made by the polar method. The release of ciprofloxacin HCl from the implants was biphasic at ≤20% w/w drug loading, and monophasic at drug loading levels ≥35% w/w. In vivo studies indicated that PLGA 50:50 implants were almost completely resorbed within 5–6 weeks. Sustained drug levels, greater than the minimum inhibitory concentration (MIC) of ciprofloxacin, up to 70 mm from the site of implantation, were detected for a period of 6 weeks.

Other new developments include biopolymeric implant materials with shape-memory for application in biomedicine. These polymers can change their shape in response to changes in environmental factors such as temperature. The shape-memory effect may be due to minute changes in polymer structure and morphology. Examples of thermoplastic materials or of the covalently cross-linked elastomers have been reported [116]. Using similar technology, it may be possible to develop materials which respond to the bacterial environment, and hence deliver antibiotics as and when required by the implanted site and the body.

In a recent study, cefoxitin has been incorporated into PLGA scaffolds [103]. Here, PLGA controls the rate of degradation whilst high molecular weight poly(lactide) (PLA) confers mechanical strength to the scaffold. An amphiphilic diblock copolymer comprising poly(ethylene glycol)-b-poly(lactide) was added to the polymer solution to encapsulate the hydrophilic cefoxitin sodium, which otherwise has poor solubility in the PLGA solvent N,N-dimethyl formamide. The drug/polymer solution was electrospun in a spinneret at 23–27 kV. As the concentration of cefoxitin increased, the scaffolds changed from a bead and string morphology, attributed to insufficient stretching of the polymer jet, to a fine fibrous structure of diameter 260 ± 90 nm. The spinning process did not affect the cefoxitin and following an initial burst, prolonged release was measured for up to 1 week.

15.5 Implants for Cancer Treatment

Cytotoxic chemotherapy and the resulting neutropenia create a high risk for infectious morbidity and mortality in patients with cancer. Long-term central venous catheters provide reliable access for administration of chemotherapeutic agents, blood transfusions, parenteral nutrition, and antibiotics to these patients, but they are also an important cause of nosocomial bloodstream infections [117–118]. Generally, as a preventative measure, patients undergoing chemotherapy treatment are given antibiotics [119–122]. There have been significant advances in the central venous access port system design; nevertheless, thrombosis and infection remain major causes to preclude access survival. The reported catheter-related infection rate was as high as 29.1% (142/488 catheters) in Benezra's early study [123]. Calle and others reported infection rates of 0.8%–6.2% among 123 jugular vein cut down patients [124]. The reported incidences of central venous thrombosis were between 0% and 16% [125–127]. Kuzel and others claimed that 5-fluorouracil may have thrombogenecity [128]. However, inclusion of heparin should prevent clotting and promote antibacterial properties. They advocated that low dose warfarin may significantly reduce the thrombosis and infection rate without changing the coagulation status [128]. The reported catheter-related infection rates ranged from 3.2% to 15% in recent studies [129,130]. Howell and others [131] studied catheter-related infection or sepsis among a cohort of 71 adult patients with cancer with long-dwelling tunneled central venous catheters for a total of 12,410 catheter days. They found that gram-positive cocci accounted for 7 of 13 catheter-related infections (5 coagulase-negative Staphylococci and 2 *S. aureus*).

Such patients would benefit from the development of slow releasing, multiple drug delivery implants. Cancer therapeutics are commonly delivered intravenously and only a small fraction of the drug reaches the target tumor. This has two consequences. Firstly, a relatively high dose of drug must be administered to obtain a therapeutic effect and secondly, nonspecific damage to healthy tissue is a commonly-occurring serious complication. The design and development of biodegradable, implantable "multiple drug" delivery systems could overcome these challenges. However, the problem remains of the susceptibility of the biodegradable device to being colonized with microbial biofilms and becoming a focus of infection [121]. This could be particularly problematic in immunocompromised patients.

Novel intratumor release by means of implanted biodegradable polymer carriers has been proposed over the past few years, and its merits of higher therapeutic index and lower toxicity have been made [132–135]. Some success has been achieved for one chemotherapy device, Gliadel Wafers, for clinical applications in patients with malignant brain tumor [135]. In contrast, there is limited development on a comparable product for radiotherapy, although some research work has been carried out on the controlled release devices of radiosensitizers in mouse tumors using polymeric rods [136]. The study concluded that intratumor release of etanidazole using a biodegradable polymer implant device, coupled together with radiation, could improve treatment outcome in tumors. Etanidazole is generally administered systemically in clinical trials. A large dosage is normally required because the drug has a short half-life, and a majority of it is depleted via excretion and cell metabolism before it reaches the targeted tumor. However, a prolonged exposure time to the drug is necessary to ensure the efficacy of the treatment, main disadvantage of the long-term devices is that they are prone to infections.

The failure in the survival rate improvement may be partly due to systemic toxicities and infection when the drug is administered over a long-term treatment schedule. The disadvantages of the systemic drug administration advocated the polymer-based drug delivery system, which releases drug locally to the tumor via polymer carriers implanted in the proximity of the tumor. Such a system offers the advantages of sustained drug release, reduced overall toxicity, higher therapeutic index, as well as improved survival rate [132]. In vitro study of the anticancer drug doxorubicin in PLGA-based microparticles was reported by Lina and others [137]. Doxorubicin, also known as adriamycin, is an anthracycline antibiotic (which has antibacterial properties) and one of the most widely-used cytostatic drugs in the field of chemotherapy [137]. It works by inhibiting the synthesis of nucleic acids within rapidly growing cancer cells thus eventually causing cell death. Doxorubicin is commonly used in the treatment of neoplastic diseases such as leukemia and various solid tumors, and has also been found to be particularly effective in the treatment of breast cancer. However, like many other anticancer drugs, it is a potent vesicant (blister agent) that may cause extravasations and necrosis at the site of injection or drug exposure [137]. If liquid infusion is used it also has side effects such as cardiotoxicity and myelosuppression which leads to a very narrow therapeutic index. Therefore, to overcome these difficulties, development of a drug encapsulated slow releasing implant would be beneficial to this group of patients.

In another study, Domb [138] reported that the injection of a mixture of the liquid biocompatible copolymer with an anticancer drug into a body solidifies on contact with body fluids to form a depot implant [139,140]. They claimed that such injectable therapeutic implants may offer a more effective and safer alternative to standard chemotherapy. They also prepared biocompatible and biodegradable liquid polyester-anhydrides by *trans*-esterification of ricinoleic acid and sebacic acid. Its mixture with paclitaxel or with *cis*-platinum injected into animal models, solidified in vivo and formed a solid implant, which released the drug in a controlled manner. It provided high local concentration of the anticancer drug, which was able to destroy malignant cells, with minimal systemic distribution [139,140]. Various physicochemical principles can be used to prepare injectable biocompatible liquids (liquid embolics) that solidify in situ to form semisolid implants. Solubility changes, responsible for precipitation of water insoluble polymers, may be used for certain therapeutic applications [138–141]. The novel technique of treating cancer by implanting polymeric drug carriers has not yet realized its intended potential, as is evident in many hospitals worldwide that have yet to adopt this technique as a mainstream viable option in tumor excision. The reasons for its delayed acceptance in medical practice are that the exact drug transport mechanism, the implications of clinical associated conditions, such as vasogenic edema, infections, and the effects of the release profiles of drug vehicles are not yet clearly understood.

15.5.1 Commercial Cancer Implants

The current state-of-the-art for sustained drug delivery includes polymeric implantable drug-filled devices where the drug is released over time by osmosis. Examples include Alza's Duros implant, which stores and releases leuprolide acetate for the treatment of prostate cancer; rods such as Zoladex [142] (produced by AstraZeneca), a continuous-release PLGA cylindrical rod which contains goserelin acetate, also for the treatment of prostate cancer, and implantable wafers such as the Gliadel Wafer, which releases carmustine for the treatment of glioma. Sandostatin LAR (Novartis Pharma) is an intramuscular injection of octreotide acetate for the treatment of acromegaly as well as gastro-entero-pancreatic neuroendocrine (GEP NE) tumors. Zoladex is presented as a biodegradable, continuous-release depot formulation 1 cm long by 1 mm in diameter which is made of poly(lactide-co-glycolide).

This depot is inserted into the body by subcutaneous injection into the anterior abdomen wall, using a syringe applicator. Effective concentrations of drug are maintained for 4 weeks for the 3.6 mg dosage, and 3 months for the 10.8 mg dosage, after which administration of a new depot is required. This implant offers a reliable and convenient method of administering therapy, with steady drug release and long periods between injections. The biocompatible and biodegradable material used does not require removal and local reactions at the site of injection are usually uncommon. An ideal drug delivery device should be fully resorbable (no extra operations) and capable of precise drug delivery in terms of dose and location. Eligard on the other hand is injected subcutaneously as a liquid and is used for the management of advanced prostate cancer. The liquid solidifies and then gradually degrades to release a continuous supply of leuprolide acetate. One-, three-, and four-month time-release formulations were approved in 2002, and a six-month formulation was approved in 2004. The variation in drug release is controlled using different copolymer ratios [143]. In a study, 120 patients were dosed with Eligard 7.5 mg for up to 6 months and injection sites were closely monitored. Whilst there were some localized adverse reactions and infections, most of them were reported as mild and were nonrecurrent over time with antibiotic treatment [143]. Infections due to these polymeric implants are currently unknown.

15.5.2 Design of Antibacterial Biodegradable Cancer Implants

Fibrous scaffolds are currently receiving attention both as a means to prevent postsurgery adhesions and also to release drugs in a site-specific manner. These could be in tubular form. Whilst surgical implantation is required, they could be used where surgery is already indicated and the drug release profile can be controlled by varying the scaffold's morphology, porosity, and composition [144,145]. Drug-encapsulated hollow fibers could also be configured in the form of a stent design which could be implanted into certain cancerous areas such as prostate cancers without any surgery. An ideal medical implant would be a single or multidrug encapsulated flexible fully resorbable device which has antibacterial properties that can be formulated as a bead filled fiber/mesh, or tubular or stent configuration (Figure 15.4a through 15.4d). The antibiotic layer could be the outer and protective layer of this multilayer fiber (Figure 15.4c). The drug could be incorporated into the walls (see Figure 15.4b) and the lumen of the device, the combination of which creates a large surface area for drug release and allows slow precise release of the drug through the walls of the fiber in a sustained manner throughout the period of fiber degradation. Alteration of the wall properties allows the production of fibers tailored to deliver the precise doses over defined time lengths for any given application.

We are developing a novel drug delivery vehicle designed for use in local sustained chemotherapy. A range of biocompatible chemotherapy drug and antibiotic-encapsulated microfibers has been developed. Antimicrobial

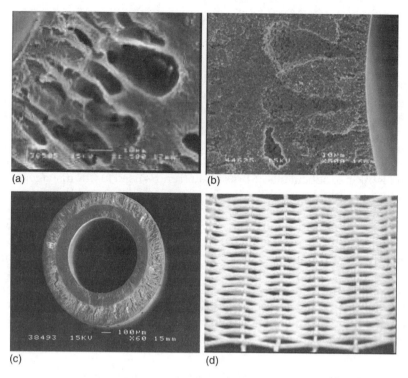

FIGURE 15.4

Drug encapsulated hollow fibers. (a) Macrocavities in fiber for liquid drug and skin to prevent high dose of drug release initially, (b) powder drugs encapsulated within walls, (c) multilayer microfibers (inner layer chemotherapy and outer layer antibiotics), and (d) drug-loaded fiber patch.

properties have been included into biomaterial design aimed at preventing biofilm formation whilst the device is intact. The examples of drug release studies with fibers system using chemotherapy drug 5-Fluorouracil (5-FU), heparin, and antibiotics are given in this section.

Figure 15.5a through 15.5c shows the heparin-loaded 85L:25 PLGA hollow fibers spun using a polymer solution of 20 wt%. This resulted in a drug permeable membrane wall with very large finger cavities on the inner layer and closely spaced external finger cavities as shown in Figure 15.5a and 15.5b. A middle sponge-like region was also present as shown in Figure 15.5c. The air gap selected during spinning also has an effect on the hollow fiber structure [144]. A larger air gap will cause a delay in the precipitation of the external surface leading to smaller external finger cavities. Decreasing the air gap results in a thinner membrane skin, which increases the degradation rate and drug permeability through the membrane.

Our studies so far have produced a range of biocompatible micro-hollow fibers with different degradation levels and diffusion properties [144,145]. To date, it has been demonstrated that it is possible to spin drug-loaded novel hollow fibers and produce unique microbeads with a range of biodegradation and diffusion properties [144], which should prove suitable

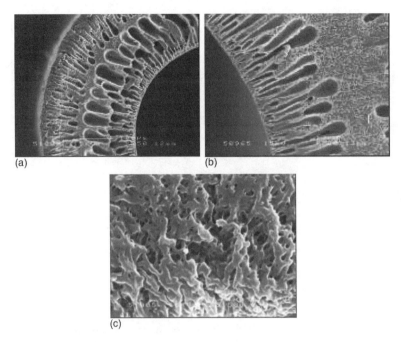

FIGURE 15.5
Heparin-loaded PLGA 85:15 fiber. (a) Cross section showing macrovoids, (b) PLA fiber with
fine pores in the center for controlled release of the drug, and (c) macroporous nature of the
65:35 PLGA for fast diffusion of the drug.

for treating cancer patients. The hollow fibers with different ratios and
combinations of drugs and polymers can be prepared by phase-inversion
technique, where the polymer is converted in a controlled manner from
liquid to solid state. The hollow fiber is formed when the polymer/solvent
solution is forced through a special orifice (the spinneret) at the same time as
an internal coagulant is pumped through the inner tube of the spinneret.
After the hollow fiber is formed, there is a period in which the fiber is
exposed to the air before it enters the nonsolvent bath where coagulation
occurs. The properties of the single and multidrug layer hollow fibers can
be controlled by manipulation of the polymer type, drug type composition/
concentration and varying the spinning parameters (orifice dimensions,
solvents to dissolve the polymer, coagulant type, and the residence time in
the air gap).

An ideal chemotherapy implant should consist of combined antibiotics
or antimicrobial agents and chemotherapy drugs and this could be a
drug-loaded multilayer hollow microfiber device or it could be drug-loaded
microbeads. Hence, release only occurs at the cancer target area. An advantage of the fiber system discussed is that it can allow the contemporaneous
release of several drugs into the target area. The heparin or antibiotic
could be loaded onto biodegradable fibers to achieve antimicrobial
activities [88,123]. Initial studies have identified optimal fiber composition

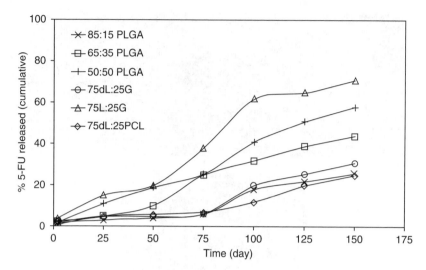

FIGURE 15.6
Pharmacokinetics of 5-FU from PLGA biopolymers.

for appropriate and controlled release of a model chemotherapy drug (5-FU). Pharmacokinetics of 5-FU and heparin release from hollow fibers (20% w/w drug loading) are given in Figures 15.6 and 15.7, respectively. A promising anti-biofilm strategy is the development of materials with antiadhesive surfaces such as heparin (Figure 15.7). Many studies show that such devices are protected against incrustation and biofilm formation for a longer period of time: 6–12 months, both in vitro and in vivo [58].

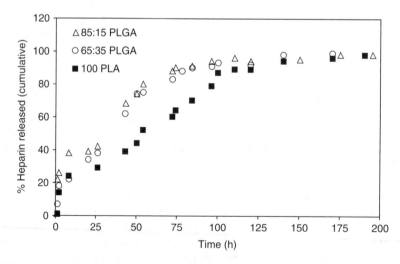

FIGURE 15.7
Release of heparin from PLA, 65:35 PLA/PCL, and 85:15 PLGA.

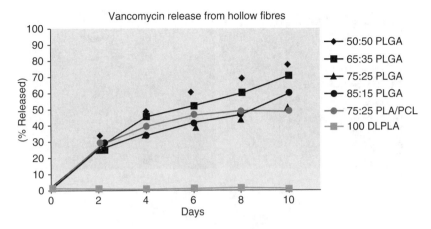

FIGURE 15.8
Vancomycin release from range of biodegradable PLGA, PLA, and PCL hollow fibers.

Figure 15.8 shows in vitro vancomycin release from PLGA, PLA, and PCL hollow fibers in phosphate buffer at 37°C. Vancomycin was chosen due to its wide antimicrobial spectrum and low MIC against gram-positive bacteria such as *S. aureus* [110]. Also, this model antibiotic is generally prescribed for cancer patients. The results in Figure 15.8 show that the release rate of vancomycin changes with PLA to PGA ratio. The higher the PLA content in the copolymer, the slower the release rate of antibiotic and higher storage modulus [146]. There is evidence that one will be able to produce antibiotic-encapsulated biodegradable fibers, beads, and capsules that can release a number of antibiotics over a long period for effective drug release against gram-positive and gram-negative bacteria [2,9,15].

Figure 15.9 shows that vancomycin-loaded 65:35 PLGA underwent extensive structural change, forming a brittle crystalline mass after 6 weeks of degradation and drug elution [145]. The surface of the structure was wrinkled with an almost "melted" appearance and consisted of large pores through which water or drug could diffuse through at a high rate (Figure 15.9b). This contraction of the polymer is due to bond scission of the amorphous phase in the polymer which causes relaxation of the macromolecules in the phase [115,146–148]. For water soluble antibiotic such as vancomycin in a hydrophobic polylactide matrix, the release mechanisms are controlled by channel diffusion, osmotic pressure, and polymer degradation. There is evidence that if the antibiotic loading is high within the polymer matrix, antibiotic particles will connect together to form channels leading to the surface of the fiber (Figures 15.9 and 15.10). This antibiotic will be released by channel diffusion [146]. 65:35 PLGA degraded quite rapidly in water, and hence is only suitable for short-term treatments (such as 5 day antibiotic treatment) or as part of a system for long-term drug delivery.

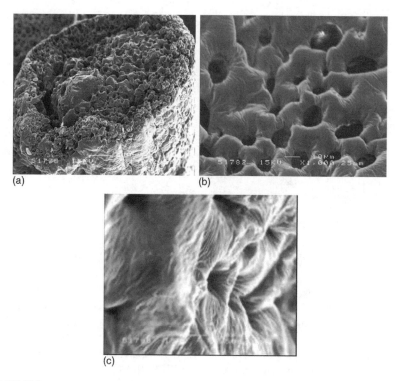

FIGURE 15.9

Vancomycin encapsulated 65:35 PLGA fiber after 6 weeks degradation. (a) Overview of cross section, (b) 1000 × magnification of the surface, and (c) surface 2500 × magnification.

Figure 15.10 shows the SEMs of the surface of polymer/vancomycin 3 weeks after the elution. Due to the large amounts of PCL in this copolymer, the glass transition of this polymer was low, leading to a very rubbery texture of the spun hollow fiber. However, after 3 weeks of degradation, the structure became more crystalline and brittle indicating water uptake

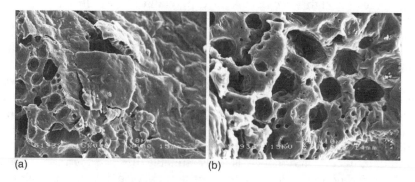

FIGURE 15.10

Vancomycin encapsulated 50:50 PLA/PCL after 3 weeks degradation. (a) surface showing osmotic cracking and (b) cross section showing large pore structures formed.

leading to plasticization and an increase in the T_g. The surface of the fiber was wrinkled and osmotic cracking had occurred as can be seen in Figure 15.10a. This structure had large pores within the cross section, quite similar to those formed in the 65:35 PLGA hollow fibers (Figure 15.10b).

In hollow fiber implants, the fibers have thin walls and large interconnecting pores and cavities that can be filled with specific drugs, Figure 15.4a and 15.4b. In the first phase after implantation, the drug released from the reservoir into cancer cells occurs in number of steps: dissolution of the drug in the polymer, diffusion of drug across the polymer membrane, and dissolution of the drugs into the external phase. Finally the biodegradation of the fibers would guarantee the complete release of the loaded drug. The kinetics of the drug release will depend on both diffusion and material degradation [72,73,121–123].

Generally with medical devices, bacteria on the device surface multiply to produce a confluent biofilm on the surface [11,12], leading to infection in the patient. Bacterial adherence and colonization are influenced by the surface characteristics and properties of a biomaterial, hence need to be investigated. One of the main design considerations is to understand bacteria attachment on polymer surfaces. Tests for bacterial attachment were carried out on the hollow fibers using an epidemic strain of methicillin-resistant *S. aureus* (EMRSA-5) and the gram-negative *P. aeruginosa* (PAO1) strains to identify the possibility of biofouling of the hollow fiber membranes. SEM micrographs of hollow fibers exposed to the strains after 1 and 18 h showed bacterial attachment on 85:15 and 50:50 PLGA (Figures 15.11 and 15.12).

It is extremely important to have a biodegradable matrix that releases antibiotics locally at the implant surface, thereby minimizing bacterial accumulation. Fabricating and formulating antibiotic-loaded devices, as well as evidence of their bactericidal properties, need to be investigated.

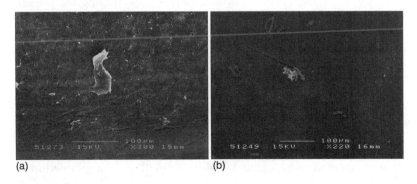

(a)   (b)

FIGURE 15.11
Fiber surface. (a) EMRSA-5 and (b) PAO1 bacteria strains, attached to the surface of 85:15 PLGA hollow fibers after an incubation time of 1 h.

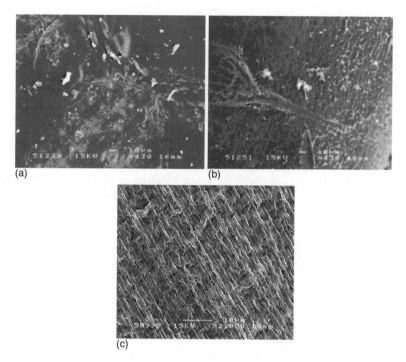

FIGURE 15.12
Hollow fiber surface. (a) 50:50 PLGA in PA01 bacteria after 1 h incubation, (b) 50:50 PLGA in PA01 after 18 h incubation, and (c) heparin-treated surface exposed to PA01 for 18 h.

Also further research is need to be conducted to find ways of treating the surface to make it less attractive to bacteria.

Heparin-loaded biofilm did not show any bacteria after an incubation time of 18 h (Figure 15.12). However, the effect of heparin on biopolymer surfaces needs to be further investigated. Our observations suggested that the released Heparin was biologically active in preventing the bacteria from permanently adhering to the substrate, and thus decreasing the possibility of biofilm-related infection.

15.6 Discussion and Future Aspects

We are beginning to understand the physiology of biofilm bacteria at the genomic and proteomic levels and we must look to these studies to yield new targets and to stimulate the design of drugs that are worth carrying. In the meantime, the adage that "prevention is better than cure" comes to mind. In the context of preventing biofilm formation on implanted medical devices, it is certainly better but no less challenging. The key has to be recognition of the problem of microbial colonization and biofilm formation at the outset of material design.

Acknowledgments

Work in the authors' laboratories is supported by the BBSRC, EPSRC, Royal United Hospital Oncology/Gynae–Oncology team and the Department of Health.

References

1. Fux, C.A. et al., Bacterial biofilms: A diagnostic and therapeutic challenge, Expert review, *Anti-Infect. Ther.*, 1, 667–683, 2003.
2. Stewart, P.S. and Costerton, J.W., Antibiotic resistance of bacteria in biofilms, *Lancet*, 358, 135–138, 2001.
3. Hall-Stoodley, L., Costerton, J.W., and Stoodley, P., Bacterial biofilms: From the natural environment to infectious diseases, *Nat. Rev. Microbiol.*, 2, 95–108, 2004.
4. Stoodley, P. et al., Biofilm as complex differentiated communities, *Annu. Rev. Microbiol.*, 56, 187–209, 2002.
5. Zimmerli, W., Trampuz, A., and Ochsner, P.E., Prosthetic-joint infections, *N. Engl. J. Med.*, 351, 1645–1654, 2004.
6. Darouchie, R.O., Treatment of infections associated with surgical implants, *N. Engl. J. Med.*, 350, 1422–1429, 2004.
7. Branda, S.S. et al., Biofilms: The matrix revisited, *Trends Microbiol.*, 13, 20–26, 2005.
8. Dankert, J., Hogt, A.H., and Feijen, J., Biomedical polymers: Bacterial adhesion, colonization, and infection, *CRC Crit. Rev. Biocompat.*, 2, 219–301, 1986.
9. Gilbert, P., Collier, P.J., and Brown, M.R.W., Influence of growth rate on susceptibility to antimicrobial agents: Biofilms, cell cycle, dormancy, and stringent response, *Antimicrob. Agents Chemother.*, 34, 1865–1868, 1990.
10. Costerton, J.W., Stewart, P.S., and Greenberg, E.P., Bacterial biofilms: A common cause of persistent infections, *Science*, 284, 1318–1322, 1999.
11. Lewis, K., Riddle of biofilm resistance, *Antimicrob. Agents Chemother.*, 45, 999–1007, 2001.
12. Mah, T.-F.C. and O'Toole, G.A., Mechanisms of biofilm resistance to antimicrobial agents, *Trends Microbiol.*, 9, 34–39, 2001.
13. Brown, M.R.W. and Smith, A.W., Antimicrobial agents and biofilms, in *Medical Implications of Biofilms*, Wilson, M. and Devine, D., (Eds.), Cambridge University Press, Cambridge, 2003, pp. 36–55.
14. Gilbert, P. et al., The physiology and collective recalcitrance of microbial biofilm communities, *Adv. Microb. Physiol.*, 46, 202–256, 2002.
15. Brooun, A., Liu, S., and Lewis, K., A dose-response study of antibiotic resistance in *Pseudomonas aeruginosa* biofilms, *Antimicrob. Agents Chemother.*, 44, 640–646, 2000.
16. Smith, A.W. and Brown, M.R.W., Biofilms, dormancy and resistance, in *Dormancy and Low-growth States in Microbial Disease*, Coates, A.R.M., (Ed.), Cambridge University Press, Cambridge, 2003, pp. 161–180.
17. Keren, I. et al., Persister cells and tolerance to antimicrobials, *FEMS Microbiol. Lett.*, 230, 13–18, 2004.
18. Keren, I. et al., Specialized persister cells and the mechanism of multidrug tolerance in *Escherichia coli*, *J. Bacteriol.*, 186, 8172–8180, 2004.

19. Roberts, M.E. and Stewart, P.S., Modelling protection from antimicrobial agents in biofilms through the formation of persister cells, *Microbiology*, 151, 75–80, 2005.
20. Drenkard, E., Antimicrobial resistance of *Pseudomonas aeruginosa* biofilms, *Microbes Infect.*, 5, 1213–1219, 2003.
21. Gilbert, P., Hodgson, A.E., and Brown, M.R.W., Influence of the environment on the properties of microorganisms grown in association with surfaces, in *Microbiological Quality Assurance: A Guide Towards Relevance and Reproducibility of Inocula*, Brown, M.R.W. and Gilbert, P., (Eds.), CRC Press, Boca Raton, FL, 1995, pp. 61–82.
22. Stewart, P.S., Theoretical aspects of antibiotic diffusion into microbial biofilms, *Antimicrob. Agents Chemother.*, 38, 1996, pp. 2125–2133.
23. Xu, K.D., McFeters, G., and Stewart, P.S., Biofilm resistance to antimicrobial agents, *Microbiology*, 146, 547–549, 2000.
24. Nichols, W.W., Biofilms, antibiotics and penetration, *Rev. Med. Microbiol.*, 2, 177–181, 1991.
25. Szomolay, B. et al., Adaptive responses to antimicrobial agents in biofilms, *Environ. Microbiol.*, 7, 1186–1191, 2005.
26. Mah, T.F. et al., A genetic basis for *Pseudomonas aeruginosa* biofilm antibiotic resistance, *Nature*, 426, 6964, 306–310, 2003.
27. Smith, A.W., Bacterial resistance to antibiotics, in *Hugo and Russell's Pharmaceutical Microbiology*, 7th ed., Denyer, S.P., Hodges, N.A., and Gorman, S.P., (Eds.), Blackwell Publishing, Oxford, UK, 2004, pp. 220–232.
28. Bagge, N. et al., Rapid development in vitro and in vivo of resistance to ceftazidime in biofilm-growing *Pseudomonas aeruginosa* due to chromosomal β-lactamase, *APMIS*, 108, 589–600, 2000.
29. Giwercman, B. et al., Induction of β-lactamase production in *Pseudomonas aeruginosa* biofilm, *Antimicrob. Agents Chemother.*, 35, 1008–1010, 1991.
30. Anderl, J.N., Franklin, M.J., and Stewart, P.S., Role of antibiotic penetration limitation in *Klebsiella pneumoniae* biofilm resistance to ampicillin and ciprofloxacin, *Antimicrob. Agents Chemother.*, 44, 1818–1824, 2000.
31. Sutherland, I.W., The biofilm matrix—An immobilised but dynamic environment, *Trends Microbiol.*, 9, 222–227, 2001.
32. Kumar, A. and Schweizer, H.P., Bacterial resistance to antibiotics: Active efflux and reduced uptake, *Adv. Drug Del. Rev.*, 57, 10, 1486–1513, 2005.
33. Hassett, D.J. et al., Anaerobic metabolism and quorum sensing by *Pseudomonas aeruginosa* biofilms in chronically infected cystic fibrosis airways: Rethinking antibiotic treatment strategies and drug targets, *Adv. Drug Del. Rev.*, 54, 1425–1443, 2002.
34. Borriello, G. et al., Oxygen limitation contributes to antibiotic tolerance of *Pseudomonas aeruginosa* biofilms, *Antimicrob. Agents Chemother.*, 48, 2659–2664, 2004.
35. Werner, E. et al., Stratified growth in *Pseudomonas aeruginosa* biofilms, *Appl. Environ. Microbiol.*, 70, 6188–6196, 2004.
36. Burmolle, M. et al., Enhanced biofilm formation and increased resistance to antimicrobial agents and bacterial invasion are caused by synergistic interactions in multispecies biofilms, *Appl. Environ. Microbiol.*, 72, 3916–3923, 2006.
37. Parsek, M.R. and Greenberg, E.P., Sociomicrobiology: The connections between quorum sensing and biofilms, *Trends Microbiol.*, 13, 1, 27–33, 2005.
38. Davies, D.G. et al., Quorum-sensing genes in *Pseudomonas aeruginosa* biofilms: Their role and expression patterns, *Science*, 280, 295–298, 1998.

39. An, Y.H., Dickinson, R.B., and Doyle, R.J., Mechanisms of bacterial adhesion and pathogenesis of implant and tissue infections, in *Handbook of Bacterial Adhesion: Principles, Methods, and Applications*, An, Y.H. and Friedman, R.J., (Eds.), Humana Press Inc., Totowa, NJ, 2000, pp. 1–27.

40. Kolter, R. and Greenberg, E.P., The superficial life of microbes, *Nature*, 441, 300–302, 2006.

41. Lin, J. et al., Bactericidal properties of flat surfaces and nanoparticles derivatized with alkylated polyethylenimines, *Biotechnol. Prog.*, 18, 1082–1086, 2002.

42. Bach, A. et al., Efficacy of silver-coating central venous catheters in reducing bacterial colonization, *Crit. Care Med.*, 27, 515–520, 1999.

43. Donelli, G. et al., New polymer-antibiotic systems to inhibit bacterial biofilm formation: A suitable approach to prevent central venous catheter-associated infections, *J. Chemother.*, 14, 501–507, 2002.

44. Schierholz, J.M. et al., Efficacy of silver-coated medical devices, *J. Hosp. Infect.*, 40, 257–262, 1998.

45. Sheretz, R.J. et al., Efficacy of antibiotic-coated catheters in preventing subcutaneous *Staphylococcus aureus* infection, *J. Infect. Dis.*, 167, 98–106, 1993.

46. Tebbs, S.E. and Elliott, T.S.J., A novel antimicrobial central venous catheter impregnated with benzalkonium chloride, *J. Antimicrob. Chemother.*, 31, 261–271, 1993.

47. Donelli, G. and Francolini, I., Efficacy of antiadhesive, antibiotic and antiseptic coatings in preventing catheter-related infections: Review, *J. Chemother.*, 13, 595–606, 2001.

48. Danese, P.N., Antibiofilm approaches: Prevention of catheter colonization, *Chem. Biol.*, 9, 873–880, 2002.

49. Rachid, S. et al., Effect of subinhibitory antibiotic concentrations on polysaccharide intercellular adhesin expression in biofilm-forming *Staphylococcus epidermidis*, *Antimicrob. Agents Chemother.*, 44, 3357–3363, 2000.

50. Piozzi, A. et al., Antimicrobial activity of polyurethanes coated with antibiotics: A new approach to the realization of medical devices exempt from microbial colonization, *Int. J. Pharm.*, 280, 173–183, 2004.

51. Woo, G.L.Y. et al., Biological characterization of a novel biodegradable antimicrobial polymer synthesized with fluoroquinolones, *J. Biomed. Mater. Res.*, 59, 35–45, 2002.

52. DiTizio, V. et al., A liposomal hydrogel for the prevention of bacterial adhesion to catheters, *Biomaterials*, 19, 1877–1884, 1998.

53. Norwood, S., Hajjar, G., and Jenkins, L., The influence of an attachable subcutaneous cuff for preventing triple lumen catheter infections in critically ill surgery patients, *Surgery*, 175, 33–40, 1992.

54. Jansen, B., New concepts in the prevention of polymer-associated foreign body infections, *Int. J. Med. Micro.*, 272, 401–410, 1990.

55. Kamal, G.D. et al., Reduced intravascular catheter infection by antibiotic binding: A prospective, randomized, controlled trial, *J. Am. Med. Assoc.*, 265, 2364–2368, 1991.

56. Rodriguez, J.L. et al., Reduced bacterial adherence to surfactant-coated catheters, *Curr. Surg.*, 43, 423–425, 1986.

57. Farber, B.E. and Wolff, A.G., The use of nonsteroidal anti-inflammatory drugs to prevent the adherence of *Staphylococcus epidermidis* to medical polymers, *J. Infect. Dis.*, 166, 861–865, 1992.

58. Mermel, L.A., Stolz, S.M., and Maki, D.G., Surface antimicrobial activity of heparin-bonded and antiseptic-impregnated vascular catheters, *J. Infect. Dis.*, 167, 920–924, 1993.

59. Prahlad, A., Harvey, R.A., and Greco, R.S., Diffusion of antibiotics from a poly-tetrafluoroethylene-benzalkonium surface, *Am. Surgeon*, 47, 515–518, 1981.

60. Chinn, J.A., Ratner, B.D., and Horbett, T.A., Adsorption of baboon fibrinogen and the adhesion of platelets to a thin film polymer deposited by radio-frequency glow discharge of allylamine, *Biomaterials*, 13, 5, 322–332, 1992.

61. Samuel, U. and Guggenbichler, J.P., Prevention of catheter-related infections: The potential of new nano-silver impregnated catheter, *Int. J. Antimicrob. Agents*, 23, 75–78, 2004.

62. Furno, F. et al., Silver nanoparticles and polymeric medical devices: A new approach to prevention of infection? *J. Antimicrob. Chemother.*, 54, 1019–1024, 2004.

63. Jagur-Grodzinski, J., Polymers for tissue engineering, medical devices, and regenerative medicine, concise general review of recent studies, *Polym. Adv. Technol.*, 17, 395–418, 2006.

64. Han, S.Y. et al., Biodegradable polymer releasing antibiotic developed for drainage catheter of cerebrospinal fluid: In vitro results, *J. Korean Med. Sci.*, 20, 297–301, 2005.

65. Sternberg, K. et al., Tissue cultivation for in vitro testing of polymer-based bioabsorbable urethral stents, *Urologe*, A, 43, 1400, 2004.

66. Francolini, I. et al., Usnic acid, a natural antimicrobial agent able to inhibit bacterial biofilm formation on polymer surfaces, *Antimicrob. Agents Chemother.*, 48, 4360–4365, 2004.

67. Freiberg, S. and Zhu, X.X., Polymer microspheres for controlled drug release, *Int. J. Pharm.*, 282, 1–18, 2004.

68. Varde, N.K. and Pack, D.W., Microspheres for controlled drug delivery, *Expert Opinion. Biol. Ther.*, 4, 35–51, 2004.

69. Moses, M.A., Brem, H., and Langer, R., Advancing the field of drug delivery: Taking aim at cancer, *Cancer Cell*, 4, 337–341, 2003.

70. Sinha, V.R. and Trehan, A., Biodegradable microspheres for protein delivery, *J. Control. Release*, 90, 261–280, 2003.

71. Zhao, Z. et al., Polyphosphoesters in drug and gene delivery, *Adv. Drug Del. Rev.*, 55, 483–499, 2003.

72. Rabinow, B.E., Nanosuspensions in drug delivery, *Nat. Rev. Drug Discov.*, 3, 785–796, 2004.

73. Sinha, R. et al., Poly-caprolactone microspheres and nanospheres: An overview, *Int. J. Pharm.*, 278, 1–23, 2004.

74. Campbell, C.S.J., Lee, I.Y.S., and Perera, S., Dept. of Chemical Engineering, University of Bath., Gilby, E., Jaaback, K., and Johnson, N., Royal United Hospital, Bath., Slow release intraperitoneal chemotherapy for ovarian metastatic cancer, presented at British Gynaecological Cancer Society Annual Conference, Leeds, November 9–11, 2005.

75. Fan, L.T. and Singh, S.K., *Controlled Release a Quantitative Treatment*, Springer-Verlag, London, Berlin, 1989, pp. 3–5.

76. Chaubal, M., *Overview on drug delivery with emphasis on polymeric implants*, http://members.tripod.com/~Chaubal/review.html, accessed 9 April 2005.

77. Brannon-Peppas, L., Polymer in controlled drug delivery, *Med. Plast. Biomater.*, 4, 34–44, 1997.

78. Lewis, D.H., Controlled release of bioactive agents from lactide/glycolide polymers, in *Biodegradable Polymers in Drug Delivery*, Langer, R. and Chasin, M., (Eds.), Marcel Dekker Publishers, NY, 1990, pp. 1–41.

79. Ramchandani, M., Pankaskie, M., and Robinson, D.H., The influence of manufacturing procedure on the degradation of poly(lactide-co-glycolide) 85:15 and 50:50 implants, *J. Control. Release*, 4, 161–173, 1997.

80. Middleton, J.C. and Tipton, A.J., Synthetic biodegradable polymers as orthopedic devices, *Biomaterials*, 21, 2335–2346, 2000.

81. Hammond, J.S., Beckingham, I.J., and Shakesheff, K.M., Expert review of medical devices, *Biomaterials*, 21, 2335–2346, 2000.

82. Schlieckera, G. et al., Characterization of a homologous series of D,L-lactic acid oligomers: A mechanistic study on the degradation kinetics in vitro, *Biomaterials*, 24, 3835–3844, 2003.

83. Perera, S.P., Hollow fibres for drug delivery and tissue engineering, presented at engineering with membranes 2005, European Membrane Society: Medical and Biological Applications, Camogli, Italy, May 15–18, 2005.

84. Cutright, D.E. et al., Degradation rates of polymers and copolymers of polylactic and polyglycolic acids, *Oral Surg. Oral Med. Oral Pathol.*, 37, 142–152, 1974.

85. Miller, R.A., Brady, J.M., and Cutright, D.E., Degradation rates of oral resorbable implants (polylactates and polyglycolates): Rate modification with changes in PLA/PGA copolymer ratios, *J. Biomed. Mater. Res.*, 11, 711–719, 1977.

86. Kweona, H.Y. et al., A novel degradable polycaprolactone networks for tissue engineering, *Biomaterials*, 24, 801–808, 2003.

87. Kister, G. et al., Structural characterization and hydrolytic degradation of solid copolymers of D,L-lactide-co-1-caprolactone by Raman spectroscopy, *Polymer*, 41, 925–932, 2000.

88. Sinha, R. et al., Poly-caprolactone microspheres and nanospheres: An overview, *Int. J. Pharm.*, 278, 1–23, 2004.

89. Kwok, C.S. et al., Design of infection-resistant antibiotic-releasing polymers: I. Fabrication and formulation, *J. Controlled Release*, 62, 62, 289–299, 1999.

90. Anwar, H., Strap, J.L., and Costerton, J.W., Establishment of aging biofilms: Possible mechanism of bacterial resistance to antimicrobial therapy, *Antimicrob. Agents Chemother.*, 36, 7, 1347–1351, 1992.

91. Segreti, J. and Levin, S., The role of prophylactic antibiotics in the prevention of prosthetic device infections, *Infect. Dis. Clin. of N. Am.*, 3, 2, 357–370, 1989.

92. Kwok, C.S., Horbett, T.A., and Ratner, B.D., Design of infection-resistant antibiotic-releasing polymers: II. Controlled release of antibiotics through a plasma-deposited thin film barrier, *J. Controlled Release*, 62, 301–311, 1999.

93. Kwok, C.S., Ratner, B.D., and Horbett, T.A., Infection-resistant biomaterials releasing antibiotics through a plasma-deposited thin film, *ACS Polym. Prepr.*, 38, 1, 1077, 1997.

94. Kwok, C. et al., Infection resistant biomaterials releasing antibiotics into the interfacial zone, in *23rd Proc. Int. Symp. Control. Release Bioactive Mater.*, 23, 230–231, 1996.

95. BIOSPAN segmented polyurethanes fact sheets, The Polymer Technology Group Inc., Emeryville, CA, 1995.

96. Kanellalopoulou, K. and Giamarellos-Bourboulis, E.J., Carrier systems for the local delivery of antibiotics in bone infections, *Drugs*, 59, 1223–1232, 2000.

97. Castro, C. et al., Ciprofloxacin implants for bone infection, in vitro/in vivo characterization, *J. Control. Release*, 93, 341–354, 2003.

98. Yenice, I. et al., In vitro/in vivo evaluation of the efficiency of teicoplanin-loaded biodegradable microparticles formulated for implantation to infected bone defects, *J. Microencapsul.*, 20, 705–717, 2003.

99. Yenice, I. et al., Biodegradable implantable teicoplanin beads for the treatment of bone infections, *Int. J. Pharm.*, 242, 271–275, 2002.

100. Gursel, I. et al., In vitro antibiotic release from poly(3hydroxybutyrate-co-3-hydroxyvalerate) rods, *J. Microencapsul.*, 19, 153–164, 2002.

101. Hasegawa, M. et al., High release of antibiotic from a novel hydroxyapatite with bimodal pore size distribution, *J. Biomed. Mater. Res.*, 70B, 332–339, 2004.

102. Yoo, J.Y. et al., Effect of lactide/glycolide monomers on release behaviours of gentamicin sulfate-loaded PLGA discs, *Int. J. Pharm.*, 276, 1–9, 2004.

103. Kim, K. et al., Incorporation and controlled release of a hydrophilic antibiotic using poly(lactide-co-glycolide)-based electrospun nanofibrous scaffolds, *J. Control. Release*, 98, 47–56, 2004.

104. (a) Mincheva, R. et al., Hydrogels from chitosan crosslinked with poly(ethylene glycol) diacid as bone regeneration materials, e-Polymers 2004, 058, (b) Hoemann, C.D. and Hurtig, M., Chitosan-glycerol phosphate/blood implants improve hyaline cartilage repair in ovine microfructere defects, *J. Bone Joint Surg. Am.*, 87A, 2671–2686, 2005.

105. Alt, V. et al., Effect of glycerol-L-lactide coating polymer on bone ingrowth of BFGF-coated hydroxyapatite implants, *J. Control. Release*, 99, 103–111, 2004.

106. Kelly, H.M. et al., Formulation and preliminary *in vivo* dog studies of a novel drug delivery system for the treatment of periodontitis, *Int. J. Pharm.*, 274, 167–183, 2004.

107. Miyazaki, S. et al., Percutaneous absorption of indomethacin from pluronic F 127 gels in rats, *J. Pharm. Pharmacol.*, 47, 455–457, 1995.

108. Esposito, E. et al., Comparative analysis of tetracycline-containing dental gels: Poloxamer- and monoglyceride-based formulations, *Int. J. Pharm.*, 142, 9–23, 1996.

109. Veyries, M.L. et al., Controlled release of vancomycin from poloxamer 407 gels, *Int. J. Pharm.*, 192, 183–193, 1999.

110. Lee, S.H. et al., Regional delivery of vancomycin using pluronic F-127 to inhibit methicillin resistance *Staphylococcus aureus* (MRSA) growth in chronic otitis media in vitro and in vivo, *J. Control. Release*, 96, 1–7, 2004.

111. McLaren, A.C., Alternative materials to acrylic bone cement for delivery of depot antibiotics in orthopaedic infections, *Clinic. Orthopaed. Relat. Res.*, 427, 101–106, 2004.

112. Schiller, C. et al., Geometrically structured implants for cranial reconstruction made of biodegradable polyesters and calcium phosphate/calcium carbonate, *Biomaterials*, 25, 1239–1247, 2004.

113. (a) Akmaz, I. et al., Biodegradable implants in the treatment of scaphoid non-unions, *Int. Orthopaed.*, 28, 261–266, 2004, (b) Leiggener, C.S. et al., Influence of copolymer composition of polylactide implants on cranial bone regeneration, *Biomaterials*, 27, 202–207, 2006.

114. Williams, D.F., An introduction to materials in medicine, in *Biomaterials Science*, Ratner, B.D. et al., (Eds.), Academic Press, New York, book review, 2004, *Biomaterials*, 26, 24, 5093, 2005.

115. Ramchandani, M. and Robinson, D., In vitro and in vivo release of ciprofloxacin from PLGA 50:50 implants, *J. Cont. Release*, 54, 2, 167–175, 1998.

116. Xie, G., Tan, X.L., and Gao, Z.Q., An interference-free implantable glucose microbiosensor based on use of a polymeric analyte-regulating membrane, *Frontiers Biosci.*, 10, 1797–1801, 2005.

117. Lowell, J.A. and Bothe, A., Venous access, preoperative, operative, postoperative dilemmas, *Surg. Clin. N. Am.*, 7, 1231–1246, 1991.

118. Mirro, J. Jr. et al., A prospective of Hickman/Broviac catheters and implantable ports in pediatric oncology patients, *J. Clin. Oncol.*, 7, 214–222, 1989.
119. Ross, M.N. et al., Comparison of totally implanted reservoirs with external catheters as venous access devices in pediatric oncologic patients, *Surg. Gynecol. Obstet.*, 167, 141–144, 1988.
120. Mueller, B.U. et al., A prospective randomized trial comparing the infections complications of the externalized catheters versus a subcutaneously implanted device in cancer patients, *J. Clin. Oncol.*, 10, 1943–1948, 1992.
121. D'Antonio, D. et al., Comparison of ciprofloxacin, ofloxacin and pefloxacin for the prevention of the bacterial infection in neutropenic patients with haematological malignancies, *J. Antimicrob. Chemother.*, 33, 837–844, 1994.
122. Pasquale, M.D., Compell, J.M., and Magnant, C.M., Groshong versus Hickmen catheters, *Surg. Gynecol. Obstet.*, 174, 408–410, 1992.
123. Benezra, D. et al., Prospective study of infections in indwelling central venous catheters using quantitative blood cultures, *Am. J. Med.*, 85, 495–498, 1988.
124. Calle, J.P. et al., Totally implanted device for long-term intravenous chemotherapy: Experience in 123 adult patients with solid neoplasm, *J. Surg. Oncol.*, 62, 273–278, 1996.
125. Kock, H.J. et al., Implantable vascular access systems: Experience in 1500 patients with totally implanted central venous port systems, *World J. Surg.*, 22, 12–16, 1998.
126. Pegues, D. et al., Comparison of infections in Hickman and implanted port catheters in adult solid tumor patients, *J. Surg. Oncol.*, 49, 156–162, 1992.
127. Salem, R.R., Ward, B.A., and Ravikumar, T.S., A new peripherally implanted subcutaneous permanent central venous access device for patients requiring chemotherapy, *J. Clin. Oncol.*, 11, 2181–2185, 1993.
128. Kuzel, T.B. et al., Thrombogenecity of intravenous 5-fluorouracil alone or in combination with cisplatin, *Cancer*, 65, 885–889, 1990.
129. Muller, B.U., Skelton, J., and Dallender, D., A prospective randomized trial comparing the infectious and noninfectious complications of an externalized catheter versus a subcutaneously implanted devices in cancer patients, *J. Clin. Oncol.*, 10, 1943–1948, 1992.
130. Schlapp, M. and Friess, W., Collagen/PLGA microparticle controlled composites for local delivery of gentamicin, *J. Pharm. Sci.*, 92, 2145–2151, 2003.
131. Howell, P.B. et al., Risk factor for infection of adult patents with cancer who have tunnelled central venous catheters, *Cancer*, 75, 1367–1375, 1995.
132. Fung, L.K. et al., Chemotherapeutic drugs released from polymers: Distribution of 1,3-bi(2-chloroethyl)-1-nitrosurea in the rat brain, *Pharm. Res.*, 13, 671–682, 1996.
133. Wang, F.J., Lee, T.K.Y., and Wang, C.H., PEG modulated release of etanidazole from implantable PLGA/PLLA discs, *Biomaterials*, 23, 3555–3566, 2002.
134. Guney, O. and Akgerman, A., Synthesis of controlled-release products in supercritical medium, *AIChE J.*, 48, 4, 856–866, 2002.
135. Brem, H. and Langer, R., Polymer-based drug delivery to the brain, *Sci. Med.*, 3, 2–11, 1996.
136. Donald, T. et al., Radiosensitization of a mouse tumor model by sustained intra-tumoral release of etanidazole and tirapazamine using a biodegradable polymer implant device, *Radiother. Oncol.*, 53, 77–84, 1999.
137. Lina, R., Nga, L.S., and Wang, C.H., In vitro study of anti-cancer drug doxorubicin in PLGA-based microparticles, *Biomaterials*, 26, 4476–4485, 2005.

138. Domb, A.J., Polymeric carriers for regional drug therapy, presented at VI International Symposium on Frontiers in Biomedical Polymers, Granada, June 16–18, 2005.
139. Sliwniak, R. and Domb, A.J., Macrolactones and polyesters from ricinoleic acid, *Biomacromolecules*, 6, 1679–1688, 2005.
140. Sliwniak, R. and Domb, A.J., Lactic acid and ricinoleic acid based copolyesters, *Macromolecules*, 38, 5545–5553, 2005.
141. Jordan, O. et al., Biomaterials for injectable therapeutic implants, *Chimia*, 59, 353–356, 2005.
142. Gibson, J.W. and Tipton Arthur, J., Patent Appl. WO2005067889, Polymeric implants, preferably containing a mixture of PEG and PLG, for controlled release of active agents, preferably GNRH, 2005.
143. Eligard Sanofi Aventis, http://www.eligard.com/index.asp
144. Perera, S.P., GB/EU Patent (pending) Appl. 0514026.4 and 054072.8, Hollow Fibres, 2005.
145. Perera, S.P., GB/EU Patent Appl. 0522569.3, Bio-compatible Drug Delivery Device, 2005.
146. Liu, S.-J. et al., A novel solvent-free method for the manufacture of biodegradable antibiotic-capsules for a long-term drug release using compression sintering and ultrasonic welding techniques, *Biomaterials*, 26, 4662–4669, 2005.
147. Lamprecht, A. et al., Microsphere design for the colonic delivery of 5-fluorouracil, *J. Control. Release*, 90, 313–322, 2003.
148. Harper, E. et al., Enhanced efficacy of a novel controlled release paclitaxel formation (PACLIMER delivery system) for local-regional therapy of lung cancer tumour nodules in mice, *Clin. Cancer Res.*, 5, 4242–4248, 1999.

16

Infections of Intracardiac Devices

George A. Yesenosky, Scott W. Sinner, Steven P. Kutalek, and Allan R. Tunkel

CONTENTS

16.1   Introduction

In the past two decades, the number of implanted intracardiac devices has increased markedly. The greatest area of growth has been with the use of cardiac pacemakers and implantable cardioverter defibrillators (ICDs). Between the years 1990 and 2002, approximately 2.25 million pacemakers and 416,000 ICDs were implanted in the United States alone [1]. The use of prosthetic heart valves, left ventricular assist devices (LVADs), and other prosthetic intracardiac devices has likewise increased over time. There have been widely varying reports of the rates of infection of these devices, ranging from as low as 1% in some studies of ICDs [2] to as high as 85% in one study of LVADs implanted for longer than 2 weeks [3]. Even with some estimates of the decreasing incidence of infection of these devices, the total number of infections may well continue to rise, in parallel with their increasing use. In this chapter, we will review the current state of knowledge regarding infections of intracardiac prosthetic devices, highlighting the epidemiology, etiology, clinical features, and management of these infections, as well as the changes that might be seen with these infections in the future.

16.2 Cardiac Pacemakers and Implantable Cardioverter Defibrillators

16.2.1 Epidemiology and Etiology

Pacemaker and implantable cardioverter defibrillator (ICD) systems consist of a generator, typically located in a surgically created pocket below the subcutaneous or subfascial tissue in the chest or abdominal wall, and subcutaneous leads that enter the endovascular system and interact with the generator through electrodes attached to the endocardium. Leads are occasionally positioned from an endovascular site within the coronary sinus (an epicardial structure), and leads or patches with direct epicardial attachments are utilized; these epicardial leads are tunneled in the subcutaneous tissues [4]. The incidence of pacemaker and ICD system infections has been reported to be as high as 19.9% [5]. The overall incidence of pacemaker and ICD infections, however, has decreased with the advent of transvenous leads and improved surgical techniques, although the incidence of infection may still be as high as 7% [6–8]. Pacemaker and ICD infections can be divided into early (0–28 days after device implantation), late (28–364 days after device implantation), and delayed (>364 days after device implantation).

Infections of the generator pockets are more common and clinically tend to appear earlier than infections involving the endovascular portions of the leads. This may be secondary to contamination of the generator pocket at the time of implant or generator change [9]. Lead infections, that tend to develop at least 1 month after insertion, can result from contiguous spread from an infected generator, or direct seeding of the endovascular portions of the leads [6,9–11]. Tissue erosion and local lesions over the device pocket may serve as a portal of entry for organisms, although hematogenous device seeding also can occur [9]. Contiguous spread from infected leads can infect the myocardium and cardiac valves; however, pacemaker endocarditis is rare (0.1%–3.1% of all pacemaker infections) and is defined by continuous bacteremia with evidence of vegetation located on the endovascular portion of device lead [7,10]. Transient bacteremia from catheter-related infections and other sources have been documented to cause seeding of pacemaker leads [12]. Late primary infections of leads are significantly less common than pocket infections, because as the leads become endothelialized, they are less likely to become infected; an inert coating over the leads may also diminish the likelihood of bacterial adherence.

Factors associated with an increased risk of pacemaker infection include skin necrosis over the generator, trauma at the generator site, bacteremia from a source other than the infected pacemaker, dermatologic conditions involving the skin overlying the generator, prolonged surgery, and replacement or repositioning of the generator or electrodes [6,7,10,11]. An increased incidence for infection is also noted in patients with coronary artery disease, coronary artery bypass surgery, atrial fibrillation, chronic underlying illnesses such as diabetes mellitus, end-stage renal disease, malignancy, malnutrition, use of corticosteroids or other immunosuppressive agents, and the use of anticoagulants, because hematoma formation following implantation is also considered to be a risk factor for subsequent infection [9,13]. The microbial etiologies of pacemaker and ICD infections are often from organisms that exist as normal skin flora, including certain organisms with an affinity for prosthetic material [12]. Staphylococci are the most common organisms isolated, with *Staphylococcus aureus* and *Staphylococcus epidermidis* accounting for approximately 75% of the cases [14]. Staphylococci seem to be particularly capable of adhering to and forming microcolonies on the polyethylene or silicone rubber sheaths covering the electrodes [12]. Other organisms isolated include *Corynebacterium* species, streptococci, enterococci, micrococci, HACEK organisms (noninfluenzae *Haemophilus* species, *Actinobacillus actinomycetemcomitans*, *Cardiobacterium hominis*, *Eikenella corrodens*, and *Kingella kingae*), various Enterobacteriaceae, and *Pseudomonas* species. Fungi (*Candida* and *Aspergillus* species) and nontuberculous mycobacteria have also been reported to cause pacemaker infections [7,9,10,15,16]. Other bacteria and fungi have rarely been reported (Table 16.1).

16.2.2 Clinical Presentation and Diagnosis

The clinical presentation of pacemaker infection varies with the site infected, the causative microorganism, and whether or not there is concomitant bacteremia [6,7,10,11]. Infection can involve the generator pocket, the subcutaneous pacemaker leads, the endovascular leads, or a combination. Patients with generator pocket infections frequently present with pain, tenderness, swelling, and erythema over the generator site. The skin overlying the generator may be adherent or erode, and purulent discharge may be present (Table 16.2). Erythema tracking along the path where the leads enter the vasculature may also be seen. In one series of 75 patients with pacemaker infection, 76% presented with pocket infection; erosion of part of the pacing system through the overlying skin occurred in 24% of cases [17]. Patients with infection caused by *S. aureus* tend to present more rapidly with more acute symptoms than patients with infection caused by *S. epidermidis*, who generally present later and may have a more indolent course related to the low virulence of this organism.

Patients with bacteremia complicating pacemaker infection are generally acutely ill with fever, chills, arthralgias, and leukocytosis. However, in one series, the overall incidence of bacteremia in patients with pacemaker infection was 33%, despite 60% of these patients having systemic symptoms [13]. Pacemaker and ICD endocarditis is rare and is defined as persistent

TABLE 16.1

Microbiology of 250 Consecutive Pacemaker and ICD Infections Treated with Device Explant

Organism	Bloodstream Isolates, n (%)	Pocket Isolates, n (%)
Coagulase-negative staphylococci	20 (8.0)	82 (32.8)
MSSA	37 (14.8)	45 (18.0)
MRSA	31 (12.4)	19 (7.6)
Micrococcus	1 (0.4)	5 (2.0)
VSE	3 (1.2)	5 (2.0)
VRE	1 (0.4)	0
Group B streptococcus	3 (1.2)	0
Peptostreptococcus	0	1 (0.4)
Viridans streptococci	0	1 (0.4)
Propionibacterium acnes	0	8 (3.2)
Pseudomonas aeruginosa	1 (0.4)	7 (2.8)
Enterobacter	1 (0.4)	3 (1.2)
Escherichia coli	2 (0.8)	2 (0.8)
Klebsiella	0	3 (1.2)
Proteus	0	3 (1.2)
Citrobacter	0	1 (0.4)
Serratia	0	1 (0.4)
Candida	2 (0.8)	2 (0.8)
Mycobacterium fortuitum	1 (0.4)	1 (0.4)
Polymicrobial infections	1 (0.4)	37 (14.8)

MSSA = methicillin-susceptible *Staphylococcus aureus*, MRSA = methicillin-resistant *S. aureus*, VSE = vancomycin-sensitive enterococci, VRE = vancomycin-resistant enterococci. Numbers total to >100% due to some patients who had concomitant bloodstream and pocket infections, and due to the 38 polymicrobial infections.

TABLE 16.2

Clinical Findings of Patients with Pacemaker and ICD Pocket Infections

Signs and Symptoms	Percentage of Patients with Pacemakers (n = 87)	Percentage of Patients with Defibrillators (n = 36)	Percentage of All Patients (n = 132)
Device pocket erythema	48	69	55
Device pocket warmth	16	39	23
Device pocket pain	46	78	55
Device pocket erosion	32	31	32
Drainage from device pocket (nonpurulent)	36	58	42
Drainage from device pocket (purulent)	21	28	23
Device pocket swelling	29	53	36
Fever	29	28	29
Chills	22	22	22
Malaise	21	22	12

Source: Adapted from Chua, J.D. et al., *Ann. Intern. Med.*, 133, 604, 2000.

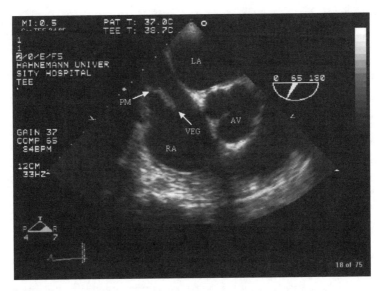

FIGURE 16.1 (See color insert following page 210)
Transesophageal echocardiographic view of a vegetation on a pacemaker lead. LA = left
atrium, PM = pacemaker lead, AV = aortic valve, VEG = vegetation, RA = right atrium.

bacteremia in patients with evidence of a vegetation on the device lead
[7,10,18,19]. It is frequently difficult to make a definitive diagnosis of lead
related endocarditis. Among native cardiac tissues, the tricuspid valve is
most commonly involved, but infection may also develop on the right
atrial or ventricular mural endocardium. Other evidence of endocarditis
includes anemia, hematuria, embolic manifestations, and a new valvular
murmur. New or changing murmurs were only reported in 13% of patients
in one review [18], and a murmur may be absent despite tricuspid
valve involvement. Transesophageal echocardiography (TEE) should be
performed to look for valvular and lead vegetations (Figure 16.1); the
sensitivity of this technique ranges from 91% to 96% [19,20].

The diagnosis of ICD epicardial lead infection is often difficult [6,11,20,21].
Chest radiography may show epicardial patch deformity. Computed
tomographic (CT) scanning may demonstrate fluid collections or epicardial
patch distortion, which may indicate infection in the appropriate clinical
setting. Gallium scanning has also been used [9,22]; increases in tracer uptake
around the generator and electrodes suggest that the infection involves the
entire system. Indium-111 white blood cell scanning may be more useful
because labeled polymorphonuclear leukocytes quickly accumulate at sites
of acute infection; this modality has a higher specificity than gallium scanning.

16.2.3 Therapy

Following implantation of cardiac pacemakers, the generator pocket and
intravascular and intracardiac leads become encased in dense layers of

endothelialized fibrous tissue so that treatment of infection is complicated by lack of penetration of antimicrobial agents into the avascular matrix, necessitating removal of some or all of the components to eradicate infection. Empiric antimicrobial therapy must include an antistaphylococcal agent (e.g., vancomycin); if the patient is extremely ill, coverage for gram-negative organisms may be added pending culture results. Antimicrobial therapy can then be further tailored based on final culture results.

The need for explantation therapy for all cardiac device infections is controversial [6,7,9–11]. Therapeutic decisions depend on the site of infection, evidence of spread from the pocket site, the presence of associated bacteremia, and the infecting microorganism. There are reports of successful treatment of generator pocket infections, especially with less virulent organisms such as *S. epidermidis*, using local debridement and irrigation systems supplemented with administration of systemic antimicrobial therapy. Most investigators agree, however, that these measures are inadequate for pocket infections, if there is associated erosion of overlying skin. The most widely advocated approach for generator pocket infection is removal of the entire system, which leads to the highest cure rates, even in the absence of a documented lead infection [23–25]. A less successful option is partial removal of the system to include removal of the generator and retunneling without removal of the sterile proximal electrodes. One study retrospectively reviewed culture results from portions of leads in 105 patients admitted for local inflammatory findings, impending erosion, and erosion or overt infection who underwent subsequent device removal and lead extraction. The authors identified that 79% of the cultures of the intravascular portions of the leads were positive with a potential risk of progressing to systemic infection [25]. In another study of 38 patients with infected antiarrhythmic implantable devices (21 of which were cardiac pacemakers and 17 of which were ICDs), patients were stratified to three treatment regimens: one group received intravenous antimicrobial therapy and relocation of the device without removal, and two groups underwent removal of the system with differing durations of antimicrobial therapy [26]. Treatment failure was observed in 100% of patients without device removal. Lead infections generally require removal of the entire system. Functionless pacer electrodes left in place in the face of infection carry a significant risk, with a mortality rate of 25% in one study [27].

The vast majority (98%) of pacemaker and defibrillator systems are completely transvenous, although utilization of epicardial leads and patches or subcutaneous electrodes is occasionally required [9]. Treatment of epicardial lead associated infection generally requires total generator and patch/lead system removal [28]. This may be problematic in patients with infected epicardial patches, because thoracotomy is required for removal; infection of epicardial patches may be suppressed with antimicrobial therapy, but cure requires removal. In patients who cannot tolerate removal of an entire epicardial system, reimplantation of the generator at a new site may be considered in those with normal radiologic studies, absence of bacteremia, infection caused by coagulase-negative staphylococci, absence of purulence in the generator pocket, and no ongoing sepsis [29]. Several reports have documented salvage of the ICD system following a pocket infection if the mediastinum is not involved [30,31]. Any patient who does not undergo complete extraction of the device should be monitored closely for relapse.

In cases of pacemaker and ICD infections with associated bloodstream infection, most investigators recommend removal of all pacemaker components to eradicate infection [6]. There are reports of patients with infected transvenous pacemakers and bloodstream infection who have been cured with antimicrobial therapy alone [17,23]. This may be a consideration provided that the bacteremia can be cleared with antimicrobial therapy, fungemia is not present, echocardiography is negative, and there is no evidence of septic embolization to the lung [10]. However, in one review of 26 patients with pacemaker infection and sustained bacteremia, all patients with staphylococcal bacteremia required removal of the entire pacing device; of the six patients with infections caused by nonstaphylococcal bacteria, five were cured with antimicrobial therapy alone [32]. In another study in patients with bacteriologically proven cardiac device endocarditis, medical treatment without removal of the pacemaker or ICD system was performed on seven patients. One hundred percent developed relapses of the endocarditis [16]. In those patients with documented vegetation on endovascular leads or right-sided cardiac structures, percutaneous lead removal may still be performed. In one study of nine patients with echocardiographically documented vegetations measuring 10–38 mm, transvenous lead removal was performed and five patients (55%) developed evidence of pulmonary embolism. All made a complete recovery with antimicrobial treatment and anticoagulation [33].

Removal of cardiac pacemakers and ICDs presents many difficulties [7,10,11]. Removal of the leads may not be possible by percutaneous traction, and damage of the vasculature or myocardium or tearing of the tricuspid valve may result as the leads become imbedded over time. Patients whose leads cannot be easily extracted may require cardiotomy with direct visualization [34], which also had the potential for associated morbidity and mortality. However, there are several techniques now available for removing an entire pacemaker system that does not require cardiotomy [35,36]. Within several months of implantation, extraction may be accomplished by simple manual traction from where the lead exits the venous system. Should removal be required more than a few months after implantation, the use of locking stylets should be the method of choice, because the force of traction is transferred to the lead tip, the usual site of fixation. Advances in lead technology have resulted in the utilization of expanded polytetrafluoroethylene (Gore-Tex, W.L. Gore & Associates, Inc., Newark, Delaware) coating of the endovascular defibrillator coils to inhibit tissue ingrowth and ease future lead removal. Advances in extraction equipment have included sheaths with tips powered by electrocautery or laser excision to aid in the dissection of leads from the endovascular surface [14]. Since the introduction of the locking stylet and powered extraction sheaths, open heart surgery for removal of infected pacemaker and ICD leads is rarely required.

Patients often require a new pacemaker system before discharge from the hospital [11]. The timing of pacemaker or defibrillator reimplantation in this setting is an ongoing area of controversy. Some authors have recommended reimplantation at a new site in patients without bacteremia at the time of explantation [37]. Alternatively, either temporary or no pacing for several days after explantation can be done before insertion of a new system; this may be preferable with infection caused by virulent organisms (e.g., *S. aureus*), if explantation is done during the acute illness, or if an entire new system is to be placed.

The duration of antimicrobial therapy for patients with pacemaker infections must be individualized [6,7,10,11]. Patients with infection limited to the generator pocket and no bacteremia can be treated with a less intensive regimen (parenteral therapy followed by oral therapy) with surgical debridement and local wound care of the infected pocket [38]. The duration of parenteral antimicrobial therapy should be at least 6 weeks if any foreign material remains; lifelong suppressive therapy may be necessary in some cases. Ten to 14 days of therapy is adequate if all hardware is removed and there is no bacteremia; 4 weeks of therapy is recommended if all hardware is removed and bacteremia was present. Any patient who does not undergo complete extraction of the device should be monitored closely for relapse. If relapse occurs, patients should have all material removed and receive an additional 6 weeks of antimicrobial therapy. The optimal antibiotic management of patients with these infections requires further study.

16.3 Left Ventricular Assist Devices

16.3.1 Epidemiology and Etiology

With the increasing incidence of congestive heart failure and the relatively limited supply of donors for heart transplantation, the use of LVADs as either a bridge to transplantation or as destination therapy has been a valuable option for certain patients. LVADs pose unique problems in their potential for infection, both from a biomechanical and immunologic point of view. They have a large surface area, are constructed of various materials, traverse several tissue planes and organ spaces, often include a driveline that permanently breaks the cutaneous barrier to infection, and usually cannot be removed until the time of transplantation (Figure 16.2) [39,40]. Additionally, the LVAD induces an immunodeficient state by leading to excess programmed cell death of T-cells [41–43]. This T-cell immunodeficiency can be demonstrated in patients by examining responses to intradermally

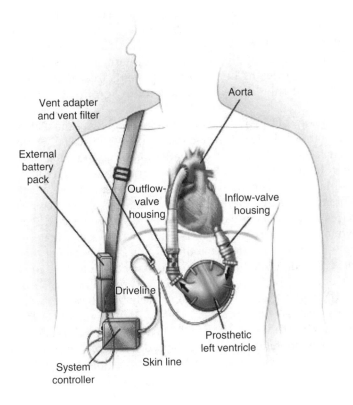

FIGURE 16.2 (See color insert following page 210)
Schematic of components of one type of left ventricular assist device.

injected antigens and measuring T-cell proliferative responses [44,45], and may predispose to certain types of infection, including fungal infections. Finally, many patients who receive an LVAD require prolonged hospitalization, often in an intensive care setting, increasing their risk of nosocomial infection.

The incidence of LVAD infection is significantly higher than other intracardiac devices, with reported figures ranging from 13% to 85% [46–49]. Some of this variation in reporting is due to differences in how infections are counted, with some studies including all infections that occurred during LVAD use, whether or not the device was thought to be the true source of the infection. Another important variable, however, is the time of LVAD support; in one study of patients with LVAD support time exceeding 30 days, 100% of the patients developed infection, with more than half developing bacteremia or fungemia [50], although infections were rather loosely defined as "any positive culture." LVAD infection may present as one or more of three basic syndromes. Driveline infections involve the cutaneous exit site of the electrical power line. LVAD pocket infections include skin and subcutaneous infections in the pocket that holds the actual pump. LVAD endocarditis involves bacterial or fungal infection of the valves or other internal surfaces of the device. Attributable mortality from these infections is quite high, with one long-term study showing sepsis causing 17 of the 41 deaths in a group of patients treated for heart failure with an LVAD, but only 1 of 54 deaths in a control group of patients treated medically [51].

The microbiology of LVAD infections differs somewhat from that of implantable antiarrhythmic devices (cardiac pacemakers and ICDs). Staphylococci are still seen frequently, but gram-negative bacteria and fungi are isolated with greater frequency, and polymicrobial infections are common. These differences may be explained, in part, by the preferential and specific colonization of biomaterials by different organisms; the large polymer surface area of the devices may predispose to infection by *S. epidermidis* and *Pseudomonas* species, whereas *S. aureus* has a predisposition for metallic material. In one large study of 214 patients (representing 17,831 LVAD-days of therapy), an overall incidence of LVAD-associated bloodstream infection of 49% was reported. Coagulase-negative staphylococci were isolated in 33 patients, *S. aureus* and *Candida* species in 19 patients each, and *Pseudomonas aeruginosa* in 16 patients. The risk of death was 10.9-fold higher in patients with fungal infection vs. those with no infection [52]. On average, *Candida* infections have been found to occur later than those caused by staphylococci, which is a common observation in other studies of nosocomial infection [47].

16.3.2 Clinical Presentation and Diagnosis

Clinical presentation of LVAD infection varies, depending on the anatomic location(s) from which the infection arises. Driveline infections are the most common and involve cutaneous inflammatory changes and drainage at the exit site of the driveline. LVAD pocket infections can be more complicated and severe, and present as cellulitis, with or without abscess in the cavity containing the pump. The least common and most concerning infection is endocarditis, which typically presents with fever and positive blood cultures, and can lead to mechanical incompetence of the device and embolic sequelae [47,48].

Any patient with an LVAD who develops bacteremia should be assumed to have an LVAD infection until proven otherwise, but it is often difficult to determine the site or extent of infection in a patient with an LVAD who is bacteremic. There are no standard radiographic criteria to determine the site of infection. Ultrasonography and CT are used frequently, although their sensitivity and specificity in the detection of infection have not been systematically studied and appear to be limited. There is at least one report of immunoscintigraphy, utilizing Tc-99 m labeled anti-NCA 95 antigranulocyte antibodies, being used to localize a staphylococcal LVAD endocarditis to the device outflow tract, allowing for a targeted surgical approach to curing the infection [53].

16.3.3 Therapy

Treatment of infections associated with LVADs must be individualized. Depending on the site of infection, the degree of illness of the patient, and potentially on Gram stain results from infected material, empiric antimicrobial therapy should include coverage against methicillin-resistant staphylococci, often against gram-negative bacilli, and sometimes also against fungi. More specific therapy can then be targeted to final culture results. Suppressive antimicrobial therapy until the time of transplantation is often required, with frequent relapses noted if therapy is stopped without device removal. Isolated driveline infections are typically managed with systemic antimicrobial therapy, surgical debridement of the exit site as needed, and a consideration of antimicrobial suppression until the time of transplantation. Aggressive surgical debridement of the exit site has been advocated by Pasque and others [54], based on the theory that the mechanism for prolonged infection of the driveline site is neo-epithelialization of the drainage tract, which prevents antimicrobial agents from reaching the infected tissues in adequate concentrations. LVAD pocket infections are managed similarly, with debridement or drainage of abscess cavities as needed. Management of LVAD endocarditis can be more problematic, owing to a potentially large area of infected prosthetic material, persistent bloodstream infection, the potential for embolic complications, and the presence of more systemic toxicity in these patients (which can then further compromise their marginal cardiac function). One prospective study has described four different considerations in managing LVAD endocarditis: LVAD replacement, cardiac transplantation, LVAD explantation without transplantation, and indefinite antibiotic suppression [55].

Infection of an LVAD, even when deep-seated, has not been shown to be a contraindication to eventual transplantation, even when considering the immunosuppression that follows transplantation. In one review of 20 patients with LVAD infections, 8 patients developed a total of 15 infections; 7 of the patients had infections that could be directly related to the LVAD [48]. In this series, all patients underwent medical therapy and local wound debridement or drainage was indicated; all infected patients were successfully bridged to surgery with no evidence of recurrence of infection at 16 months posttransplant. In another series of 33 patients with LVAD infections, 18 patients developed bloodstream infections and, of those, 12 were successfully bridged to transplantation [47]; 9 of the 12 patients with bloodstream infections had positive device cultures at the time of explantation. A larger, more recent, review examined 174 patients who underwent device implantation at a single center over 7 years [56]; 32 patients developed a device infection (12 driveline infections, 18 infections of the pump or device pocket, and two cases of LVAD endocarditis). The success rate of bridging to transplant and the one, three, and five-year survival did not differ significantly from the patients without infection. One final study did show decreased survival in patients who were bridged from LVAD support to transplant, if infection occurred while the LVAD was in place [57]. In the subset of patients with bloodstream infection, there was a higher rate of persistent bloodstream infection and a trend toward reduced posttransplant survival if the bloodstream infection was directly related to the LVAD itself (as defined by a positive culture of the device at explantation). The differences in successful bridging, clearance of infection, and survival among these groups are likely related to underlying morbidity and length of LVAD support, as well as local expertise in managing these patients. Prolonged antimicrobial therapy, even after device removal, is probably a significant contributor to success in treating these infections.

16.4 Prosthetic Cardiac Valves

16.4.1 Epidemiology and Etiology

Prosthetic valve endocarditis (PVE) accounts for 10%–20% of all cases of infective endocarditis. The long-term cumulative incidence of PVE among all valve recipients ranges from 3% to 6%, when patients are followed for up to 5 to 10 years after valve placement [58–60]. In some series, the incidence of endocarditis in the first postoperative year after implantation of a mechanical cardiac valve is higher than that for a tissue prosthesis; these rates become comparable after 5 years [59–63]. While it is not a frequent complication of valve replacement, valve infection remains a difficult problem and carries significant morbidity and mortality.

PVE is classified as either early or late, depending upon when symptoms appear; early PVE occurs within 60 days of insertion of the prosthetic valve, and late PVE occurs more than 60 days after valve insertion, with inter-mediate cases sometimes defined as occurring from 60 days to 1 year after implantation. The epidemiology and the likely pathogenesis differ between the early and late forms of PVE, as does the microbiology (Table 16.3). In both early and late PVE, host-tissue factors, including damaged endocardial tissue and platelet aggregation, predispose to development of infective endocarditis. Early PVE generally reflects valve contamination arising in the perioperative period, either from direct inoculation at the time of surgery or via hematogenous spread from a distant focus [64]. The most common organisms causing early PVE are coagulase-negative staphylo-cocci, and epidemiologic and microbiologic studies have linked many of these cases to intraoperative contamination, some occurring in clusters or epidemics [65,66]. Early-onset PVE has also been linked to bacteremia arising from invasive monitoring and support devices such as the cardiac bypass pump, and from surgical wound infections. In addition to staphylo-cocci, a variety of gram-negative organisms, streptococcal species, *Coryne-bacterium* species, and fungi are seen with higher frequency than in patients with native valve endocarditis (NVE) [64]. A recent European study has also

TABLE 16.3

Microbiology of Prosthetic Valve Endocarditis by Time after Implantation

Organism	Incidence among All Cases of PVE, %		
	<2 months	2–12 months	>12 months
Streptococci	2	9	31
Enterococci	8	12	11
Staphylococcus aureus	20–25	12	18
Coagulase-negative staphylococci	30–35	32	10–12
Corynebacteria	6	<5	3
HACEK group	Rare	Rare	6
Other gram-negative bacilli	12	3	6
Fungi	7	12	1
Polymicrobial/miscellaneous	2	6	5
Culture-negative	<5	<5	<5

Source: Adapted from Karchmer, A.W., in *Principles and Practice of Infections Disease*, Mandell, G.L., Bennett, J.E., and Dolin, R. (eds.), Churchill Livingstone, Philadelphia, 2000, 903–917.

HACEK = noninfluenzae *Haemophilus* species, *Actinobacillus actinomycetemcomitans*, *Cardiobacterium hominis*, *Eikenella corrodens*, and *Kingella kingae*.

showed that *Coxiella burnetii* endocarditis is significantly associated with intracardiac prosthetic material [67].

The microbiologic etiology of late PVE has classically resembled that of NVE, but with a slightly higher rate of infection caused by *S. epidermidis*, gram-negative bacilli, and fungi [64]. Transient bacteremia unrelated to the initial surgical procedure, often occurring in the community setting, is presumably the most common pathogenesis in late PVE. One long-term, single-center study of PVE showed significant changes in the epidemiology of both early and late PVE over time [68]; the overall incidence of early PVE decreased, and the incidence of late PVE remained stable. Among early cases, gram-negative bacilli became increasingly rare. *S. aureus* and enterococci superseded the viridans streptococci as the most common causes of late PVE, with significantly more cases of late PVE being hospital acquired. Hospital acquisition of late PVE was associated with a higher mortality. Other studies have also shown that various nosocomial bloodstream infections in patients with valvular prostheses are significantly associated with PVE [69–71]. *S. aureus* bacteremia results in a particularly high rate of PVE (51%), with a 62% mortality noted among those with definite endocarditis in one study [71].

16.4.2 Clinical Presentation and Diagnosis

Most clinical features of PVE are similar to those of NVE [64]. The manifestations may result from processes related to the heart valve and adjacent native cardiac tissues, persistent bacteremia and metastatic infection, septic or bland embolization to any organ, and circulating immune complexes. The most common clinical features that suggest the diagnosis of PVE are fever, new cardiac abnormalities, and embolic phenomena. The presence of unexplained fever in a patient with a prosthetic valve should always raise the clinical suspicion of infective endocarditis, and serial blood cultures should be obtained. Particularly in prosthetic aortic valves during the first postoperative year, there is a high rate of paravalvular spread of the infection, resulting in higher frequencies of new or changing regurgitant murmurs, congestive heart failure, persistent fever despite antimicrobial therapy, and new cardiac conduction disturbances than are seen in NVE [64,72]. As the time after surgery increases (i.e., in late PVE), the clinical features of PVE are more similar to those in patients with NVE. In terms of extracardiac complications, almost 40% of all PVE patients experience at least one systemic embolus; 20% to 40% of PVE cases are complicated by embolic stroke, central nervous system hemorrhage, or other neurologic complications [73,74]. A recent study specifically investigated, whether enterococcal PVE might have any unique features [75]. They found that patients with enterococcal PVE presented less often with demonstrable vegetations or a new regurgitant murmur, but more often with myocardial abscess, than patients with enterococcal NVE, but still showed a trend toward lower mortality than patients with nonenterococcal PVE.

In the early postoperative period, the diagnosis of PVE is often difficult because other nosocomial events can complicate the interpretation of bacteremia. Patients should be considered to have PVE if they develop sustained bacteremia caused by relatively avirulent organisms (e.g., *S. epidermidis*) or if there is no apparent extracardiac source of the bacteremia [11]. Transient bacteremia may indicate PVE as well, though this finding is not compatible with the traditional definition of endocarditis [76]. It has been suggested that patients with transient postoperative gram-negative bacteremia associated with an obvious extracardiac focus, and without clinical evidence of infective endocarditis, need not be placed on prolonged therapy for PVE [77].

In the absence of prior antimicrobial therapy, approximately 90% of patients with PVE will have positive blood cultures [11]. Patients with PVE caused by *Candida* species may have intermittent fungemia or negative cultures. Patients with fastidious microorganisms such as *Legionella* species, *C. burnetii*, *Mycoplasma hominis*, *Bartonella* species, nontuberculous myco-bacteria, and *Aspergillus* and other molds may have negative blood cultures; histological and molecular analysis of the vegetation may be necessary to make a diagnosis and guide proper antimicrobial therapy [78].

Echocardiography is essential to aid in the diagnosis of PVE and is useful for demonstrating vegetations, prosthesis dehiscence, paravalvular abscess, fistula formation, and pericardial effusion. Though TEE and transthoracic echocardiography (TTE) are sometimes complimentary in assessing a particular patient, TEE has shown superiority in examining mitral valve prostheses, detecting smaller vegetations, and evaluating complications such as dehiscence, abscess, and fistulization [79]. The overall sensitivity of TEE in demonstrating PVE is 96%, vs. only 82% for TTE [80,81]. Whereas, TEE findings of mobile vegetation, valve dehiscence, pseudoaneurysms, fistulae, or moderate to severe paravalvular regurgitation are strongly suggestive of the diagnosis of PVE, an isolated finding of mild paravalvular regurgitation has been shown to have a poor positive predictive value (only 6% in one study) for the diagnosis [82].

16.4.3 Therapy and Outcomes

Antimicrobial therapy for PVE requires parenteral administration of bactericidal agents for 6 weeks or longer [64]. Stable patients, particularly those who have received recent antimicrobial therapy, may have antimicrobials withheld until an organism has been isolated from blood cultures; unstable patients should have broad-spectrum empiric therapy, started as soon as blood cultures are collected. Empiric antimicrobial therapy should include vancomycin plus gentamicin, with consideration of adding broader spectrum gram-negative coverage in early postoperative cases. Recommendations for specific therapy against the common etiologic agents are available from the American Heart Association [83]. One major difference from treatment of NVE concerns patients with staphylococcal PVE; in these cases, therapy should include an appropriate β-lactam or vancomycin plus rifampin for 6 weeks, with gentamicin added for the first 2 weeks of antimicrobial therapy. The addition of rifampin is recommended based on a significant amount of in vitro, animal model, and clinical data that have demonstrated its usefulness in killing staphylococci adherent to foreign material [84–86]. Fungal PVE has traditionally been very difficult to cure with antimicrobial therapy alone. The standard of therapy has been amphotericin B with a consideration of combination therapy with flucytosine, and also the possibility of long-term suppressive therapy with an oral antifungal [87,88]. In one patient with *Candida glabrata* PVE in whom intolerance to amphotericin B developed, the combination of fluconazole and caspofungin was successfully used [89].

Many patients with PVE will benefit from combined antimicrobial and surgical therapy (i.e., valve replacement). Indications for cardiac surgery include the following [7,11,64,72,90]:

1. Medically refractory heart failure caused by prosthesis dysfunction (incompetence or obstruction)
2. Invasive and destructive paravalvular infection (partial valve dehiscence, new or progressive conduction system disturbances, fever persisting >10 days during appropriate antimicrobial therapy, purulent pericarditis, significant abscess, sinus of Valsalva aneurysm or intracardiac fistula)
3. Uncontrolled bacteremia despite antimicrobial therapy
4. Infection caused by selected organisms (e.g., fungi, *S. aureus*, *S. lugdunensis*, early-onset infection caused by coagulase-negative

staphylococci, aminoglycoside-resistant enterococci, other multi-drug resistant organisms, *Pseudomonas* species)

5. Relapse after appropriate antimicrobial therapy
6. Recurrent arterial emboli
7. Culture-negative PVE unresponsive to antimicrobial therapy

Despite appropriately aggressive antimicrobial and surgical therapy, the diagnosis of PVE still carries a significant mortality. Factors predicting mortality have included advanced age, occurrence of early PVE, severe comorbid illness, renal failure, stroke, moderate to severe valve regurgitation, severe heart failure, and infection with *S. aureus* or *Candida* species [91–95]. Patients with complicated PVE and those with staphylococcal infection have retrospectively been shown to have the greatest benefit from valve replacement surgery, in terms of reduction of mortality [91]. In one recent study of patients with aortic valve endocarditis, occurrence of a paravalvular abscess in a subset of patients ($n = 67$) was not shown to increase mortality; among patients with abscess, however, *S. aureus* infection was an independent predictor of mortality [95].

16.5 Future Directions

The numbers of procedures that result in the utilization of intracardiac prosthetic material will undoubtedly continue to increase. Technological advances have allowed for the development of percutaneously placed intracardiac devices to correct defects in the intraatrial and intraventricular septum, and occlusion devices for the left atrial appendage. The incidence of infection in these devices has not been well described, but appears to be low [96].

Expanded indications for implantable antiarrhythmic devices have resulted in a marked increase in the number of devices being implanted. Improved surgical techniques may allow for reductions in the number of devices which subsequently become infected. One study found that the utilization of fibrin sealant before wound closure in device pockets of patients receiving anticoagulant therapy reduced the incidence of hematoma formation, with a potential reduction in the likelihood of infection [97]. Alterations in the surgical techniques for LVAD implantation which reduce infection have additionally been described. One study identified that with the addition of a Dacron (Invista, Wichita, Kansas) graft covering of the upper surface of an LVAD pump, the incidence of pocket infection was reduced from 33.3% to 11.1% [98]. Alterations in the surfaces of implanted materials may offer additional mechanisms to reduce the likelihood of infection; the use of antibiotic coated materials has also been employed. The utilization of a gentamicin-releasing copolymer on pacemaker leads reduced the incidence of infection in an animal model [99].

Ongoing research into the development of vaccines for the prevention of infection is an exciting advancement in clinical medicine. The development of a staphylococcal vaccine may prove to be a valuable adjunct for the prevention of infections in patients with intracardiac prosthetic material. A recent study in patients with end-stage renal disease showed encouraging results [100].

References

1. FDA Releases Results of Study on Defibrillator and Pacemaker Malfunctions (news release, Rockville, MD: U.S. Food and Drug Administration, September 16, 2005).
2. Moss, A.J. et al., Prophylactic implantation of a defibrillator in patients with myocardial infarction and reduced ejection fraction, *N. Engl. J. Med.*, 346, 877, 2002.
3. Sivaratnam, K. and Duggan, J.M., Left ventricular assist device infections: Three case reports and a review of the literature, *Am. Soc. Artif. Intern. Organs J.*, 48, 2, 2002.
4. Marchlinski, F.E. et al., The automatic implantable cardioverter-defibrillator: Efficacy, complications, and device failures, *Ann. Intern. Med.*, 104, 481, 1986.
5. Bluhm, G., Pacemaker infections: A clinical study with special reference to prophylactic use of some isoxazolyl penicillins, *Acta. Med. Scand.*, Suppl., 699, 1, 1985.
6. Kearney, R.A., Eisen, H.J., and Wolf, J.E., Nonvalvular infections of the cardio-vascular system, *Ann. Intern. Med.*, 121, 219, 1994.
7. Heimberger, T.S. and Duma, R.J., Infections of prosthetic heart valves and cardiac pacemakers, *Infect. Dis. Clin. N. Am.*, 3, 221, 1989.
8. Lai, K.K. and Fontecchio, S.A., Infections associated with implantable cardio-verter defibrillators placed transvenously and via thoracotomies: Epidemiology, infection control, and management, *Clin. Infect. Dis.*, 27, 265, 1998.
9. Karchmer, A.W. and Longworth, D.L., Infections of intracardiac devices, *Cardiol. Clin.*, 21, 253, 2003.
10. Wade, J.S. and Cobbs, C.G., Infections of cardiac pacemakers, in *Current Clinical Topics in Infectious Diseases*, Remington, J.S. and Swartz, M.N. (eds.), McGraw Hill, New York, 1988, pp. 44–61.
11. Sentochnik, D.E. and Karchmer, A.W., Cardiac infections, in *A Practical Approach to Infectious Diseases*, 5th ed., Betts, R.F., Chapman, S.W., and Penn, R.L. (eds.), Lippincott Williams & Wilkins, Philadelphia, 2003, pp. 372–402.
12. Baddour, L.M. et al., Nonvalvular cardiovascular device related infections, *Circulation*, 108, 2015, 2003.
13. Chua, J.D. et al., Diagnosis and management of infections involving implantable electrophysiologic cardiac devices, *Ann. Intern. Med.*, 133, 604, 2000.
14. Verma, A. and Wilkoff, B.L., Intravascular pacemaker and defibrillator lead extraction: A state of the art review, *Heart Rhythm*, 1, 739, 2004.
15. Joly, V. et al., Pacemaker endocarditis due to *Candida albicans*: Case report and review, *Clin. Infect. Dis.*, 25, 1359, 1997.
16. del Rio, A. et al., Surgical treatment of pacemaker and defibrillator lead endocarditis: The impact of electrode lead extraction on outcome, *Chest*, 124, 1451, 2003.

17. Lewis, A.B. et al., Update on infections involving permanent pacemakers: Characterization and management, *J. Thorac. Cardiovasc. Surg.*, 89, 758, 1985.
18. Arber, M. et al., Pacemaker endocarditis: Report of 44 cases and review of the literature, *Medicine*, 73, 299, 1994.
19. Klug, D. et al., Systemic infection related to endocarditis on pacemaker leads: Clinical presentation and management, *Circulation*, 95, 2098, 1997.
20. Victor, F. et al., Pacemaker lead infection: Echocardiographic features, management, and outcome, *Heart*, 81, 82, 1999.
21. Goodman, L.R. et al., Complications of automatic implantable cardioverter defibrillators: Radiographic, CT, and echocardiographic evaluation, *Radiology*, 170, 447, 1989.
22. Kelly, P.A. et al., Postoperative infection with the automatic implantable cardioverter defibrillator: Clinical presentation and use of the gallium scan in diagnosis, *PACE*, 11, 1220, 1988.
23. Harjula, A., Jarvinen, A., and Virtanen, K.S., Pacemaker infections—treatment with total or partial pacemaker system removal, *Thorac. Cardiovasc. Surg.*, 33, 218, 1985.
24. Choo, M.H. et al., Permanent pacemaker infections: Characterization and management, *Am. J. Cardiol.*, 48, 559, 1981.
25. Klug, D. et al., Local symptoms at the site of pacemaker implantation indicate latent systemic infection, *Heart*, 90, 882, 2004.
26. Molina, J.E., Undertreatment and overtreatment of patients with infected antiarrhythmic implantable devices, *Ann. Thorac. Surg.*, 63, 504, 1997.
27. Rettig, G. et al., Complications with retained transvenous pacemaker electrodes, *Am. Heart J.*, 98, 587, 1979.
28. Wunderly, D. et al., Infections in implantable cardioverter defibrillator patients, *PACE*, 13, 1360, 1990.
29. Spratt, K.A. et al., Infections of implantable cardioverter defibrillators, *Clin. Infect. Dis.*, 17, 679, 1993.
30. Taylor, R.L. et al., Infection of an implantable cardioverter defibrillator: Management without removal of the device in selected cases, *PACE*, 13, 1352, 1990.
31. Lee, J.H. et al., Salvage of infected ICDs: Management without removal, *PACE*, 19, 437, 1996.
32. Camus, C. et al., Sustained bacteremia in 26 patients with a permanent endocardial pacemaker: Assessment of wire removal, *Clin. Infect. Dis.*, 17, 46, 1993.
33. Meier-Ewert, H.K., Gray, M.E., and John, R.M., Endocardial pacemaker or defibrillator leads with infected vegetations: A single-center experience and consequences of transvenous extraction, *Am. Heart J.*, 146, 339, 2003.
34. Neiderhäuser, U. et al., Infected endocardial pacemaker electrodes: Successful open intracardiac removal, *PACE*, 16, 303, 1993.
35. Myers, M.R., Parsonnet, V., and Bernstein, A.D., Extraction of implanted transvenous pacing leads, *Am. Heart J.*, 121, 881, 1991.
36. Frame, R., Brodman, R.F., and Furman, S., Infections of nonthoracotomy ICD leads: A note of caution, *PACE*, 16, 2215, 1993.
37. Byrd, C., Management of implant complications, in *Clinical Cardiac Pacing*, Ellenbogen, K., Kay, G., and Wilkoff, B.L. (eds.), W.B. Saunders, Philadelphia, 2000, pp. 669–694.
38. Karchmer, A.W., Infections of prosthetic valves and intravascular devices, in *Principles and Practice of Infectious Diseases*, 5th ed., Mandell, G.L., Bennett, J.E., and Dolin, R. (eds.), Churchill Livingstone, Philadelphia, 2000, pp. 903–917.

39. Goldstein, D.J., Oz, M.C., and Rose, E.A., Implantable left ventricular assist devices, *N. Engl. J. Med.*, 339, 1522, 1998.

40. Portner, P.M. et al., Implantable electrical left ventricular assist system: Bridge to transplantation and the future, *Ann. Thorac. Surg.*, 47, 142, 1989.

41. Erren, M. et al., Immunologic effects of implantation of left ventricular assist devices, *Transplant Proc.*, 33, 1965, 2001.

42. Schuster, M. et al., Induction of T-cell apoptosis by polyurethane biomaterials used in left ventricular assist devices is dependent on calcineurin activation, *Transplant. Proc.*, 33, 1958, 2001.

43. Ankersmit, H.-J. et al., Quantitative changes in T-cell populations after left ventricular assist device implantation: Relationship to T-cell apoptosis and soluble CD95, *Circulation*, 100 (Suppl. II), II-211, 1999.

44. Itescu, S. et al., Immunobiology of left ventricular assist devices, *Prog. Cardiovasc. Dis.*, 43, 67, 2000.

45. Rotenburger, M. et al., Immune response in the early postoperative period after implantation of a left-ventricular assist device system, *Transplant. Proc.*, 33, 1955, 2001.

46. Holman, W.L. et al., Infection during circulatory support with ventricular assist devices, *Ann. Thorac. Surg.*, 68, 711, 1998.

47. McCarthy, P.M. et al., Implantable LVAD infections: Implications for permanent use of the device, *Ann. Thorac. Surg.*, 61, 359, 1996.

48. Fischer, S.A. et al., Infectious complications in left ventricular assist device recipients, *Clin. Infect. Dis.*, 24, 18, 1997.

49. Mekonsto-Dessap, A. et al., Nosocomial infections occurring during receipt of circulatory support with the paracorporeal ventricular assist system, *Clin. Infect. Dis.*, 35, 1308, 2002.

50. Holman, W.L. et al., Infections during extended circulatory support: University of Alabama at Birmingham experience 1989 to 1994, *Ann. Thorac. Surg.*, 61, 366, 1996.

51. Rose, E.A. et al., Long-term use of a left-ventricular assist device for end-stage heart failure, *N. Engl. J. Med.*, 345, 1435, 2001.

52. Gordon, S.M. et al., Nosocomial bloodstream infections in patients with implantable left ventricular assist devices, *Ann. Thorac. Surg.*, 72, 725, 2001.

53. de Jonge, K.C. et al., Diagnosis and management of left ventricular assist device valve-endocarditis: LVAD valve replacement, *Ann. Thorac. Surg.*, 70, 1404, 2000.

54. Pasque, M.K. et al., Surgical management of Novacor drive-line exit site infections, *Ann. Thorac. Surg.*, 74, 1267, 2002.

55. Oz, M.C. et al., Bridge experience with long-term implantable left ventricular assist devices: Are they an alternative to transplantation? *Circulation*, 95, 1844, 1997.

56. Morgan, J.A. et al., Device related infections while on left ventricular assist device support do not adversely impact bridging to transplant or posttransplant survival, *ASAIO J.*, 49, 748, 2003.

57. Poston, R.S. et al., LVAD bloodstream infections: Therapeutic rationale for transplantation after LVAD infection, *J. Heart Lung Transplant.*, 22, 914, 2003.

58. Glower, D.D. et al., Determinants of 15-year outcome with 1,119 standard Carpentier-Edwards porcine valves, *Ann. Thorac. Surg.*, 66, S44, 1998.

59. Arvay, A. and Lengyel, M., Incidence and risk factors of prosthetic valve endocarditis, *Eur. J. Cardiothorac. Surg.*, 2, 340, 1988.

60. Rutledge, R., Kim, J., and Applebaum, R.E., Actuarial analysis of the risk of prosthetic valve endocarditis in 1,598 patients with mechanical and bioprosthetic valves, *Arch. Surg.*, 120, 469, 1985.
61. Bloomfield, P., Wheatley, D.J., and Prescott, R.J., Twelve-year comparison of a Bjork-Shiley mechanical heart valve with porcine bioprosthesis, *N. Engl. J. Med.*, 324, 573, 1991.
62. Grover, F.L. et al., Determinants of the occurrence of and survival from prosthetic valve endocarditis, *J. Thorac. Cardiovasc. Surg.*, 108, 207, 1994.
63. Hammermeister, K.E., Sethi, G.K., and Henderson, W.G., A comparison of outcomes eleven years after heart valve replacement with a mechanical valve or prosthesis, *N. Engl. J. Med.*, 328, 1289, 1993.
64. Baddour, L.M. and Wilson, W.R., Infections of prosthetic valves and other cardiovascular devices, in *Principles and Practice of Infectious Diseases*, 6th ed., Mandell, G.L., Bennett, J.E., and Dolin, R. (eds.), Churchill Livingstone, Philadelphia, 2005, Chapter 75, pp. 1022–1044.
65. Boyce, J.M. et al., A common-source outbreak of *Staphylococcus epidermidis* infections among patients undergoing cardiac surgery, *J. Infect. Dis.*, 161, 493, 1990.
66. van den Broek, P.J. et al., Epidemic of prosthetic valve endocarditis caused by *Staphylococcus epidermidis, Br. Med. J.*, 291, 949, 1985.
67. Barrau, K. et al., Causative organisms of infective endocarditis according to host status, *Clin. Microbiol. Infect.*, 10, 302, 2004.
68. Rivas, P. et al., The impact of hospital-acquired infections on the microbial etiology and prognosis of late-onset prosthetic valve endocarditis, *Chest*, 128, 764, 2005.
69. Fang, G. et al., Prosthetic valve endocarditis resulting from nosocomial bacteremia: A prospective, multicenter study, *Ann. Intern. Med.*, 119, 560, 1993.
70. Nassar, R.M. et al., Incidence and risk of developing fungal prosthetic valve endocarditis after nosocomial candidemia, *Am. J. Med.*, 103, 25, 1997.
71. El-Ahdab, F. et al., Risk of endocarditis among patients with prosthetic valves and *Staphylococcus aureus* bacteremia, *Am. J. Med.*, 118, 225, 2005.
72. Calderwood, S.B. et al., Prosthetic valve endocarditis: Analysis of factors affecting outcome of therapy, *J. Thorac. Cardiovasc. Surg.*, 92, 776, 1986.
73. Ismail, M.B. et al., Prosthetic valve endocarditis: A survey, *Br. Heart J.*, 58, 72, 1987.
74. Tornos, P. et al., Late prosthetic valve endocarditis: Immediate and long-term prognosis, *Chest*, 101, 37, 1992.
75. Anderson, D.J. et al., Enterococcal prosthetic valve endocarditis: Report of 45 episodes from the International Collaboration on Endocarditis-merged database, *Eur. J. Clin. Microbiol. Infect. Dis.*, 24, 665, 2005.
76. Durack, D.T., Lukes, A.S., and Bright, D.K., New criteria for diagnosis of infective endocarditis: Utilization of specific echocardiographic findings, *Am. J. Med.*, 96, 200, 1994.
77. Sande, M.A. et al., Sustained bacteremia in patients with prosthetic cardiac valves, *N. Engl. J. Med.*, 286, 1067, 1972.
78. Bayer, A.S. et al., Diagnosis and management of infective endocarditis and its complications, *Circulation*, 98, 2936, 1998.
79. Daniel, W.G. et al., Improvement in the diagnosis of abscesses associated with endocarditis by transesophageal echocardiography, *N. Engl. J. Med.*, 324, 795, 1991.
80. Khandheria, B.K., Transesophageal echocardiography in the evaluation of prosthetic valves, *Am. J. Cardiac Imag.*, 9, 106, 1995.

81. Daniel, W.G. et al., Comparison of transthoracic and transesophageal echo-cardiography for detection of abnormalities of prosthetic and bioprosthetic valves in the mitral and aortic positions, *Am. J. Cardiol.*, 71, 210, 1993.

82. Ronderos, R.E. et al., Are all echocardiographic findings equally predictive for diagnosis in prosthetic endocarditis? *J. Am. Soc. Echocard.*, 17, 664, 2004.

83. Wilson, W.R. et al., Antibiotic treatment of adults with infective endocarditis due to streptococci, enterococci, staphylococci, and HACEK microorganisms, *JAMA*, 274, 1706, 1995.

84. Karchmer, A.W., Archer, G.L., and Dismukes, W.E., *Staphylococcus epidermidis* causing prosthetic valve endocarditis: Microbiologic and clinical observations as guides to therapy, *Ann. Intern. Med.*, 98, 447, 1983.

85. Archer, G.L. et al., Efficacy of antibiotic combinations including rifampin against methicillin-resistant *Staphylococcus epidermidis*: In vitro and in vivo studies, *Rev. Infect. Dis.*, 5, S538, 1983.

86. Chuard, C. et al., Successful therapy of experimental chronic foreign body infection caused by methicillin-resistant *Staphylococcus aureus* by antimicrobial combinations, *Antimicrob. Agents Chemother.*, 35, 2611, 1991.

87. Ellis, M.E. et al., Fungal endocarditis: Evidence in the world literature, 1965–1995, *Clin. Infect. Dis.*, 32, 50, 2001.

88. Melgar, G.R. et al., Fungal prosthetic valve endocarditis in 16 patients: An 11 year experience in a tertiary care hospital, *Medicine* (Baltimore), 76, 94, 1997.

89. Lye, D.C. et al., *Candida glabrata* prosthetic valve endocarditis treated success-fully with fluconazole plus caspofungin without surgery: A case report and literature review, *Eur. J. Clin. Microbiol. Infect. Dis.*, 24, 753, 2005.

90. Yu, V.L. et al., Prosthetic valve endocarditis: Superiority of surgical valve replacement versus medical therapy only, *Ann. Thorac. Surg.*, 58, 1073, 1994.

91. Habib, G. et al., Prosthetic valve endocarditis: Who needs surgery? A multi-centre study of 104 cases, *Heart*, 91, 954, 2005.

92. Benjamin, D.K. Jr. et al., *Candida* endocarditis: Contemporary cases from the international collaboration of infectious endocarditis merged database (ICE-mD), *Scand. J. Infect. Dis.*, 36, 453, 2004.

93. Chirouze, C. et al., Prognostic factors in 61 cases of *Staphylococcus aureus* pros-thetic valve endocarditis from the international collaboration on endocarditis merged database, *Clin. Infect. Dis.*, 38, 1323, 2004.

94. Castillo, J.C. et al., Long-term prognosis of early and late prosthetic valve endocarditis, *Am. J. Cardiol.*, 93, 1185, 2004.

95. Anguera, I. et al., Clinical characteristics and outcome of aortic endo-carditis with periannular abscess in the international collaboration on endocarditis merged database, *Am. J. Cardiol.*, 96, 976, 2005.

96. Calachanis, M., Infective endocarditis after transcatheter closure of a patent foramen ovale, *Cath. Cardiovasc. Intervent.*, 63, 251, 2004.

97. Milic, D.J. et al., Prevention of pocket related complications with fibrin sealant in patients undergoing pacemaker implantation who are receiving anticoagu-lation treatment, *Europace*, 7, 374, 2005.

98. Arusoglu, L. et al., A novel method to reduce device-related infections in patients supported with the HeartMate device, *Ann. Thorac. Surg.*, 68, 1875, 1999.

99. van Wachem, P.B., Tissue reactions to bacteria-challenged implantable leads with enhanced infection resistance, *J. Biomed. Mater. Res.*, 41, 142, 1998.

100. Robbins, J.B. et al., *Staphylococcus aureus* types 5 and 8 capsular polysaccharide-protein conjugate vaccines, *Am. Heart J.*, 147, 593, 2004.

17

Intrauterine Devices: Infections and Biofilms

Ben M. J. Pereira and Vikas Pruthi

CONTENTS

17.1 Introduction

Regulation of fertility and family planning is now considered a human right that is basic to human dignity. To facilitate this, science and technology has made rapid strides and it is now possible for couples to choose from a variety of contraceptive methods. These include injectables, implants, female and male sterilization, intrauterine devices (IUDs), oral contraceptives, condoms, spermicides, diaphragms, and cervical caps. However, it is apparent that the development of contraceptive methods is greatly biased toward women since it is easier to target a single egg released cyclically, in a span of one month, than to deal with millions of sperm released at each ejaculation. In western countries, new generations of contraceptives like nasal sprays and dermal patches have gained popularity in preventing unwanted pregnancies. There is no doubt that men and women would like to use a method of their choice which is effective and gives them satisfaction, while safeguarding against avoidable, negative health effects. Nevertheless, in stable relationships, it is the woman who generally volunteers or is forced to take the responsibility to adopt appropriate measures to deal with conception or contraception. Today, the mind-set of an average Asian couple is to delay the first child, space the second, and stop the third.

17.2 Contraceptive Options for Women and Preference for IUDs among Asian Women

The choice of a contraceptive method is based on several considerations: the most important being safety, effectiveness, reliability, and reversibility. In addition to this, social, economic, and cultural factors also play a decisive role [1–7]. As far as the Asian population is concerned, though there is lack of exposure and experience to other modern contraceptive methods, education and counseling have encouraged couples to seek oral pills and IUDs as a means of family planning [8]. The outcome of a small survey of 1578 couples using various forms of contraception sampled from the state of Uttaranchal in North India is summarized in Table 17.1.

The data in this survey clearly establishes the wide acceptance of IUDs for birth control primarily, because of its long-term effectiveness, reversibility, and most important of all, the sense of privacy. Thus, the decision to use IUD can safely prevent pregnancy for several years. It is now well known that the pregnancy rate with these devices in situ is 0.4%–2.8% women per year, depending on the device and age of the patient [9,10]. In the urban areas, IUD is popular since it protects women from unplanned pregnancies

TABLE 17.1

Choice of Contraceptive Use in the Uttaranchal State of North India

Contraceptive Option	$n = 1578$	Users (%)
Injectables	8	0.51
Implants	47	2.98
Sterilization (female)/hysterectomy	331	20.97
IUDs	459	29.09
Oral contraceptives	410	25.98
Spermicides	0	—
Diaphragms and cervical caps	0	—
Natural methods (BT charts, billings method, coitus interruption, periodic abstinence)	55	3.49
Others (condoms/vasectomy of male partners)	268	16.98

during the peak years of their professional career or until the completion of their education, up to the time they are ready for motherhood. The added advantage is that since IUDs are very effective, there is increased sexual enjoyment without the worry of pregnancy.

In other countries like Norway, parous women who are either married or unmarried but living in a stable relationship commonly use IUDs [11]. The Demographic and Health Survey of 2005 found that 56% of women used some form of family planning or the other and of these, 63% used IUDs while 27% used oral contraceptives [12]. In India, health care providers also promote the use of IUDs since it can be inserted with ease soon after childbirth and does not interfere with breast-feeding or sex life. The greatest advantage of IUD is its reversibility i.e., when women have their IUDs removed in the reproductive age, they become pregnant as quickly as women who have not used IUDs. Besides, the cost of IUDs is subsidized by government agencies in many primary health care centers in India. Since the insertion or removal of the IUD needs the assistance of a trained health care provider, it is convenient to monitor and keep a track of the clients so that government incentives can be offered to them.

17.3　Types of IUDs

An IUD also known as intrauterine contraceptive device (IUCD) is usually a small, flexible, plastic frame, which often has copper wire or copper sleeves on it. It is inserted into a women's uterus through the vagina, which is easily accomplished immediately after childbirth. Almost all popular brands of IUDs have one or two strings or threads attached to them. The strings hang through the opening of the cervix into the vagina so that the user can check that the IUD is still in place by touching the strings (Figure 17.1). A health care provider can remove the IUD by gently pulling on the strings with a forceps.

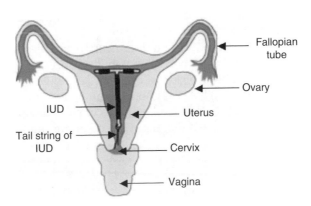

FIGURE 17.1
Diagram illustrating the positioning of IUD inside the human uterus.

There is a whole range of IUDs available in the market, which differ in respect to design, shape, size, and the kind of material used.

The initial IUDs that were developed were inert or unmedicated, made completely of plastic or stainless steel and popularly called "the loop." Later, IUDs with copper sleeves or copper wire on the plastic were introduced, like the Copper T, Tcu-380A, MLCu-375 (multiload), and Nova T. It is believed that Cu has a detrimental effect on several microorganisms like *Gardnerella vaginalis, Neisseria gonorrhea, Lactobacilli, Chlamydia*, and a large number of gram-positive anaerobes [13–16]. Polymorphonuclear cells [17] and mast cells [18] are found in larger numbers in the endometrium of Cu-IUD users. Perhaps copper ions enhance the release of two groups of toxic metabolites, reactive oxygen intermediates (ROI) and reactive nitrogen intermediates (RNI) from these cells, which might explain the bactericidal action [19–22]. However, it must be repeatedly emphasized that IUDs do not protect against sexually transmitted diseases (STDs).

Hormone-releasing IUDs such as Progestasert and levonorgestrel (LNG)-20 are used for more specific purposes. Their protective function has been attributed to impenetrable cervical mucus, endometrial changes, or reduced retrograde menstruation. However, such devices are less widely used in India. Hormonal measurements and follicular ultrasonography indicate that ovulation is generally not inhibited in women wearing hormonally active IUDs. Since the IUDs do not prevent ovulation, it is more effective at preventing intrauterine than extrauterine pregnancy. Thus, IUD users have a higher risk of ectopic pregnancy than nonusers [23]. Nevertheless, the long-term nature and reversibility of the contraceptive action meet the needs of young women and are therefore preferred.

The insertion timing of IUDs is not significantly related to any sociodemographic or reproductive history. From both physical and psychological standpoints, the ideal time for insertion appears to be immediately after abortion or parturition for several reasons [24–26]. First, the insertion of IUD is easier through a dilated cervix. Second, the demand on the hospital resources is limited to the cost of the IUD. Third, fertility can return rapidly after abortion/postpartum. Finally, women's motivation to initiate contraception is highest immediately after abortion and many patients do not return for a postabortion/postpartum examination. However, in such cases, they run the risk of perforation, expulsion, or pelvic inflammatory diseases (PIDs), particularly due to differential effects of sexual behavior [27,28]. For the postpartum period, hormone containing IUDs are not recommended, for their inhibiting effects on lactation [29]. Besides, small amounts of steroids find their way in milk and could possibly have a long-term effect on the child [30].

17.4 Mechanism of Action

Earlier, the accepted opinion was that IUDs played a predominant role after fertilization in preventing implantation [31–33]. But, now there is increasing evidence that points toward important contributions before fertilization [34–36]. The possible prefertilization mechanisms of action of the IUD include inhibition of sperm migration and viability at the level of the cervix, endometrium, and tube; slowing or speeding the transport of the ovum through the fallopian tube; and damage to and destruction of the ovum before fertilization. The contraceptive effects of copper IUDs occur predominantly before fertilization [37–39]. The postfertilization mechanism of action of IUD includes slowing or speeding the transport of the early embryo through the fallopian tube, damage to, or destruction of the early embryo before it reaches the uterus, and prevention of implantation. Thus, the potential mechanism of action of IUDs depend on whether the IUD is made of inert material or is coated with Cu/hormones [40].

17.5 Problems Associated with IUD Usage with Special Reference to Microbial Infections

Most of the problems associated with IUDs and termed as side effects (not sickness) are noticed soon after the insertion of IUDs, but gradually disappear after three months. Some pain, bleeding, or spotting may occur immediately after IUD insertion, which usually goes away in a day or two. The common side effects are longer/heavier menstrual periods, bleeding or spotting between periods, and more cramps/pain during periods. These can be tackled through the use of hormone (levonorgestrel) releasing IUDs, which provide immense health benefits. Advanced research and development has resulted in the marketing of several, small, novel intracervical/intrauterine devices. The product range is large and sometimes expulsion occurs which may go unnoticed, resulting in pregnancy. Possibly, this may be due to deviation in the insertion technique and not to faulty design of the device itself. Improper insertion of the IUD could also result in perforation (piercing) of the uterine wall.

The most common grounds for discontinuation of contraception in women irrespective of the method in use is the desire to conceive, change in patients preference, altered pattern of sexual activity, and vasectomy of the male partner [41]. In addition, the cause for early discontinuation of IUD is bleeding or pain along with genital infections [42]. The IUD can be safe and effective, if inserted under aseptic conditions in women free of cervical infection [43]. In a recent study conducted over a period of 5 years in Finland, several reasons for the premature removal of IUD were analyzed and the lowest incidence was due to infections [44]. On the contrary, infections due to IUD use are a serious cause for concern among Asian women [45,46]. The data available for the Indian population also subscribes to this view (Figure 17.2).

The most probable cause of infection is contamination at the time of IUCD insertion [47]. Even after safe insertion, there is a higher risk of infection when IUDs are promoted in areas of high STI/HIV prevalence [43]. In most cases, IUDs remain in the correct position except when the IUD is inserted

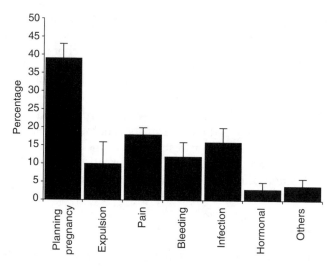

FIGURE 17.2
Causes for removal of IUD in Indian women.

soon after childbirth, but this is very rare. Some women use their fingers to check the position of the IUD strings from time to time and this may be yet another source of infection when proper hygiene is not followed.

17.6 Mode and Mechanism of Infection and Biofilm Formation on IUDs

In a recent study, the IUD removed due to symptoms of PID was investigated via both microbial culture and scanning electron microscopy [48]. A close association between the distribution of aerobic and anaerobic bacteria on the IUDs with different times in situ, was recorded. Among the two basic states of microbial cells, planktonic and sessile, it is believed that the former are important for rapid proliferation and spread into new territories, whereas the latter are slow-growing populations focused on perseverance. Candida may enter the uterine cavity from the vagina, by ascending transcervical infection using IUCD strings as a ladder [49,50]. Another study suggests that the Dalkon shield multifilament tail-string could carry bacteria cephalad by capillary action [51]. However, there are conflicting reports regarding this view [52,53]. Adherent microbes growing in consortia, known as biofilms, are present in virtually all natural and pathogenic ecosystems [54,55]. IUDs are no exception [56]. The organisms detected in the IUDs removed from Indian women are shown in Figure 17.3 and Table 17.2. These biofilms form microbial communities encased in characteristic architecture composed of polysaccharide-rich extracellular matrices and live in association with both abiotic and biotic surfaces [57].

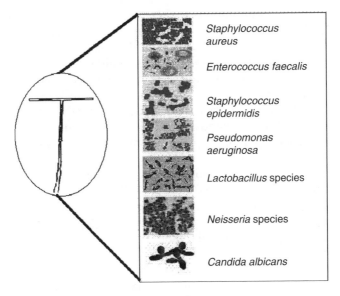

FIGURE 17.3 (See color insert following page 210)
Microflora associated with infected IUDs in Indian women.

The process of microbial adherence to surfaces is the initial event in the pathogenesis of infection and is largely dictated by a number of variables such as bacterial species, cell surface composition, nature of surface, nutrient availability, hydrodynamics, cell-to-cell communication, regulatory network, etc. This complex, dynamic process of biofilm formation starts shortly after

TABLE 17.2

Various Pathogenic Microbial Colonies Sampled from the Surfaces of the Infected IUDs

Microorganisms	Percentage Incidences
Staphylococcus aureus	28–35
Staphylococcus epidermidis	24–32
Pseudomonas aeruginosa	16–20
Escherichia coli	14–18
Lactobacillus species	16–19
Enterococcus faecalis	18–22
Candida albicans	14–16
Candida dubliniensis	12–16
Neisseria gonorrhea	4–7
Bacteroides species	3–5
Proteus species	2–4
Mycoplasmas	4–7
Gardnerella vaginalis	3–6
Chlamydia	8–10
Others	6–11

the insertion of the device into the body with conditioning of the surface by way of water, lipid, albumin, extracellular polymer matrix, or other nutrients from the surrounding environment [58]. The subsequent adherence of microbes to the surface is a reversible association mediated by different physical, chemical, and biological interactions between bacterial cell and abiotic surfaces through nonspecific interactions such as electrostatic, hydrophobic, or van der Waals forces. Microorganisms randomly arrive near the surface of the implanted medical device by several mechanisms such as direct contamination, contiguous spread, or hematogenous spread [59]. These abiotic surfaces then become an easy target for biofilm formation, as they lack innate protective coating and secretion associated with host mucosal and epithelial tissues [60].

The planktonic cells make contact with the surface of IUD either randomly or in a direct fashion via chemotaxis and motility [61–64]. Pratt and Kolter [65] elucidated the role of flagella, motility, chemotaxis, and type I pili of *Escherichia coli* in biofilm formation. Similarly, type IV pili are responsible for a form of surface associated movement known as twitching motility and have been shown to play an important role in bacterial adhesion and pathogenesis [66]. Once planktonic cells bind to the surface through exopolymeric matrix, proliferation and accumulation of monolayer cells to multilayer cell clusters start in an irreversible manner. The stages involved in the formation of biofilms on IUDs are consolidated in Figure 17.4.

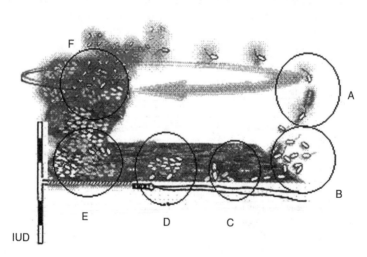

FIGURE 17.4
Hypothetical stages involved in the formation of biofilms on IUDs. (A) Organic molecules produced by microbial cells condition the IUD. (B) Microbial cells are transported/migrated to the surface of the medical device. (C) Initial adhesion of microbial cells takes place. (D) Firm attachment of microbial cells is established. (E) Biofilms begin to grow and mature. (F) Biofilm fragments detach and the cycle either repeats or the detached fragments migrate into adjoining areas leading to pelvic inflammatory diseases.

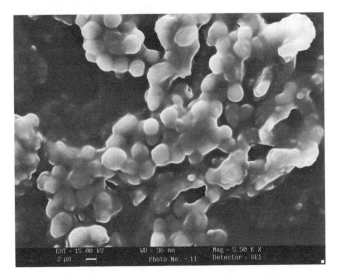

FIGURE 17.5
Scanning electron micrograph of biofilm recovered from an infected IUD.

The high molecular weight, extracellular, polymeric matrices formed are generally composed of a mixture of protein, polysaccharides, and nucleic acid, which remain under the control of gene cluster, as reported by Davies and Geesey [67], during their study on the regulation of the alginate biosynthesis gene in *Pseudomonas aeruginosa*. In the biofilm, algC is expressed at levels approximately 19 times greater than that in planktonic cells. Slime production in *P. aeruginosa* was also considered to be essential in cementing bacterial cells together in the biofilm structures together with trapping and retaining nutrients for biofilm formation. During this phase, coordinated activities of different genes related with switching of microorganisms from planktonic to sessile form, under different environmental conditions, also take place.

Under suitable conditions, microorganisms adherent to the device surface eventually develop into an adherent biofilm of a highly structured and cooperative consortia of microbes. Scanning electron microscope (Figure 17.5) and confocal laser scanning microscope (Figure 17.6) studies in our laboratory show that biofilms recovered from infected IUDs typically consist of pillar-like mushroom structures of microbial microcolonies, separated by fluid-filled channels and void. It appears that such a configuration allows convective flow that transports nutrients and oxygen from the interface to reach all level of the biofilm and diffuse toxic waste products [55,68–73]. In fact, the basic "design" of the mushroom-like microcolonies with intervening water channels mimics the primitive circulating system of higher organisms [74]. Steep oxygen gradient formed by the cells near the surface of the biofilm microcolony further adds to the structural and metabolic heterogenic

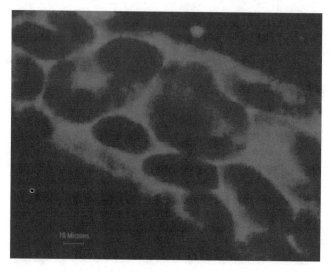

FIGURE 17.6 (See color insert following page 210)
Confocal laser scanning microscope (CLSM) picture of biofilm formed on IUD string as visualized by fluorescent staining with propidium iodide showing fluid-filled channels.

environment within the biofilm [75–77]. This provides microcolonies with the capability to resist stress, whether from host defense systems or antimicrobial therapy [78].

Recent evidence indicates that biofilm formation might be regulated at the level of population density-dependent gene expression, controlled by quorum sensing (QS) or cell-to-cell signaling [79] along with functions as diverse as motility, virulence, sporulation, antibiotic production, DNA exchange, and development of multicellular fruiting body formation [3,80,81]. It is hypothesized that the QS system makes use of two components: a small diffusible signaling molecule known as the autoinducer (AI) and a positive transcriptional activator or R-protein. At low cell density the AI is produced at a basal level; however, as the cell density increases, so does the concentration of AI. Once the AI level reaches a threshold, sufficient AI/R-protein complexes accumulate, encasing the activation of target genes. Among these, AI, acylhomoserine lactones (AHLs) signals in gram-negative, amino acids and short, cyclic peptide signals in gram-positive bacteria, and furanosyl borate diester known as AI-2 signals of both groups have been detected in naturally occurring biofilms [82–85].

17.7 Problems Associated with Eradication of Biofilms

In implanted, medical device-related infections, detachment of microorganisms from biofilms may represent a fundamental mechanism leading to the dissemination of infection in the host [59,86]. Detachment is determined as a function of growth phase, biofilm size, nutrient starvation conditions, loss of EPS, hydrodynamic flow, or mechanical shear forces [59,72,78]. Even the role of QS involved in control of biofilm detachment has been reported [87,88]. On the basis of magnitude and frequency of detachment, microbial aloofness from the biofilm has been categorized into two processes [86]: (a) Erosion—the continual detachment of single cells and small portion of biofilm. (b) Sloughing—the rapid massive loss of biofilms. Since these detached cell clusters will most likely possess the antimicrobial resistance and ability to evade the host immune responses—characteristics of the original biofilm—the resultant secondary infection (PID) caused by these detached cells may be particularly difficult to treat, and therefore necessitate removal of the implanted device.

17.8 IUDs and Pelvic Inflammatory Diseases

It must be borne in mind that IUDs do not protect against STDs including HIV/AIDS. Therefore, IUD may not be the right choice of contraceptive for women with recent STDs or with multiple sex partners or with sex partners who have multiple sex partners. In the mid 1980s and 1990s the prevalence of *Chlamydia trachomatis* in populations of IUD users was 3% in Sweden [89], 13% in Kenya [90], and 7% in Nigeria [4]. Screening for *N. gonorrhea* in African populations recorded 2% [4] and 4% [90] while in Norway the incidence is very low since this kind of infection has been practically eradicated [91]. Another study reported that IUD users with PID had significantly more *Fusobacterium* spp. and *Peptostreptococcus* spp. than non-IUD users with PID [92]. Statistical analysis of data collected in recent times reveals that infection in IUD users has dropped substantially. Although the prevalence of STDs may vary in different ethnic populations and geographical locations, the changed scenario can be attributed to the early detection of infection and treatment with antibiotics, before insertion of the IUD. The absolute risk of PID for any particular insertion is the presence or prevalence of gonorrhea or chlamydia, multiplied by the probability of contacting PID when an IUD is inserted with these diseases present [93]. In other words, PID is more likely to follow STD infection if women use an IUD [94]. A single infection could affect the tubal epithelium and lead to infertility or ectopic pregnancy.

Bacterial vaginosis (BV) is the most common vaginal infection among women of reproductive age and studies indicate a higher prevalence of this infection among IUD users in different countries [95,96]. In fact, they support the hypothesis that IUDs may change the vaginal flora in favor of BV-associated bacteria [97]. It may be that the presence of an IUD facilitates the ascent of cervicovaginal microorganisms into the uterus, thus predisposing women to PID.

Pelvic actinomycosis is usually reported in literature as an infection related to prolonged use of IUDs [98–101]. *Actinomyces* species are gram-positive, nonacid fast anaerobic bacteria that exhibit branching, filamentous growth. *Actinomyces israelii* is the most common subtype in human disease, a normal anaerobic commensal of the genital tract and gastrointestinal tract in both oropharynx and bowel. It has been identified in 8%–20% of women who have an IUD. Most patients are asymptomatic but up to 25% reportedly develop symptoms of pelvic infection. It is widely believed that ascending actinomyces infection derives from the perineum, vagina, cervix, or through oro- and anogenital contact. *Actinomyces* species grow as colonies that extend slowly across natural anatomical boundaries, forming abscesses and sinus tracts filed with polymorphonuclear leucocytes and surrounded by thick, indurated, fibrotic tissue [102].

17.9 Precautionary and Remedial Measures

The first major precaution that should be taken before opting for an IUD is a pelvic examination. As far as possible, all counseling and screening should be handled on the same day as the insertion. The IUD should not be inserted in pregnant women and adequate measures should be taken to ascertain this fact. Unexplained vaginal bleeding generally suggests an underlying medical condition and it is unadvisable to insert the IUD until the problem is diagnosed and treated. In case there is a history of STD/PID in the previous three months, one should refrain from having the IUD inserted even after treatment, until all signs/symptoms of the infection have disappeared. These include abnormal vaginal discharge, fever, or frequent urination with burning. Since it is a common practice to insert an IUD following childbirth, it is essential to check for puerperal sepsis/genital tract infection.

Medical device infection is associated with significant morbidity, mortality, and associated health care costs. Therefore, one follow-up visit is recommended approximately one month after insertion, since the peak incidence of PID is within this time. Thereafter, there is no need for a fixed, follow-up schedule. However, the client may be encouraged to return if there is excessive abdominal pain or pain with intercourse, infection exposure, abnormal vaginal discharge, strings missing, or if strings seem shorter or longer. At the university clinic in Turkey, a substantial number of women with IUD were diagnosed as having a pelvic abscess within a 7 year period [103]. Despite current knowledge that PID and pelvic abscess are rarely encountered in long-term IUD users, the presence of an IUD should be investigated in cases with an initial diagnosis of pelvic abscess based on clinical and ultrasonographic evaluation.

The health care costs associated with medical device infections are enormous. The difference between the antibiotic resistance of planktonic and biofilm forming microbial population recovered from the female

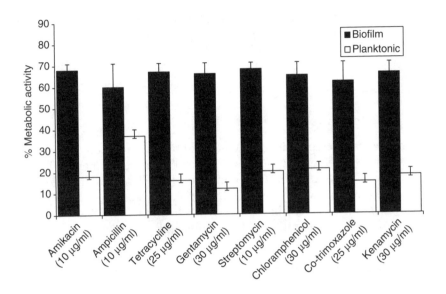

FIGURE 17.7
The effect of antimicrobial agents on planktonic and biofilm-forming microbes cultured from the reproductive tract and IUD of infected Indian women.

reproductive tract is presented in Figure 17.7. The biofilm formed due to colonization of microorganisms contains 5–50 Å thick glycoprotein matrices that protect the bacteria through a diffusion limitation process and increase their resistance to antibodies, macrophages, and antibiotics. In addition, the materials used to make the IUD can inhibit macrophage chemotaxis, phagocytes, and other processes of the immune system meant for dealing with infection. As a result, local and systemic infections may lead to insertion site abscess, septicemia, endocarditis, and gangrene of extremities. The sustained delivery of antimicrobial agents into the medical device microenvironment may prevent biofilm formation and avoid systemic side effects associated with oral or intravenous antimicrobial administration. One novel option is to coat the medical device with appropriate nanoparticles that may impart antimicrobial and anti-inflammatory properties [104,105].

17.10   Concluding Remarks

Statistical data indicates that Asian women in general and Indian women in particular regard IUDs as a safe and effective means of contraception. Nevertheless, poor sanitation conditions in the primary health centers and casual personal hygiene of IUD users are a cause for concern. Infection and related PIDs are one among several important reasons for the removal of IUDs. To receive IUDs, women should not be at an increased risk of STDs.

Rapid screening, early diagnosis and treatment of infections before insertion of IUDs will be more affordable in developing countries than treatment of complications that might result later.

Acknowledgments

Research related to regulation and control of biofilms on medical devices at our institution is being generously supported by the Department of Biotechnology, Government of India. The authors wish to thank Sunitha Pereira for editorial assistance and Professor D.R.P. Tulsiani, Department of Obstetrics and Gynecology, Vanderbilt University School of Medicine, for useful suggestions in the preparation of the manuscript.

References

1. American College of Obstetrics and Gynecology (ACOG). ACOG practice bulletin. Clinical management guidelines for obstetrician-gynecologists: Intrauterine device, *Obstet. Gynecol.*, 105, 223–232, 2005.
2. Chhabra, S., Gupte, N., and Mehta, A., Medical termination of pregnancy and concurrent contraceptive adoption in rural India, *Stud. Fam. Plann.*, 19, 244–247, 1988.
3. Hentzer, M., Givskov, M., and Eberl, L., Quorum sensing in biofilms: Gossip in slime city, in *Microbial biofilms*, Ghannoum, M. and O'Toole, G.A. (Eds.), ASM Press, Washington DC, pp. 118–140, 2004.
4. Lapido, C.G., Farr, G., and Otolorin, E., Prevention of IUD-related pelvic infection: The efficacy of prophylactic doxycycline at IUD insertion, *Adv. Contracept.*, 7, 43–54, 1991.
5. Villanueva, Y., Mendoza, I., and Aguilar, C., Expansion of the role of nurse auxiliaries in the delivery of reproductive health services in Honduras, Population Council, Washington, DC, 2001.
6. White, M.K., Ory, H.W., and Rooks, J.B., Intrauterine device termination rates and the menstrual cycle day of insertion, *Obstet. Gynecol.*, 55, 220–224, 1980.
7. World Health Organization (WHO), Johns Hopkins Bloomberg School of Public Health, Center for Communication Programs. Information and knowledge for optimal health (INFO), Decision-making tool for family planning clients and providers, Baltimore, INFO and Geneva, WHO, 2005.
8. Chadwani, A. and Amin-Hanjani, S., Incidence of actinomycosis associated with intrauterine devices, *J. Reprod. Med.*, 39, 585–587, 1994.
9. Jain, A.K., Safety and effectiveness of contraceptive devices, *Contraception*, 11, 243, 1975.
10. Tatum, H.J., Clinical aspects of intrauterine contraception: Circumspection 1976, *Fertil. Steril.*, 28, 3, 1977.
11. Skjeldestad, F.E., Choice of contraceptive modality by women in Norway, *Acta. Obstet. Gynecol. Scand.*, 73, 48–52, 1994.

12. Demographic and Health Surveys, STAT compiler, 2005. http://www.measuredhs.com/statcompiler/
13. Mishell, D.R. Jr. et al., The intrauterine device: A bacteriological study of the endometrial cavity, *Am. J. Obstet. Gynecol.*, 96, 119–126, 1966.
14. Fiscina, B. et al., Gonococcal action of copper in vitro, *Am. J. Obstet. Gynecol.*, 115, 86–90, 1973.
15. Elhag, K.M., Bahr, A.M., and Mubarak, A.A., The effect of a copper IUD on the microbial ecology of the female genital tract, *J. Med. Microbiol.*, 28, 245–251, 1988.
16. Mehanna, M.T.R. et al., Chlamydial serological characteristics among intrauterine contraceptive device users. Does copper inhibit chlamydial infection in the female genital tract? *Am. J. Obstet. Gynecol.*, 171, 691–693, 1994.
17. Moyer, D.L. and Mishell, D.R. Jr., Reactions of human endometrium to the intrauterine foreign body. II. Longterm effects of the endometrial histology and cytology, *Am. J. Obstet. Gynecol.*, 111, 66–80, 1971.
18. Yin, M. et al., The presence of mast cells in the human endometrium pre- and post insertion of intrauterine devices, *Contraception*, 48, 245–254, 1993.
19. Ding, A.H., Nathan, C.F., and Steuhr, D.J., Release of reactive nitrogen intermediates and reactive oxygen intermediates from mouse peritoneal macrophages: Comparison of cytokines and evidence for independent production, *J. Immunol.*, 141, 2407–2412, 1981.
20. Amla, A. et al., Active oxygen species in copper intrauterine device users, *Contraception*, 48, 150–156, 1993.
21. Nathan, C. and Hibbs, J.B. Jr., Role of nitric oxide synthesis in macrophage antimicrobial activity, *Curr. Opin. Immunol.*, 3, 65–70, 1991.
22. Pradhan, M., Gupta, I., and Ganguli, N.K., Nitrites and L-citrulline levels in copper intrauterine device users, *Contraception*, 55, 315–318, 1997.
23. Bouyer, J. et al., Risk factors for extrauterine pregnancy in women using an intrauterine device, *Fertil. Steril.*, 74, 899–908, 2000.
24. O'Hanley, K. and Huber, D.H., Postpartum IUDS: Keys for success, *Contraception*, 45(4), 351–361, 1992.
25. Welkovic, S., Post-partum bleeding and infection after post-placental IUD insertion, *Contraception*, 63, 155–158, 2001.
26. El-Tagy, A. et al., Safety and acceptability of post-abortal IUD insertion and the importance of counseling, *Contraception*, 67, 229–234, 2003.
27. Gillet, P.G., A comparison of the efficiency and acceptability of the copper-7 intrauterine device following immediate or delayed insertion after first trimester therapeutic abortion, *Fert. Steril.*, 34, 121–124, 1980.
28. Grimes, D.A., Intrauterine device and upper genital-tract infection, *Lancet*, 356, 1013–1019, 2000.
29. World Health Organization, Effects of hormonal contraceptives on breast milk composition and infant growth, *Stud. Fam. Plann.*, 19, 361–369, 1988.
30. Diaz, S. and Croxatto, H.B., Contraception in lactating women, *Curr. Opin. Obstet. Gynecol.*, 5, 812–822, 1993.
31. Corfman, P. and Segal, S., Biological effects of intrauterine devices, *Am. J. Obstet. Gynecol.*, 100, 448–459, 1968.
32. Hilgers, T.W., The intrauterine device: Contraceptive or abortifacient? *Minn. Med.*, 57, 493–501, 1974.
33. Spinnato, J.A., Mechanism of action of intrauterine contraceptive devices and its relation to informed consent, *Am. J. Obstet. Gynecol.*, 176, 503–506, 1997.

34. Ortiz, M.E. and Croxatto, H.B., The mode of action of IUDs, *Contraception*, 36, 37–53, 1987.
35. Silvin, I., IUDs are contraceptives, not abortifacients: A comment on research and belief, *Stud. Fam. Plann.*, 20, 355–359, 1989.
36. Ortiz, M.E., Croxatto, H.B., and Bardin, C.W., Mechanism of action of intrauterine devices, *Obstet. Gynecol. Surv.*, 51, S42–S51, 1996.
37. Mishell, D.R., Intrauterine devices: Mechanism of action, safety and efficacy, *Contraception*, 58, 45S–53S, 1998.
38. Dardanokman, R.T. and Burkman, R.T., The intrauterine contraceptive device: An often-forgotten and maligned method of contraception, *Am. J. Obstet. Gynecol.*, 181, 1–5, 1999.
39. Rivera, R., Yacobson, I., and Grimes, D., The mechanism of action of hormonal contraceptives and intrauterine contraceptive devices, *Am. J. Obstet. Gynecol.*, 181, 1263–1269, 1999.
40. Stanford, J.B. and Mikolajczyk, R.T., Mechanism of action of intrauterine devices: Update and estimation of post fertilization effects, *Am. J. Obstet. Gynecol.*, 187, 1699–1708, 2002.
41. Colli, E. et al., Reasons for contraceptive discontinuation in women 20–39 years old in New Zealand, *Contraception*, 59, 227–231, 1999.
42. Morrison, C.S., Sekadde-Kigondu, C., and Miller, W.C., Use of sexually transmitted disease risk assessment algorithms for selection of intrauterine device candidates, *Contraception*, 59, 97–106, 1999.
43. Steen, R. and Shapiro, K., Intrauterine contraceptive devices and risk of pelvic inflammatory disease: Standard of care in high STI prevalence settings, *Reprod. Health Matters*, 12(23), 136–143, 2004.
44. Pakarinen, P. and Luukkainen, T., Five years experience with a small intracervical/intrauterine levonorgestrel-releasing device, *Contraception*, 72, 342–345, 2005.
45. Brabin, L. et al., Reproductive tract infections, gynecological morbidity and HIV seroprevalence among women in Mumbai, India, *Bull. World Health Organ.*, 76(3), 277–287, 1998.
46. Yonemura, S. et al., Ureteral obstruction associated with pelvic inflammatory disease in a long-term intrauterine contraceptive device user, *Int. J. Urol.*, 13(3), 315–317, 2006.
47. Schweid, A.I. and Hopkins, G.B., Monilia chorionitis associated with an intra uterine contraceptive device, *Obstet. Gynecol.*, 31, 791, 1968.
48. Pal, Z. et al., Biofilm formation on intrauterine devices in relation to duration of use, *J. Med. Microbiol.*, 54, 1199–1203, 2005.
49. Delprado, W.J., Baird, P.J., and Russell, P., Placental candidiasis: Report of 3 cases with a review of the literature, *Pathology*, 14, 191–195, 1982.
50. Donders, G.G. et al., Intrauterine Candida infection: A report of four infected fetuses from two mothers, *Eur. J. Obstet. Gynecol. Reprod. Biol.*, 38, 233–238, 1991.
51. Tatum, H.J. et al., The Dalkon Shield controversy: Structural and bacteriological studies of IUD tails, *JAMA*, 231, 711–717, 1975.
52. Rivera, R., Is there an effect of the IUD strings in the development of pelvic inflammatory disease in IUD users? in *Proceedings from the Fourth International Conference on IUDs*, Bardin, C.W. and Mishell, D.R. (Eds.), Butterworth-Heinemann, Boston, MA, 1994, 171–178.
53. Ebi, K.L. et al., Evidence against tail strings increasing the rate of pelvic inflammatory disease among IUD users, *Contraception*, 53, 25–32, 1996.

54. Costerton, J.W. et al., Bacterial biofilms in nature and disease, *Annu. Rev. Microbiol.*, 41, 435–464, 1987.

55. Davey, M.E. and O'Toole, G.A., Microbial biofilms: From ecology to molecular genetics, *Microbiol. Mol. Biol. Rev.*, 64, 847–867, 2000.

56. Pruthi, V., Al-Janabi, A., and Pereira, B.M.J., Characterization of biofilm formed on intrauterine devices, *Ind. J. Med. Microbiol.*, 21, 161–165, 2003.

57. Watnick, P. and Kolter, R., Biofilm, city of microbes, *J. Bacteriol.*, 182, 2675–2679, 2000.

58. Dunne, W.M., Bacterial adhesion: Seen any good biofilm lately? *Clin. Microbiol. Rev.*, 15, 155–166, 2002.

59. Gristina, A.G., Biomaterial-centred infection: Microbial adhesion versus tissue integration, *Science*, 237(4822), 1588–1595, 1987.

60. Veenstra, G.J. et al., Ultrastruture organization and regulation of a biomaterial adhesin in *Staphylococcus epidermis*, *J. Bacteriol.*, 178, 537–541, 1996.

61. Jiang, X. and Pace, J.L., Microbial biofilms, in *Biofilms, Infection, and Antimicrobial Therapy*, Pace, J.L., Rupp, M.E., and Finch, R.G. (Eds.), CRC Press, Boca Raton, FL, pp. 3–19, 2006.

62. Carpentier, B. and Cerf, O., Biofilms and their consequences, with particular references to hygiene in the food industry, *J. Appl. Bacteriol.*, 75, 499–511, 1993.

63. Fletcher, M. and Loeb, G.I., Influence of substratum characteristics on the attachment of a marine pseudomonad to solid surfaces, *Appl. Environ. Microbiol.*, 37, 67–72, 1979.

64. O'Toole, G. and Kolter, R., Flagellar and twitching motility are necessary for *Pseudomonas aeruginosa* biofilm development, *Mol. Microbiol.*, 30, 285–293, 1998.

65. Pratt, L.A. and Kolter, R., Genetic analysis of *Escherichia coli* biofilm formation: Roles of flagella, motility, chemotaxis and type I pili, *Mol. Microbiol.*, 30, 285–293, 1998.

66. Shirtliff, M.E., Mader, J.T., and Camper, A.K., Molecular interactions in biofilms, *Chem. Biol.*, 9, 859–871, 2002.

67. Davies, D.G. and Geesey, G.G., Regulation of the alginate biosynthesis gene *algC* in *Pseudomonas aeruginosa* during biofilm development in continuous culture, *Appl. Environ. Microbiol.*, 61, 860–867, 1995.

68. Kolenbrander, P.E. and London, J., Adhere today, here tomorrow: Oral bacterial adherence, *J. Bacteriol.*, 175, 3247–3252, 1993.

69. Douglas, L.J., Candida biofilms and their role in infection, *Trends Microbiol.*, 11, 30–36, 2003.

70. De Beer, D. et al., Effect of biofilm structure on oxygen distribution and mass transport, *Biotechnol. Bioeng.*, 59, 302–309, 1994.

71. Costerton, J.W., Stewart, P., and Greenberg, E.R., Bacterial biofilms: A common cause of persistent infections, *Science*, 284, 318–322, 1999.

72. Donlan, R.M. and Costerton, J.W., Biofilms: Survival mechanisms of clinically relevant microorganisms, *Clin. Microbiol. Rev.*, 15, 167–193, 2002.

73. Jefferson, K.K., What drives bacteria to produce a biofilm? *FEMS Microbiol. Lett.*, 236, 163–173, 2004.

74. Costerton, J.W. et al., Microbial biofilms, *Annu. Rev. Microbiol.*, 49, 711–745, 1995.

75. Lewandowski, Z., Dissolved oxygen gradients near microbially colonized surfaces, in *Biofouling and Biocorrosion in Industrial Water Systems*, Geesey, G.G., Lewandoski, Z., and Flemming, H.C. (Eds.), Lewis Publishers, Boca Raton, FL, pp. 175–188, 1994.

76. De Beer, D., Stoodley, P., and Lewandowski, Z., Liquid flow in heterogeneous biofilms, *Biotech. Bioeng.*, 44, 636, 1994.
77. Walters, M.C. et al., Contributions of antibiotic penetration, oxygen limitation and low metabolic activity to tolerance of *Pseudomonas aeruginosa* biofilms to ciprofloxacin and tobramycin, *Antimicrob. Agents Chemother.*, 47, 317–323, 2003.
78. Stoodley, P. et al., Biofilms as complex differentiated communities, *Annu. Rev. Microbiol.*, 56, 187–209, 2002.
79. Davies, D.G. et al., The involvement of cell-to-cell signals in the development of a bacterial biofilm, *Science*, 280, 295–298, 1998.
80. De Kievit, T.R. and Iglewski, B.H., Quorum sensing and microbial biofilms, in *Medical Implications of biofilms*, Wilson, M. and Devine, D. (Eds.), Cambridge University Press, Cambridge, pp. 18–35, 2003.
81. Smith, J.L., Fratamico, P.M., and Novak, J.S., Quorum sensing: A primer for food microbiologists, *J. Food Prot.*, 67, 1057–1070, 2004.
82. Fuqua, C. and Greenberg, E.P., Self perception in bacteria: Quorum sensing with acylated homoserine lactones, *Curr. Opin. Mircobiol.*, 1, 183–189, 1999.
83. Novick, R.P., Regulation of pathogenicity in staphylococcus aureus by peptide-based density-sensing system, in *Cell-cell signaling in bacteria*, Dunny, G.M. and Winans, S.C. (Eds.), American Society for Microbiology Press, Washington DC, pp. 175–192, 1999.
84. Miller, M.B. and Bassler, B.L., Quorum sensing in bacteria, *Annu. Rev. Microbiol.*, 55, 165–199, 2001.
85. McLean, R.J.C. et al., Evidence of autoinducer activity in naturally occurring biofilms, *FEMS Microbiol. Lett.*, 154, 259–263, 1997.
86. Stoodley, P. et al., Growth and detachment of cell clusters from mature mixed-species biofilms, *Appl. Environ. Microbiol.*, 67(12), 5608–5613, 2001.
87. O'Toole, G., Kalpan, H.B., and Kolter, R., Biofilm formation as microbial development, *Annu. Rev. Microbiol.*, 54, 49–79, 2000.
88. Boyd, A. and Chakrabarty, A.M., Role of alginate lyase in cell detachment of *Pseudomonas aeruginosa*, *Appl. Environ. Microbiol.*, 60, 2355–2359, 1994.
89. Pap-Akeson, M. et al., Genital tract infections associated with the intrauterine contraceptive device can be reduced by inserting the threads into the uterine cavity, *Br. J. Obstet. Gynecol.*, 99, 676–679, 1992.
90. Sinei, S.K.A., Schulz, K.F., and Lamptey, P.R., Preventing IUCD-related pelvic infection: The efficacy of prophylactic doxycycline at insertion, *Br. J. Obstet. Gynecol.*, 97, 412–419, 1990.
91. Skjeldestad, F.E. et al., IUD users in Norway are at low risk for genital *C. trachomatis* infection, *Contraception*, 54, 209–212, 1996.
92. Viberga, I. et al., Microbiology profile in women with pelvic inflammatory disease in relation to IUD use, *Infect. Dis. Obstet. Gynecol.*, 13(4), 183–190, 2005.
93. Shelton, J.D., Risk of clinical pelvic inflammatory disease attributable to an intrauterine device, *Lancet*, 357, 443, 2001.
94. Mohllajee, A.P., Curtis, K.M., and Peterson, H.B., Does insertion and use of an intrauterine device increase the risk of pelvic inflammatory disease among women with sexually transmitted infection? A systematic review, *Contraception*, 73, 145–153, 2006.
95. Calzolari, E. et al., Bacterial vaginosis and contraceptive methods, *Int. J. Gynecol. Obstet.*, 70, 341–346, 2000.
96. Josoef, M.R. et al., High rate of bacterial vaginosis among women with intrauterine devices in Manado, Indonesia, *Contraception*, 64, 169–172, 2001.

97. Haukkamaa, M. et al., Bacterial flora of the cervix in women using an intrauterine device, *Contraception*, 36, 527–534, 1987.

98. Nayar, M. et al., Incidence of actinomycetes infection in women using intrauterine contraceptive devices, *Acta. Cytol.*, 29(2), 111–116, 1985.

99. Tendolkar, U. et al., *Actinomyces* species associated with intrauterine contraceptive devices and pelvic inflammatory disease, *Ind. J. Pathol. Microbiol.*, 36(3), 238–244, 1993.

100. Kalaichelvan, V., Maw, A.A., and Singh, K., Actinomyces in cervical smears of women using intrauterine device in Singapore, *Contraception*, 73(4), 352–355, 2006.

101. Kayikcioglu, F. et al., Actinomyces infection in the female genital tract, *Eur. J. Obstet. Gynecol.*, 118, 77–80, 2005.

102. Henderson, S.R., Pelvic actinomycosis associated with an intrauterine device, *Obstet. Gynecol.*, 41, 726–732, 1973.

103. Tanir, H.M. et al., Pelvic abscess in intrauterine device users, *Eur. J. Contracept. Reprod. Health Care*, 10(1), 15–18, 2005.

104. Furno, F. et al., Silver nanoparticles and polymeric medical devices: A new approach to prevention of infection? *J. Antimicrob. Chemother.*, 54(6), 1019–1024, 2004.

105. Morrison, M.L. et al., Electrochemical and antimicrobial properties of diamond-like carbon-metal composite films, *Diamond Relat. Mater.*, 15, 138–146, 2006.

18

Antimicrobial Urinary Catheters

David J. Stickler and Nora A. Sabbuba

CONTENTS

18.1 Introduction

Indwelling bladder catheters are the most commonly deployed prosthetic medical devices [1]. It is estimated that worldwide over 100 million are used annually. They have many applications in modern medicine. In acute care in hospitals, they have important roles monitoring and controlling urine production from unconscious patients, and facilitating repair of the urethra after surgical procedures such as prostatectomy. In chronic care, they are used for the long-term management of urinary retention and incontinence in the elderly and in patients disabled by strokes, spinal injury, or neuropathies such as multiple sclerosis. While these devices perform essential functions, unfortunately they also pose serious threats to the health and welfare of many patients. As long ago as 1958, Paul Beeson was warning of the dangers of indwelling bladder catheterization [2]. In his famous editorial

"The Case against the Catheter" he concluded that "...the decision to use this instrument should be made with the knowledge that it involves the risk of producing a serious disease which is often difficult to treat." These warnings were echoed some 30 years later when Calvin Kunin made the point that it is really quite extraordinary that although sophisticated technological advances have been made in the development of prosthetic devices in many other areas of medicine, we are still not able to perform the apparently simple task of draining the urine from patient's bladders without putting them at risk of serious illness [3]. The catheters that are used in such large numbers today have changed little in design from the devices introduced into urological practice by Dr. Fredrick Foley in the 1930s.

In this chapter, we will review the uses and prevalence of indwelling catheters, consider the infections that are associated with their use, review attempts that have been made to incorporate antimicrobial agents into the catheters to prevent these infections, look at the special problems associated with the care of patients enduring long-term catheterization, and finally discuss some of the novel strategies that have been suggested recently for controlling the infections and their complications.

18.2   Uses and Prevalence of Urinary Catheters

The Foley catheter is a simple flexible hollow tube generally manufactured from either latex or silicone polymer. The tip of the device, which is located in the bladder, has eyelets through which the urine drains. The other end of the device has a connector to a drainage tube and urine collection bag (Figure 18.1). A valve at the connector end allows the injection of sterile water through an inflation line in the catheter wall, filling a balloon that is located just below the eyeholes. The inflated balloon ensures that the catheter is not easily expelled from the bladder. It also seals the bladder sphincter so that urine does not normally leak along the outside of the catheter. The balloons are generally inflated with about 10 mL of water. Catheters come in a variety of sizes ranging from about 10–15 in. long with external diameters of 0.25–0.5 in. Cross-sections of these catheters reveal the central channel through which the urine flows and the lumen of the inflation line (Figure 18.1). It is clear that the thick walls of these catheters make the internal diameter of the urine drainage channel considerably smaller than the external dimensions.

Prevalence studies have provided data on the enormous extent to which these devices are used in both hospital and community care. Jepsen and others [4], for example, carried out a multicenter study in eight European countries and reported that 11% of hospital patients were undergoing indwelling bladder catheterization. Warren and coworkers [5] found that 9.4% of women and 6.4% of men in a population of nursing home residents in Maryland were undergoing this form of bladder management. A study

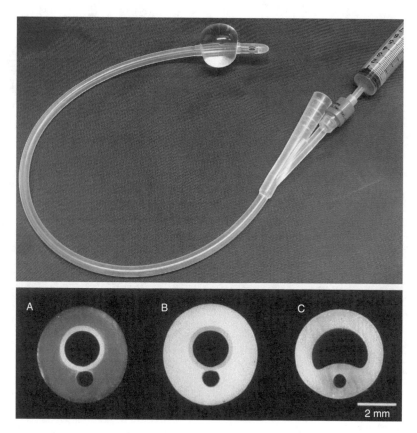

FIGURE 18.1
An all-silicone Foley catheter with an inflated retention balloon. Also shown are cross-sections of a silicone-coated latex catheter (A), a hydrogel-coated latex catheter (B), and an all-silicone catheter (C). (From Stickler, D.J. et al., *Urol. Res.*, 31, 306, 2003. With permission.)

from Denmark showed that 13.2% of hospital patients, 4.9% of patients in nursing homes, and 3.9% of patients in home care were catheterized [6]. A recent study of patients over the age of 65 receiving home care in 11 different countries in Europe reported that 11.5% of men and 3.3% of women had indwelling bladder catheters [7].

18.3 Catheter-Associated Urinary Tract Infections

Catheters constitute a convenient way to drain urine from the bladder, but unfortunately they also provide access for bacteria from a heavily contaminated external skin site to a vulnerable body cavity. The placement of the catheter also undermines the important basic defenses of the urinary tract against infection. The periodic filling and emptying cycle of the normal

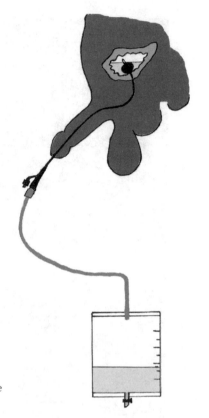

FIGURE 18.2
An indwelling urethral catheter with its urine drainage system.

urinary tract, which is so important in flushing out any microorganisms that may contaminate the urethra, does not take place in the catheterized bladder. A sump of residual urine forms in the bladder below the catheter eyeholes. The urine then trickles through the catheter, its drainage system, and into a collection bag (Figure 18.2). Thus, when bacteria gain access to the catheterized bladder, they are provided with a continually replenished reservoir of growth medium. Contamination of this reservoir results in the proliferation of even small inocula and bacterial populations rapidly exceed 10^5 cfu/mL [8]. The risk of infection is related to the length of time the catheter is in place and to the standard of catheter care. Most patients catheterized for less than a week should escape infection, but it is inevitable for those catheterized for periods longer than 4 weeks, even with modern drainage systems and meticulous care [9,10]. The number of catheters used is so large that catheter-associated urinary tract infection (UTI) is the most common of the infections that are acquired by patients in healthcare facilities [3,11].

The routes by which bacteria infect the catheterized bladder are well established. The insertion of the catheter may carry organisms from the urethra into the bladder. Good technique should ensure that this accounts for only a small proportion of infections. Organisms colonizing the periurethral skin can migrate to the bladder along the outside of the catheter through the mucoid film that forms between the mucosal surface of the urethra and the catheter [12–15]. Bacteria contaminating the urine drainage system can also gain access to the bladder through the lumen of the catheter. The outlets of the urine collection bags can become contaminated as health-care workers move from patient to patient to empty the bags. The regular opening of the taps to drain the urine facilitates the passage of contaminants into the bag. Bacterial multiplication in the urine can result in the build up of large populations of contaminating organisms. From these reservoirs of infection, bacteria can be carried through the drainage tube to the catheter and even into the bladder. This can happen when the bag is raised to the level of the bladder or above, as patients are transported around healthcare facilities [16]. Even in the absence of this sort of mismanagement, bacteria can actively migrate through the drainage tube, and catheter to the bladder [17,18]. A study by Nickel and coworkers [19], for example, demonstrated that *Pseudomonas aeruginosa* could migrate along a catheter surface against gravity and the flow of urine. They found that following contamination, bacteria adhere to the surfaces of the drainage tube or catheter. These bacteria multiply rapidly and develop into a firmly attached biofilm embedded in a protective polysaccharide matrix that they secrete around themselves. Electron microscopy suggested that the biofilm then spreads gradually ascending the luminal surface. They also found individual adherent bacteria ahead of the advancing biofilm front. They suggested that plank-tonic cells suspended in the urine column could leapfrog ahead of the biofilm, perhaps facilitated by turbulence created when the urine flow meets the biofilm front. This was termed "saltatory" movement allowing some cells to establish adherent microcolonies ahead of the ascending biofilm, which then develop and coalesce with the main body of the ascending biofilm. They concluded that only with the development of a biomaterial that effectively inhibits bacterial adherence we will be able to reduce the rate of catheter-associated infection.

Some workers have tried to determine the relative importance of the various routes of infection. Daifuku and Stamm [17], for example, conducted a prospective study on patients in an intensive care unit, from the first day of catheterization until the catheter was removed. They reported that in 18 females who acquired infection 14 had antecedent rectal colonization by the infecting strain and 12 had antecedal periurethral colonization. In contrast, in men only 5 of 17 episodes were preceded by rectal or urethral colonization. They concluded that in women the main route is by fecal strains colonizing the periurethral skin and then migrating through the urethra along the outside of the catheter. In men, the majority of infections were considered to arise by cross-infection and intraluminal migration. In a much larger prospective study, Tambyah and others [20] reported no significant differences in the pathogenic mechanisms for men or women. The probable mechanism of infection was identified as extraluminary in 66% of the cases and 34% were intraluminary. Earlier, Gillespie and others

[21] had pointed out that the relative importance of the various routes of infection will vary with the standard of catheter care. In units where careful management secures the closed drainage system, and bag-emptying procedures are performed according to strict guidelines, the external route will be the most important. Inwards where high standards are not maintained, the spread of organisms to the bladder from contaminated drainage systems will be the more significant and rapid route. The blocking of one of the routes of course merely allows other routes to become predominant. It is, thus, quite clear that strategies for the development of new technologies to prevent these infections must be designed to block all possible routes of entry [20,22].

The initial infections that occur after the first catheterization are usually by single bacterial species such as *Staphylococcus epidermidis*, *Enterococcus faecalis*, or *Escherichia coli* [13,20,21,23]. The longer the catheter remains in situ, the greater is the variety of organisms that accumulate in the bladder urine. Patients undergoing long-term catheterization become infected with complex mixed communities of mainly Gram-negative nosocomial species. Organisms such as *Providencia stuartii*, *Proteus mirabilis*, *Ps. aeruginosa*, *Morganella morganii*, and *Klebsiella pneumoniae* become particularly common and stable members of these communities [24,25]. These polymicrobial infections are extremely difficult to eliminate with antibiotic therapy, while the catheter remains in place [24]. It is common practice not to treat these infections with antibiotics unless clinical symptoms suggest that the kidneys or bloodstream have become involved. The mere colonization of the bladder urine is not considered to be clinically significant.

While it is true that most catheterized patients with bacteriuria are asymptomatic for most of the time, they are at risk from a range of complications, which can severely threaten their well-being. An important prospective study of 1540 patients in nursing homes in Ohio [26] clearly demonstrated significantly higher morbidity and mortality rates in catheterized patients than in matched noncatheterized controls. Logistic regression analysis of the data, adjusted for six other factors that had been shown to be independent risk factors for mortality (age, activities of daily living, cardiac disease, cancer, diabetes, and skin condition), showed that in comparison to noncatheterized patients, catheterized patients were three times more likely to die within a year. In addition, 279 pairs of patients, one catheterized the other not, matched for the six risk factors listed above, were selected for further analysis. The catheterized patients were found to have spent significantly more time in hospital over the 12 months (means of 10.4 compared to 2.8 days) and received significantly more antibiotics (average treatment for 35.1 days compared to 13.2 days). In both these cases, the mean values were significantly different ($P < 0.01$).

18.4 Complications of Catheter-Associated UTI

In addition to local periurinary infections such as urethritis, periurethral abscess, and prostatitis, patients undergoing long-term catheterization are also prone to kidney and bladder stones [27]. These can vary from hard crystalline bodies to diffuse aggregates of crystals in an organic matrix. Their formation is a consequence of chronic infection with urease producing bacteria that generate the conditions under which calcium and magnesium phosphate crystals form in the urine [28].

Symptomatic UTI is also a common complication in catheterized patients. Ouslander and others [29], for example, reported that over a prospective study period of a year, 80% of 54 male patients undergoing long-term catheterization had at least one episode. In a similar prospective study, Warren and coworkers [30] reported an incidence of 1.1/100 days of catheterization of febrile episodes of urinary origin in women. Although clinical symptoms are not necessarily reliable indicators that the urinary infections have progressed to the kidneys or the bloodstream in patients with catheters [16,31], there is evidence that kidney infections are common in this group. A prospective clinicopathological study involving autopsies on 75 nursing home residents [32], for example, demonstrated that acute inflammation of the renal parenchyma was present in 38% of patients who had a urinary catheter at the time of death compared to 5% of noncatheterized patients ($P = 0.004$).

Several surveys have revealed that 1%–4% of patients undergoing catheterization will develop symptomatic bloodstream infections [33,34]. Although this is a small percentage, the number of catheterized patients is so large that catheter-associated UTI is a major cause of nosocomial bacteremias [35]. A study of a large population of nursing home patients, for example, revealed that while only 5% of patients had catheters, this minority group experienced 40% of the Gram-negative bacteremias [36]. An additional problem being that these infections are commonly with antibiotic-resistant Gram-negative pathogens such as *Pv. stuartii, Klebsiella, Proteus, Enterobacter,* and *Ps. aeruginosa* [37,38].

In the case of long-term patients, many of the complications are triggered by catheter encrustation. These crystalline deposits can cover the retention balloon and obstruct the eyelet and the lumen of the catheter. They can cause trauma to the bladder mucosa and to the urethra on catheter withdrawal. The flow of urine through the catheter is blocked and under these circumstances, the urine can leak around the outside of the catheter and patients become incontinent. Alternatively, urine is retained within the bladder causing its painful distension. Reflux of infected urine to the kidneys can then precipitate serious symptomatic episodes such as pyelonephritis, septicemia, and endotoxic shock [39]. Several surveys have reported that about half of long-term patients will experience this complication [40–42].

18.5 Attempts to Control Catheter-Associated UTI

Over the years, there have been many attempts to prevent catheter-associated UTI with apparently rational strategies to block the routes of infection.

These include routine cleansing of the periurethral area with antibacterials, applying sponges soaked with antiseptic to the urethral meatus, application of antiseptic ointments to the periurethral skin, irrigation of the urethra with antibacterial solutions, and regular washing of the bladder with antibacterial solutions. These have all proved to be ineffective in preventing infection [39,43]. Paradoxically, the more comprehensive the attempts have been to put up antibacterial barriers, the less effective the strategies have been in preventing infection [44]. A dramatic example to illustrate this point was the attempt at a comprehensive policy to prevent infections in catheterized patients using the antiseptic chlorhexidine [45]. This policy involved preparing the periurethral skin with chlorhexidine; a gel containing the antiseptic was then instilled into the urethra to lubricate the passage of the catheter into the bladder. The catheter–metal junction was cleansed daily with chlorhexidine solution after which a cream containing the antiseptic was applied to the periurethral skin. Chlorhexidine was also included in the urine drainage bags every time they were emptied. The result was that with all the conceivable routes of infection blocked, an extensive outbreak occurred with a chlorhexidine-resistant multidrug-resistant strain of *Pr. mirabilis* involving at least 90 patients. The problem was only resolved when the policy was abandoned [46,47].

Given all the difficulties in preventing and controlling catheter-associated UTI using antibacterials in the daily care of catheterized patients and the complications caused by the formation of bacterial biofilms on catheters, it is not surprising that attempts have been made to prevent infection and bacterial colonization by incorporating antibiotics or biocides into the catheter material. An early attempt at such a strategy was reported by Butler and Kunin [48]. As a result of their experience, the authors found that in most cases bacteria appeared in the bladder of catheterized patients at the same time as they were found in the urine in the drainage bag, and these authors considered that the main route of infection was periurethral contamination extending along the outside of the catheter. They developed a policy in which catheters were impregnated with antimicrobial agents and their passage into the bladder was lubricated by a gel containing a mixture of polymyxin and benzalkonium chloride. Three varieties of catheter were studied: nonimpregnated catheters, impregnated catheters with tetra-methyl-thiuramdisulfide, and catheters impregnated with a proprietary agent of the cyclic thiohydroxamic group. The criterion of infection used in the study was bacteriuria $>10^5$ cfu/mL of urine. The antibacterial lubricant had no effect on the rate of infection ($P > 0.05$). Most patients were catheterized for less than 7 days. In those undergoing longer-term catheterization the infection rate in both groups was 100% by day 28. In the catheter study, no significant differences were observed in the rates of bacteriuria for the three types of catheter over the 7 days observational period ($P > 0.05$). It was also reported that the antibacterial activity of these catheters was not detectable after they had been soaked in normal saline for 24 h. Catheters removed from patients after 24–48 h had also lost their antibacterial activity. This early attempt at infection control with antibacterial catheters highlights one of the central problems of the strategy, i.e., maintaining activity over the lifetime of the catheter.

The desirable specifications of an infection-resistant catheter material were listed by Guggenbichler and coworkers [49]. The "gold standard" infection-resistant catheter material should possess a broad spectrum of antimicrobial activity and be resistant to colonization by microbial biofilms. The activity should endure for the lifetime of the device and should not be reduced by contact with body fluids. In addition, and most importantly, it should not select for and spread resistance to antibiotics and other antimicrobials.

The failure to achieve these aims in the 40 years since Kunin's early attempts is testimony to the intrinsic difficulties involved in achieving these aspirations. As a partially implanted device, the urinary catheter is constantly under threat from contamination of the drainage system and from the rich microbial skin flora at the insertion site. So, protection needs to be for the lifetime of the catheters. While catheters in place for 2–3 days should not be problematic, those many catheters that can be *in situ* for 2–3 months offer major challenges. The strategy has to deal with contamination from a wide range of organisms including many multidrug-resistant nosocomial pathogens. The urinary catheter insertion site has an abundant diverse microbial flora that presents a formidable challenge to the antimicrobial agents [15,50].

The catheter most likely to succeed in preventing infection is one from which a broad-spectrum agent elutes into the surrounding fluid and attacks any contaminating organisms before they can reach the bladder and before they can attach to the surface and assume the biofilm-resistant phenotype [51,52]. It is also important that the active agent should elute from both the external and internal surfaces of the catheter if it is to block the various routes of infection and prevent the development of biofilm. It is difficult to conceive how a firmly attached agent could prevent organisms ascending the urethra in a film of mucous between the catheter and the urethral mucosa or in the columns of urine that form in the catheter and its drainage system. The rapid coating with proteins and other organic molecules from urine that forms a conditioning film [53,54] can mask the properties of a nondiffusing antibacterial surface. In addition, although cells that initially attach to such a surface might be killed, the debris from these dead pioneering cells can provide attractive sites for subsequent colonization shielded from the antibacterial in the catheter [55].

During the years since Butler and Kunin's early attempt [48] to develop antibacterial catheters, despite considerable research activity to try and produce biomaterials resistant to bacterial biofilms, surprisingly few such catheters have been developed to the stage at which they have been tested in clinical trials.

18.6 Silver Catheters

The most popular agent that has been used in antimicrobial urinary catheters is silver. On the face of it, silver-coated or impregnated materials should satisfy many of the requirements of the gold standard. Silver has a very broad spectrum of activity including a wide range of nosocomial organisms such as antibiotic-resistant staphylococci, enterococci, Gram-negative enteric bacilli, *Ps. aeruginosa*, and *Candida albicans* [56]. However, while silver-containing materials have been exploited in many prosthetic devices, their use in urinary catheters remains controversial [57]. Doubts have been expressed about the activity of silver ions in urine due to neutralization by chloride ions [58]. Nevertheless, silver has been presented in several formats on urinary catheters, as silver oxide, silver citrate, submicron particles of metallic silver, silver electrodes, and films of silver alloy.

Kumon and coworkers [59] developed a silver coating for silicone catheters that contained a mixture of lecithin, silver citrate, and liquid silicone. The idea of the lecithin was to produce a hydrophilic surface and to modulate the release of silver ions. Experiments in a modified Robbins device showed that over 24 h test periods, this material resisted colonization by *Ps. aeruginosa*. Further tests in a laboratory model in which *E. coli* was able to colonize and migrate over sections of control silicone catheters demonstrated that the migration of this test organism was inhibited by catheters coated in the silver gel. Clinical trials are apparently underway with these catheters and we await the publication of the results.

Guggenbichler and others [49] described a technology that achieved the distribution of billions of submicron particles of metallic silver evenly through a polyurethane matrix. Polyurethanes are hygroscopic and attract water and electrolytes. Interaction of electrolytes with the silver particles throughout the matrix results in the slow release of silver ions, which concentrate in an aqueous film close to the surface of the material. Catheters manufactured from this material released 0.5 μg of silver per 24 h and it was calculated that continuous elution of silver from these materials could be expected for several months. These Erlanger silver catheters were primarily intended for use as central venous catheters, but the application of this technology to urinary catheters is being considered [60].

An alternative way to deliver silver ions to urine is by applying direct electric currents to catheters that have been fitted with silver electrodes. This iontophoresis of silver has been shown to generate antimicrobial activity in urine [61,62]. Recently, it has also been shown to inhibit catheter encrustation in bladder models infected with *Pr. mirabilis* [63].

Johnson and coworkers [64] examined the ability of a silver oxide–coated all-silicone catheter to prevent catheter-associated UTI. This catheter had been coated on both the external and luminal surfaces with a thin layer of silicone elastomer containing micronized silver oxide. In a prospective clinical trial on 482 unselected patients undergoing acute bladder catheterization in a general hospital, they found that the overall incidence of UTI (defined as two consecutive urine cultures with $>10^2$ cfu/mL of the same organism, or a single culture with $>10^5$ cfu/mL) was 10% in both the control and test groups. The median length of catheter placement was 4 days in the control catheter group and 3 days in the patients who had received the silver catheter. Analysis of the data from a subgroup of patients (females not receiving antibiotics) showed a significant reduction ($P = 0.04$) in UTI in the silver catheter group.

In a later study [65], 1309 hospitalized patients due to be catheterized for 24 h or longer were randomly assigned to receive either a silicone catheter coated with 5% silver oxide on the outer surface or a standard silicone elastomer–coated latex catheter. Infection, defined as bacteriuria ($>10^3$ cfu/mL), developed in 11.4% of the silver catheter group and 12.9% in the control group ($P = 0.45$). The patients who received the silver catheter had a meantime of onset of bacteriuria of 4.5 days compared to 6.1 days in the control group. In contrast to the observations of Johnson and others [64] no difference was found between the two catheter groups in women not receiving antibiotics. In men who did not receive antibiotics, the rate of bacteriuria was significantly higher with the silver catheter than with the control (29.4% compared to 8.3%, $P = 0.02$). The evidence, thus, indicates that these silver oxide catheters have little if any protective effect, even for short-term catheterization.

The antimicrobial catheter that has undergone the most extensive testing in clinical trials is the hydrogel silver alloy latex-based device marketed by C.R. Bard as their Bardex IC catheter. In this device, metallic silver is chemically anchored to the latex surface in a coating of gold and palladium, which apparently allows an extended slow release of silver ions from both the internal and external surfaces. The external hydrogel layer gives the catheter its lubricity and allows the release of the silver ions to exert their antibacterial activity in the surrounding fluids [66].

The reports of the early clinical trials on these devices claimed dramatic reductions in the incidence of bacteriuria in patients undergoing short-term catheterization. The first study published as a letter to *Lancet* [67] reported that a prospective randomized trial involving 102 patients showed that after 9 days catheterization, the incidence of bacteriuria (defined as 100 cfu/mL) was reduced from 34% in patients using a standard catheter to 12% in patients using the silver-coated catheters ($P < 0.001$). The second trial from this group [68] reported the incidence of bacteriuria (this time defined as 10^5 cfu/mL) in 120 patients randomized to receive either a Teflonized latex catheter or a silver alloy catheter. After 6 days, bacteriuria was reported in 6 of the 60 (10%) patients who had the silver catheter and 22 (37%) of the 60 who had the Teflon catheter ($P < 0.01$). The third trial conducted by this group [69] investigated the incidence of bacteriuria ($>10^5$ cfu/mL) in 90 patients who had been randomly assigned to receive either a noncoated latex catheter (NC), a hydrogel-coated latex catheter (HC), or a silver alloy/hydrogel-coated catheter (SHC). Of the 30 patients in each group,

50% of the NC group, 33% of the HC group, and 10% of those in the SHC group were bacteriuric after 5 days. Statistical analysis revealed a significant difference in these rates between the NC and SHC groups ($P < 0.002$) but not between the SHC and HC groups ($P < 0.06$). A serious problem with this study, however, was that urine samples were collected from the drainage bag rather than by aspiration from the catheter.

It was 9 years before the next report of clinical trials appeared as full papers in peer-reviewed journals. Verleyen and coworkers [70] conducted two randomized prospective studies in which bacteriuria was considered to be 10^5 cfu/mL and urine samples were taken by suprapubic puncture. In the first of these trials, male patients (only 27 in total) were fitted after prostatectomy, with either a Bard IC catheter or a silicone catheter for 14 days. After 14 days, there was no significant difference in the number of patients with bacteriuria in the two sets of patients (50% with the silver group and 53.3% in the silicone group). In the second trial, patients in a urology ward randomly received either a latex catheter (101 patients) or the Bard IC catheter (79 patients). The results showed a significant delay in the onset of bacteriuria. On day 5, only 6.3% of patients in the silver catheter group had bacteriuria compared to 11.9% in the latex catheter group ($P < 0.003$). The authors' concluded from these trials that the silver-coated catheter significantly delays the onset of bacteriuria compared to the basic latex catheter in short-term catheterization. In medium-term catheterization, the performance of the silver catheter is equivalent to that of the silicone catheter in preventing bacteriuria. The choice of the control catheters in these two trials is curious. A better control to determine the effect of the silver coating would have been the HC (the Bard Biocath).

Another study that failed to include the hydrogel-coated latex-based catheter as a control examined the adhesion of urinary pathogens onto catheter sections *in vitro* [71]. The tests measured the attachment of radio-labeled cells onto all-silicone, hydrogel/silver-coated latex, and hydrogel/silver-coated silicone catheter materials.

Catheter sections were incubated with cell suspensions in a minimal medium. After 2 and 18 h, the sections were then removed and "rinsed vigorously" before being placed in scintillant to determine the amount of radioactivity that had been retained on their surfaces. With the exception of tests with *Ps. aeruginosa* (which was said to have bound to the uncoated-cut ends of the sections) significantly more radioactivity was recovered from the silicone sections than from the silver-coated sections. It has to be said, however, that the rinsing techniques used in these kinds of experiments are difficult to control and standardize. An alternative explanation of these observations is that the cells adhered but bound less firmly to the hydrogel/silver coating. These could then have been preferentially removed by the vigorous rinsing. As biomaterials rapidly become coated in conditioning films that modify their surface properties when exposed to body fluids, it would also have been more appropriate if these tests were performed in urine rather than a minimal medium. The test period of 18 h is also short compared to the life span of many catheters. It is thus not possible to come to any firm conclusions from this study about the effect of the hydrogel/silver alloy coating of latex-based catheters on bacterial adherence.

A large-scale study by Karchmer and others [72] compared the infection rates in hospitalized patients given the hydrogel/silver alloy–coated or silicone-coated latex catheters. In the University of Virginia Hospital, wards rather than patients were randomized into two groups. During the first 6 months of the study, wards in group 1 were supplied with the silver-coated catheters, and group 2 with the silicone-coated catheters. After a 1 month washout period in which all wards were stocked with the silicone-coated catheters, crossover was instituted, group 1 wards receiving the silicone-coated and group 2 the silver-coated catheters. Surveillance for nosocomial infections was performed by infection control practitioners using the CDC criteria [73]. These definitions register both symptomatic UTI and asymptomatic bacteriuria. Analysis of the data showed that 154 infections were recorded among the 13,945 (1.1%) patients fitted with the silver catheter, compared to 189 of the 13,933 (1.36%) patients having the silicone-coated catheter ($P = 0.07$). When the data was expressed as infection rates per patient days, infections on the wards using silver catheters were calculated to be 2.66 infections per 1000 patients days, compared to 3.35 per 1000 days in the control wards ($P = 0.04$). Secondary bloodstream infections complicated the care of 14 of the 343 patients with UTI. The rates of these being 0.04% in patients on units supplied with the silver catheters and 0.07 on units receiving the silicone-coated catheters ($P = 0.42$). Estimates of the annual cost savings that might be achieved by the use of the silver catheter rather than the silicone-coated catheter ranged from $14,456 to $573,293. While this is perhaps the most convincing of the studies showing a beneficial effect of the silver catheters, it would have been more illuminating if the authors had distinguished between symptomatic UTI and asymptomatic bacteriuria.

Several authors considered that their studies do not provide evidence that the hydrogel/silver-coated latex catheter gives greater protection against infection. Thibon and others [74], for example, came to this conclusion after performing a randomized, prospective, double-blind, multicenter trial to compare the silver catheters with all-silicone standard catheters. Urine samples from 109 patients given the standard catheter and 90 with the silver catheter were analyzed daily for up to 10 days after catheter insertion. Unlike many other studies, the two groups were comparable with no significant difference regarding sex, age, weight, antibiotic treatment at time of catheter insertion, and duration of catheterization. UTI was defined as $>10^5$ cfu/mL and $>10^5$ leucocytes/mL. There was no significant difference in the infection rates recorded in those using the silver catheter (10%) and in the control group (11.9%).

The working party (Niël-Weise et al.) that provides guidelines for infection prevention in healthcare in the Netherlands evaluated the evidence in the literature on the efficacy of silver-coated catheters [75]. They assessed each study for five basic requirements of a good quality controlled clinical trial, awarding each a score from 0 to 5. The two trials by Verleyen and coworkers [70] scored 0 and those by Liedberg and coworkers [68,69] scored 1, all being designated of poor quality. Only the trial conducted by Thibon and coworkers [74], which had concluded that the silver alloy catheter had no protective effect, was considered to be of good quality (scoring 5). The fundamental problems with these trials led the working party to the conclusion that there was insufficient evidence to recommend the use of the silver alloy catheters.

A previous meta-analysis [76] of the trials on the silver alloy catheters had come to similar conclusions about the quality of the studies. Despite their acknowledgment that meta-analyses based on methodologically poor studies generally show a benefit to an intervention, and more rigorous studies do not, the authors proceeded to pool all the data without attaching more importance to studies of good quality. Their tentative conclusion was that the silver alloy catheters were significantly more effective than the silver oxide-coated devices and that they may be worth the extra cost in terms of preventing a common nosocomial infection. Niël-Weise and others [75] were critical of this review, expressing the opinion that meta-analysis should be based on clinically homogeneous studies of high methodological quality.

The subsequent literature on the silver alloy latex catheter has not resolved the controversy over its efficacy. Rupp and others [77], for example, based their conclusions that the silver catheter reduces catheter-associated UTI on a study that used historical control data to assess its effect. They reported that over a 2 year period during which the silver catheter was used, the incidence of catheter-associated UTI was 2.6/1000 catheter days. This compared to the rate in the previous 2 years of 6.13/1000 catheter days ($P = 0.002$). Their definition of infection included both symptomatic UTI and asymptomatic bacteriuria, 43% of the infections recorded being asymptomatic. They calculated estimates of the annual cost savings that might ensue from the use of the silver catheter, obtaining values that ranged from $5,811 to $535,452. It was encouraging, however, that all the urinary isolates from the period when the silver catheter was used seemed to have remained susceptible to silver, exhibiting minimum inhibitory concentrations (MICs) in Mueller–Hinton agar of equal to or less than 16 μg/mL. A similar study by Lai and Fontecchio [78] using historical baseline data concluded that the introduction of the silver alloy catheter into a university hospital resulted in a nonsignificant reduction in catheter-associated UTIs and only a modest cost-saving.

Two economic models assessing the potential effect of using silver alloy catheters have appeared in the literature. Saint and coworkers [79] used decision-analytical software to evaluate the care-costs of a simulated group of 1000 hospitalized patients on short-term catheterization with either a silver alloy catheter or a noncoated urinary catheter. Their analysis predicted that the use of the silver catheter would produce savings of

$4 per patient. Plowman and others [80] constructed a model to estimate the economic effect of introducing the silver catheter for short-term use into clinical practice. They concluded that to cover the cost of the silver alloy catheter, reductions of 14.6% in the incidence of infections in medical patients and 11.4% in surgical patients would be required.

The growing concern over the allergy problems associated with the use of latex products has led some hospitals to adopt "latex free" policies. In response to this, the silver alloy coating has also been applied onto all-silicone catheters. The technical difficulties of binding the silver alloy to silicone surfaces were overcome by improved pretreatment and primer coating of the all-silicone catheters. Srinivasan and others [81] recently reported a large-scale prospective trial of these silver/silicone catheters at Johns Hopkins Hospital in Baltimore, one of the hospitals that has adopted a goal of using no-latex products. During a 14 month baseline period, all-silicone catheters (Lubri-Sil, CR Bard) were used throughout the hospital. During the 10 month intervention period, the catheter supply for the units involved in the study was changed to silver alloy/hydrogel-coated silicone catheters (Lubri-Sil IC, CR Bard), which are coated on both the external and internal surfaces. Adult inpatients who had indwelling catheters for >48 h were enrolled in the study. The criterion of infection was bacteriuria $>10^5$ cfu/mL urine with no more than two species of organisms. Analysis of the data showed that in the silver catheter group (1165 patients) there were significantly more men, a larger percentage of catheters had been inserted in medical wards and the median duration of catheterization was 4 days compared to 5 days in the nonsilver group (1871 patients). A total of 334 cases of UTI were recorded, 116 in the silver group (rate of 9.6 per 100 catheters) and 218 in the nonsilver group (rate of 11.65 per 100 catheters). There was no significant difference in these rates ($P = 0.15$) or when the data was expressed as rate of UTI per 1000 catheter days (14.29 compared to 16.15, $P = 0.29$). The median length of catheterization before the onset of a UTI was 4 days in each group. There were no significant differences between the two groups with respect to the types of organisms recovered from their urine, *Candida, E. coli, Ent. faecalis,* and *Ps. aeruginosa* being predominant. During the study, seven urinary catheter-associated bloodstream infections were detected in the nonsilver group compared to nine in the silver group. The bloodstream infection incidence rate per 1000 catheter days were not significantly different, the rate being 1.11 in the silver group and 0.52 in the nonsilver group ($P = 0.06$). The authors came to the conclusion that the use of the silver-coated catheters neither delayed the onset of UTI nor was protective against UTI. The higher incidence rate of bloodstream infections in the silver group although not statistically significant is of concern. These data should be treated with caution, however, because the trial was not designed to assess the incidence of these secondary infections. The value of the data from this trial is also limited by the fact that the study groups were not identical. The confounding factors, such as more men in the silver group and the duration of catheterization being shorter in the silver group, should, however, have biased the results toward showing a beneficial effect of the silver catheters.

18.7 Minocycline- and Rifampicin-Impregnated Catheters

Darouiche and others [82] argued that the use of antiinfective catheters impregnated with a combination of antimicrobial agents has potential for reducing UTI. They suggested that the combination of minocycline and rifampicin could be ideal for this purpose. This combination has synergistic and broad-spectrum activity against almost all potential urinary pathogens, including Gram-negative bacilli, Gram-positive cocci, and yeasts. Neither of the agents is generally used to treat UTI. Another factor that may be useful in preventing the development of resistance is that they have different modes of action, minocycline inhibiting bacterial protein synthesis and rifampicin inhibiting DNA-dependent RNA polymerase.

All-silicone Foley catheters were impregnated with the antibiotic combination and a clinical study conducted with the help of spinal cord-injured patients to determine the spectrum of antimicrobial activity and its durability. A total of 30 impregnated catheters were placed via the urethra into patients' bladders. Catheters were removed at various times up to 21 days and the residual antimicrobial activity was assessed by measuring the zones of inhibition produced by 1 cm segments of the catheters against lawns of test bacteria. The activity was shown to persist for 7 days against all of an impressive list of urinary tract pathogens. By day 14, however, there was little activity against *C. albicans* and by day 21 the segments had also lost activity against strains of *E. coli*, *K. pneumoniae*, *Ent. cloacae*, and *Citrobacter diversus*. Activity against important species such as *Ps. aeruginosa*, *Pr. mirabilis*, and *Ent. faecalis*, however, was retained for the full 21 days. Chemical analysis showed that the total amounts of the agents bound to each of the catheters were 22.3 mg of minocycline and 16.4 mg of rifampicin. Most of both the agents had eluted by day 7. By day 21, little minocycline could be detected and rifampicin levels in the silicone had fallen to about 25% of its original concentration. No adverse effects were observed in any of the patients.

The main objective of antimicrobial impregnation of bladder catheters is to reduce bacterial adherence to and migration along the surfaces of the catheters. A subsequent *in vitro* study by the same group [83] therefore examined the ability of the minocycline/rifampicin catheters to inhibit bacterial migration along the external surface of the catheter. In a simple laboratory model of the catheterized urinary tract, the test organisms (*E. coli*, *K. pneumoniae*, *Ent. faecalis*, *Ps. aeruginosa*, and *C. albicans*) took 2–5 days to migrate over the all-silicone control catheters into the bladder. In contrast, the test organisms took 9–34 days to migrate over the minocycline/rifampicin catheters.

A clinical trial [84] subsequently examined the efficacy of these catheters in preventing UTI in patients who had catheters after radical prostatectomy. In a prospective trial conducted in five academic medical centers, patients were randomized to receive either control all-silicone catheters or the silicone catheters impregnated with minocycline and rifampicin. The catheters were *in situ* for 14 days and urine samples were collected by aseptic needle puncture from the sampling port after clamping the drainage tube. The criterion of infection was defined as bacteriuria at $>10^4$ cfu/mL of urine. Patients who received the antimicrobial catheters had significantly lower rates of bacteriuria than those in the control group, both at day 7 (15.2% vs. 39.7%) and day 14 (58.5% vs. 83.5%). When the data from day 14 was analyzed by category of organism, this difference was shown to be due to a reduction in Gram-positive bacteriuria. The rates of Gram-positive bacteriuria in patients who received the antimicrobial catheter vs. the control catheter were 7.1% compared to 38.2% ($P < 0.001$). However, the two groups had very similar rates of Gram-negative bacteriuria (46.4% vs. 47.1%, $P = 1.0$) and candiduria (3.6% vs. 2.9%, $P = 1.0$). The authors concluded that the results of their trial were "quite promising" but they made the point that these might well not apply to patients who require long-term drainage, such as patients with spinal cord injury, elderly, and nursing homes residents. As minocycline and rifampicin are generally less active against Gram-negative than Gram-positive bacteria, the heavy contamination of the urethral meatus of long-term patients by Gram-negative bacteria suggests that these antimicrobial catheters are unlikely to be effective in these patients. Presumably because of the limited effect of these catheters in preventing only Gram-positive bacteriuria in patients undergoing short-term catheterization, these catheters were not developed further and were not marketed by the manufacturers.

18.8   Nitrofurazone Catheters

Nitrofurazone is a nitrofuran, chemically related to nitrofurantoin a drug that has been used to treat UTIs and as an oral prophylactic to prevent catheter-associated UTI [85]. It is a broad-spectrum agent with activity against a wide range of Gram-positive and Gram-negative organisms. The chemical nature of nitrofurazone allows its combination with silicone to form a stable biomaterial, which has been used by Rochester Medical Corporation (Stewartville, Minnesota) to manufacture Foley catheters. Laboratory tests on these catheters were most encouraging. Sustained release of active concentrations of nitrofurazone was achieved for at least 30 days *in vitro*. The activity of nitrofurazone was tested against isolates of a range of species that are commonly responsible for catheter-associated UTI [86]. MIC determinations showed that nitrofurazone was very similar to nitrofurantoin in its activity. Staphylococci, *E. coli*, enterococci, and some isolates of *Citrobacter* and *Klebsiella* sp. were sensitive to concentrations of 2–16 mg/L. Isolates of *Enterobacter, Serratia* sp., and *Pr. mirabilis* exhibited a degree of resistance (MICs 32–64 mg/L). The intrinsic resistance of *Pseudomonas* sp., however, was confirmed, not even the highest concentrations tested (128 mg/L) being effective against these organisms. This pattern of activity was also shown when sections of the impregnated catheters were incubated on lawns of test species on Mueller–Hinton agar plates. Large inhibition zones (up to 20 mm in diameter) were found around the catheter segments against most (75%) of the isolates tested. The pseudomonads and many of the strains of *Serratia, Proteus*, and *Enterobacter*, however, were not inhibited by the catheter segments.

The authors considered that these nitrofurazone catheters might prevent the migration of uropathogens along the external surface of the catheter from the insertion site into the bladder. In this way, these catheters might prevent or at least delay the initiation of infection. It was conceded, however, that as certain of the Gram-negative organisms (*Serratia, Proteus*, and *Pseudomonas* spp.) were not inhibited by the nitrofurazone matrix, it could not be expected to prevent all catheter-associated infections. As these resistant species are much more commonly associated with infections in long-term catheters, it was suggested that these nitrofurazone catheters would be appropriate for short-term use in acute patients in hospitals.

A subsequent study from this group [87] funded by the Rochester Medical Corporation compared the activity of the nitrofurazone containing catheter with that of the silver hydrogel-coated latex catheter against multidrug-resistant organisms that are typically responsible for catheter-associated UTI. The list of organisms tested did not include any of the species shown to be resistant to the nitrofurazone catheters in the earlier paper. The six species selected for study were *E. coli, K. pneumoniae, Citrobacter freundii, Staphylococcus aureus*, coagulase-negative staphylococci, and *Enterococcus faecium*. Recent multidrug-resistant isolates of these species were tested against segments of the two types of catheter in the zone of inhibition test. The nitrofurazone segments produced inhibition zones against all of the multidrug-resistant isolates of the test species, with the exception of vancomycin-resistant *Ent. faecium*. A subset of strains were then tested against sections of both catheter types. Inhibition zones were not visible around the silver catheters on lawns of *E. coli, K. pneumoniae, Cit. freundii*, or *Ent. faecium*. It was only against *Staph. aureus* that zones were visible. The durability of the *in vitro* activity of the catheters was also assessed against the test organisms by daily serial transfer of catheter segments to fresh agar plates containing the test organisms. The silver catheters lost the little activity they had shown against staphylococci by day 2, whereas the nitrofurazone catheter material retained most of its activity until day 3. By day 5, however, it was only showing very slight activity against just one isolate of *E. coli*. This was a surprising result as other workers had claimed that the NFC released nitrofurazone at greater than 50 mg/L for 44 days [88].

Johnson and coworkers [87] suggested that their findings showed that the emergence of multidrug resistance among the range of important uropathogens was not likely to undermine the efficacy of the nitrofurazone catheter. They also argued that nitrofurazone resistance was unlikely to develop as a result of the use of these catheters. They did not consider the possibility that these catheters would select for the intrinsically resistant *Pseudomonas*, *Proteus*, *Serratia*, and *Candida*. They concluded that another important implication of their study was that the nitrofurazone was substantially more effective than the silver catheters at providing antibacterial activity in the vicinity of the catheter. Not only were the silver catheters less potent but their activity waned rapidly in comparison to the nitrofurazone catheters. These claims for superior performance of the nitrofurazone catheters in these *in vitro* tests should of course be examined by comparative clinical trials. Such trials have yet to be performed [89].

The efficacy of the nitrofurazone catheters compared to control silicone catheters has been studied in three clinical trials. Each study examined the ability of the catheter to prevent infections over short periods. A study by Maki and others [90], which has not appeared as a peer-reviewed paper but only as a conference abstract, compared the rates of bacteriuria (10^3 cfu/mL of urine) in patients undergoing short-term catheterization (1–7 days). In a randomized prospective first trial, the rate of bacterial infection in 170 patients in the antibacterial catheter group was 2.4% compared to 6.9% in the 174 patients in the group receiving a control all-silicone catheter. This difference had an associated P value of 0.07. When *Candida* infections were included in the calculations, the respective figures for the infection rates were 4.7% and 8.0%. In this case, the difference was far from significant ($P = 0.271$).

A multicenter study [88] was conducted in South Korea on 177 patients undergoing short-term catheterization in five university hospitals. The criterion of infection employed was $>10^3$ cfu/mL. Of the 95 patients in the silicone control catheter group (median length of catheterization 3.9 days), 19 (20% not the 22.4% quoted in the paper) were reported as infected. The incidence of infection in the nitrofurazone group (mean duration of catheterization 4.4 days) was 15.2% (14 of the 92 becoming infected). This difference was not significant ($P = 0.223$). When the analysis was limited to just those patients who had been catheterized for between 5 and 7 days, however, the corresponding figures were 16/38 infected in the controls and 12/58 in the nitrofurazone group. The incidence of bacteriuria in the 5–7 days group was, however, significantly lower in the experimental group compared with the control group ($P = 0.026$). Surprisingly, the most common organisms isolated from the control group were *Pseudomonas* species. On the basis of scanning electron micrographs of catheters removed from patients after 7 days, showing less dense bacterial colonization on the nitrofurazone catheters, it was concluded that the nitrofurazone catheter inhibited catheter-associated UTI by inhibiting the formation of biofilm on the catheter. The results of this second trial cannot be said to provide strong evidence for the claim that the nitrofurazone catheter can lower the incidence rate of UTI.

A third trial [91] tested the ability of the nitrofurazone catheters to prevent bacteriuria ($>10^5$ cfu/mL) in postoperative orthopaedic and trauma patients. In this case, silicone-coated latex catheters were used for the control group. There were just 50 patients in each group, six cases of bacteriuria were reported in the control group compared to none in the nitrofurazone group ($P = 0.028$). The median length of catheter placement was 7.9 days in the test group and 7.2 days in the control group.

It is interesting that the authors of reviews of the clinical trials that have been performed on the silver-coated catheters all complained that the differences between the definition of infection used by the various groups confound attempts to assess and compare their efficacy [57,76,92]. It is clear that the same criticism could be made of the trials on the nitrofurazone catheters.

18.9   General Conclusions on Clinical Trials on Catheters

Some authorities consider that coating catheters with the antibacterials silver or nitrofurazone have "engineered out" the risk of infection in catheterized patients [37,63]. Others are less enthusiastic, for example, in the recent comprehensive Cochrane review, Brosnahan and others [93] concluded that although the data from clinical trials on hospitalized adults undergoing short-term catheterization suggests that the use of silver alloy catheters might reduce the risk of infection, the evidence is not strong and the trials are generally of poor quality.

Trautner and coworkers [94] have been even more critical and expressed the view that clinical trials using bacteriuria, at whatever level, as the criterion for infection are fundamentally flawed. They argued that in order to attempt a rational assessment of the evidence in the literature concerning the prevention of catheter-associated UTI, it is important to appreciate the distinction between UTI and asymptomatic bacteriuria. The mere presence of bacteria in the urine of asymptomatic patients is distinct from the situation in which bacteria are present in the urine and the patient exhibits symptoms indicating that invasion of the urinary tract has occurred. More than 90% of cases of catheter-associated bacteriuria are asymptomatic and most do not progress to UTI. It would clearly be more impressive if clinical trials showed that the various antimicrobial catheters could significantly reduce the incidence of pyelonephritis or bloodstream infections. The difficulty here is that the diagnosis of symptomatic UTI in catheterized patients is problematic [31,38]. When a catheterized patient has a fever even in the absence of localizing urinary tract symptoms, because the urine culture is invariably positive, the episodes are usually described as symptomatic UTI. A prospective study using serological criteria to define

UTI identified only one-third of such episodes in long-term patients as originating from a urinary source [95]. There are other difficulties; of course, in performing such trials on symptomatic UTI, because of the lower incidence of these conditions [96], they would have to be on a much larger scale than those performed to date. The distinct advantage of bacteriuria as the criterion of infection is that it is simple to measure.

Another important issue raised by Trautner and others [94] is the common extrapolation from the clinical trials that measure to prevent bacteriuria during short-term catheterization (<14 days) will also be effective in persons undergoing permanent or long-term catheterization (>30 days). While they concluded that silver alloy catheters temporarily delay the onset of bacteriuria in short-term catheterization (<14 days), they argued that the use of systemic antimicrobial prophylaxis can have the same effect. This strategy is not generally used because the benefits of preventing largely asymptomatic bacteriuria are dubious and the risk of selecting resistant flora is high [37]. In summary, the use of the more expensive silver-coated catheters is not supported by quality data. Resistance to silver could be a problem with widespread use [94]. These concerns of course also apply to other agents such as nitrofurazone.

The systematic review by Johnson and others [89] of the evidence from clinical trials on the currently marketed antimicrobial catheters concluded that both nitrofurazone and silver alloy–coated catheters seem to reduce the development of asymptomatic bacteriuria in comparison to latex or silicone control catheters, during short-term use. The trials, however, provide little or no data on the effect of these devices on symptomatic UTI, morbidity, secondary bloodstream infection, or mortality rates. They concluded that the lack of data on these clinically meaningful endpoints and cost effectiveness of these catheters means that it is not possible to make definitive recommendations to decision makers. The clinical benefit and cost savings have yet to be demonstrated directly in a randomized trial with any of these devices in any patient population.

There are other important issues relating to these trials on antimicrobial catheters. Surprisingly, little attention has been paid to the bacteriology. Surely there are lessons to be learned from a study of the nature of the organisms that cause the infections in patients fitted with the antimicrobial catheters. In the trial reported by Darouiche and others [84], for example, 58.6% of patients receiving the antibiotic catheter developed bacteriuria. How did these organisms manage to infect and colonize the urinary tract? Were they colonizing the catheters? Were they resistant to the concentrations of the antibiotics that had eluted into the urine? An understanding of how they manage to survive could give valuable insights into ways of improving the efficacy of the antimicrobial catheter strategy.

A technical issue relating to these studies is that in none of these studies were steps taken to neutralize the activity of the antimicrobials in the urine once the samples had been collected. In the cases of minocycline, rifampicin, and nitrofurazone, substantial concentrations of drugs elute from the catheter into the urine. While the drainage tube is clamped to allow aspiration of urine from the catheter, further leaching of the drugs can occur. What then happens to bacterial populations that might be present in the bladder urine during exposure to these drugs on subsequent transit of the sample to the laboratory? If the organisms are sensitive to the agents, will they survive in the sample bottle before analysis? It is particularly interesting that the study by Al-Habdan and others [91] in which none of the patients receiving the nitrofurazone catheter were deemed to have bacteriuria, the highest population density (10^5 cfu/mL) was used as the criterion of infection. So are these studies recording the effect of the antibacterial agents in the urine in the period between sampling and analysis, rather than the prevention of bacteriuria in the bladder urine? These doubts must be settled by the use of suitable controls.

18.10 Antimicrobial Catheters Yet to Be Tested in Clinical Trials

While many other attempts have been made to develop catheters that deliver antimicrobial agents, the literature generally records their performance in laboratory tests or animal models. The antiseptic chlorhexidine digluconate was chosen by Richards and others [97] as a broad-spectrum agent for catheter impregnation. It was dispersed into a silicone rubber elastomer and this material was compression molded to produce Foley catheters. Discs of the material containing 1%–4% chlorhexidine were fabricated and used to examine the release characteristics and the toxicity. No signs of mucosal irritation were recorded on vaginal implantation of the discs into rabbits. The release of the antiseptic from the polymer was assessed by placing discs in 30 mL aliquots of phosphate buffered saline at 37°C. The suspending medium was replaced daily for up to 28 days and subsequently assayed for chlorhexidine by high performance liquid chromatography. The data showed that the bulk of the antiseptic was released in the first 24 h. In the case of the 1% material, the rate of release declined from 2670 μg/cm^2 in day 1 to only 150 μg/cm^2 in day 2, and was barely detectable by day 9. A similar profile was shown with the 4% discs, but residual activity was detected at around 100 μg/cm^2 from days 20 to 28. The data is expressed in such a way that it is difficult to calculate the concentration of chlorhexidine in the eluate. It would have been more illuminating if the authors had determined the concentration of chlorhexidine released per milliliter as urine flowed through catheters. It is clear, however, that there was little control of the release of the agent, the polymer losing most of its activity within the first 24 h. There is a basic problem with the choice of chlorhexidine as an antibacterial agent for urinary catheters, in that it has limited activity against *Pr. mirabilis*, the organism that is the major cause of complications in catheterized patients [41,55].

Gaonkar and coworkers [98] also used chlorhexidine in catheters but combined it with other agents to reduce the risk of inducing resistant-bacteria. They developed a method for incorporating silver sulfadiazine, triclosan, and chlorhexidine into both latex and silicone-based catheters. Sections of these catheters exhibited greater activity than sections of silver alloy latex catheters or nitrofurazone silicone catheters in zone inhibition tests on agar against a wide range of urinary tract pathogens. They also devised a simple laboratory model to simulate the migration of bacteria over the outer surfaces of catheters from the urethral meatus to the bladder. In this system, the test bacteria (*Staph. aureus, Staph. epidermidis, E. coli, Ent. faecalis,* and *Ps. aeruginosa*) were all able to migrate over control latex and silicone catheters into the bladder within 2 days. Their antimicrobial catheters, however, prevented migration for periods ranging from 3 days for *Ps. aeruginosa* to 24 days for *Staph. aureus*. The silver alloy catheter had little effect on bacterial migration. The nitrofurazone catheter was able to delay *E. coli* but was not so effective against staphylococci. Chemical analysis of catheters removed from the models at various time intervals showed that the chlorhexidine and triclosan were released steadily over a 20 day period and that by the end of the experiment, approximately 50% of each agent was retained in the catheters. Their conclusion was that catheters impregnated with the combination of chlorhexidine, silver sulfadiazine, and triclosan may provide superior long-term protection against infection than silver or nitrofurazone catheters, with an attendant low risk of the development of antimicrobial resistance.

Malcolm and others [99] reported that the rate of release of antibacterials from silicone elastomers could be controlled by varying the concentration of the cross-linking agent used in the synthesis of the polymer. This group had developed an ingenious method for producing self-lubricating silicones. The silicone elastomers used in the manufacture of catheters have relatively high coefficients of friction and the resulting lack of lubricity can cause pain and tissue damage on insertion and removal. The use of novel silicone cross-linking agents produced inherently self-lubricating silicones with very low coefficients of friction. These silicones were produced by the condensation cure of a poly (dimethylsiloxane) with a specifically synthesized cross-linker derived from the lubricious oil, oleyl alcohol. Oleyl alcohol is formed as a product of the cross-linking reaction in the polymer synthesis and is consequently distributed throughout the elastomer matrix. Slow diffusion of oleyl alcohol through the matrix and its persistence at the surface of the material results in a highly lubricious silicone biomaterial, allowing the fabrication of devices that can be inserted and removed easily. These polymers were also capable of delivering antibacterial agents to the surrounding medium. Metronidazole was incorporated as a model drug into these self-lubricating elastomers, and it was found that the rate of metronidazole flux from the polymer could be manipulated by varying the concentration of the cross-linker. These self-lubricating silicones loaded with appropriate agents could well find applications in the manufacture of urinary catheters.

Cho and others [100] developed a gentamicin-releasing coating for Foley catheters. Silicone-coated latex catheters were treated on the inside and outside with polyethylene-covinyl acetate (EVA) containing gentamicin sulfate. They were able to load 1.05 mg of the antibiotic into each catheter. *In vitro* tests showed that in day 1 about 0.2 mg was released from the catheter. Over the next 7 days there was a steady release of around 0.05 mg/day. The authors conceded that for most of the study period, the resulting concentrations in the urine flowing through the catheters will be well below the MICs of gentamicin for uropathogenic species. In rabbit animal models, these catheters were reported to reduce the incidence of bacteriuria (defined as >100 cfu/mL). After catheterization for 5 days, all of the ten rabbits in the control group were found to be infected while only four of the ten in the gentamicin group had bacteriuria. Scanning electron microscopy revealed the presence of thick biofilms on the control catheter surfaces, while the gentamicin-releasing catheters had only patchy bacterial colonization. The bacterial flora of the rabbit urethral meatus is unlikely to present a challenge equivalent to the drug-resistant nosocomial species that often cause the infections in humans in healthcare facilities. Nevertheless, the authors concluded that their new gentamicin-releasing catheter may be useful for controlling infection in patients undergoing short-term catheterization. They failed to examine the drug susceptibilities of the organisms that caused the infections in the test group and they ignored the potential problem of developing resistance to an important broad-spectrum antibiotic, which is used to treat the sort of life-threatening septicemias that can develop from UTIs.

In a further paper [101], this group reported that by dipping catheters in mixtures of EVA with polyethylene oxide (PEO) containing gentamicin sulfate, they were able to produce coatings from which 70%–90% of the antibiotic eluted over 7 days. In all cases, however, only low concentration of the drug was eluting after 3 days. The ability of the various coated catheters to inhibit bacterial growth was tested using an agar diffusion assay. Catheter segments were placed on inoculated agar surfaces and incubated at 37°C. They were removed and placed onto fresh inoculated plates every 24 h for 7 days. The inhibition zones were only apparent around *E. coli* and *K. pneumoniae* for the first 2–3 days. In the cases of *Proteus vulgaris*, *Staph. aureus*, and *Staph. epidermidis*, inhibition zones persisted for 7 days but were very much smaller after day 4. They again suggested that these catheters would be suitable for use for preventing infections in patients undergoing short-term (up to 7 days) catheterization. The fact that after the first few days many contaminating organisms would be exposed to subinhibitory concentrations of this important drug of course adds to the potential danger of developing resistant organisms.

The same group has developed catheters coated in norfloxacin. Silicone-treated latex catheters were dip coated with EVA- and PEO-based solutions containing norfloxacin. Their aim was to use their coating technology to achieve continuous delivery of antibiotics over 30 days and thus suppress infections in patients undergoing long-term catheterization [102]. In agar diffusion tests, sections of the coated catheters were found to produce inhibition zones against *E. coli, K. pneumoniae,* and *Pr. vulgaris* but not against *Staph. epidermidis.* This activity persisted for at least 10 days. The prospect of generating resistance to the powerful fluoroquinolone drugs was not considered. Surely, the lesson of history is that the profligate use of antibiotics has led to the enormous difficulties we now face in treating many infections. Hopefully, catheter manufacturers and the regulatory authorities will take note of the views that have been expressed, for example, by Guggenbichler and others [49] "We firmly believe that the impregnation of polymers with antibiotics for prophylactic purposes violates one of the basic principles of antimicrobial therapy and is therefore contraindicated."

Several other ingenious techniques have been used to control the release of antimicrobials from biomaterials, which might conceivably have applications in urology. For example, coating a polyurethane/ciprofloxacin matrix with thin films of poly(*n*-butyl methacrylate) by radiofrequency glow discharge plasma deposition, achieved control over the elution of the antibiotic from the polymer matrix. The usual initial flush of ciprofloxacin was eliminated and release rates then approached zero-order kinetics for at least 5 days [103].

Pugach and coworkers [104] developed a catheter coating in which ciprofloxacin was encapsulated in a liposomal gelatin matrix. Liposomes are useful drug delivery systems, as these can be formulated to carry a range of active compounds and a degree of control over their release can be achieved. In this way, all-silicone catheters were loaded with up to 30 mg of the drug. The rate of efflux from the catheter was tested *in vitro* into saline and found to be around 0.2 $\mu g/cm^2/h$ and was shown to be sustained for at least 7 days. The ability of the catheters to prevent infection was assessed in an animal model in which rabbits were fitted with all-silicone catheters or silicone catheters that had been coated with gelatin hydrogel or silicone catheters coated in the ciprofloxacin containing liposomal hydrogel. A genetically marked strain of *E. coli* was applied to the urethral meatus of each rabbit twice daily for 3 days to ensure perimeatal colonization by an important urinary tract pathogen. Urine samples were collected for bacteriological analysis over the 7 day experimental period. Bacteriuria (not defined) was recorded in 6 of the 7 (86%) rabbits fitted with a control silicone catheter, 2 of the 4 rabbits (50%) fitted with catheters coated in the gelatin hydrogel, and 6 of the 11 (55%) rabbits fitted with the ciprofloxacin-coated catheters. This surprising result suggested that the hydrogel coating was just as effective at preventing infection as the ciprofloxacin coating and suggests that more work needs to be done if this approach is to proceed to clinical investigations.

Coating catheters with a single antibiotic of course is an invitation to the contaminating microbial flora to develop resistance. The liposome technology could, however, be exploited to deliver combinations of agents that might minimize the resistance problem. Another potential advantage of liposomes is that they offer the prospect of the triggered release of agents. Liposomes can be formulated so that they release their loads under specific conditions such as at particular temperatures or pHs [105,106]. Even more interesting is that enzyme-activated liposomes can be produced, which release their contents in response to phospholipase, trypsin, inflammatory enzymes, hemolysins, and other bacterial enzymes [107]. There is thus the exciting prospect that agents could be retained on the catheter and only released when the system is challenged by bacterial contamination.

An infection-responsive drug release system has been developed by Woo and others [108] in which a fluoroquinolone was used as a monomer in a polyurethane backbone. The system then took advantage of the activity of enzymes released during the inflammatory process to trigger degradation of the polymer and release of the drug. The enzyme cholesterol esterase, which is specifically released from inflammatory cells, was shown to degrade the polymer and release ciprofloxacin. After a slow initial release, concentrations of up to 2 μg/mL were produced in the surrounding medium for up to 30 days. In a variation of this approach, Tanihara and coworkers [109] produced an infection-responsive drug release system by synthesizing a polymer—drug conjugate in which gentamicin was bound to a carboxylated poly(vinyl alcohol) hydrogel by a thrombin-sensitive peptide linker. This material released gentamicin over 24 h when exposed to *Staph. aureus* infected wound fluid. They suggested that combinations of drugs could be delivered in this way. Further that by the use of specific linkers sensitive to proteinases produced by particular bacterial species, it could be possible to achieve the release of different spectra of antibiotics from the same material dependent on the nature of the infecting species. Coating catheters with these bioresponsive polymers may have a role in controlling UTI.

Hydrogels have been used as catheter coatings to increase the lubricity of the surface and facilitate the insertion of the devices through the urethra. A hydrogel is a network of hydrophilic polymers that can swell in water and hold a large amount of water within their structure. A three-dimensional network is formed by cross-linking polymer chains. Cross-linking can be provided by covalent bonds, hydrogen bonds, van der Waals interactions, or physical entanglements. Some hydrogels can undergo reversible volume/phase transitions or gel/sol phase transitions in response to small changes in environmental conditions. These environmentally sensitive gels have been called intelligent or smart hydrogels and have been used to control the delivery of drugs from surfaces [110]. Gels have been synthesized that can respond to temperature, electric fields, light, pressure, sound or magnetic fields, pH, and the presence of certain ions and specific molecules. The enormous potential of these smart hydrogels has yet to be exploited in the development of novel catheters. A catheter coated with such a hydrogel, which responded to ammonium ions or alkaline conditions and delivered antibacterials or urease inhibitors, for example, might well have a place in controlling the problem of catheter encrustation.

Hydrogels have also been used in another ingenious strategy to avoid the uncontrolled dissipation of antibacterials from prosthetic devices. Kwok and others [111] developed an ultrasound responsive biocompatible coating for controlled antibiotic release. The coating is a drug-loaded poly (2-hydroxyethyl) methacrylate (pHEMA) hydrogel. This material is then coated with layers of ordered methylene chains that create a self-assembled membrane barrier to the passive release of the drug. When the system is exposed to ultrasound, the methylene chain barrier is disturbed and a pulse of antibiotic is released. When the ultrasound is turned off, the barrier reassembles. The concept of using ultrasound to induce controlled drug delivery is attractive, as at low power levels it is noninvasive and has been used effectively in other areas of medical treatment and diagnosis.

Norris and others [112] developed and tested such a system that delivers ciprofloxacin. In vitro experiments in flow-cells revealed that before the application of ultrasound, there was a low level of ciprofloxacin (0.1–0.3 $\mu g/mL$) released from the hydrogel into the effluent. When the ultrasound was applied there was a rise of up to 1.8 $\mu g/mL$ in the effluent concentration and when the ultrasound was turned off the ciprofloxacin concentration returned to background levels. A flow cell was also used in experiments to assess the ability of *Ps. aeruginosa* to form biofilms on glass slides coated with this material. Tests were performed on hydrogels with and without ciprofloxacin and with and without exposure to ultrasound in a 43 kHz ultrasonic bath for 20 min daily. Biofilm formed on all the surfaces but was significantly reduced on the ciprofloxacin-loaded hydrogels with ultrasound induced drug delivery compared to the various controls. There is obviously a lot of work needed to refine this technology, but in principle it offers the prospect of medical devices such as catheters having an internal reservoir of antibacterial agents. If infection becomes associated with the device then with externally applied ultrasound, antibiotic release could be triggered from the device without the need for systematic antibiotics or surgical intervention. More effective control over the passive release of sublethal concentrations of drugs from this coating, which occurs in the absence of ultrasound, would be important to avoid the risk of selecting antibiotic-resistant strains. An issue with the application of such a technique to indwelling bladder catheters is that as partially implanted devices, these are under the threat of infection from the bacterial flora at their insertion sites throughout their entire life span. Such a technique might, however, be appropriate to deal with situations when particularly important pathogens such as *Pr. mirabilis* infect the urinary tract. Under these circumstances, the triggering of antibiotic release could be reserved until the urine turns alkaline and the threat of catheter blockage is imminent.

Antibacterials in hydrogel coatings on bladder catheters will of course be released into urine. It is important to realize that the surrounding medium can have a profound effect on the release of drugs from hydrogels. Jones and others [113], for example, reported that the release of fusidic acid from a hydrogel was significantly reduced when artificial urine replaced a buffer as the test medium. The effect was attributed to the formation of the poorly soluble calcium salt of fusidic acid in the urine.

18.11   The Catheter Encrustation Problem

Given the difficulties of maintaining sterile urine in long-term catheterized patients, an alternative approach might be to try and manipulate the composition of the urinary flora rather than trying to eliminate it completely. This strategy requires of course the identification of the organisms that it would be important to exclude from the catheterized tract. A prime candidate for such a policy would surely be *Pr. mirabilis*. This species is not a significant pathogen of the normal urinary tract. It is not usually among the species that first invade the catheterized tract [23], but the longer a patient is catheterized the more likely *Pr. mirabilis* is to appear in the urine, and eventually in permanently catheterized individuals, it can infect up to 40% of patients [114]. There is no doubt that infections with this organism can undermine the quality of life and seriously threaten the health of catheterized patients.

The complications all stem from the ability of this species to produce urease. The activity of this enzyme generates ammonia from urea and the urine becomes alkaline. As the pH rises in any sample of urine, there comes a point at which calcium and magnesium phosphates crystallize from solution. This point has been termed the nucleation pH and is characteristic for any particular sample of urine [115,116]. So as the activity of the bacterial urease elevates the pH toward the nucleation pH, crystalline material starts to form in the urine and complications inevitably follow. The accumulation of the calcium and magnesium salts around the catheter eyelets and on the luminal surfaces eventually blocks the flow of urine from the bladder (Figure 18.3), with potentially disastrous consequences for the patient. All types of catheter are vulnerable to encrustation in this way and there are currently no effective methods for its prevention [117–119].

The material that encrusts catheters has been shown to resemble infection-induced bladder and kidney stones in its crystalline structure, which are mainly composed of a mixture of struvite, magnesium ammonium phosphate, and a poorly crystalline form of apatite, a hydroxylated calcium phosphate in which a variable proportion of the phosphate groups are replaced by carbonate [120,121]. Scanning electron microscopy showed that large numbers of bacilli were always present throughout the material (Figure 18.4) [122]. Bacteriological analysis confirmed that urease-producing species, predominantly *Pr. mirabilis*, were consistently recovered from encrusted catheters [123]. It has become clear therefore that the material that encrusts and blocks catheters is a crystalline bacterial biofilm. If we are to control this problem, we need to understand just how these crystalline biofilms develop.

FIGURE 18.3
Scanning electron micrographs showing a freeze fractured cross-section of a patient's catheter that had blocked. The biofilm community was composed of *Proteus mirabilis* and *Staphylococcus aureus*. The catheter had been *in situ* for 8 weeks. (From Stickler, D.J., *Eur. Urol. Update*, 5, 1, 1996. With permission.)

The catheters in current use provide attractive, unprotected sites for bacterial attachment and are particularly vulnerable to encrustation and blockage. As the contaminated urine flows through the catheter, cells are trapped by the crevices and surface irregularities left by the manufacturing processes [124]. This occurs especially around the rims of the catheter eyelets, which are crudely engineered and are highly irregular in nature. The eyelets are also vulnerable as they are always exposed to contaminated residual urine in the bladder. Attached to a surface, bathed in a constant gentle flow of a nutrient-rich medium, the adherent bacterial populations

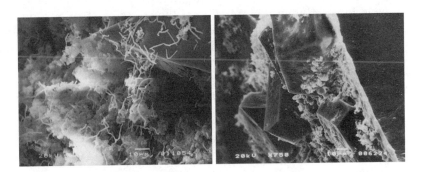

FIGURE 18.4
Scanning electron micrographs of crystalline bacterial biofilms colonizing catheter surfaces. The large crystals are struvite and the poorly crystalline material is a form of apatite. Large numbers of bacilli (*Proteus mirabilis*) are also visible.

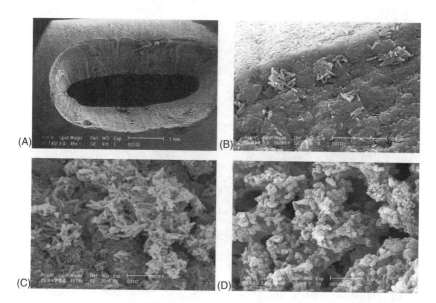

FIGURE 18.5

Scanning electron micrographs showing the colonization of a hydrogel-coated latex Foley catheter by *Proteus mirabilis* biofilm. (A) The rough irregular surface around the catheter eyelet, (B) the development of microcolonies of bacilli after 4 h, (C) at 6 h the biofilm is starting to spread over the catheter surface and produce apatite, and (D) the catheter biofilm after 20 h incubation in the bladder model. (Modified from Stickler, D.J., et al., *Urol. Res.*, 31, 306, 2003. With permission.)

thrive. The cells divide to form microcolonies (Figure 18.5). These produce exopolysaccharide that cements their attachment to the surface and constitutes the matrix of the biofilm that forms as the colonies merge and spread down the catheter lumen. Due to the action of the urease, the pH of the urine and the biofilm rises and at the nucleation pH for that urine, the crystals will form in the urine and the developing biofilm [125,126]. Although the urease is the main force driving the formation of the crystalline biofilms, there is evidence that the bacterial exopolysaccharide capsule also has a role in the process. *In vitro* studies have shown that it attracts magnesium ions and produces a gel that stabilizes crystal growth, thus, accelerating crystal formation [127,128].

While the biofilm is colonizing the catheter, aggregates of crystals continue to form in the urine and these also become incorporated into the growing biofilm. The result is that these crystalline biofilms can build up rapidly and block either the eyeholes or the lumen, usually just below the balloon region. The formation of crystalline aggregates in urine and the coaggregates that these aggregates form with cells can also initiate biofilm development on surfaces to which the bacterial cells do not readily attach. A recent study [129] identified surfaces to which *Pr. mirabilis* cells failed to attach as long as the urine remained acid. When the pH of the suspending urine was allowed to rise above its nucleation pH, however, macroscopic aggregates composed of cells and apatite formed, which then settled on the surface and initiated biofilm formation. Further attachment of cells and crystals and cell–crystal coaggregates ensured the development of the catheter blocking biofilm.

Solely selecting biomaterials for catheter manufacture with surface properties likely to resist adherence by single bacterial cells will not be a totally effective strategy for preventing the formation of crystalline *Pr. mirabilis* biofilms. To stop the bacterial colonization of devices in the urinary tract infected with *Pr. mirabilis*, it will be necessary to prevent the crystallization of apatite and struvite in the urine. Thus, if we are to use antibacterial catheters to inhibit encrustation, it is essential that the agent elutes into the urine and prevents the rise in urinary pH induced by the bacteria.

Studies in laboratory models of the catheterized bladder have shown that both the silver alloy–coated catheter and the nitrofurazone catheters are vulnerable to encrustation and blockage by *Pr. mirabilis* (Figure 18.6) [117,130,131]. This is not surprising as the silver does not diffuse well into the surrounding medium and nitrofurazone has limited activity against *Pr. mirabilis* [86]. Clinical trials to establish whether antibacterial catheters can

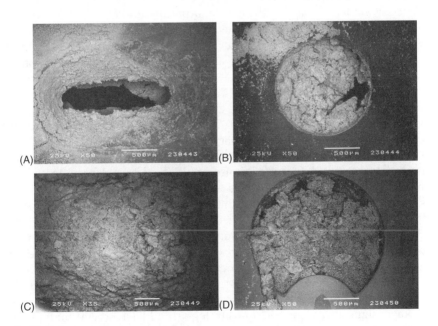

FIGURE 18.6
Scanning electron micrographs showing catheters blocked by *Proteus mirabilis* after incubation in a model of the catheterized bladder. Crystalline material is shown blocking the eyehole (A) and lumen (B) of a hydrogel/silver alloy–coated latex catheter, (C) shows an occluded eyehole, and (D) blocked lumen of a nitrofurazone coated catheter. (These micrographs were kindly provided by Robert Broomfield, Cardiff University.)

prevent catheter encrustation and blockage have not been performed, which is strange given the clinical importance of the complication. The critical endpoint of such studies could be simply defined as "catheter blockage due to encrustation." It would be straightforward to identify and uncontroversial, as it is clearly clinically significant.

Given the problems of maintaining the release of effective concentrations of antibacterials for the lifetime of long-term catheters, Bibby and coworkers [132] put forward the suggestion that the retention balloon could be used as a reservoir for large amounts of agent and the membrane of the balloon might ensure the controlled release over long periods. It was also suggested that in a patient, the release of agent could be monitored and if it dropped to ineffective levels it could simply be replenished by refilling the balloon without disturbing the catheter. These workers were able to show that mandelic acid diffused through the catheter balloon and calculated that they could achieve concentrations of around 0.1 mg/mL in patients' urine. Unfortunately, mandelic acid is not very active against *Pr. mirabilis* or other urinary pathogens, being bactericidal in urine at concentrations of around 5 mg/mL [133].

In contrast to mandelic acid, the biocide triclosan is very active against *Pr. mirabilis*. The MIC of this agent against strains of *Pr. mirabilis* isolated from encrusted catheters has been reported to be 0.5 µg/mL [134]. Using a laboratory model supplied with artificial urine and infected with *Pr. mirabilis*, Stickler and others [135] demonstrated that triclosan can diffuse through the balloons of all-silicone catheters, prevent the rise in urinary pH, and inhibit crystalline biofilm formation on the catheters. In these experiments, control catheters with their retention balloons inflated in the usual way with water were blocked within 24 h, whereas the test catheters inflated with triclosan (10 mg/mL in 5% w/v polyethylene glycol) drained freely and showed little signs of encrustation at the end of a 7 day experimental period. Subsequent experiments showed that this strategy was also effective with latex-based catheters and when artificial urine was replaced with pooled human urine. Triclosan was found to have impregnated the whole length of the all-silicone catheters, but not the latex-based devices [136]. The conditions used in these experiments were intended to simulate those in which a catheter is introduced into a bladder that is heavily colonized by *P. mirabilis* (10^8 cfu/mL) in urine at a pH of around 8.5. A concentrated urine and a low flow rate of urine to the bladder were also employed to simulate the low fluid intake characteristic of many elderly patients undergoing long-term catheterization. The observation that encrustation is inhibited under these severe experimental conditions that normally produce rapid catheter blockage suggests that the triclosan strategy should extend substantially the life span of catheters in patients prone to encrustation problems.

While the effects of triclosan on the bladder epithelium have not been tested, this antibacterial is commonly used in soaps, hand washes, deodorants, toothpastes, and mouthwashes. Its safety has been established through extensive acute and long-term toxicity, carcinogenic and teratology studies and the USA Food and Drug Administration have approved its use in oral care products [137]. As with any intervention with an antibacterial agent, the development of resistance is a concern. In the case of triclosan, however, despite extensive use in many preparations for over 30 years there has been little sign that the clinical or domestic use has led to the generation of resistant organisms [138–140]. Nevertheless, in clinical trials and any subsequent clinical application of this strategy, it will be important to monitor the urinary flora for signs of resistance to triclosan.

Chemical analysis demonstrated that loading the retention balloons with triclosan (10 mg/mL in 5% w/v polyethylene glycol) results in the daily diffusion of around 115 μg of the agent into the urine. Assuming this diffusion rate is maintained, the antibacterial activity should persist in the urine for well over the current maximum life span of 12 weeks for each long-term catheter. The strategy was also found to inhibit biofilm formation by *E. coli*, *K. pneumoniae*, and *Pv. stuartii*. Not all of the species that can cause catheter-associated UTI are sensitive to triclosan, however [141]. *Ps. aeruginosa*, *M. morganii*, and *Serratia marcescens* (MICs of triclosan >100 mg/L) were able to colonize the triclosan-impregnated silicone catheters. These species despite being urease positive in laboratory bacterial identification tests, however, do not seem to be able to produce alkaline urine or crystalline biofilms [142,143].

The plan for the use of triclosan in this way is not to routinely replace water with triclosan to fill catheter balloons, but only to use triclosan when the urine becomes colonized by *Pr. mirabilis*. The idea is, thus, not to use the triclosan prophylactically but as an intervention when encrustation becomes a problem. In this way, we would hope to reduce the risk of selecting resistant species. If the triclosan strategy could be successfully transferred from the laboratory to the clinic, it could bring major improvements in the bladder management of many elderly and disabled people. The method does not disturb the integrity of the closed drainage system and does not require the manufacture of novel catheters. There is also the possibility that it could be used to deliver other agents including antibiotics. The recent study by Carlsson and others [144], for example, showed that it was possible to deliver nitric oxide through catheter balloons by inflating the balloons with a mixture of sodium nitrite and ascorbic acid. We have subsequently tested many other antibacterials and found that few are capable of diffusion through the balloons (unpublished data). There is surely scope, however, to manipulate the composition of the polymers used in the manufacture of catheter balloons to modify their diffusion properties.

In 1988, Calvin Kunin posed the rhetorical question "Can we build a better urinary catheter?" His conclusion was—of course we can, but catheter manufacturers have been reluctant to invest in research and development [3]. The design of catheters in current use is driven not by the requirements for effective patient care but by manufacturer's requirements for convenient production processes and demands of health authorities for inexpensive devices. As a result, the catheters available today with their roughly engineered, irregular surfaces around the eyelets and their narrow central channels are readily colonized and blocked by crystalline bacterial biofilm. The development of catheters with larger internal diameters and smoother surfaces, especially at the rims of the eyeholes, would substantially reduce the complications associated with the current devices. The morbidity associated with the currently available devices undermines the welfare of many individuals and the costs of dealing with the complications are not acceptable. The medical device industry must be persuaded to take up Kunin's challenge [124].

Preventing colonization of the bladder urine of patients undergoing long-term catheterization is a tall order. It is unlikely that any strategy involving antimicrobial agents will provide a permanent solution to the problem. Trautner and Darouiche [145] suggested that it may be sensible to avoid the use of antibacterials in preventative strategies. They have argued that the inevitable failure to suppress bacteriuria in these cases is related to the facility with which the numerous, diverse, and frequently multiresistant organisms from the bowel flora can gain access to the catheterized tract. The resistance problem is also compounded by the ease with which exchange of drug resistance genes occurs between the mixed populations of bacteria in the catheter biofilm. In view of these factors, they concluded that trying to rid the catheterized bladder of all microbial florae is futile and merely promotes the selection of resistant organisms. As an alternative approach, they suggested that colonization of the catheterized urinary tract by benign bacteria might prevent or interfere with colonization by pathogenic species. This bacterial interference strategy would thus avoid the use of antimicrobial agents and the attendant risk of resistance.

Intravaginal inoculation of *Lactobacillus casei* and *Lactobacillus fermentum* has been reported to reduce symptomatic coliform UTI in women who had previously suffered from acute, noncatheter-associated infections [146]. Unfortunately, these lactobacilli are not likely candidates for the prevention of catheter-associated UTI as they seem unable to colonize urine and are lost within 48 h of instillation into the bladder [147]. Hull and coworkers [148], however, identified a nonpathogenic strain of *E. coli* 83972 that could colonize the urinary tracts of catheterized spine-injured patients. This organism had been isolated from the urine of a girl with stable asymptomatic bacteriuria, and was not able to express the virulence factors typical of pyelonephritic strains of *E. coli*. Preliminary tests prior to controlled clinical trials have suggested that inoculation of the bladders of spine-injured patients with this strain might reduce the rate of symptomatic UTI [149]. In vitro experiments have also demonstrated that catheters that had been colonized by *E. coli* 83972 impeded colonization by a pathogenic strain of *E. coli* and by *Pv. stuartii* and *C. albicans* [150]. This group has also explored whether natural microbial defense systems, such as bacteriocins, might be feasible as a means of preventing catheter-associated UTI. In vitro tests demonstrated that the presence of a colicin producing *E. coli* K12 on the surface of catheter sections completely prevented colonization by a colicin-susceptible strain [151]. The long-term aim of this group seems to be developing a nonpathogenic strain of *E. coli* genetically modified to simultaneously express bacteriocins active against a range of uropathogens. It would be of great interest if such an interference strategy could be found to be effective against *Pr. mirabilis*.

18.12 Conclusions

An overall view of the literature must inevitably lead to the conclusion that the strategy of impregnating or coating Foley catheters with antimicrobials has not yet succeeded in controlling the infection-induced problems that complicate the care of people undergoing indwelling bladder catheterization. In short-term patients, while we might be able to reduce the incidence of asymptomatic bacteriuria, there is no evidence that we can reduce the clinically more significant conditions of kidney or bloodstream infections. In long-term patients, we are not able to prevent bacteriuria, symptomatic UTI, or the encrustation and blockage of catheters.

Advances in our understanding of the pathogenesis of catheter-associated UTI have led to a realization of the central roles played by the development of bacterial biofilms on catheters. If we can inhibit catheter biofilm formation, we might be able to block the ways in which infection is initiated. If we could prevent the formation of the crystalline biofilms on catheters, we would have achieved control over a complication that undermines the health, confidence, and quality of life of many elderly and disabled people. While there are exciting prospects on the horizon, it has become very clear that attempting to block bacterial biofilm formation on catheters involves inhibiting surface colonization mechanisms that microbes have evolved over millennia as basic survival strategies in natural aquatic habitats. We need to be more ingenious than simply incorporating antibacterial compounds into the catheter materials. As we gain a deeper understanding of the mechanisms by which biofilms develop, ideas for alternative novel strategies come to light. Blocking the various adherence mechanisms used by organisms, catheter surfaces that scavenge important bacterial nutrient such as iron, or interfering with the cell-to-cell communication systems, which seem to regulate the development of some biofilms, are intriguing possibilities that might be appropriate for urinary catheters. It is important to bear in mind, however, that even if we manage to produce a potent antiinfective biomaterial suitable for urinary catheters, if given time, the ingenious rapidly evolving creatures we are dealing with will surely find ways of colonizing it. Maintaining control over the problem of catheter-associated UTI will no doubt require a continuing effort to develop novel strategies and biomaterials.

References

1. Darouiche, R.O., Device-associated infections: A macroproblem that starts with microadherence, *Clin. Infect. Dis.*, 33, 1567, 2002.
2. Beeson, P.B., The case against the catheter, *Am. J. Med.*, 24, 1, 1958.
3. Kunin, C.M., Can we build a better catheter? *N. Eng. J. Med.*, 319, 365, 1988.
4. Jepsen, O.B. et al., Urinary tract infection and bacteraemia in hospitalised medical patients—a European multicentre prevalence survey on nosocomial infection, *J. Hosp. Infect.*, 3, 241, 1982.
5. Warren, J.W. et al., The prevalence of urethral catheterization in Maryland nursing homes, *Arch. Intern. Med.*, 149, 1535, 1989.
6. Zimakoff, J. et al., Management of urinary bladder function in Danish hospitals, nursing homes and home care, *J. Hosp. Infect.*, 24, 183, 1993.
7. Sorbye, L.W. et al., Indwelling catheter use in home care: Elderly, aged 65+, in 11 different countries in Europe, *Age Ageing*, 34, 377, 2005.
8. Stark, R.P. and Maki, D.G., Bacteriuria in the catheterized patient: What quantitative level of bacteriuria is relevant? *N. Eng. J. Med.*, 311, 560, 1984.
9. Garibaldi, R.A. et al., Factors predisposing to bacteriuria during indwelling urethral catheterization, *N. Eng. J. Med.*, 291, 215, 1974.
10. Stamm, W.E., Catheter-associated urinary tract infections: Epidemiology, pathogenesis and prevention, *Am. J. Med.*, 91, suppl. 3B, 65s, 1991.
11. Hayley, R.W. et al., The nationwide nosocomial infection rate, *Am. J. Epidemiol.*, 121, 159, 1985.
12. Kass, E.H. and Schneidermann, L.J., Entry of bacteria into the urinary tracts of patients with in-lying catheters, *N. Eng. J. Med.*, 256, 566, 1957.
13. Bultitude, M.J. and Eykyn, S., The relationship between the urethral flora and urinary infection in the catheterised male, *Br. J. Urol.*, 45, 678, 1973.
14. Garibaldi, R.A. et al., Meatal colonization and catheter-associated bacteriuria, *N. Eng. J. Med.*, 303, 316, 1980.
15. Fawcett, C. et al., A study of the skin flora of spinal cord injured patients, *J. Hosp. Infect.*, 8, 149, 1986.
16. Tambyah, P.A., Catheter-associated urinary tract infections: Diagnosis and prophylaxis, *Int. J. Antimicrob. Agents*, 24S, S44, 2004.
17. Daifuku, R. and Stamm, W.E., Association of rectal and urethral colonization with urinary tract infection in patients with indwelling catheters, *JAMA*, 252, 2028, 1984.
18. Nickel, J.C., Grant, S.K., and Costerton, J.W., Catheter-associated bacteriuria, an experimental study, *Urology*, 26, 369, 1985.
19. Nickel, J.C., Downey, J., and Costerton, J.W., Movement of *Pseudomonas aeruginosa* along catheter surfaces: A mechanism in pathogenesis of catheter-associated infection, *Urology*, 34, 93, 1992.
20. Tambyah, P.A., Halvorson, K.T., and Maki, D.G., A prospective study of pathogenesis of catheter-associated urinary tract infections, *Mayo Clin. Proc.*, 74, 131, 1999.
21. Gillespie, W.A. et al., Does addition of disinfectant to urine drainage bags prevent infection in catheterised patients? *Lancet*, 1, 1037, 1983.
22. Kunin, C.M., The drainage bag additive saga, *Infect. Control*, 6, 261, 1985.
23. Matsukawa, M. et al., Bacterial colonization on intraluminal surface of urethral catheter, *Urology*, 65, 440, 2005.

24. Clayton, C.L., Chawla, J.C., and Stickler, D.J., Some observations on urinary tract infections in patients undergoing long-term bladder catheterisation, *J. Hosp. Infect.*, 3, 39, 1982.

25. Warren, J.W. et al., A prospective microbiologic study of bacteriuria in patients with chronic indwelling urethral catheters, *J. Infect. Dis.*, 146, 719, 1982.

26. Kunin, C.M. et al., The association between the use of urinary catheters and morbidity and mortality among elderly patients in nursing homes, *Am. J. Epidemiol.*, 135, 291, 1992.

27. Stickler, D.J. and Zimakoff, J., Complications of urinary tract infections associated with devices used for long-term bladder management, *J. Hosp. Infect.*, 28, 177, 1994.

28. Griffith, D.P., Musher, D.M., and Itin, C., Urease: The preliminary cause of infection-induced urinary stones, *Invest. Urol.*, 13, 346, 1976.

29. Ouslander, J.G., Greengold, B., and Chen, S., Complications of chronic indwelling urinary catheters among male nursing home patients: A prospective study, *J. Urol.*, 138, 1191, 1987.

30. Warren, J.W. et al., Fever, bacteraemia and death as complications of bacteriuria in women with long-term urethral catheters, *J. Infect. Dis.*, 155, 1151, 1987.

31. Galloway, A. et al., Serial concentrations of C-reactive protein as an indicator of urinary tract infection in patients with spinal cord injury, *J. Clin. Path.*, 39, 851, 1986.

32. Warren, J.W., Muncie, H.L., and Hall-Craggs, M., Acute pyelonephritis associated with bacteriuria during long-term catheterization: A prospective clinicopathological study, *J. Infect. Dis.*, 158, 1341, 1988.

33. Krieger, J.N., Kaiser, D.L., and Wenzel, R.P., Urinary tract etiology of bloodstream infections in hospitalized patients, *J. Infect. Dis.*, 148, 57, 1983.

34. Bryan, C.S. and Reynolds, K.L., Hospital acquired bacteremic urinary tract infection: Epidemiology and outcome, *J. Urol.*, 132, 494, 1984.

35. Saint, S., Clinical and economic consequences of nosocomial catheter-related bacteriuria, *Am. J. Infect. Control*, 28, 68, 2000.

36. Rudman, D. et al., Clinical correlates of bacteremia in a veterans administration extended care facility, *J. Am. Geriatr. Soc.*, 36, 726, 1988.

37. Maki, D.G. and Tambyah, P.A., Engineering out the risk of infection with urinary catheters, *Emerging Infect. Dis.*, 7, 1, 2001.

38. Nicolle, L.E., Catheter-related urinary tract infection, *Drugs Ageing*, 22, 627, 2005.

39. Kunin, C.M., *Urinary Tract Infections: Detection, Prevention and Management*, 5th ed., Williams and Wilkins, Baltimore, 1997, Chapter 9.

40. Kunin, C.M., Chin, Q.F., and Chambers, S., Formation of encrustations on indwelling catheters in the elderly: A comparison of different types of catheter materials in "blockers" and "non-blockers," *J. Urol.*, 138, 899, 1987.

41. Mobley, H.T.L. and Warren, J.W., Urease-positive bacteriuria and obstruction of long-term catheters, *J. Clin. Microbiol.*, 25, 2216, 1987.

42. Kohler-Ockmore, J. and Feneley, R.C.L., Long-term catheterisation of the bladder: Prevalence and morbidity, *Br. J. Urol.*, 77, 347, 1996.

43. Stickler, D.J. and Chawla, J.C., The role of antiseptics in the management of patients with long-term catheters, *J. Hosp. Infect.*, 10, 219, 1987.

44. Stickler, D.J., Susceptibility of antibiotic-resistant Gram-negative bacteria to biocides: A perspective from the study of catheter biofilms, *J. Appl. Microbiol.*, 92, suppl. 163S, 2002.

45. Southampton Control of Infection Team, Evaluation of aseptic techniques and chlorhexidine on the rate of catheter-associated urinary tract infection, *Lancet*, 1, 89, 1982.
46. Walker, E.M. and Lowes, J.A., An investigation into in vitro methods for the detection of chlorhexidine resistance, *J. Hosp. Infect.*, 6, 389, 1985.
47. Dance, D.A.B. et al., A hospital outbreak caused by a chlorhexidine and antibiotic resistant *Proteus mirabilis*, *J. Hosp. Infect.*, 10, 10, 1987.
48. Butler, H.K. and Kunin, C.M., Evaluation of polymyxin lubricant and impregnated catheters, *J. Urol.*, 100, 560, 1968.
49. Guggenbichler, J.P. et al., A new technology of microdispersed silver in polyurethane induces antimicrobial activity in central venous catheters, *Infection*, 27, suppl 1, S16, 1999.
50. Kunin, C.M. and Steele, C., Culture of the surfaces of urinary catheters to sample urethral flora and study the effect of antimicrobial therapy, *J. Clin. Microbiol.*, 21, 902, 1985.
51. Danese, P.N., Antibiofilm approaches: Prevention of catheter colonization, *Chem. Biol.*, 9, 873, 2002.
52. Knudsen, B.E., Chew, B.H., and Denstedt, J.D., Drug-eluting biomaterials in urology: The time is ripe, *BJU Int.*, 95, 726, 2005.
53. Ohkawa, M. et al., Bacterial and crystal adherence to the surfaces of indwelling urethral catheters, *J. Urol.*, 143, 717, 1990.
54. Santin, M. et al., Effect of the urine conditioning film on ureteral stent encrustation and characterization of its protein composition, *Biomaterials*, 20, 1245, 1999
55. McLean, R.J.C., Stickler, D.J., and Nickel, J.C., Biofilm mediated calculus formation in the urinary tract, *Cells Mater.*, 6, 165, 1996.
56. Russell, A.D., Antimicrobial activity of silver, *Prog. Med. Chem.*, 31, 351, 1994.
57. Darouiche, R.O., Anti-infective efficacy of silver-coated medical prostheses, *Clin. Infect. Dis.*, 29, 1371, 1999.
58. Schierholz, J.M. et al., Antiinfective and encrustation-inhibiting materials—myth and facts, *Int. J. Antimicrob. Agents*, 19, 511, 2002.
59. Kumon, et al., Catheter-associated urinary tract infections: Impact of catheter materials on their management, *Int. J. Antimicrob. Agents*, 17, 311, 2001.
60. Rösch, W. and Lugauer, S., Catheter-associated infections in urology: Possible use of silver-impregnated catheters and the Erlanger silver catheter, *Infection*, 27, suppl 1, S74, 1999.
61. Davis, C.P. et al., Electrode and bacterial survival with iontophoresis in synthetic urine, *J. Urol.*, 147, 1310, 1992.
62. Shafik, A., The electrified catheter: Role in sterilizing urine and decreasing bacteriuria, *World J. Urol.*, 11, 183, 1993.
63. Chakravarti, A. et al., An electrified catheter to resist encrustation by *Proteus mirabilis* biofilm, *J. Urol.*, 174, 1129, 2005.
64. Johnson, J.R. et al., Prevention of catheter-associated urinary tract infection with a silver oxide-coated urinary catheter: Clinical and microbiologic correlates, *J. Infect. Dis.*, 162, 1145, 1990.
65. Riley, D.K. et al., A large randomized clinical trial of a silver-impregnated urinary catheter: Lack of efficiency and staphylococcal superinfection, *Am. J. Med.*, 98, 349, 1995.
66. Davenport, K. and Keeley, F.X., Evidence for the use of silver-alloy-coated urethral catheters, *J. Hosp. Infect.*, 60, 298, 2005.

67. Lundeberg, T., Prevention of catheter-associated urinary tract infections by the use of silver-impregnated catheters, *Lancet*, 2, 1031, 1986.
68. Liedberg, H. and Lundberg, T., Silver alloy coated catheters reduce catheter-associated bacteriuria, *Br. J. Urol.*, 65, 379, 1990.
69. Liedberg, H., Lundberg, T., and Ekman, P., Refinements in the coating of urethral catheters reduces the incidence of catheter-associated bacteriuria, *Eur. Urol.*, 17, 236, 1990.
70. Verleyen, P. et al., Clinical application of the Bardex IC Foley catheter, *Eur. Urol.*, 36, 240, 1999.
71. Ahearn, D.G. et al., Effects of hydrogel/silver coatings on *in vitro* adhesions to catheters of bacteria associated with urinary tract infections, *Curr. Microbiol.*, 41, 120, 2000.
72. Karchmer, T.B. et al., A randomized crossover study of silver-coated urinary catheters in hospitalized patients, *Arch. Intern. Med.*, 160, 3294, 2000.
73. Garner, J.S. et al., CDC definitions for nosocomial infections, *Am. J. Infect. Control*, 16, 128, 1988.
74. Thibon, P. et al., Randomized multi-centre trial of the effects of a catheter coated with hydrogel and silver salts on the incidence of hospital-acquired urinary tract infections, *J. Hosp. Infect.*, 45, 117, 2000.
75. Niël-Weise, B.S., Arend, S.M., and van den Broek, P.J., Is there evidence for recommending silver-coated urinary catheters in guidelines? *J. Hosp. Infect.*, 52, 81, 2002.
76. Saint, S.J. et al., The efficacy of silver alloy-coated urinary catheters in preventing urinary tract infection: A meta-analysis, *Am. J. Med.*, 105, 236, 1998.
77. Rupp, M.E. et al., Effect of silver-coated urinary catheters: Efficacy, cost-effectiveness and antimicrobial resistance, *Am. J. Infect. Control*, 32, 445, 2004.
78. Lai, K.K. and Fontecchio, S.A., Use of silver-hydrogel urinary catheters on incidence of catheter-associated urinary tract infections in hospitalized patients, *Am. J. Infect. Control*, 30, 221, 2002.
79. Saint, S. et al., The potential clinical and economic benefits of silver alloy urinary catheters in preventing urinary tract infection, *Arch. Intern. Med.*, 160, 2670, 2000.
80. Plowman, R. et al., An economic model to assess the cost and benefits of the routine use of silver alloy coated urinary catheters to reduce the risk of urinary tract infections in catheterized patients, *J. Hosp. Infect.*, 48, 33, 2001.
81. Srinivasan, A. et al., A prospective trial of a novel silicone-based, silver-coated Foley catheter for the prevention of nosocomial urinary tract infections, *Infect. Control Hosp. Epidemiol.*, 27, 38, 2006.
82. Darouiche, R.O. et al., Antimicrobial activity and durability of a novel antimicrobial-impregnated bladder catheter, *Int. J. Antimicrob. Agents*, 8, 243, 1997.
83. Darouiche, R.O., Safar, H., and Raad, I.I., In vitro efficacy of antimicrobial-coated bladder catheters in inhibiting bacterial migration along catheter surface, *J. Infect. Dis.*, 176, 1109, 1997.
84. Darouiche, R.O. et al., Efficacy of antimicrobial-impregnated bladder catheters in reducing catheter-associated bacteriuria: A prospective randomized multi-centre clinical trial, *Urology*, 54, 976, 1999.
85. Lambert, H.P. and O'Grady, F.W., *Antibiotic and Chemotherapy*, 6th ed., Churchill Livingstone, Edinburgh, 1992, pp. 181–184.

86. Johnson, J.R., Berggren, T., and Conway, A.J., Activity of a nitrofurazone matrix urinary catheter against catheter-associated uropathogens, *Antimicrob. Agents Chemother.*, 37, 2033, 1993.
87. Johnson, J.R., Delavari, P., and Azar, M., Activities of a nitrofurazone-containing catheter and a silver hydrogel catheter against multi-drug resistant bacteria characteristic of catheter-associated urinary tract infection, *Antimicrob. Agents Chemother.*, 43, 2990, 1999.
88. Lee, S.J. et al., A comparative multicentre study on the incidence of catheter-associated urinary tract infection between nitrofurazone-coated and silicone catheters, *Int. J. Antimicrob. Agents*, 24S, S65, 2004.
89. Johnson, J.R., Kuskowski, M.A., and Wilt, T.J., Systematic review: Antimicrobial catheters to prevent catheter-associated urinary tract infection in hospitalized patients, *Ann. Intern. Med.*, 144, 116, 2006.
90. Maki, D.G., Knasinski, V., and Tambyah, P.A., A prospective investigator-blinded trial of a novel nitrofurazone-impregnated urinary catheter, [Abstract] *Infect. Control Hosp. Epidemiol.*, 18, P50, 1997.
91. Al-Habdan, I. et al., Assessment of nosocomial urinary tract infections in orthopaedic patients: A prospective and comparative study using two different catheters, *Int. Surg.*, 88, 152, 2003.
92. Schierholz, J.M. et al., The efficacy of alloy-coated urinary catheters in preventing urinary tract infection, *Am. J. Med.*, 107, 534, 1999.
93. Brosnahan, J., Jull, A., and Tracy, C., Types of urethral catheters for management of short-term voiding problems in hospitalised adults, *Cochrane Database Syst. Rev. Issue*, 1, 1, 2004.
94. Trautner, B.W., Hull, R.A., and Darouiche, R.O., Prevention of catheter-associated urinary tract infection, *Curr. Opin. Infect. Dis.*, 18, 37, 2005.
95. Orr, P. et al., Febrile urinary infection in the institutionalized elderly, *Am. J. Med.*, 100, 71, 1996.
96. Tambyah, P.A. and Maki, D.G., Catheter-associated urinary tract infection is rarely symptomatic: A prospective study of 1497 catheterized patients, *Arch. Intern. Med.*, 160, 678, 2000.
97. Richards, C.L. et al., Development and characterization of an infection inhibiting urinary catheter, *ASAIO J.*, 49, 449, 2003.
98. Gaonkar, T.A., Sampath, L.A., and Modak, S.M., Evaluation of the antimicrobial efficacy of urinary catheters impregnated with antiseptics in an in vitro urinary tract model, *Infect. Control Hosp. Epidemiol.*, 24, 506, 2003.
99. Malcolm, R.K. et al., Controlled release of a model antibacterial drug from a novel self-lubricating silicone biomaterial, *J. Control. Rel.*, 97, 313, 2004.
100. Cho, Y.H. et al., Prophylactic efficacy of a new gentamicin-releasing urethral catheter in short-term catheterized rabbits, *Br. J. Urol. Int.*, 87, 104, 2001.
101. Cho, Y.W. et al., Gentamicin-releasing urethral catheter for short-term catheterization, *J. Biomater. Sci. Polym. Ed.*, 14, 963, 2003.
102. Park, J.H. et al., Norfloxacin-releasing urethral catheter for long-term catheterization, *J. Biomater. Sci. Polym. Ed.*, 14, 951, 2003.
103. Kwok, C.S., Horbett, T.A., and Ratner, B.D., Controlled release of antibiotics through a plasma-deposited thin film barrier, *J. Control. Rel.*, 62, 301, 1999.
104. Pugach, J.L. et al., Antibiotic hydrogel coated Foley catheters for prevention of urinary tract infection in a rabbit model, *J. Urol.*, 162, 883, 1999.
105. Kono, K., Thermosensitive polymer-modified liposomes, *Adv. Drug Delivery Rev.*, 53, 307, 2001.

106. Maeda, M., Kumano, A., and Tirrell, D.A., H^+-induced release of contents of phosphatidylcholine vesicles bearing surface-bound polyelectrolyte chains, *J. Am. Chem. Soc.*, 110, 7455, 1988.

107. Meers, P., Enzyme-activated targeting of liposomes, *Adv. Drug Delivery Rev.*, 53, 265, 2001.

108. Woo, G.L.Y., Mittleman, M.W., and Santerre, J.P., Synthesis and characterization of a novel biodegradable antimicrobial polymer, *Biomaterials*, 21, 1235, 2000.

109. Tanihara, M. et al., A novel microbial infection-responsive drug release system, *J. Pharm. Sci.*, 88, 510, 1999.

110. Qiu, Y. and Park, K., Environment-sensitive hydrogels for drug delivery, *Adv. Drug Delivery Rev.*, 53, 321, 2001.

111. Kwok, C.S. et al., Self-assembled molecular structures as ultrasonically-responsive barrier membranes for pulsatile drug delivery, *J. Biomed. Mater. Res.*, 57, 151, 2001.

112. Norris, P. et al., Ultrasonically controlled release of ciprofloxacin from self-assembled coatings on poly(2-hydroxyethyl methacrylate) hydrogels for *Pseudomonas aeruginosa* biofilm prevention, *Antimicrob. Agents Chemother.*, 49, 4272, 2005.

113. Jones, D.S., Andrews, G.P., and Gorman, S.P., Characterization of crosslinking effects on the physicochemical and drug diffusional properties of cationic hydrogels designed as bioactive urological biomaterials, *J. Pharm. Pharmacol.*, 57, 1251, 2005.

114. Mobley, H.L.T., Virulence of *Proteus mirabilis*. In *Urinary Tract Infections: Molecular Pathogenesis and Clinical Management*, Mobley, H.L.T. and Warren, J.W., Eds., ASM Press, Washington, 1996, pp. 245–269.

115. Choong, S. et al., Catheter-associated urinary tract infection and encrustation, *Int. J. Antimicrob. Agents*, 17, 305, 2001.

116. Suller, M.T.E. et al., Factors modulating the pH at which calcium and magnesium phosphates precipitate from human urine, *Urol. Res.*, 33, 254, 2005.

117. Morris, N.S., Stickler, D.J., and Winters, C., Which indwelling urethral catheters resist encrustation by *Proteus mirabilis* biofilms? *Br. J. Urol. Int.*, 80, 58, 1997.

118. Capewell, A.E. and Morris, S.L., Audit of catheter management provided by district nurses and continence advisors, *Br. J. Urol.* 71, 259, 1993.

119. Getliffe, K.A. and Mulhall, A.B., The encrustation of indwelling catheters, *Br. J. Urol.*, 67, 337, 1991.

120. Cox, A.J. and Hukins, D.W.L., Morphology of mineral deposits on encrusted urinary catheters investigated by scanning electron microscopy, *J. Urol.*, 142, 1347, 1989.

121. Hedelin, H. et al., The composition of catheter encrustations, including the effect of allopurinol treatment, *Br. J. Urol.*, 56, 250, 1984.

122. Cox, A.J., Hukins, D.W.L., and Sutton, T.M., Infection of catheterised patients: Bacterial colonization of encrusted Foley catheters shown by scanning electron microscopy, *Urol. Res.*, 17, 349, 1989.

123. Stickler, D.J. et al., *Proteus mirabilis* biofilms and the encrustation of urethral catheters, *Urol. Res.*, 21, 407, 1993.

124. Stickler, D.J. et al., Why are Foley catheters so vulnerable to encrustation and blockage by crystalline bacterial biofilm? *Urol. Res.*, 31, 306, 2003.

125. Kunin, C.M., Blockage of urinary catheters: Role of microorganisms and constituents of urine on the formation of encrustations, *J. Clin. Epidemiol.*, 42, 835, 1989.

126. Morris, N.S., Stickler, D.J., and McLean, R.J.C., The development of bacterial biofilms on indwelling urethral catheters, *World J. Urol.*, 17, 345, 1999.
127. Clapham, L. et al., The influence of bacteria on struvite crystal habit and its importance in urinary stone formation, *J. Crystl. Growth*, 104, 475, 1990.
128. Dumanski, A.J. et al., Unique ability of *Proteus mirabilis* capsule to enhance mineral growth in infectious urinary calculi, *Infect. Immun.*, 62, 2998, 1994.
129. Stickler, D.J. et al., Observations on the adherence of *Proteus mirabilis* onto polymer surfaces, *J. Appl. Microbiol.*, 100, 1028, 2006.
130. Morris, N.S. and Stickler, D.J., Encrustation of indwelling urethral catheters by *Proteus mirabilis* biofilms growing in human urine, *J. Hosp. Infect.*, 39, 227, 1998.
131. Morgan, S.D. and Stickler, D.J., The development of crystalline *Proteus mirabilis* biofilms on Foley catheters, *Clin. Microbiol. Infect.*, 11, suppl 2, 541, 2005.
132. Bibby, J.M., Cox, A.J., and Hukins, D.W.L., Feasibility of preventing encrustation on urinary catheters, *Cells Mater.*, 2, 183, 1995.
133. Rosenheim, M.L., Mandelic acid in the treatment of urinary tract infections, *Lancet*, I, 1032, 1935.
134. Stickler, D.J., Susceptibility of antibiotic-resistant Gram-negative bacteria to biocides: A perspective from the study of catheter biofilms, *J. Appl. Microbiol. Symp.*, 92, suppl, 163S, 2002.
135. Stickler, D.J., Jones, G.L., and Russell, A.D., Control of encrustation and blockage of Foley catheters, *Lancet*, 361, 1435, 2003.
136. Jones, G.L. et al., A strategy for the control of catheter blockage by crystalline *Proteus mirabilis* biofilm using the antibacterial agent triclosan, *Eur. Urol.*, 48, 838, 2005.
137. Jones, R.D. et al., Triclosan: A review of its effectiveness and safety in healthcare settings, *Am. J. Infect. Control*, 28, 184, 2000.
138. Sreenivasan, P. and Gaffar, A., Antiplaque biocides and bacterial resistance: A review, *J. Clin. Periodontol.*, 29, 965, 2002.
139. McBain, A.J. et al., Exposure of sink drain microcosms to triclosan: Population dynamics and antimicrobial susceptibility, *Appl. Environ. Microbiol.*, 69, 5433, 2003.
140. Russell, A.D., Whither triclosan? *J. Antimicrob. Chemother.*, 53, 693, 2004.
141. Barganza, H.N. and Leonard, P.A., Triclosan: Applications and safety, *Am. J. Infect. Control*, 24, 209, 1996.
142. Jones, G.L. et al., Effect of triclosan on the development of bacterial biofilms by urinary tract pathogens on urinary catheters, *J. Antimicrob. Chemother.*, 57, 266, 2006.
143. Stickler, D.J. et al., Studies on the formation of crystalline bacterial biofilms on urethral catheters, *Eur. J. Clin. Infect. Dis.*, 17, 649, 1998.
144. Carlsson, S. et al., Intravesical nitric oxide delivery for prevention of catheter-associated urinary tract infections, *Antimicrob. Agents Chemother.*, 49, 2352, 2005.
145. Trautner, B.W. and Darouiche, R.O., Catheter-associated infections: Pathogenesis affects prevention, *Arch. Intern. Med.*, 164, 842, 2004.
146. Reid, G., Bruce, A., and Taylor, M., Influence of three-day antimicrobial therapy and lactobacillus vaginal suppositories on recurrent urinary tract infection, *Clin. Ther.*, 14, 11, 1992.
147. Bruce, A. and Reid, G., Intravaginal instillation of lactobacilli for prevention of recurrent urinary tract infection, *Can. J. Microbiol.*, 34, 339, 1988.
148. Hull, R. et al., Urinary tract infection prophylaxis using *Escherichia coli* 83972 in spinal cord injured patients, *J. Urol.*, 163, 872, 2000.

149. Darouiche, R.A. et al., Pilot trial of bacterial interference for preventing urinary tract infection, *Urology*, 58, 339, 2001.
150. Trautner, B.W., Hull, R.A., and Darouiche, R.O., *Escherichia coli* 83972 inhibits catheter adherence by a broad spectrum of uropathogens, *Urology*, 61, 1059, 2003.
151. Trautner, B.W., Hull, R.A., and Darouiche, R.O., Colicins prevent colonization of urinary catheters, *J. Antimicrob. Chemother.*, 56, 413, 2005.
152. Stickler, D.J., Biofilus catheters and urinary tract infections, *Eur. Urol. Update*, 5, 1, 1996.

19

Risk Factors for Postoperative Infectious Complications after Abdominal Surgery

Patrick Pessaux and Emilie Lermite

CONTENTS

19.1 Introduction

Infectious complications are the main causes of postoperative morbidity in abdominal surgery [1]. These complications have an important financial cost and are responsible for significant morbidity [2,3]. In order to reduce these complications, it is important to establish the risk factors that increase their incidence. Identification of risk factors in the perioperative period would possibly allow decreasing the rate of postoperative infectious complications. Antibiotic prophylaxis is one of the different procedures to reduce postoperative infectious complications. Controversy still persists concerning the efficacy of antibiotic prophylaxis in clean abdominal surgery (stratum 1) [4] and concerning which specific antibiotic to use [5]. Since 1992, the Centers for Disease Control and Prevention (CDC) [6] has modified the definition of surgical wound infection using the term of surgical site infection (SSI), which includes parietal and deep infectious complications. There are few studies reported in the literature taking into account global postoperative infectious complications including extra parietal abdominal infectious complications (urinary infections, intravascular catheter infections, lung infections, and late infections). The risk factors for global infectious complications and for SSI may be different [7].

19.2 Risk Factors

Age is a risk factor for postoperative complications commonly reported in literature [8,9]. In fact, age is neither a risk factor for SSI nor for one of its components (parietal infection or deep infection). Different pathophysiologic mechanisms have been proposed to explain the development of specific infectious complications. Aging leads to a progressive reduction in immunity (60% of the patients above 75 years have cutaneous anergy) [10] and organ functional deterioration [11]. Some other authors consider elderly patients as fragile with a higher prevalence of associated chronic diseases [12].

Ten percent to thirty percent of cirrhotic patients, undergoing an abdominal surgical procedure, developed postoperative bacterial infections [13,14]. Cirrhosis appears as an independent risk factor for global infectious complication, but not for SSI. This high prevalence of infectious complications in cirrhosis can be explained by the presence of various dysfunctions in the mechanisms of defense against infection [15]: a decrease in the function of the reticuloendothelial system and of granulocyte function, a decrease in complement concentrations, a change in cell mediated immunity. Changes in digestive flora and in the intestinal barrier may also play a role in the pathophysiology of bacterial infections during cirrhosis. Additionally, cirrhotic patients have often other associated disorders such as malnutrition, acute hypovolemia, hypoalbuminemia, which can worsen a preexistent immune dysfunction.

Comparison between vertical and transversal incision has been studied in seven retrospective comparative studies and in 11 prospective randomized studies regrouped in a meta-analysis [16]. None of these studies has put in any evidence of significant difference in terms of wound infection rate. It should be noted, though, that none of these studies has taken into account global postoperative infectious complications. The multivariate analysis has shown that the vertical incision is an independent risk factor for infectious complications, arguing for the advantage of transverse abdominal incision. Moreover, transverse abdominal incision is associated with less postoperative pain and less analgesic use [16]. This type of incision is related to less postoperative pulmonary complications (atelectasis, pneumonia, respiratory failure), and less evisceration and secondary incisional hernia [16]. It is important to note that the two types of incisions offer equal exposure of different intraabdominal organs [16], although the midline vertical incision allows widening section more easily and gaining 4–15 min in operative time [16]. Laparoscopy could be of interest in this setting, a technique that is not evaluated in the current report due to the small number of procedures performed by laparoscopy.

The realization of a digestive anastomosis or suture is in most cases, a necessary part of the procedure. The procedure is then classified at the least as class 2 [11], increasing the infection risk from 2% to 10%. These variables are usually known before the operation unless an accidental perforation of bowel takes place (e.g., digestive perforation during difficult adhesion lysis) or an intraoperative change in therapeutic decision due to unsuspected intraoperative findings.

Cruse [18], NASNRC [17], and Public Health Laboratory Service [19] have demonstrated the existence of a direct relationship between operative time and postoperative infectious risk. These authors showed that risk doubled with each additional operative hour. Our results are in agreement with this, as our rates of postoperative infection were respectively of 6.3%, 12.2%, and 27.7% for 1 h, 1–2 h, and more than 2 h of operation. Operative time is an unmodifiable variable. Thus, periodic reinjection of antibiotics (according to half-life) during the intervention should be favored.

Elek and Conen [20] have shown since 1957 that 6,000,000 bacteria of *Staphylococcus aureus* were needed to entail an infection in a healthy subject. An inoculum of 600 *Staphylococcus* is enough to produce the same lesion in a person with one and only suture stitch. The presence of a cutaneous infection, on the operating area, is an entrance door for bacteria to the abdominal cavity, giving birth to secondary deep infections. Out of emergency, the procedure must be performed after recovery of cutaneous infection.

Patients affected with cancer, irrespective of chemotherapy or radiotherapy, have deficits in immunity, proportional to the extension and the gravity of the disease [21,22]. The infectious risks are correlated to this immunity deficit [23].

Two risk factors more specific for parietal infectious complications have been revealed: obesity and parietal drainage. Obesity is a risk factor of wound infection already demonstrated by univariate and multivariate studies [24,25]. Although weight loss is an important long-term health goal, it is important to recognize, concerning the operative prophylaxis, that antibiotics should be adjusted to patient's body weight. In fact, Forse and others [26] have shown that by doubling the antibiotic's posology, the wound infection rate was reduced from 16.5% to 5.6%. In literature, the results are contradictory concerning the efficiency of preventive drainage of a noninfected abdominal wall [27–30]. A distinction must be made between preventive drainage of a noninfected abdominal wall and a curative drainage of an infected abdominal wall. The only indication of preventive drainage is obesity [30].

Other variables for infectious risk factors found in literature are ASA classification [31,32], ethnic group [27], hypoxia [33], hypothermia [34], albuminemia [35], glycemia [36], and cholesterolemia [37].

19.3 Prevention

The identification of risk factors, which could be avoided in the perioperative period, would possibly allow decreasing the rate of postoperative infectious complications.

Underweight reflects the nutritional state of patients [38]. A loss of weight more than 10% of usual weight has already been demonstrated to be a risk factor for postoperative infectious complications [39] due to changes in the defense system against infection [40,41]. The prevalence of malnutrition in surgery department is from 20% to 50% [42]. Nevertheless, data concerning effects of renutrition on the immune system are somewhat contradictory. Some authors have demonstrated an improvement in immune function after malnutrition [41]. Muller and others [43] have demonstrated that correction of preoperative malnutrition in patients operated for cancer reduced significantly the incidence of major postoperative complications and notably infectious complications. This perioperative renutrition is recommended and can be performed according to different modalities [44]. Eight prospective randomized trials comparing enteral to parenteral renutrition have been regrouped in a meta-analysis [45], concluding that enteral nutrition must be used, if possible.

Mechanical skin preparation is a daily nursing procedure in abdominal surgery. Though a single procedure has not yet been adopted, several evidence-based points can be found in the literature: nonshaving [46–49] or depilation of the surgical field with an electric clipper [50] gives the lowest rate of wound infection. The same is true for shaving of the operative field just before the operation [51]. The beneficial effects of depilatory cream have not been demonstrated. Finally, there are no current data dealing with the acceptability and the overall cost of depilation procedures. Some studies had demonstrated the efficacy of preoperative full-body shower wash with skin antiseptic [52,53]. The beneficial effect was better after two showers [54,55].

The main aim of abdominal drainage is to evacuate residual effusions in order to avoid their infection. Thus, every drainage has its own limits, because of obstruction and mobility. Preventive drainage appears to be a risk factor for SSI. In colorectal surgery, drainage has been demonstrated to be useless [56–58]. It can be at the origin of retrograde bacterial colonization, especially when the drainage is no more in aspiration. We discourage the placement of routine preventive abdominal drainage as per the guidelines of the French Society of Digestive Surgery (SFCD) [59]. A meticulous hemostasis without unnecessary dissections must be done in order to reduce the amount of postoperative residual collections and hematomas. The indication of postoperative anticoagulation with curative dosage must be really justified. Anticoagulation with preventive dosage that demonstrated decreasing postoperative thromboembolic complications must be privileged. Particular care must be taken for patients requiring anticoagulation with curative dosage.

Intravascular catheters and urinary catheters are the two most common causes of nosocomially acquired bloodstream infection. Endogenous infections may arise from the patient's own nose. Although the majority of bacteria are staphylococcal, gram-negative bacteria may colonize the skin around the perineum. As the majority of urinary catheters indications found to be inappropriate [60], their routinely use must be avoided.

To reduce the risk of infection and its effects, antibiotic prophylaxis is commonly used. This use of antibiotics, whilst having potential benefits in preventing postoperative infections, also carries certain risks. Perhaps the most serious of the unwanted effects is encouraging the emergence of antibiotic resistance. Other problems are the side effects, the most extreme being life-threatening anaphylactic reactions. The cost of antibiotics also needs to be taken into account when deciding whether they should be routinely used. The choice of antibiotic relies on objective criteria, the main ones being bacteriologic, pharmacologic, and individual patient criteria (age, pregnancy); past medical and surgical history and comorbidity (renal failure or liver complaint and allergy); and other prescribed treatments (drug interactions). At present, the indications of antibiotic prophylaxis are based on infectious risks of the surgical act according to the classification of the National Research Council [6]. The efficacy of antibiotic prophylaxis in clean-contaminated (or stratum 2) and contaminated surgeries (or stratum 3) [17] was demonstrated in prospective randomized studies [4,61,62]. An antibiotic prophylaxis in clean surgery (or stratum 1) with implantation of prosthetic material was also recommended [63,64]. The efficacy of antibiotic prophylaxis in clean abdominal surgery (stratum 1) [65,66], especially for inguinal hernia repair without mesh [67], continues to be controversial. In clean surgery, all patients do not have the same risk of postoperative infectious complications [68–70]. This notion of subgroup was initially reported by Haley and others [70] who demonstrated, in clean surgery, the rate of abdominal wall abscess varying from 1.1% to 15.8%. Stratification of subgroups of patients as high and low risk for infectious complications after inguinal hernia repair without mesh enables a better use of prophylaxis only to patients who really need it.

Now, risk factors for SSI and for global infectious complications are better known. Some of these factors may be modifiable before or during the procedure, allowing to prevent or to reduce the rate of postoperative complications.

References

1. Krukowski ZH, Matheson NA. Ten-year computerized audit of infection after abdominal surgery. *Br. J. Surg.* 1988; 75: 857–861.
2. Olson M, O'Connor M, Schwartz ML. Surgical wound infections. A 5-year prospective study of 20,193 wounds at the Minneapolis VA Medical Center. *Ann. Surg.* 1984; 119: 253–259.
3. McGowan JE. Cost and benefit of perioperative antimicrobial prophylaxis: Methods for economic analysis. *Rev. Infect. Dis.* 1991; 13 (suppl. 10): S879–S889.
4. Rotman N, Hay JM, Lacaine F, Fagniez PL, ARC. Prophylactic antibiotherapy in abdominal surgery. First vs third-generation cephalosporins. *Arch. Surg.* 1989; 124: 323–327.
5. Current trends in antibiotic prophylaxis in surgery. *Surgery* 2000; 128 (suppl. 4): S14–S18.

6. Centers for disease Control. CDC definitions of nosocomial surgical site infections, 1992: A modification of CDC definitions of surgical wound infections. *Infect. Control Hosp. Epidemiol.* 1992; 13: 606–608.

7. Pessaux P, Msika S, Atalla D, Hay JM, Flamant Y. Risk factors for postoperative infections complications in noncolorectal abdominal surgery: A multivariate analysis based on a prospective multicenter study of 4718 patients. *Arch. Surg.* 2003; 138: 314–324.

8. Gross PA, Rapuano C, Andrignolo A, Shaw B. Nosocomial infections: Decade-specific risk. *Infect. Control* 1983; 4: 145–147.

9. Nicolle LE, Huchcroft SA, Cruse PJE. Risk factors for surgical wound infection among the elderly. *J. Clin. Epidemiol.* 1992; 45: 357–364.

10. Schwab R, Walters CA, Weksler ME. Host defense mechanisms and aging. *Semin. Oncol.* 1989; 16: 20–27.

11. Gardner ID. The effect of aging on susceptibility to infection. *Rev. Infect. Dis.* 1980; 2: 801–810.

12. Finkelstein MS. Unusual features of infections in the aging. *Geriatrics* 1982; 37: 65–78.

13. Doberneck RC, Sterling WA, Allison DC. Morbidity and mortality after operation in nonbleeding cirrhotic patients. *Am. J. Surg.* 1983; 146: 306–309.

14. Bloch RS, Allaben RD, Walt AJ. Cholecystectomy in patients with cirrhosis. A surgical challenge. *Arch. Surg.* 1985; 120: 669–672.

15. Rimola A. Infections au cours des maladies hépatiques. In: *Hépatologie Clinique,* Ed Flammarion Médecin-Science, Paris, pp. 1272–1284.

16. Grantcharov TP, Rosenberg J. Vertical compared with transverse incisions in abdominal surgery. *Eur. J. Surg.* 2001; 167: 260–267.

17. Ad Hoc Committee on Trauma, National Research Council, Division of Medical Sciences: Organization, methods, and physical factors. *Ann. Surg.* 1964; 160 (suppl. 1): 19–31.

18. Cruse PJE, Foord R. The epidemiology of wound infection: A 10 year prospective study of 62939 wounds. *Surg. Clin. N. Am.* 1980; 60: 27–40.

19. Public Health Laboratory Service. Incidence of surgical wound infection in England and Wales: A report of the Public Health Laboratory Service. *Lancet* 1960; 2: 658.

20. Elek SD, Conen PE. The virulence of staphylococcus pyogenes for man: A study of the problems of wound infection. *Br. J. Exp. Pathol.* 1957; 38: 573–586.

21. Sample WF, Gertner HR, Chretien PB. Inhibition of phytohemagglutinin-induced in vitro lymphocyte transformation by serum from patients with carcinoma. *J. Trauma* 1971; 24: 319–322.

22. Wanebo HJ. Immunologic testing on a guide to cancer management. *Surg. Clin. N. Am.* 1979; 59: 323–347.

23. Meakins JL, Christou NV, Halle CC, MacLean LD. Influence of cancer on host defense and susceptibility to infection. *Surg. Forum* 1979; 30: 115–117.

24. Armstrong M. Obesity as an intrinsic factor affecting wound healing. *J. Wound Care* 1998; 7: 220–221.

25. Moro ML, Carrieri MP, Tozzi AE, Lana S, Greco D. Risk factors for surgical wound infections in clean surgery: A multicenter study. Italian PRINOS Study Group. *Ann. Ital. Chir.* 1996; 67: 13–19.

26. Forse RA, Karam B, MacLean LD, Christou NV. Antibiotic prophylaxis for surgery in morbidly obese patients. *Surgery* 1989; 106: 750–757.

27. Simchen E, Shapiro M, Michel J, Sacks T. Multivariate analysis of determinants of postoperative wound infection: A possible basis for intervention. *Rev. Infect. Dis.* 1981; 3: 678–682.
28. Mc Ilrath D, van Heerden J, Edis AJ. Closure of abdominal incisions with subcutaneous catheters. *Surgery* 1976; 4: 4112–4116.
29. Higson RH, Kettlewell MG. Parietal wound drainage in abdominal surgery. *Br. J. Surg.* 1978; 65: 326–329.
30. Allaire AD, Fisch J, McMahon MJ. Subcutaneous drain vs suture in obese women undergoing cesarian delivery. A prospective randomized trial. *J. Reprod. Med.* 2000; 45: 327–331.
31. Garibaldi RA, Cushing D, Lerer T. Risk factors for postoperative infection. *Am. J. Med.* 1991; 91 (suppl. 3B): S158–S163.
32. Culver DH, Horan TC, Gaynes RP, Martone WJ, Jarvis WR, Emori TG et al. Surgical wound infection rates by wound class, operative procedure, and patient risk index. National Nosocomial Infections Surveillance System. *Am. J. Med.* 1991; 91 (suppl. 3B): S152–S157.
33. Knighton DR, Halliday B, Hunt TK. Oxygen as an antibiotic. The effect of inspired oxygen on infection. *Arch. Surg.* 1984; 119: 199–204.
34. Kurz A, Sessler DI, Lenhardt R, the Study of Wound Infection and Temperature (SWIFT) Group. Perioperative normothermia to reduce the incidence of surgical-wound infection and shorten hospitalisation. *N. Engl. J. Med.* 1996; 334: 1209–1215.
35. Gibbs J, Cull W, Henderson W, Daley J, Hur K, Khuri SF. Preoperative serum albumin level as a predictor of operative mortality and morbidity. *Arch. Surg.* 1999; 134: 36–42.
36. Pomposelli JJ, Baxter JK, Babineau TJ, Pomfret EA, Driscoll DF, Forse RA et al. Early postoperative glucose control predicts nosocomial infection rate in diabetic patients. *JPEN* 1998; 22: 77–81.
37. Delgado-Rodriguez M, Medina-Cuadros M, Martinez-Gallego G, Sillero-Arenas M. Total cholesterol, HDL-cholesterol, and risk of nosocomial infection: A prospective study in surgical patients. *Infect. Control Hosp. Epidemiol.* 1997; 18: 9–18.
38. Fourtanier G, Prévost F, Lacaine F, Belghiti J, Hay JM et les association universitaire de recherche en chirurgie. *Gastroenterol. Clin. Biol.* 1987; 11: 748–752.
39. Windsor JA, Hill GL. Weight loss with physiologic impairment. A basic indicator of surgical risk. *Ann. Surg.* 1988; 207: 290–296.
40. Kahan BD. Nutrition and host defense mechanisms. *Surg. Clin. N. Am.* 1981, 61: 557–570.
41. Law DK, Dudrick SJ, Abdou NI. Immune competence of patients with protein caloric malnutrition. *Ann. Intern. Med.* 1973; 79: 545–550.
42. Corish CA, Kennedy NP. Protein-energy undernutrition in hospital-in-patients. *Br. J. Nutr.* 2000; 83: 575–591.
43. Muller JM, Brenner U, Dienst C, Pichlmaier H. Preoperative parenteral feeding in patients with gastrointestinal carcinoma. *Lancet* 1982; 1: 68–71.
44. Mariette C, Alves S, Benoist S, Bretagnol F, Mabrut JY, Slim K. Soins périopératoires en chirurgie digestive: Recommandations de la Société Française de Chirurgie Digestive (SFCD). *Ann. Chir.* 2005; 130: 108–124.
45. Moore FA, Feliciano DV, Andrassy RJ, McArdle AH, Booth FVM, Morgenstein-Wagner TB et al. Early enteral feeding, compared with parenteral, reduces

postoperative septic complications. The results of a meta-analysis. *Ann. Surg.* 1992; 216: 172–183.

46. Seropian R, Reynolds BM. Wound infection after preoperative depilatory versus razor preparation. *Am. J. Surg.* 1971; 121: 251–254.
47. Court-Brown CM. Preoperative skin depilation and its effects on postoperative wound infections. *J. R. Coll. Surg. Edinb.* 1981; 26: 238–241.
48. Mead PB, Pories SE, Hall P, Vacek PM, Davis JH, Gamelli RL. Decreasing the incidence of surgical wound infections. Validation of surveillance-notification program. *Arch. Surg.* 1986; 121: 458–461.
49. Mishriki SF, Law DJ, Jeffery PJ. Factors affecting the incidence of postoperative wound infection. *J. Hosp. Infect.* 1990; 16: 223–230.
50. Alexander J, Fisher IE, Booyanian M, Palmquist J, Morris MJ. The influence of hair-removal methods on wound infections. *Arch. Surg.* 1983; 118: 347–352.
51. Prigot A, Games AL, Nwagbo U. Evaluation of a chemical depilatory for pre-operative preparation of five hundred fifteen surgical patients. *Am. J. Surg.* 1962; 104: 900–906.
52. Byrne DJ, Phillips G, Napier A, Cuschieri A. The effect of whole body disinfection on intraoperative wound contamination. *J. Hosp. Infect.* 1991; 18: 145–148.
53. Aly R, Bayles C, Bibel DJ, Maibach HI, Orsine CA. Clinical efficacy of a chlorous acid preoperative skin antiseptic. *Am. J. Infect. Control* 1998; 26: 406–412.
54. Kaiser AB, Kernodle DS, Barg NL, Petracek MR. Influence of preoperative showers on staphylococcal skin colonization: A comparative trial of antiseptic skin cleansers. *Ann. Thorac. Surg.* 1988; 45: 33–38.
55. Byrne DJ, Napier A, Cuschieri A. Rationalizing whole body disinfection. *J. Hosp. Infect.* 1990; 15: 183–187.
56. Merad F, Hay JM, Fingerhut A, Yahchouchi E, Laborde Y, Pelissier E et al. Is prophylactic pelvic drainage useful after elective rectal or anal anastomosis? A multicenter controlled randomized trial. French Association for Surgical Research. *Surgery* 1999; 125: 529–535.
57. Merad F, Yahchouchi E, Hay JM, Fingerhut A, Laborde Y, Langlois-Zantain O et al. Prophylactic abdominal drainage after elective colonic resection and supra-promontory anastomosis. A multicenter study controlled by randomization. French Association for Surgical research. *Arch. Surg.* 1998; 133: 309–314.
58. Urbach DR, Kennedy ED, Cohen MM. Colon and rectal anastomoses do not require routine drainage: A systematic review and meta-analysis. *Ann. Surg.* 1999; 229: 174–80.
59. Mutter D, Panis Y, Escat J. Le drainage en chirurgie digestive. *J. Chir.* 1999; 136: 117–123
60. Gokula RR, Hickner JA, Smith MA. Inappropriate use of urinary catheters in elderly patients at a midwestern community teaching hospital. *Am. J. Infect. Control* 2004; 32: 196–199.
61. Lewis RT, Allan CM, Goodall RG, Marien B, Park M, Lloyd-Smith W et al. Cefamandole in gastroduodenal surgery: A controlled prospective randomized double-blind study. *Can. J. Surg.* 1982; 25: 561–563.
62. Nichols RL, Webb WR, Jones JW, Smith JW, LoCicero J. Efficacy of antibiotic prophylaxis in high risk gastroduodenal operations. *Am. J. Surg.* 1982; 143: 94–98.
63. Page CP, Bohnen JMA, Fletcher R, McManus AT, Solomkin JS, Wittmann DH. Antimicrobial prophylaxis for surgical wounds: Guidelines for clinical care. *Arch. Surg.* 1993; 128: 79–88.

64. Recommandations de la SFAR—Pour la pratique de l'antibioprophylaxie en chirurgie viscérale. *J. Chir.* 1999; 136: 211–215.
65. Knight R, Charbonneau P, Ratzer E, Zeren F, Haun W, Clark J. Prophylactic antibiotics are not indicated in clean general surgery cases. *Am. J. Surg.* 2001; 182: 682–686.
66. O'Connor LT, Goldstein M. Topical perioperative antibiotic prophylaxis for minor clean inguinal surgery. *J. Am. Coll. Surg.* 2002; 194: 407–410.
67. Andersen JR, Burcharth J, Larsen HW, Roder O, Andersen B. Polyglycolic acid, silk, and topical ampicillin. Their use in hernia repair and cholecystectomy. *Arch. Surg.* 1980; 115: 293–295.
68. Pessaux P, Atallah D, Lermite E, Msika S, Hay JM, Flamant Y et al. Risk factors for prediction of surgical site infections in clean surgery. *Am. J. Infect. Control* 2005; 33: 292–298.
69. Pessaux P, Lermite E, Blezel E, Msika S, Hay JM, Flamant Y et al. Predictive risk score for infection after inguinal hernia repair. *Am. J. Surg.* 2006; 192: 165–171.
70. Haley RW, Culver DH, Morgan WM, White JW, Emori TG, Hooton TM. Identifying patients at high risk of surgical wound infection. A simple multivariate index of patient susceptibility and wound contamination. *Am. J. Epidemiol.* 1985; 121: 206–215.

Authors

Terri A. Antonucci joined STERIS Corporation, US, in 1988 as a technical sales consultant, responsible for all initial sales of STERIS System 1. She was then responsible for beginning and maintaining all departments responsible for education and training including customer education, sales training, service training, and clinical and professional education. She serves as the technical expert for STERIS Corporation and is the professional liaison between STERIS and professional organizations. She serves as an active member and industry expert with the following organizations: AORN, SGNA, APIC, AAMI, AdvaMed, ENA, AACN, and others. Prior to her tenure at STERIS Corporation, she was employed as an operating room nurse at several hospitals in the Greater Cleveland area, where she specialized in orthopedics, cardiovascular, and general surgery.

Linda Corum has a BS and an MS in medical technology, with specialization in microbiology and immunology. She has 10 years of research experience in various laboratories in the WVU Health Sciences Center and Cancer Center Research facilities. She is working as research assistant in West Virginia University in the Department of Pathology at Morgantown, West Virginia, US.

Joseph Kirby Farrington, PhD (RM/SM), has been at Lilly for more than 2 years serving as Senior Technical Consultant Sterility Assurance and most recently as Research Advisor-Microbiology Manufacturing Science and Technology (MS&T). Before Lilly, Dr Farrington was Director of Quality Assurance and Regulatory Affairs for the Pharmaceutical Development Center from 1997–2002. His previous experience also includes a 20 year career with Schering-Plough where he held the positions of Manager Microbiology, Associate Director QC, Research Director Microbiology, and VP-Operations (Plough Labs). As well as holding adjunct positions at several universities, he is also a member of the faculty for the PDA Training and Research Institute. His areas of expertise include industrial microbiology, environmental monitoring, sterile manufacturing, pharmaceutical water systems, sterile products, antimicrobial preservative systems, laboratory automation, process validation, regulatory submissions, and training. Kirby Farrington received a BS in chemistry and biology from LaGrange College, MS in microbiology and biochemistry from Clemson University, and a PhD in pharmacy and veterinary medicine (microbiology) from Auburn University. He is a registered (RM) and specialist microbiologist (SM).

Professor David P. Fidler is a professor of law at Indiana University School of Law, Bloomington, US. He is one of the world's leading experts on international law and public health, especially with regard to infectious diseases. He is a senior scholar at the Center for Law and the Public's Health at Georgetown and Johns Hopkins universities. Professor Fidler is also an internationally recognized expert on arms control issues related to biological and chemical weapons, bioterrorism, the international legal implications of nonlethal weapons, and the globalization of baseball. He has published widely in international law, international relations, and public health journals. His books include *Biosecurity in the Global Age* (2007) (with Lawrence O. Gostin), *SARS, Governance, and the Globalization of Disease* (2004), *Stealing Lives: The Globalization of Baseball and the Tragic Story of Alexis Quiroz* (2002) (with Arturo Marcano), and *International Law and Infectious Diseases* (1999). Professor Fidler has been a visiting scholar at the University of Oxford and the London School of Hygiene and Tropical Medicine. He has also been a Fulbright New Century scholar. Professor Fidler holds a BA (magna cum laude) from the University of Kansas, an MPhil in international relations (distinction) from the University of Oxford, a JD (cum laude) from Harvard Law School, and a Bachelor of Civil Law (BCL) (first class honors) from the University of Oxford. Before joining the faculty at Indiana University, he practiced law with Sullivan & Cromwell in London and Stinson, Mag & Fizzell in Kansas City. From 1990 to 1993, he was a lecturer in the International Relations Programme at the University of Oxford. In addition to his teaching and scholarly activities, professor Fidler is very active as a consultant on international law for governmental, intergovernmental, and nongovernmental organizations.

Nancy E. Kaiser is the senior manager of Healthcare Consumable Product Technologies for STERIS Corporation, St. Louis, Missouri, US. As an employee of Merck, Bristol-Myers Squibb, and now STERIS Corporation, Kaiser has published and holds a number of patents in the areas of antimicrobial skin care and wound management products. A graduate in chemistry from Washington University, her areas of expertise are in skin care formulationand microbiology.

Daniel A. Klein is a manager of Microbiology and Clinical Affairs for STERIS Corporation, St. Louis operations, US. In this role, Dan is responsible for the microbiological and clinical testing of the consumable product line for the healthcare, defense, and life science businesses. Before joining STERIS, Dan managed a contract-testing laboratory in the Washington, DC area focusing on bacteriological, mycological, and clinical testing of products for FDA and EPA submission. Dan possesses a bachelor of science degree in microbiology from the University of Illinois and a master's degree from Washington University in St. Louis, US.

Steven P. Kutalek, MD, received his BA from The Johns Hopkins University in 1975 with a major in biophysics and his MD from the New York University School of Medicine in 1979, where he was awarded the anatomy prize. He completed an internship at The Medical College of Pennsylvania, then residency in internal medicine and cardiology fellowship at Hahnemann University School of Health Sciences, where he was chief fellow in cardiology. He completed his advanced training in cardiac electrophysiology at the Likoff Cardiovascular Institute in Philadelphia in 1985. Since 1987, he has directed the Cardiac Electrophysiology Laboratory at Hahnemann University Hospital and an accredited cardiac electrophysiology fellowship program at Drexel University College of Medicine. He has served as the Chief of the Division of Cardiology and currently as an associate chief. Beginning in 1990, he has developed a profound interest in the management of device-related complications, performing and educating fellows in the technique of transvenous lead extraction of infected and noninfected cardiac devices. He has authored or coauthored over 200 publications in the field of cardiac arrhythmias, with a strong emphasis on device management. He has developed new techniques with regard to the surgical management of the infected device patient. He has chaired the Device Database Committee of the Heart Rhythm Society and serves on the editorial board of the *Journal of Interventional Cardiology*. He is a peer reviewer for multiple cardiology and electrophysiology journals and an abstract grader for the Heart Rhythm Society.

Dr Emilie Lermite, MD, is a digestive surgeon, who earned her medical degree at the Medical University of Angers in France. She practiced in the digestive surgery department in the University Hospital of Angers. She is actually preparing a PhD thesis. She has more than 20 publications in peer-reviewed journals.

Dr Gurusamy Manivannan, PhD, MBA, has a polymer chemistry, photochemistry, and materials science background combined with business education. He has broad product development experience in medical and personal care fields. He received a PhD, from the Madras University, India, and an MBA (High-Tech) from Northeastern University, Boston, US. He has 12 years of industrial work experience in various responsibilities: senior scientist to senior manager in start-up and Fortune 500 companies—Intelligent Biocides, the Gillette Company, and STERIS Corporation. Further, he has 7 years of academic work experience in Europe, Canada, and the United States. He has more than 150 publications in various categories, such as editor and contributor of books, patents, reviews, research papers, instructor of professional courses, and invited lectures, to his credit. He is a member of the American Chemical

Society, Society for Cosmetic Chemists, and Product Development & Management Association.

Dr Gerald McDonnell, BSc, PhD, is a vice president for STERIS Corporation, based at their European headquarters in Basingstoke, United Kingdom. Dr McDonnell has a BSc (honors) in medical laboratory sciences from the University of Ulster and a PhD in microbial genetics from Trinity College, Dublin. He has worked for STERIS for over 10 years in the United States and Europe on the development, research, and support of infection prevention products and services, including cleaning, antisepsis, disinfection, and sterilization methods. His research and development focus includes cleaning, disinfection, and sterilization product development and support of existing products, including training on various aspects of decontamination. His basic research interests include infection prevention, decontamination microbiology, emerging pathogens, and the mode of action and resistance to biocides. His publications include *Antisepsis, Disinfection and Sterilization. Types, Action and Resistance* published by the American Society of Microbiology.

Lindsay Nakaishi is a native of Morgantown, West Virginia, US. She is a recent graduate of Davidson College in Davidson, North Carolina, where she received her BS in biology. She has spent the last year working with Dr Thomas at West Virginia University exploring the consequences and characteristics of biofilms in industry, medicine, and dentistry. She is currently pursuing her Masters degree in public health.

Dr Kanavillil Nandakumar holds a PhD degree from Goa University, India, and a DSc degree from Kyushu University, Japan. He is currently a faculty member at Lakehead University (Orillia campus), Canada. Before joining Lakehead, he held various scientific and research positions at Bhabha Atomic Research Centre (BARC), India; Osaka University, Kawasaki Steel Corporation, Technical Research Laboratory, and Shikoku National Research Institute in Japan; and GLIER, University of Windsor, Canada. He has been actively involved in the research on biofilms and biofouling organisms, their ecology and preventive techniques, for the last several years. His expertise also includes aquatic ecology, thermal ecology, and microbial corrosion. To his credit, he has more than 150 research publications, which include several refereed journal publications, book chapters, and patents.

Dr Jerry L. Newman obtained his BS degree in biology from Heidelberg College and his PhD in biochemistry from Ohio University. At the University of Minnesota, Hormel Institute, his postdoctoral studies focused on the physical biochemistry of biological membrane lipids. During his career he has held a variety of R&D and management positions at S.C. Johnson & Son, Johnson & Johnson Medical, Kemin Foods, and STERIS Corporation, US. During this time he was involved in the development and evaluation of numerous infection control, skin care, wound care, decontamination, and dietary supplement products and technologies. His research interests include antimicrobial technologies and the effects of various factors on skin condition of healthcare workers. He has numerous patents and published papers on topics of the formulation, skin condition, and the evaluation of antimicrobial products. He currently serves as senior director, Product Development, at STERIS Corporation, US. His team is responsible for research and development activities for the skin care and chemical decontamination businesses within STERIS.

Professor Ben M.J. Pereira obtained his PhD in endocrinology from the Dr A.L.M. Post Graduate Institute of Basic Medical Sciences, University of Madras, in 1983. After a brief stint as a research associate at Chennai, he was appointed as faculty member in the Department of Biotechnology, University of Roorkee, in 1984. He was on sabbatical to the Department of Obstetrics and Gynecology, Policlinico Umberto 1°, La Sapienza, University of Rome during 1987–1988 and developed an ELISA kit for the detection of cancer of the female genital tract. Once again he spent 1 year (1996–1997) at the Department of Obstetrics and Gynecology, Center for Reproductive Biology Research, Vanderbilt University School of Medicine, USA, contributing substantially toward the understanding of egg–sperm interaction. He is now back at the University of Roorkee, rechristened in 2001 as Indian Institute of Technology Roorkee, and continues there as professor. He has guided 12 PhD thesis, has contributed 4 book chapters, and has over 50 publications in peer-reviewed journals. Recently, he received the outstanding teacher award and the Vijay Rattan award for his services, achievements, and contributions.

Dr Semali Perera (SPP) is a senior lecturer in the Department of Chemical Engineering at University of Bath, United Kingdom. She has 14 years of experience in the fields of biomaterials, design and development of drug delivery vehicles, delivery of cancer therapeutics, membranes, advanced separations, and nanomaterials. In particular, she has detailed expertise relating to the manufacture and testing of membranes and low-pressure drop devices such as monoliths and hollow fibers. She is the inventor of a number of novel hollow fibre developments including those which are the subject of two patent applications filed by the University of Bath (2005). She gained an honors degree in chemical engineering in 1984 and a PhD from the Department of Industrial and Polymer Chemistry at Brunel University, Uxbridge, United Kingdom, in 1987. After being employed in industry for 2 years as an environmental consultant, she became a lecturer at the University of Bath.

Dr Patrick Pessaux, MD, is a digestive surgeon, who earned his medical degree at the Medical University of Angers in France. He practiced in the digestive surgery department in the University Hospital of Angers. He worked with The French Association for Surgical Research about the risk factors for infectious complications after digestive surgery. He specialized in liver surgery and liver transplantation in the Department of Hepatobiliary Surgery and Liver Transplantation, Hospital Beaujon (AP-HP, University Paris VII, Clichy, France). Now, he is with the Department of Hepatobiliary Surgery and Liver Transplantation, University Hospital of Strasbourg, France. He received a PhD for "Methodological and economical analysis of health system" from University Lyon 1 (France). He has more than 100 publications in peer-reviewed journals.

Jim Polarine is a technical service specialist at STERIS Corporation, US. He has been with STERIS Corporation for over 5 years, where his current technical focus is microbial control in clean rooms and other critical environments. Jim Polarine has lectured in North America, Europe, and Puerto Rico on issues related to disinfection and sanitation in clean rooms. He is a frequent industry speaker and has worked on publications for several scientific journals. Jim is also active on the PDA task force on cleaning and disinfection as well as the publication of documents through PDA on cleaning and disinfection. He graduated from the University of Illinois with a master of arts in biology, and is a member of the PDA, ISPE, IEST, ASM, AALAS, ASTM, and ACS. He previously worked as a clinical research coordinator with the Department of Veteran's Affairs and as an instructor at the University of Illinois.

Dr Vikas Pruthi obtained his BSc (honors) and MSc degrees from Delhi University and did his PhD from the Institute of Microbial Technology (Chandigarh, India). He was offered the post of lecturer in Guru Nanak Dev University, Amritsar, in 1997 before joining as assistant professor in the Department of Biotechnology, IIT Roorkee, Roorkee. His research interests center around medical and environmental microbiology. Presently he is working on a DBT, Government of India, project dealing with "Candida biofilms: molecular analysis of its formation and control" and also on an MHRD (Ministry of Human Resource Development), Government of India, funded project on surface active agents. He has published over 50 research articles and review papers in various journals and conferences of international repute and several popular articles in scientific magazines. He is also working on the editorial board of research journals from India.

Dr José A. Ramirez is the Global director for Open Innovation for Johnson-Diversey Inc., North American Research Center. The center focuses on technology and product development as well as technical support in the area of institutional cleaning and hygiene, emphasizing codevelopment of technologies with external partners through open innovation. Before joining JohnsonDiversey, Dr Ramirez served as a vice president of Research & Development at Virox Technologies Inc., in Mississauga, Ontario, where he led the development of various breakthrough hydrogen peroxide-based cleaning and disinfection technologies. Dr Ramirez holds master's and doctorate degrees in chemical engineering from the University of Colorado, Boulder, and an undergraduate degree in mechanical engineering from Simón Bolívar University in Caracas, Venezuela.

Dr Nancy A. Robinson received a PhD in biochemistry from Case Western Reserve University in Cleveland, Ohio, where she investigated alternative cellular energy sources. After graduation, she served 4 years active duty as a captain and principal investigator at the United States Army Research Institute of Infectious Diseases at Fort Detrick in Frederick, Maryland, investigating the mechanism of action of mycrocystin, a blue-green algal toxin. After a 4 year sabbatical at home with her two children, Dr Robinson returned to Case Western Reserve as a postdoctoral fellow where she studied the protein composition of the cornified envelope, a keratinocyte terminal differentiation product. She is currently the director of Advanced Sterilization at STERIS Corporation in Mentor, Ohio, US. She has over 8 years of experience in the development of low-temperature sterilization systems for heat-sensitive, reusable medical devices.

Dr Nora A. Sabbuba, after gaining a first class honors degree in applied biology, took up a research studentship in Dr Stickler's laboratory in the Cardiff School of Biosciences. She was awarded a PhD for a thesis entitled "Studies on the epidemiology and pathogenesis of catheter-associated *Proteus mirabilis* urinary tract infections." Subsequently, she worked as a postdoctoral research assistant in Cardiff on a programme entitled New Urological Technologies and Devices, sponsored by the Engineering and Physical Sciences Research Council of the United Kingdom. This was a collaborative project involving academic centers and medical device companies. Nora worked on the development of novel catheters and catheter valves. Currently, Nora has a postdoctoral research post in the Biomed Centre at Southmead Hospital, Bristol, where she is continuing her work on the development and testing of novel strategies to reduce the complications associated with the use of prosthetic devices in the urinary tract.

Michael G. Sarli of STERIS Corporation, US, has over 25 years experience in the aerosol and pesticide industries with responsibilities for quality, safety, and regulatory affairs. STERIS Corporation is a developer and manufacturer of FDA-regulated drugs, cosmetics, medical devices, and EPA-regulated antimicrobial products.

Professor Syed A. Sattar, MSc, Dip Bact, MS, PhD, RM (CCM), FAAM, is emeritus professor of microbiology and founding director of the Centre for Research on Environmental Microbiology (CREM) at the University of Ottawa, Canada. His research relates to understanding the influence of environmental factors on the fate of human pathogens in water, food, air, soil, municipal wastes, and on animate and inanimate surfaces and medical devices. A major part of his work includes investigations on safer and better means of environmental control of microbial pathogens, including infectious bioagents, using chemical and physical means. He is an active member of several standards-setting organizations, and microbicide test methods developed by his research team now form the basis for numerous national and international standards. He serves as an advisor to the Government of Canada and also to the World Health Organization. He has previously served the U.S. Environmental Protection Agency as a member of the FIFRA subpanel on germicide test methodology. He has delivered over 250 invited lectures internationally and has published more than 200 papers and chapters in books and conference proceedings; he has also authored one book, coedited another, and has written four lab manuals, three monographs, and numerous commissioned reviews. He is the recipient of several international fellowships and was recently awarded the Martin S. Favero Lectureship of the Association of Professionals in Infection Control & Epidemiology. Dr Sattar is a registered microbiologist of the Canadian College of Microbiologists and an elected fellow of the American Academy of Microbiology. He is also a member of the editorial boards of the *American Journal of Infection Control, Infection Control and Hospital Epidemiology*, and the *International Journal of Infection Control*.

Scott W. Sinner, MD, was born in Williamsport, Pennsylvania. After graduating magna cum laude from Franklin and Marshall College in Lancaster, Pennsylvania, with a degree in biology, he earned his medical degree at the Hahnemann University School of Medicine in Philadelphia, Pennsylvania. He completed his internal medicine internship and residency at MCP-Hahnemann University, and then spent an additional year there as chief medical resident. He then completed his fellowship in infectious diseases at the same institution, now known as Drexel University College of Medicine. For the past 3 years, Dr Sinner has been in private practice in central New Jersey. Dr Sinner has published articles on the management of bacterial and tuberculosis meningitis, and conducted research on the diagnosis and management of pacemaker infections during his fellowship. He was board certified in internal medicine in 2001 and in infectious diseases in 2004. His professional activities include memberships in the American College of Physicians, the Infectious Disease Society of America, and the American Society for Microbiology. He is on staff at Somerset Medical Center, the University Medical Center at Princeton, Saint Peter's University Hospital, and Robert Wood Johnson University Hospital in New Brunswick, US. His practice's name is ID Care, and is located in Hillsborough, New Jersey (www.idcare.com).

Dr Anthony W. Smith is the principal and dean of The School of Pharmacy, University of London. He took up his appointment in April 2006. He gained BPharm with first class honors from the University of Bath in 1983. After registration as a pharmacist, he returned to Bath to study for his PhD on "Photoenzymatic repair in *E. coli*" and he graduated in 1987. He spent

6 years at the School of Pharmacy and Pharmaceutical Sciences at Aston University in Birmingham and returned to Bath in 1993. At its broadest level, his research is aimed at understanding how pathogenic bacteria adapt and survive in the environment in which they are present. He currently investigates how pathogenic bacteria have adapted to cause infection in higher eukaryotes (animals and humans). The experimental design is to examine the interaction between bacteria and lower-order eukaryotes such as amoebae that will have coexisted together in complex biofilms for billions of years longer than their association with animals and humans. Work in progress illustrates that the paradigm of bacterial–protozoal interaction is not restricted solely to *Legionella pneumophila*, but also includes major nosocomial pathogens such as methicillin-resistant *Staphylococcus aureus* (MRSA) and *Pseudomonas aeruginosa*. A new area of collaborative study is the design of polymeric drug-releasing matrices optimized to reduce bacterial biofilm formation. The combined purpose of this work therefore is to address questions of a fundamental nature, yet from which some of the answers will contribute directly to clinical benefit.

Susan Springthorpe is the director of research at the Centre for Research on Environmental Microbiology (CREM) at the University of Ottawa, Canada. With Dr Sattar, she studies the fate and control of pathogens in a variety of natural and engineered environments, especially those related to drinking water. She is particularly interested in the interactions of pathogens with environmental microbiota as well as in the response of microorganisms to single or multiple stressors, which she believes are important, though largely neglected, areas of research. She is involved in methods and standards development at the national and international levels, and serves as leader of the Canadian delegation to ISO TC 198 on sterilization of healthcare products. She has authored more than 100 articles including reviews, book chapters, and research articles, and regularly reviews manuscripts for several leading journals in infection control, environmental microbiology, and microbial control.

Dr Kurissery R. Sreekumari holds a doctorate degree and 16 years of postdoctoral research and teaching experience in microbiology. She is currently leading the microbiology unit in ALS Laboratory Group, Thunder Bay, Canada. Before joining ALS, she was holding microbiology research positions in Lakehead University, Canada, Osaka University, Japan, and Indira Gandhi Centre for Atomic Research (IGCAR), India. Sree's areas of expertise include environmental and food microbiology, molecular biology, and microbiological quality standards and specifications. To her credit, she has more than 50 research publications including several refereed journal publications, book chapters, and patents.

Dr David J. Stickler after gaining DPhil at Oxford University (1967) for a thesis on a study of bacterial viruses, moved to the Department of Bacteriology in the Medical School of Trinity College, Dublin. It was during his time in Dublin that he became interested in the problem of catheter-associated urinary tract infections in paraplegic patients at the Irish National Spinal Unit. He moved to Cardiff in 1972 where he now holds the post of reader in medical microbiology at the Cardiff School of Biosciences, Cardiff University. Throughout his research career, his major research activities have been concerned with a study of the complications of urinary tract infections in patients undergoing bladder catheterization. He has published over 120 papers on this and related topics. He has worked on many aspects of the problem from the fundamental biology of biofilms to the practical management of catheter encrustation in patients. In recent years he has worked closely with Professor Roger Feneley in Bristol. Since 2003, he has been the director of research of the BioMed Centre at Southmead Hospital, a specialized clinical and research centre dedicated to improving the care of patients undergoing long-term bladder catheterization.

Dr Vesna Šuljagić is a physician and epidemiologist studying emerging pathogens and hospital acquired infections. She received her MD from the University School of Medicine in Belgrade and her MS and PhD from the Department of Epidemiology at the Military Medical Academy in Belgrade, Serbia. Dr Šuljagić is now head of Department of Infection Control and assistant professor in the Department of Epidemiology at Military Medical Academy. Dr Šuljagić has 11 years of professional experience in prevention and control of nosocomial infections and in applied researches in the field of epidemiology of nosocomial infections. She is also the manager of the HIV/AIDS Prevention Program in Armed Forces of Serbia. She has published over 20 papers and book chapters and mentored postgraduate students and fellow residents.

Scott V.W. Sutton, PhD, earned his BS in genetics from the University of California at Davis, and his master's and PhD in microbiology from the University of Rochester, Rochester, New York. After an NIH postdoctoral fellowship at the Medical College of Virginia, Richmond, Virginia, he went to work for Bausch and Lomb, Rochester, New York, until 1994 when he accepted a position at Alcon Laboratories, Fort Worth, Texas. Dr Sutton left Alcon Laboratories in 2004 as a director in the R&D division to accept a position as pharma consultant (Microbiology) with Vectech Pharmaceutical Consultants, Inc. There, he assisted companies as varied as start-ups, generics, and established pharmaceutical innovators with laboratory management and compliance issues, process change opportunities, project management, and implementation of rapid microbiological methodologies. Dr Sutton is an active author and speaker as well as being active in a volunteer capacity, involved with the USP Analytical Microbiology Committee of Experts since 1993. He operates the Pharmaceutical Microbiology Forum with its monthly newsletter (http://www.microbiologyforum.org) and serves as a reviewer for several peer-reviewed journals. Dr Sutton also operates an information source on the Internet, The Microbiology Network (http://www.microbiol.org), that provides services to microbiology-related user groups. This service also supports two e-mail lists, the first devoted to pharmaceutical microbiology (www.microbiol.org/pmflist.htm) and the second devoted to pharmaceutical stability (www.microbiol. org/psd-psdglist.htm).

Dr John G. Thomas' worldwide studies and travels, in an array of disciplines, have led to his unique knowledge and diversified interests that have developed into his current focus on translational studies that bridge the gap between research scientists and clinical practitioners. His journey began in Vermont with a bachelor of science at Norwich University and then to Syracuse University, where he received his master's and PhD. He completed his residency as a captain in the U.S. Army, Medical Service Corp, where he was director microbiology/virology at Letterman General Hospital in San Francisco, California. Dr Thomas has been at West Virginia University for the past 13 years as a pathology and microbiology professor for the School of Medicine, clinical professor for periodontics at the School of Dentistry, and an adjunct professor for the Graduate School, Department of Microbiology, Immunology, and Cell Biology, as well as for the School of Pharmacy. He is also the director of microbiology and virology in the Clinical Laboratories at WVU Hospitals. Although his many appointments keep him busy, they have also given him the opportunity to investigate the nature of dental and medical biofilms, particularly the microbiology of plaque and biofilm development on indwelling medical devices such as endotracheal tubes.

Allan R. Tunkel, MD, PhD, received his BS from Seton Hall University in 1976 and his PhD and MD from the University of Medicine and Dentistry of New Jersey, New Jersey Medical School in 1980 and 1984, respectively, where he was elected to membership in Alpha Omega Alpha Honor Medical Society. He completed his residency and served a chief residency at the hospital of the Medical College of Pennsylvania, and did a fellowship in infectious diseases at the University of Virginia Health Sciences Center. During his fellowship, he initiated his interest in the pathogenesis and pathophysiology of bacterial meningitis and central nervous system infections; he has authored or coauthored more than 160 publications (original articles, reviews, editorials, and book chapters) in his field of interest and is chair of the Infectious Diseases Society of America Practice Guideline Committee for Central Nervous System Infections. He was previously at Drexel University College of Medicine where he had significant roles including director of the Internal Medicine Residency Program and vice chair of the Department of Medicine. In January 2006, he became chair of the Department of Internal Medicine at Monmouth Medical Center in Long Branch, New Jersey. Dr Tunkel has received numerous honors and awards throughout his career, including six Golden Apple Awards, three dean's awards for excellence in clinical teaching, the trustees' award, two gender equity awards from the American Women's Medical Association, two excellence in teaching awards from the Internal Medicine Residents, the Christian R. and Mary F. Lindback Foundation Award for Distinguished Teaching at Drexel University, and the Laureate Award from the Pennsylvania Chapter of the American College of Physicians; he has been listed as one of the best doctors in America.

Nicole Williams received a BS in microbiology from Gannon University in Erie, Pennsylvania, US. She joined STERIS Corporation in 2000 as an associate scientist, and was responsible for validation testing of various accessories for SYSTEM 1. When Nicole left STERIS in 2006, she was the supervisor of device testing/development where she was responsible for scheduling and supervising all testing for various development projects relating to the reprocessing of endoscopes and the STERIS Device Testing Program. She also was the technical expert for STERIS Corporation and was the liaison between STERIS and many medical device manufacturers. Currently, Nicole is the Product Manager for Repairs and Reprocessing for Richard Wolf Medical Instruments Corporation in Vernon Hills, Illinois. In her free time she enjoys reading and taking walks with her dog Andy.

George A. Yesenosky, MD, was born in Johnstown, Pennsylvania, and graduated from the Medical College of Pennsylvania, US, in 1998. He completed his internship and residency in internal medicine at MCP-Hahnemann University and spent an additional year there as chief medical resident. Dr Yesenosky subsequently completed his fellowship training in cardiovascular disease and cardiac electrophysiology at the same institution, now known as Drexel University College of Medicine. He served as the chief fellow for both the cardiac electrophysiology and cardiovascular medicine training programs. He is employed with the Cardiology Consultants of Philadelphia, and is the director of cardiac electrophysiology at Jeanes Hospital in Philadelphia, Pennsylvania, and is a staff electrophysiologist at Hahnemann University Hospital where he is appointed to the academic staff of Drexel University College of Medicine.

Index

A

Ablative type paint, 273
Acanthamoeba, 111, 299
Accelerated hydrogen peroxide
    (AHP)® technology, 80
Acidic detergents, 46
Acinetobacter sp., 137, 297, 308
    baumannii, 13, 129
Acrylic bone cement
    for depot administration of
        antibiotics, 352
Active Implantable Medical Device
    Directive (90/383/EEC), 258
Active species, *in situ* generation, 77–79
Acyl homoserine lactones use, 333
Adriamycin, for chemotherapy, 355
Aerobic bacterial counts in human
    skin, 69
Aerobic plate count, 63
Aeromonas hydrophila, 200
Agar-agar (medium)
    and conventional colony growth,
        340–341
Airway mucus, of cystic fibrosis
    patients, 342
AISI type 304 stainless steel, 270–271
Alcohol-based hand rubs (ABHR), 256
Alkylating agent, 93
Alza's Duros implantation, for prostate
    cancer treatment, 356
American College of
    Gastroenterology, 178
American Gastroenterological
    Association, 178
American National Standard for Safe
    Use of Lasers in Health Care
    Facilities, 179
American Society for Gastrointestinal
    Endoscopy (ASGE), 178, 182
American Society for Testing and
    Materials (ASTM), 217, 254

Amertrap systems in biofilm formation
    prevention, 279
Amino sugar biosynthesis, 116
Amoxicillin absorption, on
    polyurethanes, 344
Ampicillin and vancomycin-resistance
    strains, 295
Anaerobic cell metabolism, 105
Antibacterial biodegradable cancer
    implants
    design of, 357–364
Antibacterial materials in biofilm
    formation prevention, 270–271
Antibiotic encapsulation, in
    matrix, 345
Antibiotic flush solutions, 326
Antibiotic lock technique, 325–326
Antibiotic resistance mechanisms, 114
Antibiotic-resistant pathogens, 142
Antibiotics, 6
    in biofilm formation prevention,
        276–277
Antifouling coatings, in biofilm
    formation prevention, 273–274
Antifungal agents, 88
"Anti-immunization insurgency", 329
Anti-infective drugs, 4, 119
Antimicrobial agents, 127
    reaction with exopolymer, 342
    role of, 322–323
    types of, 5 (*see also* Antiseptics;
        Disinfectants)
Antimicrobial chemotherapy, 6
Antimicrobial compounds, biocidal
    efficacy, 83
Antimicrobial dyes, 118
Antimicrobial impregnated
    matrices, 343
Antimicrobial peptides, 324–325
Antimicrobial preservative efficacy
    test, 63

V

Vacuum-assisted closure technique, 330
VADs, *see* Ventricular assist devices
Vancomycin encapsulation
  in 50:50 PLA = PCL, 361–363
  in 65:35 PLGA fiber, 361–362
Vancomycin-resistant enterococci (VRE),
  13, 17
VAP, *see* Ventilator-associated
  pneumonia
VAP risk reduction, 327
Vascular catheters role, 332
Vegetative bacteria, 100
Vegetative cells disinfection, 170
Ventilator-associated pneumonia, 297,
  301, 303, 306, 308, 327–328
Ventricular assist devices, 299
Ventriculoperitoneal shunts,
  298–299, 310
Vibrio sp.
  alginolyticus, 278
  parahaemolyticus, 10
Viral infections, 48
Virulence definition and factors, 66
Virulence of organisms, evaluating
  factors, 66
Voice prostheses during oral cavity,
  297–298
Voice prosthesis infection, 308

VP, *see* Ventriculoperitoneal shunts
VP shunt infections, 310

W

Water-borne cryptosporidium
  infection, 8
Water-line disinfection, 119
Weapons of mass destruction (WMD),
  26, 33
Wetting agents role in biofilm formation
  prevention, 274
Whilst surgical implantation, 357
Workplace Hazardous Materials
  Information System
  (WHMIS), 262
World Health Organization (WHO), 3, 7,
  13, 24, 31
  authority and power of, 33–34
Wound-associated medical devices,
  clinical trials, 329
Wuchereria, 111

Y

Yersinia enterocolitica, 10

Z

Zoladex, 356